DETECTION OF LOW-LEVEL OPTICAL SIGNALS

SOLID-STATE SCIENCE AND TECHNOLOGY LIBRARY

Volume 4

Aims and Scope of the Series

The aim of this series is to present monographs on semiconductor materials processing and device technology, discussing theory formation and experimental characterization of solid-state devices in relation to their application in electronic systems, their manufacturing, their reliability, and their limitations (fundamental or technology dependent). This area is highly interdisciplinary and embraces the cross-section of physics of condensed matter, materials science and electrical engineering.

Undisputedly during the second half of this century world society is rapidly changing owing to the revolutionary impact of new solid-state based concepts. Underlying this spectacular product development is a steady progress in solid-state electronics, an area of applied physics exploiting basic physical concepts established during the first half of this century. Since their invention, transistors of various types and their corresponding integrated circuits (ICs) have been widely exploited covering progress in such areas as microminiaturization, megabit complexity, gigabit speed, accurate data conversion and/or high power applications. In addition, a growing number of devices are being developed exploiting the interaction between electrons and radiation, heat, pressure, etc., preferably by merging with ICs.

Possible themes are (sub)micron structures and nanostructures (applying thin layers, multi-layers and multi-dimensional configurations); micro-optic and micro-(electro)mechanical devices; high-temperature superconducting devices; high-speed and high-frequency electronic devices; sensors and actuators; and integrated opto-electronic devices (glass-fibre communications, optical recording and storage, flat-panel displays).

The texts will be of a level suitable for graduate students, researchers in the above fields, practitioners, engineers, consultants, etc., with an emphasis on readability, clarity, relevance and applicability.

Detection of Low-Level Optical Signals

Photodetectors, Focal Plane Arrays and Systems

by

M. A. TRISHENKOV

Institute of Applied Physics,
Russian State Scientific and Production Centre "Orion",
Moscow, Russia

Translated by A. Dobroslavsky

KLUWER ACADEMIC PUBLISHERS

DORDRECHT / BOSTON / LONDON

A C.I.P. Catalogue record for this book is available from the Library of Congress.

ISBN 0-7923-4691-2

Published by Kluwer Academic Publishers,
P.O. Box 17, 3300 AA Dordrecht, The Netherlands.

Sold and distributed in the U.S.A. and Canada
by Kluwer Academic Publishers,
101 Philip Drive, Norwell, MA 02061, U.S.A.

In all other countries, sold and distributed
by Kluwer Academic Publishers,
P.O. Box 322, 3300 AH Dordrecht, The Netherlands.

Printed on acid-free paper

Printed in the Netherlands

Preface

This book is addressed to designers of photodetectors and photodetecting systems, designers of focal plane arrays, charge-coupled devices, specialists in IR technologies, designers of optoelectronic detecting, guiding and tracking systems, systems for IR direction finders, lidars, lightwave communication systems, IR imagers.

All these specialists are united by one common purpose: they are all striving to catch the weakest possible optical signal. The most important characteristic of photosensitive devices is their detectivity, which determines the lowest level of optical signal they are able to detect above the noise level. These threshold characteristics define the most important tactical and technical parameters of the entire optoelectronic system, such as its range, resolution, precision. The threshold characteristics of optoelectronic system depend on many of its components; all designers agree, however, that the critically responsible part of the system is the photodetector [1]. By the end of the 1960s the physicists and the engineers were able to overcome many obstacles and to create photodetectors (at least single-element or few-element ones) which covered all the main optical bands (0.4 ... 2, 2 ... 3, 3 ... 5, 8 ... 14 μm), carried out the detection almost without any loss (the quantum yield being as high as 0.7 ... 0.9), and reduced the noise level to the lowest possible limit. The noise level was defined either by the fluctuations of background thermal radiation (as in IR detectors), or by the shot noise of the dark current which was lower than the noise of the subsequent amplifier (as in silicon photodiodes).

However, as the technical parameters were improved almost to their theoretical limit, further progress in the theory and practice of photosensitive devices did not stop. On the contrary, late sixties and early seventies saw the onset of new breathtaking developments in the field of photodetectors, which were stimulated by the advances in microelectronics. Hybrid and solid-state technologies have made it possible to combine the photodetector and the signal processing IC in one module or even on the same crystal. This gave birth to a new class of devices. In the Russian-language literature these devices are called ***фотоприемные устройства* (*ФПУ*)** [fotopriYOMniye ustROYstva], or, literally, photo receiving devices. The English counterpart of this term, *photodetecting assembly*, commonly refers to only the tandem of photodetector and amplifier. In the present book we are going to use the term photodetecting assembly (PDA) in a broader sense, to apply to the entire class of devices with photodetectors and signal processing electronics, including focal plane arrays. The structure of modern PDAs is discussed in Ch. 1.

While a photodetector performs just one function, converting optical pulse into electrical signal, PDA also carries out subsequent operations: as a minimum, amplification, then, in some systems, filtering, and switching in multielement systems.

Today there exists a large variety of most diverse PDAs, which include other signal processing functions (for instance, comparison with the preset threshold), and auxiliary circuits which maintain operating regimes of the PDA.

Such signal processing functions have been earlier entrusted to separate electronic units of the optoelectronic system. Incorporation of these functions into the PDA itself opened new possibilities not only for microminiaturization, but also for improving the threshold characteristics (see Sect. 1.1, 1.2, 2.5). In dealing with the basic task of detection of ultimately weak optical signals, the designer of PDA must follow the guidelines of the *theory of detection*. This theory tells how to design the signal processing circuits, how to select the optimal (quasi-optimal) frequency response of the device, which will extract the signal in the best possible way from its mixture with the noise, how to formulate the rule for deciding whether or not there is a useful signal at the input of the device.

Now what is the literature that can be found on photodetectors, PDA, and the theory of detection? There is an extensive bibliography on photodetectors, on focal plane arrays, and on detection and measurement of radio signals. At the same time, there are rather few titles concerned specifically with PDAs and the theory of detection of optical signals. The list is quite short: a monograph on PDA circuit engineering [2], a reference book on Soviet-made PDAs up to 1987 [3], and N.S. Shestov's classic *Extraction of Optical Signals from Background Random Noise* [4]. The latter has become the theoretical basis for optoelectronic engineering; however, it was published as early as 1967, before the advent of PDAs, and naturally does not say anything about the detection of weak optical signals with the aid of these.

This monograph is intended to fill this gap. PDAs, as described here, are all designed to register ultimately weak radiation. This is their principal task, and this is, consequently, the purpose of the author. The main subject of the book is the theory of detection of optical signals as applied to PDAs. We discuss almost all the main types of threshold PDAs: from one-element PDAs based on photoresistors (PR), photodiodes (PD), avalanche photodiodes (APD), to multielement PDAs including focal plane arrays (FPA), charge-coupled devices (CCD), and other charge transfer devices (CTD). We take into account the particular features of PDAs: their typical noise spectra, the effect of the number of channels on the process of detection, the real shape of optical signal.

The author did his best to make the book useful for the most broad circle of designers of PDAs and optoelectronic systems, to present the material in a clear and comprehensible way, to avoid cumbersome mathematics as much as possible, laying the emphasis on the physics and physical interpretation of the results.

The contemporary theory of detection is one of the most sophisticated branches of statistical radio engineering. Because of this, we shall often rely on simplified models and skip rigorous mathematical proofs, but without compromising the physics and the precision of final results. In almost all sections we give numerical evaluations, simple approximations and diagrams which will be appreciated by the practical engineer.

Acknowledgments. This book is based on the collective experience gained by the research workers of the Russian State Research Center "Orion", the renowned leader in the field of optoelectronics, the sole affiliation of the author for over thirty years. The Russian version of the book was published in 1992 under the title *Photodetecting assemblies and CCDs. Detection of weak optical signals* (Moscow, Radio i Svyaz, 400 pp.). The first volunteers who tested out the first chapters of the book were Yu.R. Vinetsky, N.V. Kravchenko, P.E. Khakuashev, V.M. Yungerman. The author took due account of their valuable comments in preparing the final version of the book. Great help in numerical calculations came from A.M. Gutorenko, M.A. Polunin, Yu.M. Eskin.

This English-language version has been completely revised and amended. It would not have been possible without assistance and help of Executive Director of International Science and Technology Center Dr. Oles Lomacky, General Director of "Orion" professor V.T. Khryapov and his first deputy professor I.I. Taubkin (it was the latter who stimulated the course of lectures that eventually grew into this book). The manuscript was translated into English by A.S. Dobroslavsky and edited by Yu.M. Eskin. Yu.R. Vinetsky did all the drawings.

I express my heartiest gratitude to all who helped this book to see the light.

January 1997, Moscow

About the Author

Mikhail A. Trishenkov (b. 1938), Doctor of Technology, graduate of Moscow Institute of Power Engineering, is head of laboratory at the Russian State Research Center "Orion", a major optoelectronics research facility. Professor of Moscow Institute of Radio, Electronics and Automation.

A leading expert on photodetectors, Dr. Trishenkov was in charge of development of a range of high-performance photodiodes and IR linear arrays. He is the author of 120 papers, 30 inventions, author and editor of several monographs. His research is concerned with kinetics of internal photoelectric effect, avalanche multiplication in photodiodes, electric injection in CCD, processing and filtering of optical signals.

Contents

Preface v

Abbreviations and Notation xi

1. Photodetecting Assembly as Detection System for Weak Optical Signals 1

1.1. Functions and Structure of Photodetecting Assembly, CCD, and Other Charge Accumulation Devices 2
1.2. Photodetectors 23
1.3. Amplifier 59

2. Signal and Noise in Photodetecting Assemblies 66

2.1. Pulsed Optical Signal 66
2.2. Basic Notions of Noise Process 75
2.3. Photodetector Noise 102
2.4. Noises of Photodetecting Assembly 121
2.5. Minimization of Photodetecting Assembly Noises 142

3. Foundations of Theory of Optimal Filtering as Applied to Photodetecting Assemblies 167

3.1. Signal – No Signal Decision Making 167
3.2. Optimal Filter (Frequency Analysis) 181
3.3. SNR at the Output of Optimal Filter (Frequency Analysis) 190
3.4. Noise Equivalent Power, Detectivity, and Other Detection Invariants 211

4. Optimal Filtering of Photo Signals (Time-Domain Analysis) 227

4.1. Pulse Characteristic of Optimal Filter 227
4.2. Shape of Signal at the Output of Optimal Filter 248

5. Signal Processing in Real Photodetecting Assemblies . 257

5.1. Quasi-Optimal Filtering 257
5.2. Loss in SNR in Quasi-Optimal Filter 265
5.3. Broad-Band Amplification 297

6. Detection of Weak Optical Signals with Photodetecting Assemblies Based on Avalanche Photodiodes 305

6.1. Broad-Band Amplification and Quasi-Optimal Filtering in Photodetecting Assemblies Based on Avalanche Photodiodes 305
6.2. Quantization and Fluctuation of Signal 311
6.3. Photon Counting Mode of Avalanche Photodiode 344

7. Detection and Processing of Optical Signals in Matrices 357

7.1. Integration Cell As Quasi-Optimal Filter 357
7.2. Multielement Traditional-Type Photodetecting Assemblies and Focal Plane Arrays 385

8. Detection of Weak Optical Signals in Special-Purpose Optoelectronic Systems 417

8.1. Timing of Arrival of Optical Pulse 417
8.2. IR Imager 432

Conclusion 452
References 453

Abbreviations and Notation

ADC	Analog-Digital Converter
AFC	Amplitude-Frequency Characteristic
APD	Avalanche PhotoDiode
BLIP	Background Limited Infrared Photodetector
BT	Bipolar Transistor
CCD	Charge Coupled Device
CID	Charge Injection Device
CTD	Charge Transfer Device
DCS	Double Correlated Sampling
DFGA	Distributed Floating Gate Amplifier
FCC	Frequency-Contrast Characteristic
FET	Field Effect Transistor
HF	High Frequency
IC	Integrated Circuit
IR	InfraRed
LED	Light Emitting Diode
LF	Low Frequency
MCT	Mercury-Cadmium-Telluride
MIS	Metal-Insulator-Semiconductor
NEP	Noise Equivalent Power
NETD	Noise Equivalent Temperature Difference
OG	Optical Generation [mode]
PD	PhotoDiode
PDA	PhotoDetecting Assembly
PR	PhotoResistor
PSE	PhotoSensitive Element
SCR	Space Charge Region
SNR	Signal-to-Noise Ratio
SPRITE	Signal PRocessing In The Element
TDI	Time Delay and Integration [mode]
UHF	Ultra High Frequency
UV	UltraViolet
A	photoresistor cross-section area
A	frame area
A	excess (low-frequency) noise intensity
A_e	photosensitive element area
A_{im}	object image area in the focal plane

A_{ob}	object area
$\mathcal{A}$	light spot area
a	coefficient in formula for detection invariant
a_t	coefficient in the threshold charge formula
a_i	frame-superposition mode weighted coefficients
a_{ob}	part of the area of the object whose radiation is projected on to one element
a'_{ob}	image of the area a_{ob} (equal to the area of element A_e)
B_L	coefficient in the formula for signal-to-noise ratio in multi-element photodetecting device in case of predominantly low-frequency noise
B_W	coefficient in the formula for signal-to-noise ratio in multielement photodetecting device in case of predominantly white noise
B_H	coefficient in the formula for signal-to-noise ratio in multielement photodetecting device in case of predominantly high-frequency noise
b	FET channel width
C	capacitance (including the integrating cell capacitance)
C_{int}	main integrator (source) capacitance in the charge-partitioning cell
$C_{d\,int}$	drain integrating capacitance in the charge-partitioning cell
C_p	capacitance of photodiode or photodetector
C_{in}	net input capacitance of amplifier
C_a	input capacitance of amplifier (input transistor)
C_c	circuit (stray) capacitance
C_{fb}	feedback circuit capacitance
C_{ft}	first node feedthrough capacitance
C_{ox}	oxide capacitance
C_{FET}	capacitance of field-effect transistor
C_{BT}	capacitance of bipolar transistor
c	velocity of light
$\mathcal{D}$	coefficient of diffusion
$D = 1/\Phi_t$	detectivity (at fixed frequency band and photodetector area)
$D_1 = 1/\text{NEP}$	detectivity, or reciprocal unit frequency bandwidth noise equivalent power
$D^* = 1/\text{NEP}^*$	specific detectivity (per unit bandwidth and unit area)
$D^*_{BLIP} = 1/\text{NEP}^*_{BLIP}$	specific detectivity of the background limited infrared photodetector (BLIP)
d	diameter of optical spot
$\mathcal{E}$	electric field strength

$\mathcal{E}_B$	electric field strength at avalanche breakdown
$\mathcal{E}_i$	electric field strength in *i*-layer
E	optical radiation power
$E_s(t)$	optical pulse signal (time dependence of power)
$\mathrm{E}_s(f)$	spectral density of optical pulse in the frequency band from 0 to ∞.
$\tilde{\mathrm{E}}_s(f)$	spectral density of optical pulse in the frequency band from -∞ to +∞
E_{sA}	optical pulse amplitude
E_{sT}	threshold value of optical pulse corresponding to PDA output signal equal to noise
E_b	power of background optical radiation
E_g	energy band gap
E_n	rms noise value at fixed frequency bandwidth
e_{ph}	photo-emf
$e_s(t)$	relative shape of optical pulse
$\mathrm{e}_s(f)$	relative spectrum of optical pulse
e_{s1}, e_{s2}	relative spectral density of optical pulse at frequencies f_1 and f_2, respectively
$\mathrm{e}_\delta(f)$	relative frequency spectrum of δ-type signal
$e_n = \sqrt{\overline{e_n^2}}$	rms value of noise voltage (emf-noise)
e_{n2}	emf-noise of second stage (or subsequent amplifier)
$\mathcal{F}$	noise factor
f	cycle frequency
f_s	effective frequency band of signal
Δf_s	quasi-harmonic signal bandwidth
f_a	effective frequency band of amplifier
$f_{a\,opt}$	optimal effective frequency bandpass of amplifier (corresponding to maximum SNR value)
f_{lens}	focal length of lens
$\Delta f = 1/T$	frequency bandwidth corresponding to one harmonic of spectrum of realization of random function on interval T
Δf_a	frequency band of narrow-band amplifier
f_{fa}	false alarm frequency
f_{char}	characteristic frequency of false alarms
f_c	dark counting rate in photon count mode
f_0	characteristic frequency at which high-frequency noises of BP and FET-based PDAs become the same
f_n	effective noise frequency band (for arbitrary noise spectrum)

$f_{\mathrm{n\,w}}$	effective noise frequency band in case of white noise
$f_{\mathrm{n\,h}}$	effective noise frequency band in case of high-frequency noise
f_{high}	high-frequency limit
f_{low}	low-frequency limit
$f_{\mathrm{n\,s}}$	noise frequency bandwidth for calculation of signal noise
G	transistor transconductance
G_{fb}	transconductance of the feedback four-pole device
$\mathrm{H}(t)$	pulse response
$\mathrm{h}(t)$	relative pulse response
$\mathrm{h_{eq}}(t)$	equivalent pulse response (of rectangular shape whose amplitude and area are the same as those of the replaced $\mathrm{h}(t)$ response)
$h_{21\mathrm{e}}$	current transfer (gain) of bipolar transistor in common-emitter circuit
h	Planck's constant
I	current
I_{p}	photo current
$I_{\mathrm{p\,b}}$	background radiance-induced photo current (excluding equilibrium thermal radiation)
$I_{\mathrm{d\,b}}$	infrared photodetector dark current component due to equilibrium thermal radiation
$I_{\mathrm{d\,g}}$	infrared photodetector dark current component due to thermal generation, or thermal generation current
$I_{\mathrm{d}} = I_{\mathrm{d\,b}} + I_{\mathrm{d\,g}}$	dark current
$I_0 = I_{\mathrm{pb}} + I_{\mathrm{d}}$	total, or back, current
$I_{\mathrm{s}}(t),\ I_{\mathrm{s}}$	signal photocurrent
I'_{s}	initial (non-gained) photocurrent of APD signal
$I_{\mathrm{s\,A}}$	signal current amplitude
I_{in}	photodetector initial (non-gained) current
I_{out}	photodetector secondary (gained) current
I_{a}	input leakage current of amplifier (transistor)
$I_{\mathrm{f\,b}}$	feedback connection current
$I_{\mathrm{E}},\ I_{\mathrm{B}},\ I_{\mathrm{C}}$	emitter, base and collector current of bipolar transistor
I_{D}	FET drain current
I_{A}	amplitude of one-electron current pulse
I_{A}	noise pulse amplitude
$I(t)$	current random function
$\bar{I}$	mean current value
$I_1,\ I_2,\ldots,\ I_i,\ldots,\ I_N$	sequence of realizations of random current function
$I_1,\ I_2$	input and output current of transistor

I_{eq}	current of equivalent noise diode which characterizes the photodetector noise
I_n	current of equivalent noise diode which characterizes the PDA noise
$i_n = I(t) - \bar{I}$	deviation of current from the mean value (random function)
$\mathrm{I}_{nn}(f)$, $\mathrm{I}_{nnc}(f)$, $\mathrm{I}_{nns}(f)$	*n*th current realization spectra: amplitude, sine and cosine-wave components
$i_n = \sqrt{\overline{i_n^2}}$	rms value of PDA noise current
i_{na}	rms value of PDA noise current: amplifier-induced component (amplifier noise)
$i_{n\,PD}$	rms value of PDA noise current: photodetector-induced component (photodetector noise)
$i_{a2} = \sqrt{\overline{i_{a2}^2}}$	transistor output noise current
$\overline{J^2}$, $\overline{J^2(f)}$	PDA input noise power density (noise spectrum)
$\overline{J_a^2}$	spectral density of input current noise of four-pole device (amplifier, transistor)
$\overline{J_{a2}^2}$	spectral density of output current noise of four-pole device (amplifier, transistor)
$\overline{J_{am}^2}$	spectral density of noise power of amplifier from all sources related to its input
$\overline{J_h^2}$	high-frequency noise component of amplifier
$j = \sqrt{-1}$	imaginary unit
$\mathrm{j}(f)$	line spectrum of noise realization (for particular regular model)
j	ordinal number
j_d	dark current density of photodetector
j_{db}	generation current density due to thermal radiation
j_{dg}	thermal generation current density
K_{21}	amplifier transfer coefficient (absolute amplitude value)
$\mathbf{K}_{21}$	amplifier transfer coefficient (complex value)
$\mathrm{K}_{(n+1)1}$	gain of *n*-stage amplifier
K_{eff}	effective transfer coefficient of feedbacked amplifier
K_0	transfer coefficient (gain) of amplifier: maximum value at low frequencies
K_{int}	transfer coefficient of integrator (integrating cell)
K_{lens}	coefficient of collection of optical system of IR imager
K_p	internal gain of photodetector

K_{opt}	internal gain coefficient: optimal value corresponding to maximum SNR
$\mathcal{K}(\Delta t)$	correlation function
k	ordinal number
k	ratio of impact ionization coefficients
k	ratio of readout (spilling) charge to integrated charge in the incomplete charge-spilling mode
$k(f)$	relative frequency response of amplifier
κ	Boltzmann's constant
$\mathcal{L}$	loss coefficient: ratio of the real filter SNR to the optimal filter SNR
L	FET channel length
L	object distance
$L \times L$	frame size of multielement optoelectronic system
$\mathcal{L}_{im}$	object image length in the focal plane
$\mathcal{L}_{ob}$	object length
l	photoresistor length
$l \times l$	size of one element of multielement PDA
l	ordinal number
l_f	free path length of electron
$\mathcal{M}$	number of pixels (image elements)
M	multiplication coefficient (random number in case of statistical analysis)
$\overline{M}$	mean value of multiplication coefficient (the bar is dropped when statistical characteristics are not analyzed)
M_e	number of elements (channels) in PDA (matrix)
M_{opt}	multiplication coefficient: optimal value corresponding to maximum SNR
M_{qua}	multiplication coefficient: quasi-optimal value
m	ordinal number
m	number of lines per frame
m_d	number of defective channels in the matrix
N	ordinal number
N	number of photons
N_d	number of donor atoms
N_{eff}	effective number of quanta of equilibrium thermal radiation
N'	number of radiation-generated photocurrent producing pairs per unit time
N_s, $\overline{N}$	number of signal photoactive pairs and its mean value

$\{N_s\}$	mean value of the minimum number of signal photoelectrons ensuring the desired probability of target miss
$N_t = U_t/qK_0$	discriminator threshold level in terms of the number of photoelectrons
N_n	noise number
NEP	noise equivalent power in unit frequency band with fixed area of photosensitive element
NEP^*	specific noise equivalent power in unit frequency band per unit area of photosensitive element
NEP^*_{BLIP}	specific noise equivalent power of background-limited infrared photodetector
NETD	noise equivalent temperature difference
$NETD_{BLIP}$	noise equivalent temperature difference in BLIP mode
$NETD^*_{BLIP}$	specific noise equivalent temperature difference in BLIP mode
n	ordinal number
n	exponent in Miller's formula for multiplication coefficient
n	concentration of electrons
n_i	intrinsic concentration
n	ratio of signal duration to sampling period
n_{fa}	ratio of threshold voltage to rms noise voltage which determines the probability of false alarm
P	power
P_{fa}	false alarm probability
P_{cfa}	false alarm probability for each channel of the matrix
P_{tm}	target miss probability
$P_{tm\,i}$	target miss probability for ith element of the matrix
P_{err}	total probability of error
$P(N_s)$	probability of realization of optical pulse consisting of N_s photoelectrons
$p(1,x)$	probability of passage of one electron across avalanche layer of width x without multiplication
$p(i)$	probability of light spot falling on ith element of the matrix
p	concentration of holes
Q_s	cell charge caused by accumulation of signal photocurrent (signal charge)
Q_{sn}	signal charge accumulated over nth subframe time
Q_{s1}, Q_{s2}, ΔQ_s	signal charges accumulated over 1st and 2nd frames, and their difference
Q_1, Q_2, ΔQ_{12}	charges accumulated over 1st and 2nd frames and their difference
Q_{ini}	initial (preset) charge in the integrating capacitor
Q_d, Q_0	cell charge due to the dark and the total current, respectively

Q_{const}	constant charge deduced from the total charge in the subtraction technique
q	charge of electron
R	resistance
R_L	photodetector load resistance
R_{LPD}, R_{LPR}	load resistance of photodiode, photoresistance
R_p	dynamic resistance of photodiode, photodetector
R_{pSC}	short-circuit dynamic resistance of photodiode
R_{pOC}	open-circuit dynamic resistance of photodiode
R_{pAPD}	dynamic resistance of avalanche photodiode
R_s	photodiode series resistance
R_{st}	transistor input circuit series resistance
R_{in}	PDA input resistance
R_a	amplifier (transistor) input resistance
R_{a2}	amplifier (transistor) output resistance
R_d	drain circuit load resistance (amplifier node output resistance)
R_{fb}	feedback resistance
R_{ch}	FET channel resistance
R_{eq}	equivalent input resistance of feedbacked amplifier which determines input voltage
S	sensitivity of photodetector
$S_{\lambda\,max}$	peak monochromatic sensitivity of photodetector (corresponding to wavelength λ_{max})
S_0	sensitivity of initial photoconversion
S_0	direct current sensitivity
SNR	signal to noise ratio
SNR_{OF}	SNR at the output of optimal filter
SNR_1	SNR at the output of the first stage of measuring amplifier
SNR_1^{OF}	SNR at the output of the first stage of optimal measuring device
SNR_{PDA-PD}	SNR in photodiode-based PDA
$SNR_{PDA-APD}$	SNR in avalanche photodiode-based PDA
SNR_{ki}, SNR_k	SNR for individual elements of the matrix
$\overline{\Delta SNR_{mn}^2}$, $\overline{\Delta SNR_m^2}$	dispersion of SNR for its distribution with respect to the matrix elements
$s(f)$	relative sensitivity frequency response
T	absolute temperature
T	time interval

T_s	duration of optical signal
T_{05}	optical signal duration at half-amplitude
$T_{05\,qua}$	signal duration at half-amplitude at the output of quasi-optimal filter
$T_{05\,opt}$	signal duration at half-amplitude at the output of optimal filter
T_i	integration time (observation time)
$T_{i\,opt}$	optimum integration time corresponding to maximum SNR
$T_{i\,max}$	maximum allowable integration time (as limited by the cell current and capacitance)
T_{int}	interrogation time required for readout of information from one cell
T_f	frame time (time of renewal of frame information)
T_S	channel sampling period (time between two consecutive readouts from the same channel)
T_m	measurement time
T_{per}	time period between one-electron pulses
t_m	time instant at which the amplifier output signal is at maximum
$t_1, t_2,\ldots,t_i,\ldots, t_N,\ldots,t_{2N}$	sampling instants
t_0	time instant at which the optical signal is at maximum (usually, $t_0 = 0$)
t_{lag}	lag (time delay)
U	photodetector voltage
$\|U'\| = \|U\| + \Psi_c$	voltage across the space charge region in back-biased photodiode
U_{opt}	optimal photodiode bias voltage
U_B	avalanche breakdown voltage
U_{ps}	power supply voltage
U_{co}	FET cut-off voltage
$U'_{co} = U_{co} + \Psi_c$	voltage across the space charge region of FET in cut-off regime
U_d	FET drain voltage
U_t	discriminator set-off (threshold) level
$U_{t\,opt}$	optimal threshold level corresponding to the minimum total error
$U_s(t)$	signal voltage
$U_s(t)$	amplifier output signal voltage when transfer coefficient is unity ($K_0 = 1$) (relative value)
U_{sA}	signal voltage amplitude
U_{in}, U_{out}	amplifier input and output voltage

U_{fb}	feedback circuit voltage
$U_{n\,in}$	PDA input noise voltage
U_n	PDA (amplifier) output noise voltage
ΔU	integrator capacitor discharge voltage
$U_{n\,PD}$	noise voltage of photodiode-based PDA
$U_{n\,PR}$	noise voltage of photoresistor-based PDA
U_{s1}	output signal voltage of the first stage of measuring amplifier
$U_{n\,a}$	PDA amplifier noise
$u_s(t)$	relative shape of amplifier output signal (output signal corresponding to transfer coefficient, optical power and photodetector sensitivity all equal to unity, $K_0 = 1,\ E_s = 1,\ S = 1$)
$u_{s\,A} = u_s(t_m)$	relative pulse response which characterizes the decrease in photoresponse due to limited bandwidth of filter-amplifier (output signal amplitude corresponding to transfer coefficient, optical power and photodetector sensitivity all equal to unity, $K_0 = 1,\ E_s = 1,\ S = 1$)
Ψ_{DI}	detection invariant
W	thickness of space charge region of diode (photodiode)
W_i	thickness of *i*-layer
W_{av}	thickness of avalanche multiplication layer
W_{ch}	FET channel thickness
X	random quantity
X_i	random quantity value
x	coordinate (*x*-axis)
x_0 (and y_0)	coordinates of power center of optical spot
Y	random quantity
Y_i	random quantity value
y	coordinate (*y*-axis)
y_0 (and x_0)	coordinates of power center of optical spot
Z_p	impedance of photodiode (photodetector)
Z_{fb}	impedance of feedback circuit
Z_{eq}	equivalent input impedance of feedbacked amplifier
Z_a	input impedance of PDA (amplifier, transistor)
Z_{a2}	output impedance of PDA (amplifier, transistor)
Z_{in}	input impedance
α	coefficient of absorption of optical radiation
α	impact ionization coefficient for electrons
β	impact ionization coefficient for holes

2β	angular field of view of photodetector
$\Gamma_1 = f_{nw}/f_a$	ratio of the effective white noise frequency band to the effective frequency band of amplified signal
$\Gamma_2 = f_{nh}/f_a$	ratio of the effective high-frequency noise frequency band to the effective frequency band of amplified signal
γ	ratio of time constant of *RC* circuit to signal duration
γ	coefficient in the formula for SNR
ε	relative permittivity
ε_0	permittivity of vacuum
ε_e	emissivity
ε_{ph}	energy of photon
η	quantum efficiency of photodetector
η_c	general efficiency of the counting mode
η_c'	counting efficiency caused by the loss due to multiplication fluctuations
η_c''	counting efficiency caused by the loss due to limited counting time
$\Theta(f)$	relative spectral density of noise power (noise spectrum) at PDA input
$\Theta_{out}(f)$	relative spectral density of noise power (noise spectrum) at PDA output
θ	coefficient which relates $NETD^*_{BLIP}$ and D^*_{BLIP}
æ	APD noise coefficient
λ	optical radiation wavelength
λ_{max}	peak sensitivity wavelength
λ'	short-wave cut-off wavelength
λ''	long-wave cut-off wavelength
μ	mobility of charge carrier
μ_n	electron mobility
μ_p	hole mobility
ν	frequency of photon
ξ	relative temperature coefficient of background current
ρ	resistivity
ρ	density of probability
$\sigma^2 = \overline{i_n^2}$	dispersion (variance)
τ	time interval
τ	transit (diffusion) time
$\tau_h = 1/2\pi f_h$	characteristic time constant of high-frequency noise
τ_{tr}	transit time

$\tau_{\mathrm{tr}\,n}$, $\tau_{\mathrm{tr}\,p}$	transit time for electrons and holes in photoresistor
$\bar{\tau}_{\mathrm{tr}\,n}$	mean value of electron transit time in photoresistor (averaged for all transit lengths)
τ_{SCR}	transit time in the space charge region
τ_{life}	lifetime
τ_{p}	time constant of photodetector
τ_{M}	Maxwell time constant
τ_{eff}	effective time constant of amplifier
τ_{f}	time constant of integrating *RC* circuit
τ_{fd}	time constant of differentiating *RC* circuit
τ_{f1}	time constant of high-frequency (additional) integrating *RC* circuit
τ_{cor}	correlation time
τ_{n}	duration of noise pulse
τ_{th}	time between consecutive collisions of electron (due to thermal motion)
$\Phi(z)$	error function
Φ_{t}	noise equivalent power (sensitivity threshold) in fixed frequency band
$\varphi_{\mathrm{th}} = \kappa \mathrm{T}/q$	thermal potential
φ_{eff}	APD effective thermal potential
$\varphi_{\mathrm{f}}(f)$	phase response characteristic
φ_{c}	contact potential difference
ω	radian frequency
ω'	base-current optimization frequency: at this frequency the base current is adjusted so as to equalize the shot noise and the emf noise components of bipolar transistor

Chapter 1

PHOTODETECTING ASSEMBLY AS DETECTION SYSTEM FOR WEAK OPTICAL SIGNALS

A broad class of optoelectronic systems is designed for detection of weak pulsed signals [1, 5–10]. It would be easier to name those systems which use unmodulated radiation (exposure meters, some automatic systems) than to enumerate all systems working with modulated radiation. In some of the latter modulated is the source itself — such are the devices based on pulsed lasers and LEDs: geodesic laser range finders, fiber optical telecommunication systems, reflectometers (devices for locating breaks in fiber optical cables by timing the return of the pulse bounced off the faulty spot), active tracking and aiming systems (such as coordinators). In some optoelectronic systems the source is not modulated, and the task consists in defining the coordinates of the point source. The pulsed signal in such systems is caused by modulation and scanning in the system, or by the movement of the point source. Such systems include star orientation devices, IR direction finders, various kinds of passive tracking and aiming systems. Even if the radiation of the source is constant, and there is no modulation or scanning in the system, the measurement time T_m is still limited, and the optical signal may be considered as a pulse of length T_m.

In this book we are going to deal with PDAs for this class of optoelectronic systems, designed for detection of pulsed optical signals. Naturally, in many cases the optical signal must be not only detected (that is, whether or not there is an optical signal at the input of PDA), but also measured (amplitude, timing, duration). First of all, however, one needs the object of measurement. This is the reason why we start with the detection of optical signals with the aid of PDA. Particular designs of PDAs for certain measuring systems (which determine the coordinates of light spot, time of arrival of the signal, temperature difference across the object) will be treated in Ch. 7, 8.

First we shall answer the most general questions: what must be the functions and the structure of the electronic part of optoelectronic system for detection of pulsed optical signal. Then we shall consider the main structural units responsible for the magnitude of the lowest detectable signal: the photodetector and the amplifier.

1.1. Functions and Structure of PDA, CCD, and Other Charge Accumulation Devices

Functional diagram of the detection system. The first operation in the sequence is the detection, whereby the optical pulse is converted into electrical signal (the so-called photosignal). Since the broad-band noise exists in the system along with the signal, the next operation is filtering: one has to separate the signal harmonics from the noise, and cut off the frequencies dominated by the noise. In real devices the filtering must, as a rule, be accompanied by the amplification of the signal, so as to make it "transportable" (that is, stronger than the noises and strays in the subsequent stages), and suitable to serve as the input of the decision-making unit. This last unit takes the decision whether or not there is a signal at the front end of the device. The entire device that carries out detection, filtering (amplification) and "decision-making" is referred to as the detection system (Fig. 1.1a) [4].

A multichannel detection system in the first approximation can be considered as simply an array (mosaic) of one-channel detection systems. Theoretically, each channel should incorporate its own decision-making unit. However, one decision-making unit may service a number of channels via a switching unit, which connects the channels one by one, at the appropriate switching rate (commutation or sampling rate), to the common decision-making unit (Fig. 1.1a). In this way, a multichannel detection system performs yet another function called switching — or, more precisely, multiplexing, since parallel information is converted into serial information.

In the early optoelectronic systems the detection was carried out by photodetectors, and all the remaining functions of the detection system were performed by separate electronic modules [1]. The advances of microelectronics in the late sixties and early seventies have made it possible to pack into one enclosure (or even on to the same crystal) both the photosensitive element, and all or some of the electronic circuits performing other functions of the detection system: filtering, amplification, switching, and sometimes taking the decision whether or not there is the signal at the input. Thus emerged a new class of devices, photodetecting assemblies [2]. Currently there are many varieties of PDAs [3]. Let us consider the main types of these, and prove that, despite all kinds of technical differences, the structure of all PDAs is essentially the same, and corresponds to the functional diagram of the detection system shown in Fig. 1.1a.

Traditional PDAs with amplifiers. The most simple is the one-channel (one-element) PDA which incorporates photodetector and amplifier, the latter usually doubling as the filter and selecting the desired frequency band (Fig.1.1b). We distinguish the low-frequency PDAs, intended for mechanical modulation and scanning (with the frequencies typically lying between 1 Hz and 10 kHz), and the high-frequency PDAs for laser and LED sources (with the frequencies ranging from 1 to 100 MHz). The latter are based on *pn* and *pin* photodiodes (PD) and avalanche photodiodes (APD). As indicated above, such PDAs may also incorporate the decision-making unit, made, for example, in the form of discriminator (see Sect. 3.1 below).

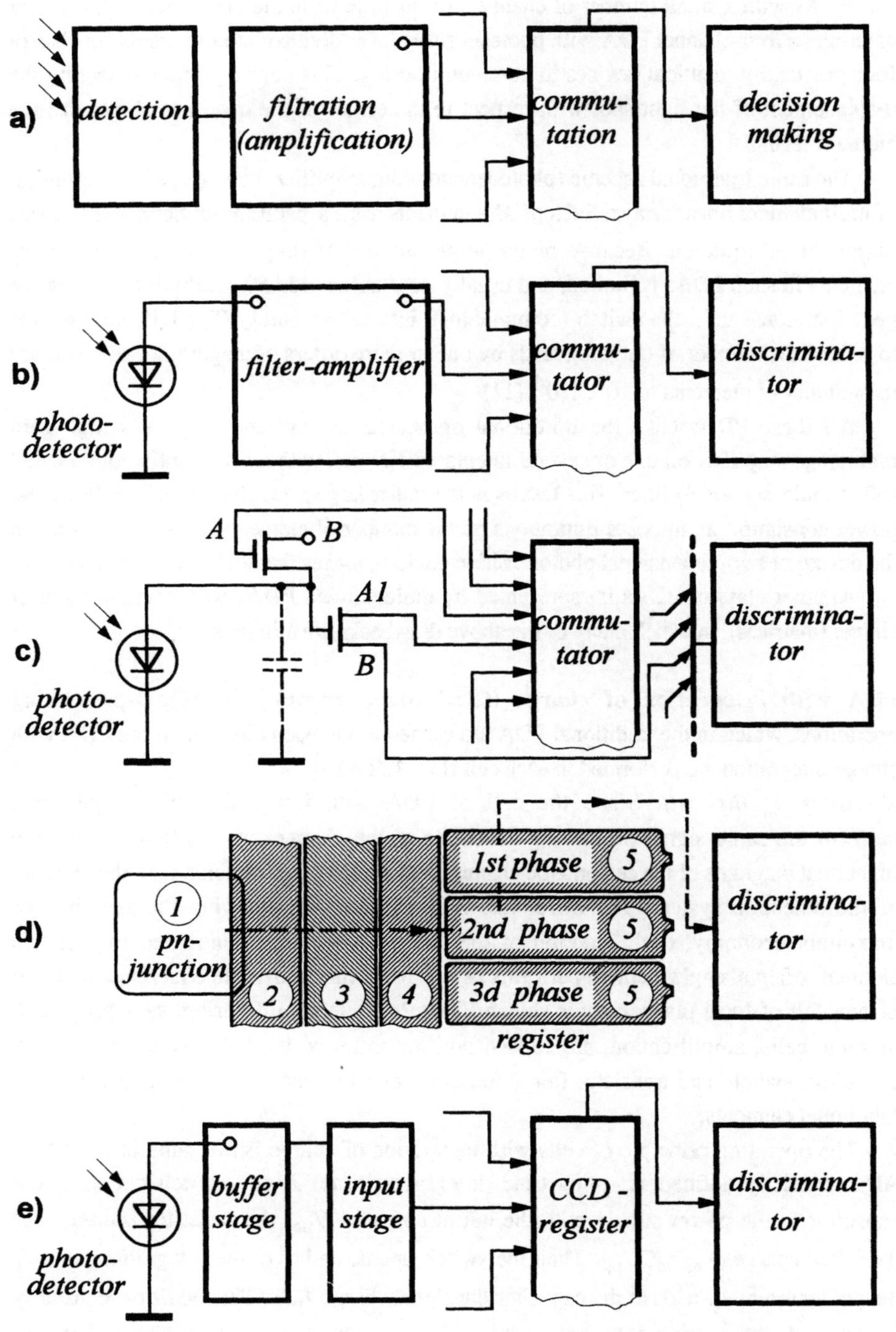

Figure 1.1: By their functional structure, PDAs are detection systems: (a) functional diagram of a detection system; (b) PDA of traditional type with amplifiers; (c) PDA with integration; (d) CCD with photodiode cells; (e) hybrid IR CCD.

PDAs with a small number of channels differ little from the one-channel PDAs. For example, a four-channel PDA with photosensitive layer divided into quadrants consists of four practically identical devices in common housing. This device permits detecting the displacement of the light spot with respect to the center of the quadrant photosensitive element (PSE).

The same traditional scheme (photodetector cum amplifier) can be used in the design of multielement linear arrays. Each of M_e elements has a separate amplifier and a separate output of information. Because of the large number of output leads, the number of elements in such PDAs is limited, and usually does not exceed 50 ... 200. Introducing the next functional unit, the switch (commutator), into the assembly (Fig.1.1b) allows one to reduce the number of outgoing leads by one or more orders of magnitude, and increase the number of elements to $10^3...10^4$ [11].

All these PDAs (like the traditional optoelectronic system) employ a high-gain multistage amplifier on one or several integrated circuits, with a sufficiently complicated (often multi-sectional) filter. This results in the rather large size, elevated heat release and power consumption, imposes limitations on the number of elements, creates problems in the design of two-dimensional photosensitive devices, makes the device more expensive.

Another class of PDAs is represented by multielement PDAs with accumulation of charge (matrices), in which many of the above drawbacks are eliminated.

PDA with integration of charge (focal plane array). All signal-processing operations, which in the traditional PDA are carried out in each channel, in the PDA with charge integration are performed in each cell (Fig. 1.1c-e).

Circuitry of the cell. Since the cell of PDA with integration of charge must perform the same signal-processing functions as the channel of traditional PDA, the functional diagrams of the cell and the channel are identical, and coincide with the diagram of the detection system. The difference is in the design: following the principle of maximum economy, each function of the detection system is entrusted to just one element; often a single element performs two or more functions. The equivalent diagram of one cell of focal plane array is shown Fig. 1.2a. Detection, filtering, switching, and, in some cells, amplification, are carried out, accordingly, by photosensitive element, capacitor, switch, and transistor (amplifier). In this way, one cell contains at least four functional elements.

The operating principle of cells with integration of charge is the same in all PDAs. After a strictly defined sampling time (integration time) T_i the switch connects the capacitor to the power supply with the initial potential U_{ps}; the capacitor accumulates the initial charge $Q_{ini} = CU_{ps}$. Then the switch opens, and over the integration time T_i the photocurrent I_p adds to the capacitor the signal charge I_pT_i. This addition is usually negative: the photocurrent discharges the capacitor by the amount I_pT_i. The switch and the capacitance together make up a dynamic filter: accumulation of charge over the time T_i between two consecutive flips of the switch forms the effective bandpass of the filter, $1/2T_i$ (for more details see Sect. 7.1).

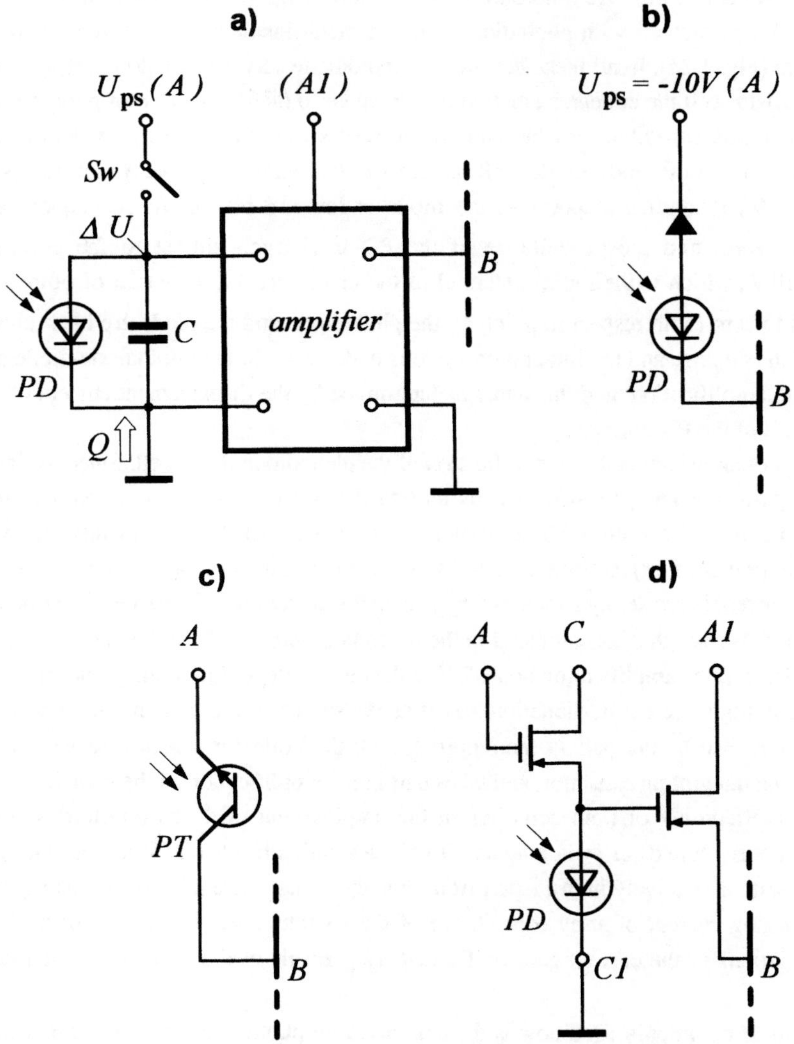

Figure 1.2: Charge integration cell is the basis of multielement PDAs: points A, A1 are connected to control buses, points B are connected to readout buses; points C, C1 are common for all cells in the matrix.

The amount of signal charge can be assessed in two ways: (1) by measuring the current though the capacitor at the instant the switch is closed. In this case the switch serves not only as the dynamic filter, but also as the switch which connects the cell to the common bus; and (2) by measuring the discharge voltage across the capacitor, $\Delta U = I_p T_i / C$. In this case the integrating capacitance must be connected to the gate of the transistor (amplifier input), and at the time of measurement the power supply voltage must be applied to the transistor, which thus serves not only as the amplifier, but also as the switch which connects the cell to the common output bus *B*.

Diagrams of the most common types of integration cells are shown in

Fig. 1.2b–d [12]. They are photodiode-diode cell (Fig. 1.2b), phototransistor cell (Fig. 1.2c), and the cell with photodiode and two metal-insulator-semiconductor (MIS) transistors (Fig. 1.2d). In all these designs the photodiode serves as photodetector (in the phototransistor cell the collector junction is equivalent to the *pn* junction of photodiode), and the junction capacitance of the photodiode serves as the integrating capacitor. The switch in the photodiode-diode cell is the diode: when negative potential (say, $U_{ps} = -10\,\mathrm{V}$) is applied to point *A*, the diode is forward-biased, its resistance falls almost to zero, and the capacitance of the PD is charged almost to the potential $U_{ps} = -10\,\mathrm{V}$, which is the initial potential of the cell. Next, the potential of point *A* is set equal to zero (with respect to point *B*), the photodiode and the diode are back-biased (the switch is open), and the integration cycle is under way. In the photodiode-diode cell there is no amplification, and the signal is determined by the displacement current of the capacitance of the PD junction.

The phototransistor cell is a modification of the photodiode-diode cell. Since the latter can be regarded as an *nppn* structure, by uniting the two *p* regions into a common thin layer we come to the structure of a bipolar *npn* transistor. At the time of measurement the switch (the emitter junction) is biased in the forward direction, and the net collector (emitter) current is greater by a factor of h_{21e} than the current which charges the collector capacitance. We see that, as compared to the photodiode-diode cell, in the phototransistor cell we also get an "amplification bonus". The degree of integration in the phototransistor cell is quite high: the emitter junction serves as the switch of the dynamic filter and the commutator, also taking part in the amplification; the collector junction serves as the detector, the integrating capacitor, and takes part in the amplification of the signal.

The shortcoming of both cells lies in the employment of diode (emitter) switch, which is poorly suited for switching small voltages and currents (since its resistance in the open state is relatively high). Apart from this, the signal current is contaminated with the recharging current of stray capacitance of the switch. Then, in the phototransistor cells, variations in the current gain coefficient h_{21e} result in the spread of sensitivity values.

The most commonly used now is the cell based on photodiode and MIS transistors, such as the one shown in Fig. 1.2d. Although the degree of integration in this kind of cell is low (each function of the equivalent diagram (Fig. 1.2a) is entrusted to a separate element, with the exception of the capacitance), all functions are performed in the best possible way. One of the transistors recharges the capacitance, serving as the switch of the dynamic filter, and does this job much better than a diode. The other transistor (amplifier) carries out the above-described procedure of reading the resulting jump of potential across the capacitance, so that the recharging current of the switch is irrelevant.

A special place among the multielement PDAs belongs to charge coupled devices (CCD) [13–18]. The CCD cell is quite simple: it is a MIS structure, or, so to say, a capacitor whose lower metallic plate is replaced by a semiconductor layer (Fig. 1.3a)*.

* In modern CCDs the upper metallic layer is replaced by the conducting polycrystalline silicon, but the term MIS (or MOS, for metal – oxide – semiconductor) structure remains in use.

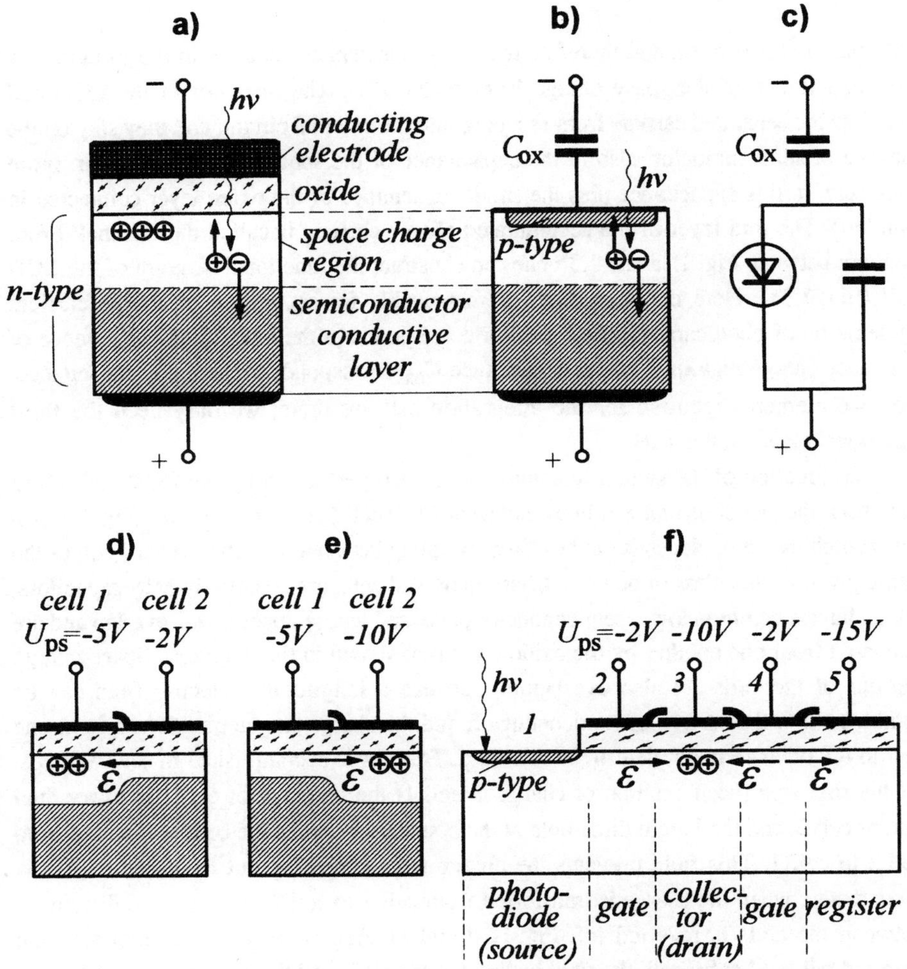

Figure 1.3: Integration of functions of photodetector, filter-integrator and commutator in the CCD cell (a – e), and separation of these functions (f).

The MIS structure operates in a nonequilibrium regime: back bias is suddenly applied to the structure, so that the majority carriers (electrons on the diagram) are repelled from the surface and form a broad subsurface space charge region. Similarly, a broad space charge region is formed in the high-resistance *n*-type region of PD, when back bias is applied to the strongly doped *p*-region (Fig. 1.3b; the PD is discussed in greater detail in Sect. 1.2). The difference, as we see, consists only in the way the back bias is applied: across the thin oxide layer in the CCD, and across the restraining *p*-type layer in the PD, so that large through currents are avoided in both cases. Similarity between the diagrams in Fig. 1.3a and 1.3b points to the "blood relationship" between the CCD cell and other integration cells. The physics is also similar: both in the PD and in the CCD there is the main space charge region, whose strong field separates the light-generated carriers. In the PD the separated charge carriers escape across the *p*-junction into the external circuit (when the latter is closed). In the integration mode of photodiode-diode cell, however, the external circuit is open, and therefore the charge carriers accumulate in the p-region (in the capacitance of the space charge layer of the PD). The insulator in the CCD cell prevents the separated carriers from escaping into the external circuit, and they stay on the surface of semiconductor, also in the capacitance of the space charge region (or, more precisely, in this capacitance plus the small capacitance of the oxide layer connected in parallel). The thin layer of the accumulated charge carriers is called the channel. From analogy between Fig. 1.3a and 1.3b one can construct the functional diagram of the CCD cell (Fig. 1.3c). Here the photodiode is responsible for separation of charge carriers (generation of photocurrent), the capacitor corresponds to the integrating capacitance of the space charge region; the input capacitance C_{ox} corresponds to the oxide capacitance. So, two elements required for the integration cell are there; we only need the third functional element, the gate.

The function of the switch (commutator) is also performed by the CCD cell. More precisely, the switch is realized in the adjacent CCD cell (MIS structure; see Fig. 1.3d,e). The switching action depends on the charge coupling between the cells, whence comes the name given to this class of devices. Charge-coupled cells are essentially MIS capacitors, which have a common lower semiconductor plate, and whose upper plates overlap and are insulated from one another by the oxide layer (not shown in the diagram). Space charge regions of the cells are also overlapping, so that a longitudinal electric field can be established on the surface of semiconductor, and the charge can be transferred from one cell to the other through the drift in this field. Transfer is accomplished in several steps. At the first step (accumulation of charge in cell 1) the potential of cell 1 is lower than that of cell 2, and the longitudinal field at the boundary between the cells is directed from cell 2 to cell 1. This field prevents the charge from spreading out from cell 1. At the second step, when the charge is transferred from cell 1 to cell 2, the potential difference between the cells is reverted (cf. Fig. 1.3d and e). Accordingly, the direction of the longitudinal field is flipped, the field is directed to cell 2, and the charge is transferred to cell 2.

CCD has the highest degree of integration: we see that all functions (detection, filtering – accumulation, commutation) can be performed by one MIS cell. In some designs, however, these functions are entrusted to different cells. For example, in matrices with interline transfer the detection and integration are carried out in a special MIS cell, and then the charge is transferred to the CCD register (which is a sequence of charge coupled MIS structures), which performs operation of commutation (charge transfer to the output bus). In CCD with PD all three functions are separated (Fig. 1.3f): detection is carried out by photodiode 1, integration by cell 3, and commutation by cell 5 which is part of the CCD register (the register is perpendicular to the plane of the drawing). The gates 2 and 4 are auxiliary: they separate the above three regions and control the movement of charge carriers between the latter. The potential values indicated on the diagram correspond to the integration regime. The potential on the surface of semiconductor under gate 2 is set to about – 2 V. The potential of photodiode at the initial state is zero, and so the longitudinal field draws out the holes from the photodiode.

This process goes on until equilibrium is established: the longitudinal field vanishes, and the photodiode is automatically biased to about the potential of the gate (–2V). Observe that the charge of the holes drawn out from the PD can be dumped from the integrator, so that it is not registered at the output. When the light generates charge carriers in the PD, the equilibrium is upset again, and the PD is biased in the forward direction. Under the action of longitudinal field the light-generated carriers move under the electrode 2 and further under the integrator electrode 3, gradually equalizing the surface potentials of the *p*–region 1 and the gate 2. The operation of switching is obvious: one has to remove the potential barrier from electrode 4 (by applying the potential of about –12 ... 13V), so as to allow the charge carriers to flow from integration cell 3 to the commutating CCD register (cell 5).

Now why is it that, despite the feasibility of building a completely integrated device, different functions in some designs of the CCDs are still entrusted to separate elements? The answer is clear: because each element, each MIS structure will cope with one specific job better than with three jobs at once. The absence of light-absorbing polycrystalline layer on top of the PD will improve the sensitivity of the photodetector. If the integration cell is separated from the switching CCD register, charge integration and commutation can be conducted in parallel. Separation of photodetector from the CCD register makes it possible to shield the register from external light, thus preventing "blurring" of the image when the charge is transferred within the register. As indicated above, in case of integration cells the preference is also given to the cell with PD and two MIS transistors: quality of signal processing is put above the utmost degree of integration.

The most important advances in the development of multielement PDAs, and, in the first place, matrix PDAs, are associated with silicon CCDs. This is due to the availability of high silicon technologies, and to the functional flexibility of CCD structures. As far as the detection of weak optical signals is concerned, the important advantage of the CCD cell over the alternative types of integration cells is that there is no initial charge on the surface (in the channel), which facilitates the signal processing procedure. While it is necessary to maintain a certain background charge (the so-called "fat zero") in CCDs with surface channel, so as to fill up the surface traps and prevent trapping of charge carriers, there is no trapping in CCDs with buried channel, and there is no need of introducing the background charge (the buried channel is located 0.3 ... 1 μm below the surface of semiconductor, which is achieved by the specially adjusted doping profile).

Finally, the last type of integration cell we are going to discuss is the ordinary photoresistor (PR) [19]. Of course, a PR performs the first function of the detection system, the detection of optical signal. The light-generated charge carriers are stored in the bulk of semiconductor. The function of the switch, which after a certain time erases the accumulated charge, is performed by the process of recombination. Because of recombination, the light-generated charge carriers only persist in the semiconductor on the average for their lifetime τ_{life}. Accordingly, we say that the integration in PR is carried out over the lifetime of charge carriers. So, owing to the recombination, the PR performs not only the detection, but also the next function of the detection system, the filtering (the bandpass of the PR is similar to that of the one-section RC-filter with the time constant τ_{life}; see Sect. 1.2, 5.1, 5.2). In the PR there is also amplification of the signal, the so-called photoresistive amplification. Switching also does not present any difficulties and is accomplished by applying the power supply voltage to photoresistor at the time of sampling. Recall that a PR is a passive device, and the current will not flow in the external circuit in the absence of power supply voltage.

All the same, despite the high degree of integration and even the availability of amplification, the PR-based detection system cannot compete with a CCD cell: in case of PR it is not possible to adjust the filter (because τ_{life} cannot be changed), frame-to-frame interference is strong, generation of light carriers usually takes place against the high background concentration of charge carriers.

How to build multielement PDAs from separate integration cells. There are two basic ways of organizing the above-discussed integration cells into the common matrix structures.

The first is based on the lattice of horizontal and vertical buses: "rows" and "columns" (Fig. 1.4a). Cells is placed at their intersections: points A (switch control points) are connected to the row buses, and points B (signal outputs) are connected to the column buses. If necessary, the buses can be twinned: for example, in the cell with PD and two MIS transistors the transistors must be controlled separately by applying appropriate pulses to points A and $A1$. Certain leads of the cells can be joined together: for example, point B in Fig. 1.2d (the drain of the switch transistor) can be made common for all cells. The matrix is operated by sending sequentially the sampling pulses to the rows, and reading column by column the outputs from the elements of the active row. To connect sequentially the column outputs to the common output bus, the lattice can be equipped with the horizontal register. Any row in the lattice can be addressed independently from the rest, and the device of this kind is appropriately termed the line-interrogation array.

All types of cells are organized in the matrix according to the same lattice principle: those shown in Fig. 1.2, and the phototransistor cells, and even the CCD cells. In the latter case the vertical bus is a MIS structure which is charge-coupled with all integrating MIS cells of the column. The charge of the active cell is transferred to this vertical bus and causes the potential jump, which is registered by the output amplifier based on a MIS transistor. This arrangement is called the charge injection device (CID), with reference to the operation of charge erasing. After completion of the readout, the potential of the cell becomes close to zero, the charge is injected into the substrate and rapidly recombines with the majority carriers (on the surface there are no majority carriers, nothing to recombine with in the space charge region). In the earlier modifications of the CID the injection of the charge into the substrate was also used for readout of the signal.

Another way of organizing the cells into the matrix is the "conveyor principle": the charge is relayed to the common output from one cell to another. CCDs seem to have an inborn aptitude towards this kind of arrangement: the CCD cells are sequentially linked together into CCD registers through charge coupling. To ensure the transfer of charge in the desired direction, each CCD cell must contain three MIS cells (or two cells with a

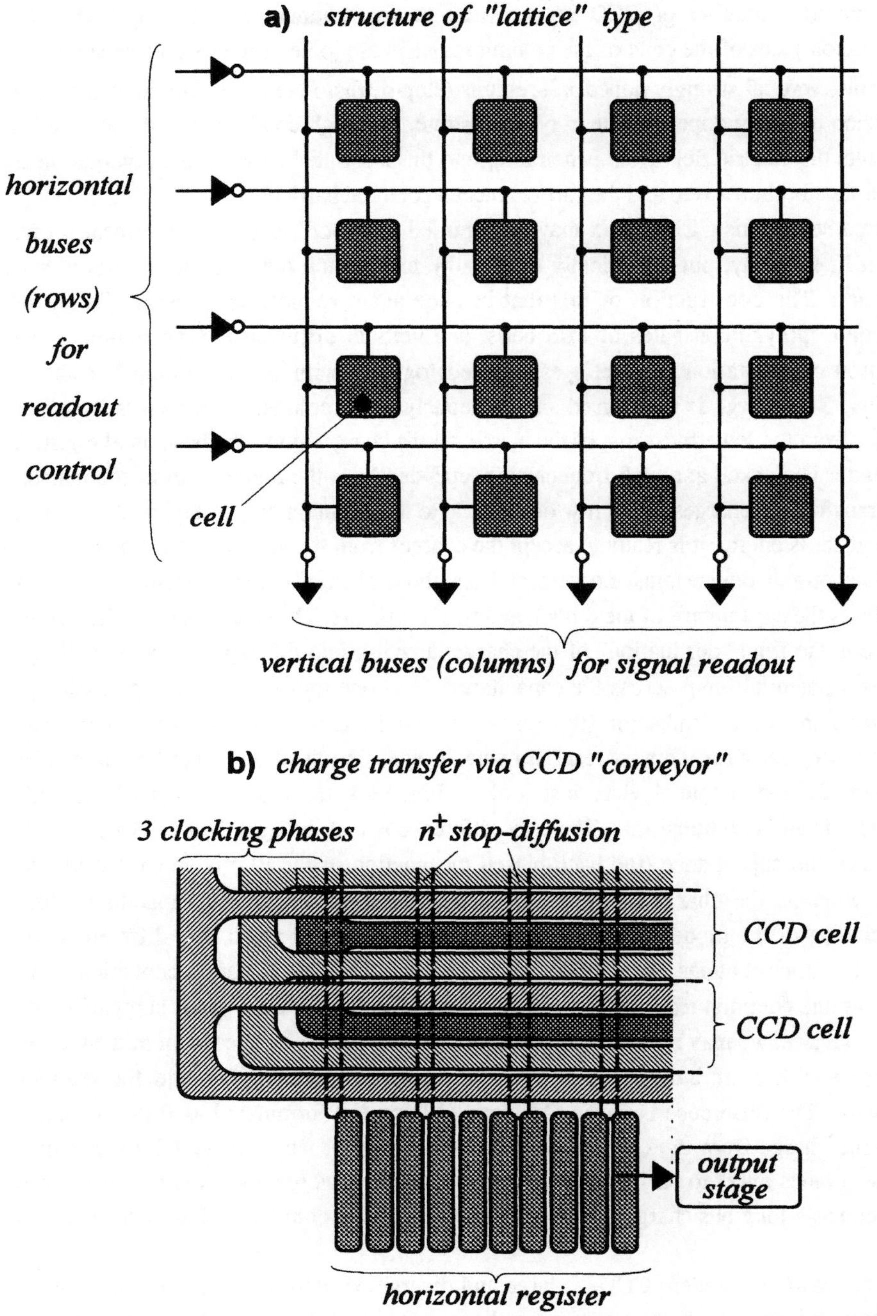

Figure 1.4: Two main principles of construction of matrices with integration: (a) lattice structure; (b) charge conveyor.

stepwise potential profile in the channel). All cells in the register are controlled by the common pulse generator (from one to four phases are usually required). The matrix is built up of a number of CCD registers, as shown schematically in Fig. 1.4b. The polysilicon gates of the cells of the column registers are joined together. To separate the columns, vertical strongly doped n^+-regions (stop-diffusion regions) are created by the diffusion of donor dope into the *n*–type substrate. The high conductivity of these zones prevents the electric field from penetrating into the substrate below, so the potential near the surface is almost zero and the carriers meet a potential barrier.

In a sense, this CCD matrix may be regarded as a modification of the linear array: take a linear array, put it vertically (Fig. 1.4b), and stretch the gates in the horizontal direction. The construction of this matrix once again reduces to a simple lattice of horizontal polysilicon gates of MIS cells, and vertical diffusion *n*–type regions. The common commutation system is established together with the additional horizontal register. The charge is transferred simultaneously in all columns downwards by clock pulses; from the lowermost row of the matrix all the charges go to the horizontal register. The latter is clocked at a high frequency independently of the regular part of the matrix, and transfers the charges of the row one by one to the common output device. As soon as the register is empty, it is ready to accept the charges from the next row, and so on.

The output device must contain at least three elements: the capacitance (which usually is the capacitance of the diode), and two transistors. The capacitance of the output device is the final "destination" of the charge: arriving into the capacitance, the charge causes a potential jump across the capacitance. First one transistor reads out this jump, and then the other transistor (the resetting switch) erases the charge, getting the capacitance ready for the next readout cycle. There is no need of giving here a special diagram for the output device: just look at Fig. 1.2d, the diagram of the cell with photodiode and two transistors. The only difference is that in Fig. 1.2d the charge which comes to the capacitance (the back-biased *pn*-junction of the PD) is generated by the light, whereas the charge in the CCD output device comes to the capacitance (the capacitance of the diode) from the CCD register. One might put it in a different way: while the lattice contains a readout device in each cell, the CCD is more "economical" and employs one common readout (output) device for all cells. Continuing our comparison of output devices, we may return to the CID. As indicated above, each column of the CID is equipped with a similar output device, consisting of the capacitor and the readout transistor. The difference between CCD and CID can be formulated as follows: in the CCD the charge from the cells goes to the output device, while in the CID the output device "comes over" to each cell of the column. In other words, the readout capacitor is designed as a long bus charge-coupled with all cells in the column, and its capacitance is large.

The family of modern CCDs is large and diversified. As indicated, the functions of detection, integration, charge transfer may be entrusted to different CCD cells. The device may include various auxiliary CCD register sections, but the conveyor principle is always employed, sometimes with branching, combining, or ringing of the charge transfer routing. For example, in Fig. 1.4b the routes of charge transfer are joined: the charges from the columns come to the common horizontal register, which is used for conveying the charge to the output device. The conveyor principle is also used in some other, less common devices, such as scanistors, "bucket-brigade devices", SPRITE (Signal PRocessing In The Element) devices [20]. The latter will be briefly discussed in Sect. 7.1.

Multielement IR PDAs with integration. Silicon, the basic material of semiconductor technologies, has been used for creating highly advanced CCDs (linear arrays of 1000, 2000, 5000 elements, matrices 500 by 500 and 1000 by 1000 pixels), which are sensitive in the spectral range from 0.5 to 1.1 μm. Considerable difficulties are usually encountered in the development of multielement PDAs (especially with a large number of elements) sensitive to IR radiation in the ranges 2 ... 3, 3 ... 5, 8 ... 14 μm (these ranges correspond to the atmospheric transparency windows). Those devices which are sensitive to the wavelengths above 1.1 μm are referred to as IR devices (IR PDA, IR CCD, IR matrices). We do not include silicon-based PDAs with intrinsic absorption into this class, although the range from 0.7 to 1.1 μm, strictly speaking, belongs to the near infrared spectrum. As far as IR CCDs and IR CIDs are concerned, the development pursues four major directions.

Silicon CCDs based on Shottky barrier. "Sensitization" of silicon to IR radiation is achieved by connecting a Shottky photodiode in parallel with the photodiode of the cell of silicon CCD (compare Fig. 1.3f and 1.5a). The Shottky photodiode (surface barrier diode) is a metal-semiconductor structure, which means that in the CCD cell an additional metallic electrode must be deposited directly on the substrate and connected electrically with the diffusion region. With specially selected metals (for instance, platinum or palladium) it is possible to form a low surface barrier ($\approx 0.2\,\mathrm{eV}$), whose field separates the carriers excited in the metal by IR radiation.

Doped silicon CCDs. Silicon is sensitized to IR radiation by introducing an impurity with low ionization energy ("shallow impurity") into the photosensitive cell of the CCD, so as to make possible the impurity absorption of IR radiation.

CCDs and CIDs based on narrow-band semiconductors. These devices employ such narrow-band semiconductors as InAs, InSb, CdHgTe (mercury-cadmium-telluride, or MCT). Theoretically, the topology of such devices could have been the same as that of silicon-based CCDs; however, the low transfer coefficient justifies (if not dictates) the use of the CID ideology in case of narrow-band materials.

Hybrid IR CCDs. These devices are a combination of a narrow-band photodetector and a silicon CCD. The photosignal is delivered to the CCD electrically, for which purpose each cell is supplemented by the injector. In its simplest modification (direct injection of photodiode signal) the injector is already known to us: it is diode 1

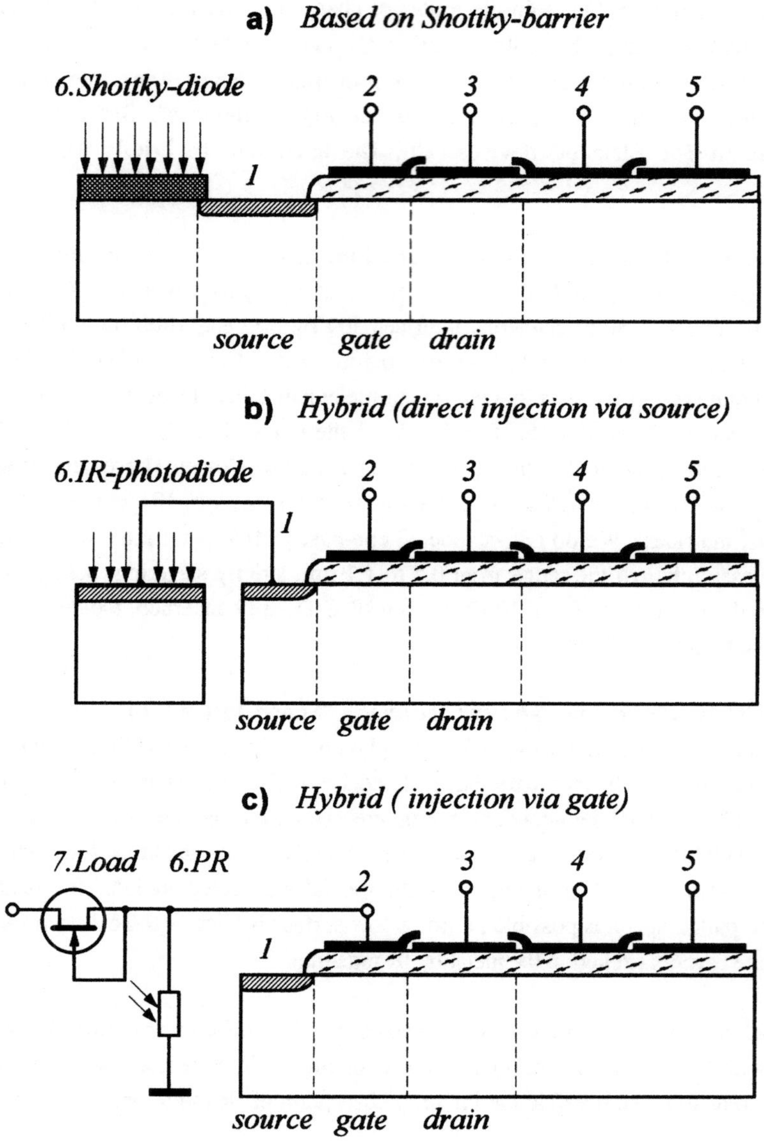

Figure 1.5: IR CCD cells: (a) based on Shottky barriers; (b) hybrid with direct injection via source; (c) hybrid with electrical injection via gate.

(photodiode) of the CCD cell in Fig. 1.3f. The IR PD is connected in parallel with this PD (Fig. 1.5b). The circuitry of the hybrid IR CCD cell, the Shottky barrier CCD, and the silicon CCD with PD is the same (compare Fig. 1.3f, 1.5a and 1.5b). In fact, the Shottky barrier CCD may be regarded as a modification of the hybrid CCD, in which the IR photodiode is constructed on the common substrate with the silicon CCD. When the photodiode is illuminated, the diffusion region is biased in the forward direction, and the carriers are injected into the integration cell. Interestingly enough, the Shottky photodiode and the CCD in the more recent designs are produced on two separate crystals, so from technical standpoint these devices do not differ from other hybrid CCDs [21].

If we take a closer look at these three designs, we shall also see that the injector is based on a MIS transistor, whose drain 3 is realized as an integration MIS cell rather than a strongly doped (diffusion) zone. Such MIS transistor is the basis of any injector: after all, there always must be the source 1 from which the charges are injected, the region where the charges are stored (drain 3), and the gate 2 which separates the two. In the circuits considered above, the signal was applied to the source. It is also possible, however, to apply the signal to the gate, which is especially convenient in case of photoresistors (Fig. 1.5c). Here, as in an ordinary amplifier, the signal voltage applied to the gate controls the drain current, which charges the capacitance of the drain rather than going to the external circuit. In other words, the diagram in Fig. 1.5c corresponds to operation into capacitive load in the drain circuit. The use of electrical rather than optical injection in the hybrid CCDs gives rise to considerable complications, which are due to the highly increased resistance of the injector at the low levels of photosignal. At low signal levels the MIS transistor operates in the sub-threshold regime, where its resistance depends not on the channel, but on the barrier between the source and the gate regions [22–24]. As will be shown in Sect. 1.2, the resistance of the barrier is

$$R = \frac{\varphi_{\text{th}}}{I} = \frac{0.007}{10^{-9} \ldots 10^{-7}} = 7 \cdot \left(10^4 \ldots 10^6\right) \text{ Ohm} \,, \tag{1.1}$$

where $\varphi_{\text{th}} = \kappa T/q$ is the thermal potential (κ is the Boltzmann constant, T is the absolute temperature, q is the charge of electron). For this estimate we have taken the value of φ_{th} at $T = 77\,\text{K}$, and the values of current I typical of the IR PD for the spectral ranges 2 ... 3 and 3 ... 5 μm, with the photosensitive area of $100 \times 100\,\mu\text{m}$. These currents are calculated in the next section (see Table 1.1). The high resistance results in the large time constant of the RC circuit of the injector: for the typical capacitance of the source circuit $C = (1 \ldots 10) \cdot 10^{-12}\,\text{F}$ the time constant is $0.7 \cdot \left(10^{-7} \ldots 10^{-4}\right)\text{s}$. If the required current injection time is of this order of magnitude or even shorter, there is no time for the injection of charge carriers. The only solution for this problem is pre-amplification. So the simplicity has to be sacrificed, and hybrid IR CCDs are equipped with pre-amplifiers (Fig. 1.1e).

In the first two types of IR CCDs (Shottky and doped CCDs), the silicon technology makes it possible to construct large matrices (100 by 100, 500 by 500 elements) [25, 26]. Recently, formats of the order of 1000 by 1000 have been realized with Schottky barrier image sensors [27]. Unfortunately, the sensitization of silicon is not good enough, and the quantum yield of such devices in the spectral range 2 ... 5 μm is as low as 0.1 ... 5%. The quantum yield of the two other types is high, since they are based on the intrinsic absorption in narrow-band semiconductors, but the number of elements is limited. To the limitations explained above we may add the following: (1) the technology of narrow-band semiconductors is far not as advanced as the silicon technology, and (2) it is difficult to join two different crystals in the hybrid CCDs. Fundamental problems in the integration IR PDAs are also associated with the large background currents produced by thermal radiation (Sect. 1.2, 7.2).

The following three directions of development of IR matrices are not related to the CCDs.

Lattice based on the cell with MIS switches. This is also a hybrid device. The IR PD is built on the narrow-band semiconductor, and is connected with the silicon crystal. This time it is not a CCD, but rather a lattice of cells based on MIS transistors. Such a cell is akin to the cell with silicon PD and MIS transistor switches shown in Fig. 1.2d. In principle, one can realize all the elements shown in the diagram on silicon, and connect the IR PD in parallel to the silicon diode (Fig. 1.6a). This hybrid cell will work, but it will not work well. The fact is that the resistance and the bias voltage of IR PD (usually based on InSb or MCT) are much lower than those of the silicon diode, so that the latter is shunted. The efficiency of charge integration is reduced. In order to separate the integrating capacitance of the silicon diode from the low-resistance IR PD, a common-gate buffer FET [28] is placed between the two (Fig. 1.6b). The photocurrent of IR PD goes through the source and the channel to the drain circuit and charges the capacitance C_i. Now this capacitance is connected to the high-resistance drain circuit, and is no longer bypassed by the low-resistance IR PD.

By introducing the buffer transistor we have actually returned to the input circuit of the hybrid IR CCD. This becomes especially clear if we redraw the electrical diagram of Fig. 1.6b as a skeleton chart (Fig. 1.6c). We see that both devices have IR PD, the diffusion source region, and the MIS switch. The differences only appear in the drain region: in the IR CCD it is a MIS cell (that is, CCD cell), and in the transistor matrix it is an ordinary diffusion region.

The designers of hybrid IR matrices are always faced with the choice between the silicon CCD and the MIS switch. As we see, the input stage of these two devices is the same, and it is the input stage that has to be matched with the IR PD and that is responsible for the input noise level (Sect. 2.4). Because of this, both solutions are currently in use. When, however, the background currents are large, as is the case, for instance, in the range 8 ... 14 μm, the matrix with MIS switches has certain advantages, since it ensures a more efficient integration of the charge. This is explained by the following two circumstances.

(1) In parallel to the diffusion drain a MIS structure with a very thin oxide layer is connected (the thickness reported in [28] is 250Å, which gives a very high value of the specific capacitance, $\sim 1.6 \cdot 10^{-7}\ \mathrm{F/cm^2}$). Such a high specific capacitance is hardly achievable in a CCD cell, because a low-loss charge transfer can only be ensured when the oxide-semiconductor boundary is well defined, which requires a rather large thickness of the oxide layer (~1700Å). For this reason the specific capacitance of the CCD drain is considerably lower ($\sim 2 \cdot 10^{-8}\ \mathrm{F/cm^2}$).

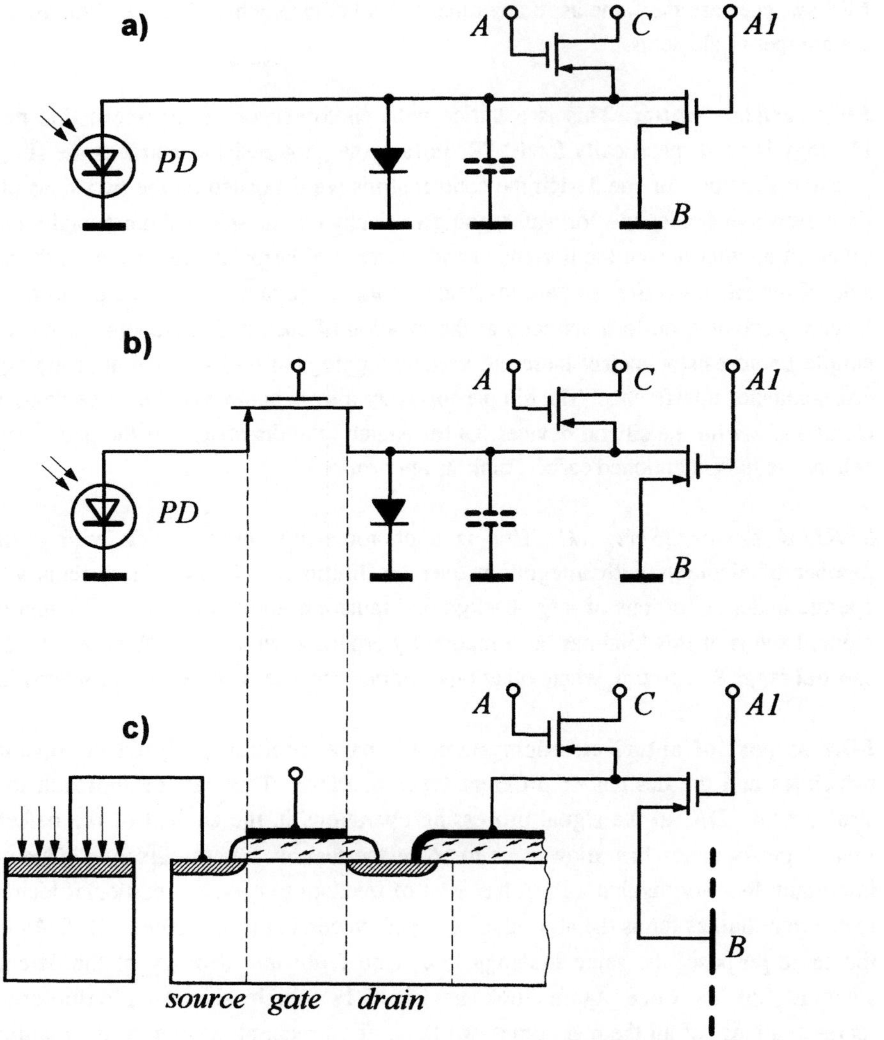

Figure 1.6: Cell of IR PDA with MIS FETs is akin to hybrid IR CCD cell.

(2) A CCD register occupies a much greater area than the vertical bus of the matrix. Because of this, there is more area available for the integrating capacitance in the matrix than in the IR CCD. On top of that, the CCD cell of the register itself has a much smaller area for the integrating capacitance C_i, which also reduces the efficiency of integration.

The lattice matrices (based in our case on MIS switches) and the CCD matrices have divided their "spheres of influence". Because of their large integrating capacitance, the lattice matrices are more advantageous in case of strong background radiation, whereas the CCD matrices are better suited for situations when the background radiation is low, since lower capacitance means less amplifier noise (see Sect. 2.5).

All the design and technological difficulties in manufacture of hybrid IR matrix with MIS switches are the same as in any other hybrid device, which imposes limitations on the number of elements.

Photoresistor matrix. This is a lattice with photoresistor in each cell; this matrix ideology is used specifically for IR PR. In fact, the photoresistor matrix is the simplest possible structure. In one design the contact strips are deposited on the two sides of the photosensitive semiconductor wafer, going vertically on one side and horizontally on the other. In another design the horizontal and the vertical strips are deposited on the same side of the silicon wafer, and are insulated by the oxide layer, while the photoresistive layer is deposited on to the silicon at the location of each cell. Of course, even in the simple photoresistor matrix there are various topological tricks aimed at reducing the cross-channel interference. The simple topology allows achieving the same density of elements as with the silicon devices. Unfortunately, the drawbacks of the photoresistor cell, as we have mentioned earlier, limit its applications.

SPRITE device (Sect. 7.1). This is a photoresistor-based device with a small number of elements with integration over the lifetime. It is used in systems which operate under conditions of large background radiation, in the time lag and integration mode. Device of this kind has been reportedly produced with the MCT material for the spectral range 8 ... 14 μm, where other types of devices meet with serious problems [29].

PDA as part of optoelectronic system. We have carefully analyzed the operating principles and the designs of different types of PDAs. This careful approach to the structure of PDA, to the signal processing operations in the context of the detection system performance has allowed us to solve the task we set to ourselves. Our most important discovery is that a PDA has a lot of freedom to choose a particular technical realization, but, as far as the algorithm of signal processing is concerned, all PDAs have the same purpose, the same ideology, the same faith: the ideology of the detection system. Let us once again look at Fig. 1.1, which shows "plenipotentiary representatives" of all the main types of PDAs. The structural diagram of the traditional PDA with amplifier (Fig. 1.1b) does not differ from the functional diagram of the detection system (Fig. 1.1a): each function of the detection system is realized in the electronic section of the PDA. Integration PDAs — both the CCDs and the cells comprised of photodetector, capacitance, switch — are intentionally shown in Fig. 1.1c,d in those modifications where all signal processing operations (detection, filtering, switching) are carried out by separate components. This helps to emphasize the commonalty of the functional diagrams of all types of PDAs. An especially instructive example is the hybrid IR CCD (Fig. 1.1e). This is one of those centaurs which may be classified as both a traditional PDA with amplifier and an integration PDA. Indeed, the photodetector and the photosignal processing circuit are produced on separate crystals, like in a traditional PDA. In some devices of this kind the gain of the preamplifier is high enough, and the preamplifier also doubles as a filter, so that the CCD is for the most part concerned with the commutation. This hybrid IR CCD does not differ from the traditional PDA even in the design. In another modification, when it is possible to join directly the photodetector and the CCD, the structural scheme and the operating principle of the device are the same as those of the CCD shown in Fig. 1.3d. Because of the universality of the functional scheme, the circuitry and even the components are also very much similar even in the seemingly different types of integration PDAs. These similarities have been more than once emphasized above. After all, the integration devices (both their cells and their output sections) are most usually built of two basic structural elements: the region of space charge (CCD cell, PD), and MIS transistors (reset, readout). This universality will help us to carry out the analysis of signal processing in the PDA in the general form.

Our study has also demonstrated that both the ordinary photodetector, equipped with appropriate optoelectronic periphery, and the various kinds of PDAs (also supplemented, if necessary, with various units) are detection systems, and consistently carry out the already known sequence of signal processing operations. As a matter of fact, this conclusion follows directly from the definition of PDA. Differences between the systems using, respectively, photodetectors and PDAs are found mostly in the design solutions. The electronic modules of optoelectronic system are, as a rule, built of discrete components and encapsulated ICs. Hybrid-film technology is used in PDAs with amplifiers, and solid state technology in PDAs with integration. Now, if the PDA has taken upon itself certain functions of the optoelectronic system, what is then the "raison d'être" of this class of devices? Or is it perhaps just the matter of shifting some of the responsibilities of the optoelectronic systems engineer to the designer of PDAs?

The answer is negative: the use of PDAs instead of photodetectors in optoelectronic systems has certain advantages, and opens up a number of essentially new possibilities.

1. First of all we should name the traditional advantages of microminiaturization, the transition to the hybrid and solid state technologies: small mass, size, power consumption, improved reliability, etc. Of greater importance, however, are the entirely new qualities brought about by the advent of PDAs.

2. To reduce the contribution of the amplifier to the total noise, the input capacitance must be minimized (Sect. 2.4, 2.5). The integral design, and especially the realization of photodetector and the first stage (FET) on the common substrate, reduce the stray capacitance and hence the total input capacitance.

3. At the threshold-level illumination the signal of photodetector is small (especially when there is no internal gain), and is easily jammed. Preamplification of the photosignal in PDAs makes the signal "transportable".

4. In a multielement photodetector the number of leads is the same as the number of elements; each of these low-signal leads has to be connected to the separate input of electronic module of the optoelectronic system. When the number of elements reaches one

hundred and more, the assembly becomes cumbersome, and stray capacitance increases. When switching is done within the PDA, the number of signal outputs can be reduced dramatically (even down to just one), which eliminates the problem.

5. In case of cooled multielement photodetectors, each element must be connected separately to the cryostat. The large number of electrical leads spoils the thermal and mechanical properties of the device, complicates the design. If the electronic signal processing circuit (amplification, filtering, commutation) has a low power consumption, and is small enough to be placed near the photosensitive element (PSE) in the focal plane, then the number of connections can be reduced by orders of magnitude, and the device will be free from the above shortcomings.

6. It is only a PDA that can ensure sufficient two-dimensional format of the image. In a photodetector there must be a separate signal lead from each element, which can be done to the formats 6 by 6, perhaps (with much effort) to 10 by 10 elements. When the number of elements is greater, it is necessary to incorporate commutation into the device, which then becomes a PDA by definition.

Since a PDA, unlike a photodetector, performs a whole sequence of operations instead of a single one (detection), and replaces sophisticated electronic modules of optoelectronic system, it should be regarded not as an electronic component, but rather as a part of the system. Generally, a PDA may include the following subsystems.

Optical system. It seems that only in the simplest one-channel PDAs the optics may consist of an inexpensive protective window. Even in one-channel threshold-level PDAs the effective aperture is increased through the use of rather complicated immersion lenses. PDAs intended for fiber optical systems may include focons for optical contact with the light guide. The highly sophisticated optical systems use multichannel cooled PDAs; the windows in such systems always are optical elements: they are part of the imaging system, and must be transparent for IR radiation (they are made of such materials as sapphire, optical ceramics, semiconductors). The background thermal radiation is eliminated by optical filters which match the spectrum of the source, and the aperture delimits the field of view. Specially designed honeycomb diaphragms can form the desired field of view in two mutually perpendicular planes. The filter and the diaphragm must be cooled; otherwise they will themselves create thermal radiation falling on the photosensitive element.

Photosensitive element. This is a proper semiconductor structure (chip or its part) which receives the light signal.

Electronic signal processing circuits. These are produced using at least the hybrid-film technology with unenclosed components. Further development followed the path of integration, reducing the length of connectors between PSE and electronic units.

As we have seen, the photosensitive elements and the electronic processing units in CCD, CID, and PDA with MIS switches are realized on one or two chips (connected with indium pins). It is not by accident that such devices are generically referred to as focal plane arrays (FPA). These key words summarize the result of development in technology and design: the electronic unit is brought as close to PSE as possible, and are located in practically the same plane. This is the best solution to realize all six advantages of PDA listed above.

The systems considered above (optical, photosensitive, electronic) comprise the signal processing path (at optical or radio frequencies). To support their operation, PDA must also include a number of auxiliary systems.

Cooling system [2, 5, 30-31]. A thermoelectric cooling system is sometimes used even in silicon PDAs (single-element threshold-level, and CCDs), although these devices can and often do work at the ambient temperature. The dark current of PSE can be reduced by lowering its temperature down to –10...+10°C. The working temperatures of the intrinsic semiconductor photodetectors for the spectral ranges 2 ... 3, 3 ... 5, 8 ... 14 μm usually lie between 77 and 190 K. The cooling is done through microscopic heat exchangers, incorporated in the PDA. These heat exchangers may operate from both open-cycle sources (high-pressure gas cylinders), and from closed-cycle networks (throttle-expansion cryogenic machines). Less common are radiation and thermoelectric cooling systems. Especially sophisticated helium and neon cooling systems are used for doped photodetectors, whose working temperatures are below 77 K.

Auxiliary electronics. These devices include secondary power sources for photosensitive elements and amplifiers, systems for maintaining the operating regime of the APD, whose operating voltage increases with increasing temperature. Clock pulse generators are necessary for PDAs with internal switching, and especially for the integration PDAs. Usually a CCD matrix requires from 10 to 15 pulse buses, and in hybrid CCDs this number may be as high as 20 or 25, since the clock pulses are also necessary for controlling the input circuits. Special clock pulse ICs are available for some types of CCDs; they can also be realized on the common wafer with the CCD. A PDA may also incorporate various sensors (for measuring the temperature of the PSE, for life diagnostics), as well as electronic control circuits for the cooling system.

Design. All the above subsystems must be packaged into a single unit, which is then placed in the optoelectronic system and adjusted to the accuracy of 10 ... 100 μm. Some integrated circuits of PDAs (for instance, CCDs) are assembled in the relatively simple standard capsules. More often, however, the housing for PDA corresponds to the complexity of the device. For example, the design of PDAs for optical fiber systems (receiving modules) includes optical connectors for plugging into the line. Especially complicated is the design of cooled PDA, which is a combination of semiconductor technology, radio engineering, optics and cryogenics. The design calculations include thermal balance, elimination of mechanical resonances at working frequencies, reliability study. Of special importance is the junction between PSE and further electronic modules. In the IR linear array, when the cooled PSE is placed in the cryostat, while the preamplifiers remain outside, the critical section is the electrical connection between them. In the hybrid devices special attention has to be paid to the junction between the PSE crystal (which usually is a narrow-band semiconductor) and the silicon wafer. Heat sinks are necessary for the electronic modules, especially for the multichannel amplifiers. All systems must follow the principle "do no harm to thy neighbor": clock pulse oscillators and cooling systems must not induce stray currents in the signal circuits; the electronic modules must not release too much heat, so as not to harass the cooling system.

From arguments developed above it follows that PDA is a complex system, and its design must rely on the advances in many engineering disciplines. The overall design of the PDA, the choice of the structural scheme, the design of each subsystem (photodetector, electronics, layout, optics) must all be directed towards achieving the common "supergoal": the detection of weak optical signals. This is exactly what our book, volume number one in the library of the PDA designer, is about.

Summary. *In this book we consider PDAs for a broad class of optoelectronic systems for registering low-level pulsed optical signals. This signal must be registered by the detection system, which is a device for detection, amplification, filtering (separation from the noise), commutation in multielement devices, and, finally, deciding whether or not there is a useful signal. In the earlier systems the detection was carried out by photodetector, and further processing was entrusted to the electronic modules of the optoelectronic system. The advances in microelectronics have made it possible to create PDAs capable of performing not only the detection, but also all or some of the other functions of the detection system. The replacement of photodetector by PDA is not only a step forward in microminiaturization. PDAs are distinguished by the entirely new qualities: the ultimately low input capacitance and noise, transportability of the reinforced signal, the reduced number of information outputs owing to internal commutation, the possibility of creating two-dimensional (matrix) devices. Although the modern PDAs come in many varieties, they are all built (in whole or in part) according to the common functional diagram of the detection system. The main difference is in the method of filtering: in the traditional PDA with amplifier the amplifier serves as the filter for the signal; in the integration PDA the effective bandpass* $1/2T_i$ *is formed by the integrating capacitance and the switch which erases the charge in* T_i *intervals. The cell of any matrix with charge integration contains a photodetector, an integrating capacitance, and a switch for erasing the charge (sometimes there also is a readout transistor). The constituents of the cell may be integrated into a single element (for example, a CCD element).*

A modern PDA is a complex system, which may include optical, semiconductor, electronic subsystems. The structure of the PDA must be designed from the principles of the theory of detection of pulsed optical signal. The foundations of this theory as applied to PDAs are laid forth in the present book.

1.2. Photodetector

Detection. This operation, the conversion of optical signal into electric signal, is the first one carried out by the detection system. Detectors in PDAs are photoelectric semiconductor sensors (photodetectors), based on the intrinsic photoeffect. The intrinsic photoeffect works on the principle of photon count: each absorbed photon creates an electron-hole pair in the semiconductor*. The photodetector must count these pairs. The most simple and straighforward method of counting is realized in photodiode. We know that the space charge region (SCR) with a strong electric field arises in the *pn*-junction (Fig. 1.7a). Coming into the SCR (either through diffusion from the quasineutral *n* and *p* regions, or being generated directly therein), the electron-hole pairs are broken by the field: the positively charged holes move along the field towards the *p* region, and the negatively charged holes move against the field towards the *n* region. Photocurrent $I_{\rm p}$ starts in the external circuit if the latter is closed.

Assume that within a certain observation period T the photodetector has been struck by one photon with the energy $\mathrm{h}\,\nu = \mathrm{h}\,c/\lambda$, where h is Planck's constant, ν the frequency, c the speed of light, λ the wavelength. Then the mean optical radiation power is $E = \mathrm{h}\,\nu/T$. Ideally, this photon produces one electron-hole pair, which after separation supplies into the external circuit one electron with the charge q, and the mean current is thus q/T. The coefficient of conversion (the ratio of the output photocurrent to the incident optical radiation power) is called the sensitivity of photodetector,

$$S = \frac{I_{\rm p}}{E} = \frac{q/T}{\mathrm{h}\,\nu/T} = \frac{q}{\mathrm{h}\,\nu} = \frac{q\lambda}{\mathrm{h}\,c}\,. \tag{1.2}$$

Assume now that the ideal photodetector has received N photons within the time T. Then the photocurrent is increased N-fold ($I_{\rm p} = Nq/T$), and so is the optical radiation power ($E = N\mathrm{h}\,\nu/T$), but their ratio (sensitivity S) remains unchanged. In real photodetectors, however, some of the incident quanta are reflected, some other pass through the crystal without being absorbed. Even some of the absorbed quanta are not lucky enough, and the electron-hole pairs generated by them recombine before reaching the junction. So it is only part of the incident quanta, $N' = \eta N$, that produce pairs (which we are going to call photoactive pairs) which are separated in the SCR and cause photocurrent in the external circuit. We see that the photocurrent is only increased by the factor of N', and the expression for the sensitivity of real photodetectors contains the coefficient $\eta = N'/N$:

$$S = \frac{I_{\rm p}}{E} = \frac{N'q/T}{N\mathrm{h}\,\nu/T} = \frac{N'}{N}\frac{q}{\mathrm{h}\,\nu} = \eta\frac{q\lambda}{\mathrm{h}\,c}\,. \tag{1.3}$$

* In case of short-wave radiation, when the energy of photon is greater than the width of forbidden band, one photon may generate two pairs. Then it is said that the intrinsic quantum yield is equal to two [32].

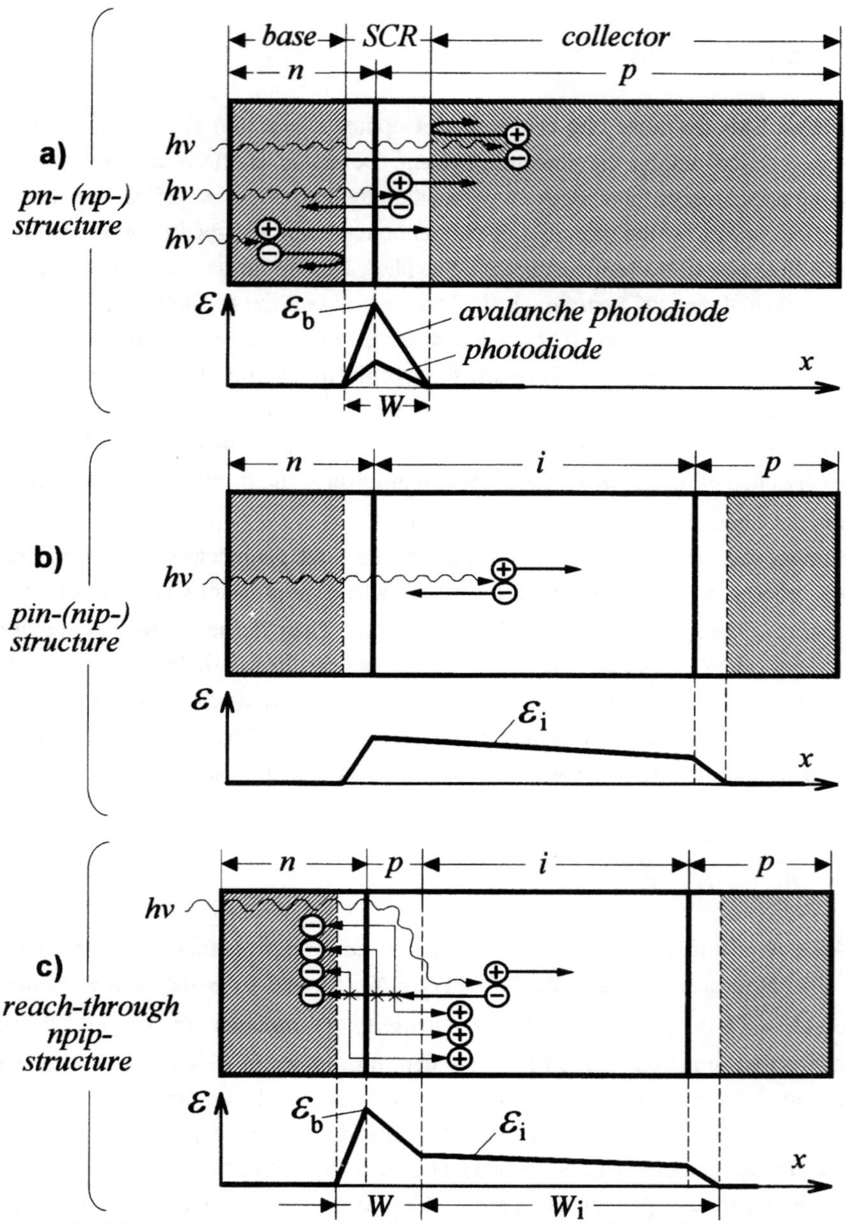

Figure 1.7: Main structures of photodiodes and their electric fields diagrams: (a) *pn* (*np*) structure; (b) *pin* (*nip*) structure; (c) *npip* structure; x denotes ionizing collision of primary electron.

The coefficient η characterizes the performance of photodetector as a counter of quanta of radiation, and is referred to as the quantum efficiency of the detector.

The conversion of optical radiation power E into the photocurrent I_p takes place in any photodetector. In the general case, however, the photocurrent in (1.3) is actually the primary photocurrent I'_p caused by the charge of photoactive carriers generated in the semiconductor in unit time. At the output of the photodetector we are dealing with the secondary current I_p. In the photodiode these two currents are the same, $I_p = I'_p$, but in the general case they are linked through the coefficient of proportionality, $I_p = K_p I'_p$. This coefficient is called the internal gain of photodetector, $K_p > 1$.

Main types of photodetectors used in PDAs. Widely employed in PDAs are photodiodes (PD), avalanche photodiodes (APD), photoresistors (PR). In CCDs, as we know from Sect. 1.1, the radiation is detected by the MIS structure. Phototransistors (PT) and other types of detectors are less common [2, 33, 34].

Photodiodes. According to their structure and properties, the photodiodes may be classified into low-frequency and high-frequency devices. The structure of the low-frequency PD is already known to us (Fig. 1.6a): it is an ordinary *pn* (or *np*) junction, with the thickness of the base layer (the illuminated front layer) being, as a rule, of the order of 1 ... 10 μm. It is easy to calculate the thickness of the space charge region W. Usually it is a narrow layer of the relatively high-resistance collector substrate which has lost the mobile carriers. The remaining unbalanced impurity atoms (for instance, the donors N_d) create the space charge qN_dW. According to electrostatics, this charge gives rise to the field $\mathcal{E} = qN_dW/\varepsilon\varepsilon_0$, where ε is the permittivity of semiconductor, and ε_0 is the permittivity of vacuum. Since the charge is distributed uniformly, the field changes smoothly from zero to the above maximum value (Fig. 1.7a), and so its mean strength across the layer is $\mathcal{E}/2$. Accordingly, the voltage drop across the layer is

$$U' = \frac{\mathcal{E}}{2} W = \frac{qN_d}{2\varepsilon\varepsilon_0} W^2 . \tag{1.4}$$

Here we have taken into account the fact that the voltage drop across the back-biased junction is created by both the external voltage $|U|$ and the intrinsic contact potential difference of the junction φ_c, so that $U' = |U| + \varphi_c$. From (1.4) we can express the thickness of SCR through the voltage drop across the *pn*-junction:

$$W = \sqrt{\frac{2\varepsilon\varepsilon_0 U'}{qN_d}} = \sqrt{\frac{2(12...16)\cdot 8.86\cdot 10^{-14}(1...10)}{1.6\cdot 10^{-19}\left(10^{14}...10^{16}\right)}} \approx 0.4...13\ \mu\text{m}. \tag{1.5}$$

This typical thickness of SCR corresponds to the typical values of ε, U, N_d.

The long-wave radiation (near the long-wave sensitivity limit) penetrates deeply into the collector region (the substrate), because the coefficient of absorption is low. The pairs produced here diffuse to the SCR from the large enough depth, defined by the diffusion length L. This is a relatively slow process whose time constant is of the order of the lifetime of the carriers in the collector (typically, from 1 to 100 μs). The inertiality of PD will be discussed later. Devices with such inertia are suitable for optoelectronic systems with mechanical modulation of the signal at 1 ... 10 kHz.

The length of LED and laser pulses (~ 1 ... 100 ns), however, is much smaller than the above time constant, and their detection requires using special high-frequency PD. The depth of absorption of short-wave radiation is small (for instance, the wavelengths ranging from 0.5 to 0.7 μm are absorbed in silicon at the depth of 1 to 5 μm); therefore, the high-frequency PD for this spectral range may have the same *pn* structure, only with a thinner base layer (from 0.5 to 2 μm thick), to ensure small enough passage time of the carriers. Much more often used, however, are semiconductor lasers and LEDs with the wavelengths 0.8 ... 0.95 μm (for lightwave communications systems) and solid state lasers operating at 1.06 μm (for optical range finders and some other devices). Such radiation penetrates deep into the bulk of silicon: to the depth of 10 ... 50 μm at $\lambda = 0.8\ldots0.95$ μm, and 700 μm at $\lambda = 1.06$ μm. Fast and efficient collection of charge carriers from such a large depth can only be ensured by establishing a very strong withdrawing field — as strong as $\geq 10^4$ V/cm, so that the speed of charge carriers is close to saturation (the numerical values given here and further in this section relate to silicon, which is the main material for photodetectors in the spectral range from 0.4 to 1.1 μm. This led to the development of high-frequency PD having a specially designed *pin* structure (Fig. 1.7b), in which *n* and *p* regions are separated by a high-resistance *i*–layer. Since the concentration of dope in this layer is low ($< 10^{12}$ cm^{-3}, which is by two to four orders of magnitude less than in *n* and *p* regions), this layer behaves almost as an insulator: when the layer is depleted, the charge of the impurity carriers is low and does not shield the field. Thanks to this, it is possible to extend the SCR throughout the entire *i*-layer. The thickness of the latter W_i is selected equal to a few absorption depths: $W_i = 20\ldots60$ μm for $\lambda = 0.8\ldots0.95$μm, and $W_i = 200\ldots500$ μm for $\lambda = 1.06$ μm (in the last case the radiation is reflected from the back side of the collector, and the path length of the radiation $2W_i \approx 400\ldots1000$ μm is of the order of the length of absorption). This thickness of SCR and the required field strength correspond to the following values of working voltage:

$$\begin{aligned} \lambda = 0.8\ldots0.9\mu\text{m}: \quad & U \cong \mathcal{E}W_i = 10^4(2\ldots6)\cdot10^{-3} = 20\ldots60 \text{ V}; \\ \lambda = 1.06\mu\text{m}: \quad & U \cong \mathcal{E}W_i = 10^4(2\ldots5)\cdot10^{-2} = 200\ldots500 \text{ V}. \end{aligned} \tag{1.6}$$

The broad depleted *i*-layer "spreads" the plates of the capacitor (the conducting *p* and *n* regions) further apart, and the specific capacitance of the *pin* structure is very small (smaller by one or two orders of magnitude than that of the normal *pn* junctions), which is another attractive feature of this construction (numerical evaluation of the capacitance will be given below).

Light-generation and separation of charge carriers in the MIS structure proceed in the same way as in the photodiode, as indicated in the previous section (see Fig. 1.3a, b). The difference is that the MIS structure, unlike the PD, does not have the base layer.

Other types of photodetectors used in PDAs are characterized by a certain amount of internal gain, and so they partly perform the next function of the detection system, amplification of the signal.

Avalanche PD. Amplification can be obtained in the ordinary *pn* structure. This requires applying a voltage high enough to create the highest possible field strength ($\mathcal{E} \approx (3...5)\cdot 10^5$ V/cm) in the SCR. The primary carrier, generated by light or by heat, accelerates in the strong field of the SCR and acquires the energy high enough to produce a new pair. The incipient secondary carriers are also capable of causing impact ionization, and so an avalanche-like process is under way. This avalanche process can be controlled, so as to get a yield of exactly M secondary carriers for each primary carrier. Quantity M is called the multiplication coefficient. In other words, the sensitivity of the avalanche photodiode is M times the sensitivity of the ordinary PD.

Controlling the avalanche, however, is not an easy task. In the first place, the technology must ensure a highly homogeneous junction. Otherwise, the tiniest inhomogeneities in the photosensitive element will give rise to strong local fields which cause premature local breakdowns (microplasmas); then it is not possible to obtain a uniform multiplication throughout the element. Also, it is necessary to prevent premature breakdowns along the perimeter of the element, which is accomplished by introducing specially designed guard rings. Secondly, the operating voltage of the APD must be maintained with a very high precision. The required stability of operating voltage U can be estimated by the known empirical Miller formula for the multiplication coefficient M as function of operating voltage U:

$$M = \frac{1}{1-\left(\frac{U}{U_B}\right)^n}, \tag{1.7}$$

where, as a rule, $n = 2...5$, U_B is the avalanche breakdown voltage, at which the multiplication factor tends to infinity. The avalanche breakdown voltage can be estimated from the above values of breakdown field strength $\mathcal{E}_B \approx (3...5)\cdot 10^5$ V/cm. Since in the abrupt *pn* junction the field strength falls linearly from its maximum value $\mathcal{E}_B$ to zero (Fig. 1.7a), the voltage drop must be calculated on the basis of the mean value $\mathcal{E}_B/2$. We have already quoted also the typical thicknesses of the SCR (0.4...13 μm); it should be noted that in a broader SCR the probability of impact ionization increases, and the accelerating field is weaker. So,

$$U_B = \frac{\mathcal{E}_B}{2} W_i \approx \frac{5\cdot 10^5}{2} 0.4\cdot 10^{-4} ... \frac{3\cdot 10^5}{2} 13\cdot 10^{-4} = 10...180 \text{ V}. \tag{1.8}$$

For the typical values of the multiplication coefficient $M = 10...200$, Miller's formula is simplified. Taking advantage of the fact that U is close to U_B, we replace U/U_B by $1-(U_B - U)/U_B$, carry out expansion and retain the first term, getting as a result

$$M=\frac{1}{1-\left(1-\frac{U_B-U}{U_B}\right)^n}\approx\frac{1}{1-\left(1-n\frac{U_B-U}{U_B}\right)}=\frac{U_B}{n(U_B-U)} . \tag{1.9}$$

Hence the required precision of U is

$$\frac{U_B-U}{U_B}=\frac{1}{nM} ;\ \frac{\delta U}{U_B}=\frac{\delta M}{nM^2} ;$$
$$\frac{\delta U}{U}\approx\frac{\delta U}{U_B}=\frac{1}{nM}\frac{\delta M}{M}=\frac{1}{(2...5)(10...200)}\cdot 0.1=5\cdot 10^{-3}...1\cdot 10^{-4} . \tag{1.10}$$

We see that the relative stability of the operating voltage must be higher by a factor of nM than the required relative stability of the multiplication coefficient $\delta M/M$, and the operating voltage must be kept constant to within a fraction of a percent if, for instance, $\delta M/M=0.1$. The high requirements to the stability of U, which are the more rigorous the higher the coefficient M, impose a serious limitation on the achievable magnitude of this coefficient. The situation is even more complicated by the fact that the breakdown voltage increases with the temperature, so that one has to keep track of the temperature variations [35].

Now, the silicon avalanche PD with *pn* structure has the same drawbacks as the non-avalanche type: the large time constant for the penetrating radiation. It might seem that, by analogy with the PD designed for the detection of short laser pulses with $\lambda > 0.8$ µm, the solution lies in the use of a *pin* (*nip*) avalanche structure. However, the working voltage of such structure is high enough even in the non-avalanche mode (see (1.6)), and in the avalanche mode it becomes unacceptably high: for a relatively thin *i*–layer of 20 ... 60 µm we get

$$U\approx U_B\approx \mathcal{E}_B W_i=5\cdot 10^5\cdot 2\cdot 10^{-3}...3\cdot 10^5\cdot 6\cdot 10^{-3}=1000...1800\ \text{V} . \tag{1.11}$$

Because of this, the attempts to design a *pin* APD were soon discontinued [36]. The modern "laser" silicon APD have a special *npip* structure [37, 38], which permits reducing considerably the working voltage as compared to the *pin* APD. This structure is a hybrid, a serial connection of the above *np* and *nip* (or, more precisely, *ip*) structures (cf. Fig. 1.7c,a,b). Multiplication and absorption in this construction are separated: the diode *np* structure (its SCR region) carries out the multiplication, and the *i*–region is entrusted with the absorption. The field strength profile in the *npip* structure (Fig. 1.7c) clearly shows the way used to reduce the working voltage of the device by localizing the high field strength $\mathcal{E}_B$ to the narrow region ($W=1...2$ µm). Then the voltage U is

$$U\approx \mathcal{E}_B W+\mathcal{E}_i W_i=\frac{3\cdot 10^5}{2}\cdot(1...2)\cdot 10^{-4}+10^4\cdot(1...1.5)\cdot 10^{-2}\approx 110...180\ \text{V} . \tag{1.12}$$

The working voltage may be increased to 400 ... 600 V to facilitate multiplication at elevated temperatures and to compensate for technological allowances in the manufacture of the *npip* structure.

The hybrid *npip* structure inherits the virtues of its parents: the moderate operating voltage from the *np* structure, and the small time lag for deeply penetrating radiation from the (*n*)*ip* structure. There also are new properties, not encountered in the parents: positive, like the better tolerance to variations of the working voltage (which are damped in the high-resistance *i*-layer), and negative, like the high sensitivity to the precise doping of the avalanche *p*–layer (the *npip* structure may degenerate into a *nip* structure in case of underdose, and into an *np* structure in case of overdose). The *npip* APDs are known under different names: APD with separated regions of absorption and multiplication (as follows directly from arguments developed above), APD with penetration (since the space charge and the field penetrate into the high-resistance *i*–layer).

Photoresistor. Another common type of photodetector with internal gain is the ordinary photoresistor (PR). As in the PD, the absorbed quantum generates in the PR one electron-hole pair, thus increasing the concentration of mobile charge carriers (the effect of photoconductivity). Unlike the PD, photoresistor is a passive device, and it has no region of intrinsic field which separates the carriers. Registration of photocarriers requires applying external bias to create the field. Then the electrons start drifting to the positively biased electrode and escape into the external circuit. The principle of neutrality requires that another electron should replace the escaped one. This new electron will traverse the photoresistor (the entire length l from one contact to the other) within the transit time

$$\tau_{\mathrm{tr}\,n} = \frac{l}{v_n} = \frac{l}{\mu_n \mathcal{E}} = \frac{l^2}{\mu_n U} \,, \tag{1.13}$$

where $U = \mathcal{E} l$ is the voltage across the contacts; v_n and μ_n are the velocity and the mobility of the electron. Similarly, for the transit time of the hole we have

$$\tau_{\mathrm{tr}\,p} = \frac{l^2}{\mu_p U} \,, \tag{1.14}$$

where subscript p marks off the parameters of the hole.

So, when the bias voltage is applied to the PR, and a photoactive pair is generated, electrons and holes start being supplied into the external circuit (in addition to the dark current), on the average, over the time intervals $\tau_{\mathrm{tr}\,n}$ and $\tau_{\mathrm{tr}\,p}$. The light-generated pair persists for its lifetime τ_{life} (the time before recombination). Within this time the external circuit will receive $\tau_{\mathrm{life}}/\tau_{\mathrm{tr}\,n}$ electrons and $\tau_{\mathrm{life}}/\tau_{\mathrm{tr}\,p}$ holes, and so the number of secondary carriers in the external circuit per one primary pair (that is, the gain of photoresistor) is

$$K_{\mathrm{p}} = \frac{\tau_{\mathrm{life}}}{\tau_{\mathrm{tr}\,n}} + \frac{\tau_{\mathrm{life}}}{\tau_{\mathrm{tr}\,p}} = \frac{\tau_{\mathrm{life}}}{l^2} U\left(\mu_n + \mu_p\right) . \tag{1.15}$$

Let us emphasize once again that expression (1.15) holds as long as the carriers recombine only in the bulk. In real single-crystal PR the minority carriers recombine also on the contact, and thus only persist no longer than the time taken to get from the place of birth to the contact. Assume, for the sake of definiteness, that the minority carriers are holes. When the PR is uniformly illuminated, the path length of the hole to the contact may vary from zero (if the hole is generated just near the contact) to the complete length of photoresistor l (for holes generated near the opposite contact). Accordingly, the mean path length is $l/2$, and the mean transit time is

$$\bar{\tau}_{\mathrm{tr}\,p} = \frac{l}{2v_p} = \frac{l^2}{2\mu_p U} . \tag{1.16}$$

If $\tau_{\mathrm{life}} >> \bar{\tau}_{\mathrm{tr}\,p}$, the calculated mean transit time $\bar{\tau}_{\mathrm{tr}\,p}$ must be taken for the mean lifetime of the pair, and then τ_{life} in (1.15) must be replaced by $\bar{\tau}_{\mathrm{tr}\,p}$:

$$K_{\mathrm{p}} = \bar{\tau}_{\mathrm{tr}\,p} \frac{U\left(\mu_n + \mu_p\right)}{l^2} = \frac{l^2}{2\mu_p U} \frac{U\left(\mu_n + \mu_p\right)}{l^2} = \frac{1}{2}\left(\frac{\mu_n}{\mu_p} + 1\right) . \tag{1.17}$$

The physical meaning of this formula is quite simple: while the slow hole drifts from the place of generation to the contact, the fast electrons traverse the entire photoresistor approximately $\mu_n/2\mu_p$ times. Equation (1.17) defines the ultimate gain of a single-crystal photoresistor with recombination on the contact, which can be achieved with a high bias voltage. Hence also follows the requirement to the material: the mobility of the majority carriers must be much higher than that of the minority carriers.

Phototransistor. This device with internal gain may be regarded as the simplest solid state IC which combines a PD with transistor amplifier. In a bipolar transistor the PD is primarily represented by the back-biased collector junction (the emitter junction is also sensitive to radiation, but its contribution to the total photocurrent is, as a rule, small because of the small area of the junction). The photocurrent of collector-base junction goes to the base and is amplified by a factor of $h_{21\mathrm{e}}$ (recall that $h_{21\mathrm{e}}$ is the current gain of bipolar transistor in the common emitter circuit). In the junction field-effect transistor (FET) the photocarriers are separated in the gate region, and the resulting photocurrent flows through the load R_{L} in the external gate circuit, producing the voltage drop across the load of $I_{\mathrm{p}}R_{\mathrm{L}}$. This voltage controls the output current, causing it to increase by $GI_{\mathrm{p}}R_{\mathrm{L}}$ (where G is the transconductance of the transistor), and so the gain of the phototransistor is equal to the gain of the stage GR_{L}.

As will be shown in Sect. 2.5, in most applications the bipolar transistors are inferior to FETs in terms of their threshold characteristics. But then, in place of a FET it is usually more advantageous to use a hybrid tandem consisting of a photodiode and a FET manufactured on different chips. This makes it easier to optimize each of the elements, and to avoid technological complications. For this reason phototransistors are commonly used in automation systems rather than in detectors intended for low-level (threshold) signals. All the same, there is one respect in which the hybrid pair is not as good as the FET: the stray capacitance of the assembly may become important even though the capacitance of the elements (PD and FET) themselves is small. This disadvantage was especially noticeable in the earlier designs. Such situation occurs in photodetectors for lightwave communication systems, where the diameter of the photosensitive element may be as small as 80 ... 200 μm. These systems use a special solid-state device which competes even with an APD, and may be conventionally referred to as a transistor. It is a *pin* FET and contains a *pin* PD and a FET; the pair is realized on the common chip but is topologically separated. The advanced technology allows optimizing the structure of each of the two elements. Elimination of stray capacitance and the small size of the sites have allowed reducing the total input capacitance down to 0.1 ... 0.5 pF, which is especially important for the threshold characteristics at high frequencies (Sect. 2.5).

Sensitivity and internal gain. The expression for sensitivity (1.3) has already been obtained in connection with the principle of photodetection; let us now rewrite it in a form more convenient for calculations:

$$S = K_p S_0 \,, \quad S_0 = \frac{\eta q \lambda}{\mathrm{h} c} = \frac{\eta \lambda}{1.24} \,, \tag{1.18}$$

where λ is measured in μm, and S in A/W.

Here we have substituted the numerical values of q, h, c; S_0 is the sensitivity of the primary conversion; the coefficient of internal gain K_p assumes the following values:

$K_p = 1$	for photodiode;
$K_p = M$	for avalanche photodiode;
$K_p = h_{21e}$	for bipolar phototransistor;
$K_p = \dfrac{\tau_{life}}{\tau_{tr\,n}} + \dfrac{\tau_{life}}{\tau_{tr\,p}}$	for photoresistor.

It has often been maintained that the higher the internal gain (and hence the sensitivity), the better the photodetector: since its main purpose is the conversion of optical signal into electrical signal, the slope of this conversion (sensitivity) is the criterion of quality. Sometimes it was also held that even a low quantum yield is not too bad, because the losses at the primary conversion can be compensated by the increased gain. Such opinions, however, are erroneous. The fact is that in the low-level PDAs we are concerned not with the signal itself, but with the mixture of signal and noise (Sect. 3.1), which also depends on the amplification. The analysis of signal-to-noise ratio, which is carried out in detail in Sect. 3.3, brings one to the following conclusion. The main criterion of performance of optical detector is its quantum efficiency η; there is a certain optimal value of internal gain K_{popt} which need not and should not be exceeded, since excessive gain worsens the signal-to-noise ratio.

The first statement can be rather easily proved for photodetectors with a large level of noise. Usually these are photodetectors operating in the so-called background limited mode, when, along with the optical signal E_{s}, the device receives a considerable amount of background radiation E_{b}. Prevalent in such case are the shot noises of the primary photocurrent I_{pb} due to the background radiation, $i_{\mathrm{n\,PD}} = \sqrt{2qI_{\mathrm{pb}}f_{\mathrm{n}}}$, where q is the charge of electron, f_{n} is the noise band (Sect. 2.3, 5.2). Any subsequent amplification (including the internal gain of photodetector), be it even a perfectly zero-noise process, cannot reinforce the signal more than the noise, since the harmonics of the signal and the noise are indistinguishable. Accordingly, the ratio of signal $K_{\mathrm{p}}S_0E_{\mathrm{s}}$ to noise $K_{\mathrm{p}}i_{\mathrm{n\,PD}}$ at the output of photodetector is not improved by amplification:

$$\mathrm{SNR} = \frac{I_{\mathrm{s}}}{i_{\mathrm{n\,PD}}} = \frac{K_{\mathrm{p}}S_0E_{\mathrm{s}}}{K_{\mathrm{p}}\sqrt{2qS_0E_{\mathrm{p}}f_{\mathrm{n}}}} \propto \sqrt{S_0} \propto \sqrt{\eta}\ . \tag{1.19}$$

Hence it follows that the criterion of quality of photodetector is its quantum efficiency, and no amount of internal gain can make up for the losses at primary conversion (because the coefficient K_{p} in (1.19) for SNR cancels out).

The same applies to photodetectors with low initial noise $i_{\mathrm{n\,PD}}$ which is less than the noise of amplifier $i_{\mathrm{n\,a}}$. In this case the noise of amplifier must be added to the noise of photodetector (which now is not determined by the background) in the denominator of (1.19):

$$\mathrm{SNR} = \frac{I_{\mathrm{s}}}{\sqrt{K_{\mathrm{p}}^2 i_{\mathrm{nPD}}^2 + i_{\mathrm{n\,a}}^2}} = \frac{K_{\mathrm{p}}S_0E_{\mathrm{s}}}{\sqrt{K_{\mathrm{p}}^2 i_{\mathrm{nPD}}^2 + i_{\mathrm{na}}^2}}\ ; \tag{1.20}$$

$$K_{\mathrm{p}}i_{\mathrm{nPD}} << i_{\mathrm{n\,a}}\ ,\quad \mathrm{SNR} = \frac{K_{\mathrm{p}}S_0E_{\mathrm{s}}}{i_{\mathrm{na}}} \propto K_{\mathrm{p}}S_0 \propto K_{\mathrm{p}}\eta\ ; \tag{1.21}$$

$$K_{\mathrm{p}}i_{\mathrm{nPD}} >> i_{\mathrm{n\,a}}\ ,\quad \mathrm{SNR} = \frac{S_0E_{\mathrm{s}}}{i_{\mathrm{n\,PD}}} \propto S_0 \propto \eta\ . \tag{1.22}$$

In writing (1.21) we assumed that the small noise of photodetector is due to the dark current of thermal generation $i_{\mathrm{nPD}} = \sqrt{2qI_{\mathrm{d\,g}}f_{\mathrm{n}}}$, and is not related to the sensitivity.

Because of this, the dependence of SNR on η is even stronger than in (1.19).

From (1.20) it becomes clear why we need the internal gain and what its optimal value is, given that the noise of photodetector is small. Internal gain is needed to "withdraw" this noise from the noise of the amplifier (or, as they say, to suppress the noise of the amplifier, to reduce its share in the total noise of the PDA). The value of K_p is optimal when the noise of the photodetector exceeds the noise of the amplifier — that is, when K_p satisfies the condition $K_{popt} i_{n\,PD} = (1...3) i_{n\,a}$. Further increase in the gain is not worthwhile from the standpoint of signal-to-noise ratio (see (1.22)). When the initial noise $i_{n\,PD}$ is low, the optimal value K_{popt} may be so large as to be not achievable in practice, and then (1.21) holds. The magnitude of K_p is limited by the growing instability (as we have seen in the case of APD), increasing nonlinearity, by the narrowing dynamic range, stronger temperature dependence of sensitivity. In order to increase the value of K_p in photoresistors, one has either to increase the lifetime, or to reduce the transit time by applying a larger bias (see (1.15)). This will either increase the time constant, or bring about excessive noise and increased likelihood of breakdown because of the high bias voltage. The inherent flaw of the high internal gain is the associated noise level. By contrast to what we have assumed in (1.19) – (1.22), the noise of photodetector may grow faster than the signal as the internal gain increases, as it happens in case of APD (Sect. 2.3). Because of the multiplication-associated noise, the signal-to-noise ratio in APD may only fall: it is always worse than the SNR in a PD without amplification (Sect. 2.5, 6.1). The SNR in a PD, however, is "thing in itself": in order to apprehend the SNR, one must hook up an amplifier to the photodetector, and the amplifier will introduce its own noise, worsening the overall signal-to-noise ratio. Only when we consider the PDA as a whole, and analyze the SNR in the entire device, we understand that the internal gain is, after all, a good thing. We also see the disadvantages of the internal gain (additional noise, deterioration of performance), and conclude that its allowable magnitude is limited. Because of this, further amplification is still necessary. The presence of internal gain does not, as a rule, slacken the requirements to the internal noise of the amplifier: the higher $i_{n\,a}$ is, the larger value of $K_{popt} \approx i_{n\,a}/i_{n\,PD}$ is required — and we already know the drawbacks of the high K_{popt}. It follows that each element of the PDA should "mind his own business": the photodetector must detect the signal with the lowest possible losses, and the amplifier must amplify the signal within the desired band and to the desired level, introducing as little noise as possible. It must be said that the modern PDAs do their job well, registering the photons practically without any loss. Reduced losses by reflection and recombination, the use of the above-described *pin* and *npip* structures with a broad absorbing *i*-layer and strong field, the use of broad-band base layer in heterostructures, owing to which the radiation is absorbed directly in the SCR (in the narrow-band collector) — these and other measures have allowed raising the quantum efficiency η as high as 40 ... 95%. The values close to the theoretical limit (80...95%) have been obtained primarily with the silicon PDs at the wavelengths from 0.7 to 0.95 μm. Somewhat lower values of η are obtained in silicon photodetectors optimized for the wavelength of 1.06 μm (20...60%), because of the incomplete

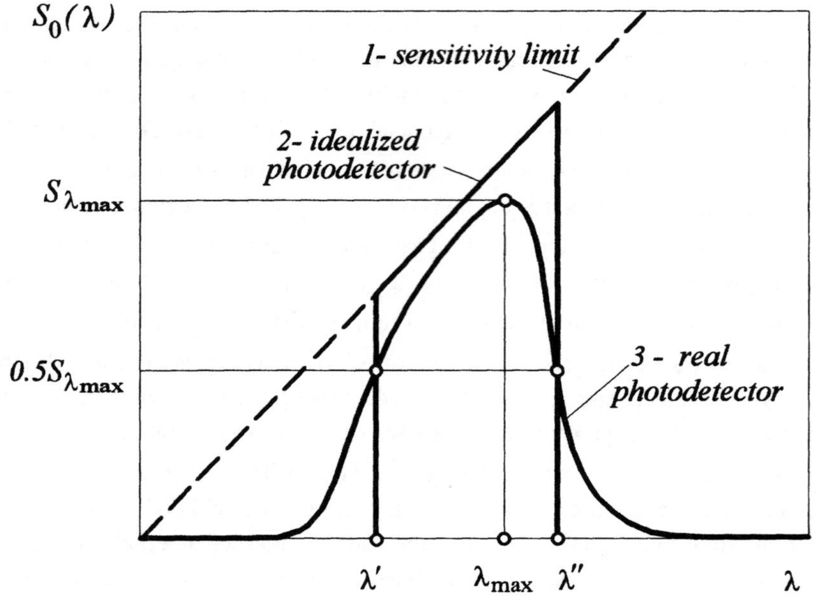

Figure 1.8: Spectral curve of photodetector: "incomplete" triangle ("incomplete" trapeze): (1) sensitivity limit; (2) idealized photodetector; (3) real photodetector.

absorption, and in IR PDs (40...80%), because of the additional losses in optical elements. The quantum efficiency is also lower in CCDs (40...60%), because of the absorption in polysilicon electrodes.

Spectral sensitivity. As follows from (1.18), the sensitivity of the ideal photodetector with unit quantum yield to monochromatic radiation, $S_0 = \lambda/1.24$, is a fundamental constant and does not depend on the material of the photodetector. It only depends on the number of quanta per one watt of power, and is thus proportional to the wavelength. The material (the width of its forbidden band E_g) defines the long-wave limit λ'' of fundamental (intrinsic) absorption: photogeneration does not take place if the energy of photon $h\nu$ is less than E_g. In case of extrinsic conduction, when the photon causes the carrier from the impurity level to the valence band or the conduction band, the long-wave limit λ'' is defined by the energy gap between the impurity level and the relevant band. From the short-wave side the sensitivity is theoretically not limited; if required, it may be restricted by optical filters. So, the spectral characteristic $S_0(\lambda)$ of such idealized photodetector is triangular or trapezoidal (curves 1, 2 in Fig. 1.8). In real photodetectors the spectral sensitivity is, of course, smoothed (curve 3): usually it falls off monotonously to both sides of the maximum (which occurs at the wavelength λ_{max}) because of the decrease in quantum efficiency: in the short-wave range ($\lambda < \lambda_{max}$) the depth of absorption is small, and the carriers recombine in the bulk of the base and on its surface, whereas in the long-wave ($\lambda > \lambda_{max}$) range the depth of penetration is large, and the absorption is incomplete. The short-wave (λ') and the long-wave (λ'') limits of the

Table 1.1. Main spectral ranges: material of photodetector, its currents and resistances

Spectral Range	I	II	III		IV	V		
Wavelength, μm	0.4...1.1	1...2	2...3		3...5	8...10	8...12	8...14
Material	Si	Ge	InAs	PbS	InSb	MCT ($x = 0.219$)	MCT ($x = 0.206$)	MCT ($x = 0.194$)
Type of photodetector	PD	PD	PD	PR	PD	PD	PD	PR
T, K (design value)	293	293	77	190	77	77	77	77
n_i, cm^{-3}	$1.7\cdot10^{10}$	$2.3\cdot10^{13}$	$4.5\cdot10^{4}$	—	$3\cdot10^{9}$	$7\cdot10^{12}$	$3\cdot10^{13}$	$8\cdot10^{13}$
τ_i, s	$10^{-3}...10^{-1}$	$\approx 3\cdot10^{-6}$	10^{-6}	—		$2\cdot10^{-8}..10^{-6}$		—
$n(p)$, cm^{-3}	$10^{14}...10^{17}$	$10^{15}...10^{16}$	10^{13}	$\approx 10^{17}$	$10^{14}...10^{15}$	$\approx 5\cdot10^{15}$		$(5\cdot10^{14})$
τ_n, s	$10^{-6}...10^{-4}$	$10^{-6}...10^{-5}$	$\sim 10^{-8}$	—	10^{-9}	$3\cdot10^{-8}$		$3\cdot10^{-8}$
(τ_n), s	—	—	$(\approx 10^{-6})$	$(10^{-3}...10^{-2})$	$(10^{-7}...10^{-6})$	—		$(1.5\cdot10^{-6})$
μ_n, cm^2/V·s	1300	3800	65000	—	620000		$(1...2)\cdot10^{5}$	
(μ_p), cm^2/V·s	(500)	(1820)	(350)	(5...50)	(600)		(≈ 500)	
j_{dg}, A/cm^2	$3\cdot10^{-10}$... ...10^{-7} (SCR)	$(1...3)\cdot10^{-4}$ (neutral region)	$< 10^{-8}$ (SCR)	—	$\approx 10^{-8}...10^{-6}$ (SCR)	$(0.3...3)\cdot10^{-3}$ (neutral region)	$\approx 10^{-2}$ (neutral region)	—
n_{eff}, cm^{-2}·s^{-1}	$1.36\cdot10^{2}$	$2.26\cdot10^{10}$	$3.74\cdot10^{13}$	$3.74\cdot10^{13}$	10^{16}	$2.4\cdot10^{17}$	$5.4\cdot10^{17}$	$8.5\cdot10^{17}$
j_d, A/cm^2	$3\cdot10^{-10}$... ...10^{-7}	$(1...3)\cdot10^{-4}$	$6\cdot10^{-6}$	—	$1.6\cdot10^{-3}$	$3.9\cdot10^{-2}$	$8.7\cdot10^{-2}$	—
j_{db}, A/cm^2	$2.2\cdot10^{-17}$	$3.6\cdot10^{-9}$	$6\cdot10^{-6}$	$6\cdot10^{-6}$	$1.6\cdot10^{-3}$	$3.9\cdot10^{-2}$	$8.7\cdot10^{-2}$	0.14
PSE size, mm	0.5dia	0.5dia	0.1×0.1	0.1×0.1	0.1×0.1	0.1×0.1	0.1×0.1	0.1×0.1
R_{pSC}, Ω	10^{8}... ...$4\cdot10^{10}$	(0.4... ...1.3)$\cdot10^{5}$	$\gtrsim 10^{10}$	$10^{5}...5\cdot10^{6}$	$7\cdot(10^{7}...10^{9})$	$2.3(10^{4}...10^{5})$	$7\cdot10^{3}$	25...50
R_{pOC}, Ω	same	same	$7\cdot10^{7}$		$3\cdot10^{5}$	$(0.7...1)\cdot10^{4}$	$3\cdot10^{3}$	

spectral range of sensitivity of real photodetectors are usually defined at a specified level of sensitivity (usually 0.1 or 0.5 of the maximum value) and are referred to as the cut-off wavelengths.

What must be the spectral range of a photodetector? The majority of optoelectronic systems may be divided with regard to their operating spectral band into five ranges (Table 1.1). The first two ranges (0.4 ... 1.1 μm and 1 ... 2 μm) correspond to the high-temperature natural sources (the Sun, the stars, other very hot bodies), and the most common lasers and LEDs: solid-state (1.06; 0.63 μm) and semiconductor (0.69; 0.8...0.95; 1.3; 1.55 μm). The limits of the first range are determined by the sensitivity of silicon, the basic material of semiconductor electronics. The remaining three spectral ranges (III – V) correspond to thermal radiation of slightly heated bodies (300...800°C) or bodies at room temperature. The limits of these ranges are determined by the windows of transparency of the atmosphere; range V contains the maximum of spectral density of radiation at room temperature. The long-wave limit of this range usually varies from 10...14 μm, and this range in our table is split into three subdivisions.

Photodetectors are capable of covering the entire spectrum of optical radiation; the listing of all varieties of photodetectors and semiconductor materials sensitive in the ranges I – V is the task of specialized monographs and handbooks [5, 33, 34, 39]. Table 1.1 lists the main and the most technologically streamlined photosensitive materials used by the majority of PDAs. In some applications these are challenged by other materials. For the first two ranges such are the binary, ternary, quaternary compounds A_3B_5: GaAs, GaP, InAs, GaSb, AlAs, and their solid solutions, like GaInAsP and AlGaAsSb [39–41]. For example, GaAs and AlGaAs can be used in photodetectors for registration of picosecond pulses of radiation at 0.8...0.95 μm. As compared with silicon, the transit length in this case is shorter (owing to the high coefficient of absorption), and the mobility is higher. Photodiodes based on the broad-band material GaP ($E_g = 2.26$ eV) are highly sensitive to the UV radiation, and can operate at very high temperatures (up to 300...400°C). The spectral curve of GaPAs-based photodiodes is similar to the visibility curve of human eye, which is useful for exposure meters. In lightwave communication systems operating at 1.3 and 1.55 μm the germanium-based PD are being replaced by PD based on GaInAs – InP (or AlInAs – InGaAs - InP) heterostructures. The dark current of such PDs ($\sim 10^{-8}$ A/mm^2) is about two orders of magnitude less than that of the silicon PDs. Illumination through a broad-band "window" (InP substrate) ensures absorption directly in the SCR within the hetero-junction and at low depth, which reduces losses and quickens the response.

The spectral range determines the entire technical design of the PDA. First of all, this concerns the type and the material of the photodetector. The low-level IR photodetectors for the ranges III – V require cooling, which governs the construction principles of the PDA. In the forthcoming sections we shall see that different spectral ranges mean different currents and different levels of photodetector noise, and hence different electronic signal processing. Finally, the spectral range usually defines the design of PDA to match a specific optoelectronic system: PDAs of the first two ranges are most commonly employed in laser and LED systems, in star navigation systems, guiding systems, exposure meters. PDAs in the three IR ranges are intended primarily for lidars and IR imaging systems.

Response time. In describing the main types of PDs (low-frequency and high-frequency) we had to touch upon the mechanism which determines the "inertiality" of the device, since the required promptness of response defines the choice of the structure of PD.

When the carriers are generated in the quasineutral n and p-regions of base and collector, the response of the PD is limited by the time of diffusion of these carriers to the junction. The lag is especially large when the carriers are generated in the collector far away from the junction, and so depth of absorption $1/\alpha$ is larger than the diffusion length. Because of the recombination, the lifetime of carriers τ_{life} in the semiconductor is limited. In the collector the recombination is supplemented by another mechanism: the "withdrawal" of carriers by the pn junction. As soon as the pair approaches the collector, it is separated by the field, and the carriers escape to the external circuit. Because of this, the photocarriers in the collector disappear at a rate twice as high as that in the neutral semiconductor (photoresistor), and their effective lifetime is $\tau_{\text{life}}/2$. As the carriers disappear, the photocurrent also disappears, and so the constant $\tau_{\text{life}}/2$ should be definitive for the time lag of withdrawal of carriers from the collector. The solution of equations of transfer gives about the same value for the time constant of photocarrier ($\sim \tau_{\text{life}}/\sqrt{3}$). The typical values of τ_{life} in a PD vary within $10^{-6}...10^{-3}$ s.

As the depth of absorption $1/\alpha$ decreases, the carriers are generated closer to the junction, and their transport to the SCR and separation become faster. The time of collection of carriers generated in the base region can also be small if the base is thin enough. As we know, within the time t the carriers diffuse to the distance $L \approx \sqrt{Dt}$, where D is the coefficient of diffusion. Accordingly, the longest time of drift of carriers generated far from the junction (at the distance of $1/\alpha$ in the collector, or W in the base) can be found from the following relations:

$$L = \sqrt{Dt} \approx \frac{1}{\alpha}, \quad t \approx \frac{1}{\alpha^2 D}; \tag{1.23}$$

$$\sqrt{Dt} \approx W, \quad t \approx \frac{W^2}{D}. \tag{1.24}$$

The time of drift of carriers generated directly near the junction is zero, so the average transit time is

$$\tau_{\text{tr}} \approx \frac{1}{2}\left(\frac{1}{\alpha^2 D} + 0\right) = \frac{1}{2\alpha^2 D} \tag{1.25}$$

in the collector, and

$$\tau_{tr} \approx \frac{1}{2}\left(\frac{W^2}{D}+0\right)=\frac{W^2}{2D}\approx\frac{\left(10^{-4}...10^{-3}\right)^2}{2\cdot(13...44)}\approx 10^{-10}...4\cdot 10^{-8}\text{ s} \qquad (1.26)$$

in the base. The numerical estimate in (1.26) has been done for a silicon PD.

The quickest response is achieved when the radiation is absorbed in the SCR, so that the mechanism of transport is drift rather than diffusion, and when the field is strong, so that the velocity of carriers v is close to its saturation value (for silicon, $v \approx 6\cdot 10^6$ cm/s at $\mathcal{E} > 2\cdot 10^4$ V/cm). Then the time of transit across the entire space charge region reaches its minimum,

$$\tau_{SCR}=\frac{W}{v}=\frac{(1...500)\cdot 10^{-4}}{6\cdot 10^6}\approx 2\cdot 10^{-11}...10^{-8}\text{ s} . \qquad (1.27)$$

The transport of carriers in an *npip* APD may be divided in two stages. First the primary electrons generated in the *i*-region drift within the time τ_{tr} to the avalanche layer, where they are multiplied (Fig. 1.7c). Then the *M*-fold-multiplied package of secondary carriers (holes) drifts back to the *p*-substrate. The drift of the charge package in the SCR induces a current in the external circuit which is proportional to the magnitude of the charge. Since the primary charge is small, we may neglect its induced current and assume that the induced current in the external circuit is entirely caused by the drift of the secondary carriers. Because of this, the duration of the pulse front in an *npip* APD is of the order of the time constant τ_{SCR} (like in a *pin* PD); the kinetics of this device is distinguished by the time lag of the same order (τ_{SCR}).

In a CCD the time of collection (diffusion in the collector, drift in the SCR) is much less than the time of accumulation.

In a photoresistor the lag is uniquely determined by the lifetime τ_{life}, whose typical values for the most common materials (PbS, MCT) fall within $10^{-6}...10^{-2}$ s.

In a phototransistor the lag is, as a rule, due to the amplification process rather than to the collection and separation of carriers, and is primarily determined by the time constant of the emitter RC circuit, which for small currents in the open-base regime may vary within $10^{-6}...10^{-3}$ s.

Now are we happy with such response? As a rule, the response is satisfactory provided that the choice of the photodetector is correct. The typical duration of the optical pulse varies within $10^{-4}...1$s in systems with mechanical scanning or modulation, and within $\sim 10^{-9}...10^{-6}$ s in laser and LED systems. The above numerical estimates indicate that in most cases the photodetector may be regarded as an inertialess device, so that the profile of the signal photocurrent exactly replicates the profile of the incoming optical pulse. In some situations, however — when the laser pulse is very short, or the lifetime is very long — the inertiality of the detector may be noticeable. Then the sensitivity of the photodetector may be represented as $S(f) = S_0 s(f)$, where S_0 is the direct current sensitivity, and $s(f)$ is the relative frequency response. The particular form of $s(f)$ depends on the operating principle of the photodetector; as a rule, this function can be approximated by the frequency response of the simplest single-section filter:

$$s(f) = \frac{1}{1 + j\omega\tau_p}, \tag{1.28}$$

where j is the imaginary unit, and ω is the radian frequency. The photodetector time constant τ_p is close to the characteristic times given above. The exact relations between these quantities are found from the analysis of equations of transfer.

Formally, the frequency response $s(f)$ may be considered as the response of the first section of the filter. Subsequent filtering stages must introduce the necessary correction, shaping the desired frequency response with due account for the initial $s(f)$.

The sluggishness of the photoresistor can be a blessing in disguise, as we already know from Sect. 1.1. In the photoresistor matrix the integration over the lifetime (transform (1.28)) is the main operation of filtering, and so the PR performs almost all function of the detection system: detection, reinforcement, filtering, and even switching, given the appropriate lattice structure of the matrix. We also pointed to the disadvantages of such universality: it is not possible to "tune up" the filter, and the time constant $\tau_p = \tau_{life}$ is inseparable from other parameters (amplification). To this we may also add such disadvantages as the predetermined shape of the frequency response: after all, the desired frequency response of the filter may be different from (1.28); because of the decline of $s(f)$ at the high frequencies the contribution to the noise from the subsequent stages increases, especially if there is little if any amount of internal amplification. So we have to repeat that "everyone should mind one's own business": the photodetector will detect the signal without distortion (which is well done by the modern devices), the amplifier will amplify the signal, and the filter will shape the desired frequency response.

These parameters characterize the processing of the signal by the photodetector. Let us now consider the dark current of the photodetector, which is responsible for the noise.

Dark current. When the voltage is applied to the photodetector, it causes not only the useful signal photocurrent, but also, unfortunately, a certain constant current. Under nonequilibrium conditions created by the bias, the carriers are generated not only by the light (photons), but also by the heat (phonons). The rate of thermal generation of pairs in unit volume is n_i/τ_i (in the SCR), and p_n/τ_p, n_p/τ_n (in the quasineutral n and p-regions). Here n_i is the concentration of intrinsic carriers, p_n and n_p are the concentrations of the minority carriers; τ_i, τ_p, τ_n are, respectively, the lifetimes of the carriers in the i-region, holes in the n-region, and electrons in the p-region. Given the volumes of the regions where the carriers are generated (Fig. 1.7a), the thermal generation current is

$$I_{dg} = qA_e W_b \frac{p_n}{\tau_p} + qA_e W \frac{n_i}{\tau_i} + qA_e L_n \frac{n_p}{\tau_n} \; . \tag{1.29}$$

Numerical calculation of the dark current is an ungrateful task: for a given semiconductor material it is only n_i and $D = \varphi_{th}\mu$ (the diffusion coefficient D is required for calculating $L = \sqrt{D\tau}$) that may be regarded as constants; other parameters, such as the concentrations $n_p = n_i^2/p$, $p_n = n_i^2/n$, the lifetimes, vary from one specimen to another over a very broad range. Then, there are too many mechanisms which add to the dark current: apart from those already named, one could mention tunneling, leakage through surface and bulk channels, which is especially difficult to assess, and the like [33, 34, 42]. Let us disregard these contributions, since in theory (and often in practice, in properly designed and properly manufactured specimens) they can be eliminated. From [14, 42–45] one may get an idea of the typical values of parameters for calculating the currents (see Table 1.1). These values have been used for calculating with the aid of (1.29) the dark current densities $j_{dg} = I_{dg}/A_e$.

Dark current component due to equilibrium thermal radiation (background current). The current I_{dg} in electronic devices is called saturation current (it flows when the back bias is of the order of several thermal potentials). In photodetectors we call it the dark current. But now, what is darkness? Even if we shield off the photosensitive element from all external sources with a non-transparent screen, the photosensitive element will still receive the equilibrium thermal radiation from the screen itself, and thus it is not possible to create "absolute darkness" (unless we reduce the temperature down to absolute zero). For this reason it is usual to include into the dark current of IR photodetectors, apart from the thermal generation current, also the photocurrent caused by the outside equilibrium thermal radiation corresponding to the ambient temperature (normally, T = 293 K).

From the outset we should warn the reader that the background component of the dark current in IR photodetectors of ranges III – V is usually the predominant one, and its noise sets the limit of the weakest detectable signal (Sect. 3.4). This is why we are going to pay a lot of attention to this component of the current.

Calculation of background current in idealized model. This model assumes that the surroundings emit blackbody radiation at a certain constant temperature T. Then, unlike the thermal generation current, the background current can be rigorously calculated: our task is similar to the task of the photodetector, and consists in counting the quanta of thermal radiation falling on the element. By Planck's law,

in unit time,
in unit wavelength range,
from unit area of photosensitive element,
into the hemisphere (space angle 2π steradian),

the number of emitted quanta is [46]

$$n(\lambda) = \frac{2\pi c}{\lambda^4\left(\exp\frac{\mathrm{h}\,c}{\kappa\,\mathrm{T}\,\lambda} - 1\right)} . \tag{1.30}$$

At equilibrium, the number of emitted and falling quanta $n(\lambda)$ is the same. In the cooled IR photodetectors the equilibrium does not exist: the working temperature of photosensitive elements is much lower than the background temperature, and therefore the emission by the element itself may be neglected. Equation (1.30) gives the density of the flux of quanta falling on the photosensitive element (from the entire hemisphere heated to the temperature T). Integrating the density over the spectral range from 0 to λ'', we find the effective number of thermal quanta registered by the photodetector:

$$n_{\text{eff}} = \int_0^{\lambda''} n(\lambda)d\lambda = \int_0^{\lambda''} \frac{2\pi c d\lambda}{\lambda^4\left(\exp\frac{\mathrm{h}\,c}{\kappa\,\mathrm{T}\,\lambda} - 1\right)}$$

$$= 2\pi c\left(\frac{\kappa\,\mathrm{T}}{\mathrm{h}\,c}\right)^3 \int_0^{\lambda''} \frac{d\left(\frac{\kappa\,\mathrm{T}\,\lambda}{\mathrm{h}\,c}\right)}{\left(\frac{\kappa\,\mathrm{T}\,\lambda}{\mathrm{h}\,c}\right)^4\left(\exp\frac{\mathrm{h}\,c}{\kappa\,\mathrm{T}\,\lambda} - 1\right)} . \tag{1.31}$$

Now we introduce dimensionless variables

$$y = \frac{\kappa\,\mathrm{T}\,\lambda}{\mathrm{h}\,c} = \frac{1}{x}, \quad y'' = \frac{\kappa\,\mathrm{T}\,\lambda''}{\mathrm{h}\,c} = \frac{1}{x''}, \tag{1.32}$$

and use them to transform the integral in (1.31) to the form

$$\mathrm{In}_1 = \int_0^{y''} \frac{dy}{y^4\left(\exp\frac{1}{y} - 1\right)} = \int_{y''}^{0} \left(\frac{1}{y}\right)^2 \frac{d\frac{1}{y}}{\exp\frac{1}{y} - 1} = \int_{x''}^{\infty} \frac{x^2 dx}{\exp x - 1} . \tag{1.33}$$

This integral is one of those which are often encountered in IR technology. In this book we shall deal with such integrals more than once, which justifies taking a closer look at the integrals of this kind.

Integrals associated with thermal radiation. Calculations of various physical quantities related to the blackbody radiation and Planck's distributions of energy, number of photons and their fluctuations, radiation heat conduction, etc., lead to the integrals of the form

$$\mathrm{In}_1(n) = \int_{x''}^{\infty} \frac{x^n dx}{\exp x - 1} \ , \tag{1.34}$$

$$\mathrm{In}_2(n) = \int_{x''}^{\infty} \frac{x^n dx}{(\exp x - 1)^2} \ . \tag{1.35}$$

The first of these we have just encountered in our calculation of n_{eff} (in this case, $n = 2$).

The technique of taking the integrals $\mathrm{In}_1(n)$, $\mathrm{In}_2(n)$ is universal. In general, they cannot be reduced to quadratures; our trick will consist in finding the particular solutions at $x'' >> 1$ and $x'' \to 0$, and then joining them to obtain a formula for arbitrary x'' (arbitrary long-wave limit λ'').

When $x'' >> 1$, the unity in the denominator in the integrand of (1.34), (1.35) may be dropped (Wien's approximation [47]); then

$$\mathrm{In}_1(2) \approx \mathrm{In}_2(2) \approx \int_{x''}^{\infty} x^2 \exp(-x) dx = \left(x^2 + 2x + 2\right)\exp(-x)\Big|_{\infty}^{x''}$$

$$= \left(x''^2 + 2x'' + 2\right)\exp(-x'') \ . \tag{1.36}$$

We have written the solution for $n = 2$: first, we shall need it for calculating the background current, and secondly, the reader will readily check the solution by comparing the derivative of the solution with the integrand. By induction, this solution can be easily extended to arbitrary values of n. From (1.36) we see that the solution is a modified integrand, in which the power term x^n is supplemented by all its derivatives:

$$\mathrm{In}_1'(n) = \mathrm{In}_2'(n) \approx \int_{x''}^{\infty} x^n \exp(-x) dx =$$

$$= \left(x''^n + nx''^{(n-1)} + n(n-1)x''^{(n-2)} + \ldots + n!\right)\exp(-x'') \ . \tag{1.37}$$

Now what are the values assumed by x'' in the spectral range of interest (up to 14 μm)? According to (1.32), the value of this argument at the normal background temperature of 293 K is

$$x'' = \frac{\mathrm{h}\,c}{\kappa \mathrm{T} \lambda''} = \frac{6.62 \cdot 10^{-34} \cdot 3 \cdot 10^{10}}{1.38 \cdot 10^{-23} \cdot 293 \cdot 1.4 \cdot 10^{-3}} = 3.51 > 1 \,,$$

$$\exp 3.51 = 33.4 >> 1 \,. \tag{1.38}$$

We see that the exponential term is indeed much greater than one, and the analytical expression (1.37) obtained in Wien's approximation is quite good for calculating n_{eff} and other parameters related to Planck's distribution at $\lambda'' < 14\,\mu\text{m}$. For the sake of universality, however, let us proceed further with our analysis: it will be useful for the case of hot background, and for photodetectors in the far infrared ($\lambda'' > 20\,\mu\text{m}$).

For $\lambda'' \to \infty$, $x'' \to 0$, the integrals (1.34), (1.35) can be found from tables [48]:

$$\mathrm{In}_1(n) = \int\limits_0^\infty \frac{x^n dx}{\exp x - 1} = n! \sum_{m=1}^{\infty} \frac{1}{m^{n+1}} \,, \tag{1.39}$$

$$\mathrm{In}_2(n) = \int\limits_0^\infty \frac{x^n \exp x dx}{(\exp x - 1)^2} = n! \sum_{m=1}^{\infty} \frac{1}{m^n} \,. \tag{1.40}$$

Let us now see, what is the value of the integral given by our first approximate solution (1.37) at $\lambda'' \to \infty$. Then $x'' = 0$, the exponential term is equal to one, and the polynomial is reduced to just the constant term:

$$x'' = 0 \,, \quad \mathrm{In}_1'(n) = \mathrm{In}_2'(n) = n! \,. \tag{1.41}$$

This approximate expression differs from the exact formulas (1.39), (1.40) in that the latter have the n-factorial multiplied by the sum, which, according to [48], rapidly tends to unity as n increases:

$$\sum_{m=1}^{\infty} \frac{1}{m^2} = \frac{\pi^2}{6} = 1.645 \,; \quad \sum_{m=1}^{\infty} \frac{1}{m^3} = 1.202 \,;$$

$$\sum_{m=1}^{\infty} \frac{1}{m^4} = \frac{\pi^4}{90} = 1.082 \,; \quad \sum_{m=1}^{\infty} \frac{1}{m^5} = 1.037 \,. \tag{1.42}$$

So, the formula (1.37) is even better than we thought: not only it holds for $\lambda'' < 14\,\mu\text{m}$ (for which it was initially derived) — it can also be used as the first approximation of $\mathrm{In}_1(n)$, $\mathrm{In}_2(n)$ at arbitrary value of λ''! The error of (1.37) only becomes noticeable when $\lambda'' \to \infty$ while $n = 2$ or 3; then the difference between the sum in (1.42) and unity is 20...65%. To justify the approximation (1.37), let us observe in anticipation that these 20...65% are the error in the calculation of background currents and noise power $\overline{i_n^2}$, whereas the error in the rms noise $\sqrt{\overline{i_n^2}}$ will be one half of this value: $\sqrt{1.645} = 1.283$, and $\sqrt{1.202} = 1.096$ (see Sect. 2.2, 2.3, 3.4). Even this error may be reduced dramatically by joining the solutions (1.37) and (1.39), (1.40): we only need to introduce the missing sum into (1.37):

$$\mathrm{In}_1(n) = \left(x''^n + n x''^{(n-1)} + \ldots + n! \sum_{m=1}^{\infty} \frac{1}{m^{n+1}} \right) \exp(-x'') \ , \tag{1.43}$$

$$\mathrm{In}_2(n) = \left(x''^n + n x''^{(n-1)} + \ldots + n! \sum_{m=1}^{\infty} \frac{1}{m^n} \right) \exp(-x'') \ . \tag{1.44}$$

Numerical calculations using (1.34), (1.35), (1.43), (1.44) indicate that the error of (1.43), (1.44) in the entire range of λ'' does not exceed 2...4%. As a matter of fact, the first approximation (1.37) (without multiplying the constant term by the sum) can be safely used when $\lambda'' < 14\,\mu\mathrm{m}$, or when $n > 3$: the error is then also within 2%. Thus, for all the above integrals we have obtained analytical approximations which are convenient both for analyzing the dependence on λ'' and T, and for getting numerical estimates with a pocket calculator.

Calculation of background current in idealized model (continued). Now we substitute the integral (1.43) and the sum (1.42) into the expression for the effective number of quanta (1.31):

$$n_{\mathrm{eff}} = 2\pi c \left(\frac{\kappa \mathrm{T}}{\mathrm{h} c} \right)^3 \mathrm{In}_1(2) = 2\pi c \left(\frac{\kappa \mathrm{T}}{\mathrm{h} c} \right)^3 \left(x''^2 + 2x'' + 2.404 \right) \exp(-x'')$$

$$= \left(2\pi \frac{\kappa^3}{\mathrm{h}^3 c^2} 2.404 \right) \mathrm{T}^3 \left(0.416 x''^2 + 0.83 x'' + 1 \right) \exp(-x'') \ . \tag{1.45}$$

Introducing the modified Stefan – Boltzmann constant

$$\sigma' = 2\pi \frac{\kappa^3}{\mathrm{h}^3 c^2} 2.404 = 2\pi \frac{\kappa^3}{\mathrm{h}^3 c^2} \left(2 \sum_{m=1}^{\infty} m^{-3} \right) = 1.52 \cdot 10^{11}\,\mathrm{cm}^{-2} \cdot \mathrm{s} \cdot \mathrm{K}^{-1} \ , \tag{1.46}$$

we rewrite (1.45) as

$$n_{\mathrm{eff}} = \sigma' \mathrm{T}^3 \left(0.416 x''^2 + 0.83 x'' + 1 \right) \exp(-x'') \ . \tag{1.47}$$

Hence follows the known Stefan – Boltzmann law for radiation over the entire spectrum:

$$x'' = 0 \ , \ \ n_{\text{eff}} = \sigma' \mathrm{T}^3 \ . \tag{1.48}$$

Observe that the constant σ' and the law $\sigma' \mathrm{T}^3$ relate to the radiation in terms of *the number of quanta*. They should not be confused with the more commonly known Stefan – Boltzmann constant σ and *the blackbody energy flux density law* $W_{\text{eff}} = \sigma \mathrm{T}^4$. The energy flux grows faster with the temperature than the number of quanta, since the mean quantum energy is proportional to the temperature, $\mathrm{h}\,\overline{\nu} = a\kappa \mathrm{T}$. Calculation yields the following value for the coefficient of proportionality a:

$$\mathrm{h}\,\overline{\nu} = a\kappa \mathrm{T} \ , \ \ a = \frac{3! \sum\limits_{m=1}^{\infty} m^{-4}}{2! \sum\limits_{m=1}^{\infty} m^{-3}} = 3 \cdot \frac{1.08}{1.202} = 2.7 \ . \tag{1.49}$$

Using this value of $\mathrm{h}\,\overline{\nu}$, we may write the following relationship between the flux of energy and the number of quanta:

$$W_{\text{eff}} = \mathrm{h}\,\overline{\nu} n_{\text{eff}} = (a\kappa \mathrm{T})\sigma' \mathrm{T}^3 = (a\kappa\sigma')\mathrm{T}^4 = \sigma \mathrm{T}^4 \ , \tag{1.50}$$

where

$$\sigma = a\kappa\sigma' = \frac{3! \sum\limits_{m=1}^{\infty} m^{-4}}{2 \sum\limits_{m=1}^{\infty} m^{-3}} \kappa \left(2\pi \frac{\kappa^3}{\mathrm{h}^3 c^2} 2 \sum_{m=1}^{\infty} m^{-3} \right) = 2\pi \frac{\kappa^4}{\mathrm{h}^3 c^2} 3! \sum_{m=1}^{\infty} m^{-4}$$

$$= 2\pi \frac{\kappa^4}{\mathrm{h}^3 c^2} 3! \frac{\pi^4}{90} = 5.67 \cdot 10^{-12}\ \mathrm{W \cdot cm^{-2} \cdot K^{-4}} \ . \tag{1.51}$$

In the ideal photodetector the number of light-generated photoactive pairs is equal to the number of falling quanta. Accordingly, to find the dark current density j_{db} (measured in A/cm^2) we only have to multiply n_{eff} (1.47) by the charge of electron,

$$j_{\text{db}}(\lambda'') = q n_{\text{eff}} = q\sigma' \mathrm{T}^3 \left(0.416 x''^2 + 0.83 x'' + 1 \right) \exp(-x'') \ . \tag{1.52}$$

If the spectral sensitivity of photodetector is limited from both the long-wave (λ'') and the short-wave (λ') sides, this formula needs to be corrected as follows:

$$j_{db}(\lambda',\lambda'') = j_{db}(\lambda'') - j_{db}(\lambda') \,, \tag{1.53}$$

since the number of quanta in the range between λ' and λ'' is equal to the number of quanta in the range between 0 and λ'' minus the number of quanta in the range between 0 and λ'. Usually, however, $j_{db}(\lambda'') >> j_{db}(\lambda')$, and so $j_{db}(\lambda',\lambda'') \approx j_{db}(\lambda'')$.

The calculated values of n_{eff}, j_{db} for all our spectral ranges I – V are presented in Table 1.1. As the long-wave limit moves from 3 to 10μm, the effective number of quanta increases as much as four orders of magnitude, since the density of thermal radiation $n(\lambda)$ grows sharply. For the three far-infrared subranges (8...10, 8...12 and 8...14μm) the dependence of n_{eff} on λ'' is not that strong: these subranges are matched to the maximum of thermal radiation, and they receive a considerable share of the total number of quanta emitted by a black body. According to the Stefan – Boltzmann law (1.48), (1.46), this number is

$$x'' = 0 \,, \quad n_{eff} = \sigma' T^3 = 3.82 \cdot 10^{18} \cdot \left(\frac{T}{293}\right)^3 \left[cm^{-2} \cdot s^{-1}\right] , \tag{1.54}$$

and so the density of background current is

$$j_{db} = qn_{eff} = 1.6 \cdot 10^{-19} \cdot 3.82 \cdot 10^{18} \cdot \left(\frac{T}{293}\right)^3 = 0.612 \cdot \left(\frac{T}{293}\right)^3 \left[\frac{A}{cm^2}\right] . \tag{1.55}$$

Comparing the above expression for the ultimate number of quanta with the values of n_{eff} from Table 1.1, we may specify precisely the share of quanta registered by photodetector at $\lambda'' = 14\,\mu m$ in the total number of quanta: this share amounts to 22%. The density of current in such photodetector is about one-fourth of the maximum.

Comparison between background and thermal generation components of the dark current. As follows from Table 1.1, the dark current $j_d = j_{dg} + j_{db}$ of photodetectors of the first two spectral ranges is really a dark current: such photodetectors are practically insensitive to thermal radiation with the room temperature, and their dark current is caused by thermal generation, $j_d \approx j_{dg}$. The calculation of j_{db} is only a matter of academical interest. On the contrary, the current of thermal generation in photodetectors sensitive to radiation with $\lambda > 2\mu m$ with the parameters given in the table is less than the current caused by the equilibrium background thermal radiation, $j_d \approx j_{db} = qn_{eff}$ (of which the reader has already been informed). Because of this, the ultimate density of dark current in such photodetectors may be regarded as a fundamental quantity which does not depend on the parameters of photodetector and is determined by the number of falling quanta of thermal radiation in the given spectral range from λ' to λ''.

Temperature dependence of thermal background dark current. We have already found the function $j_{db}(T)$: it is given by (1.37), (1.52). The temperature of the natural background (that is, the temperature around us and the photodetector) usually varies from –40 to +40°C, or 233...313 K. This a relatively narrow range of $\pm\Delta T$ as compared with the mean temperature $T_0 = 273\,K$, and therefore the functional dependence may be linearized using the expansion

$$j_{db}(T) \cong j_{db}(T_0) + \left.\frac{dj_{db}}{dT}\right|_{T_0} \Delta T = j_{db}(T_0)\left[1 + \frac{1}{j_{db}(T_0)}\left.\frac{dj_{db}}{dT}\right|_{T_0} \Delta T\right] . \qquad (1.56)$$

Then our task is reduced to finding the relative temperature coefficient ξ, which, according to (1.52), is

$$\xi = \frac{1}{j_{db}(T_0)}\left.\frac{dj_{db}}{dT}\right|_{T_0} = \frac{1}{n_{eff}}\frac{dn_{eff}}{dT} . \qquad (1.57)$$

We already have the expressions (1.31) – (1.33), necessary for calculating n_{eff}. Let us use then once again, dropping the constant coefficients and retaining only those terms which depend on λ and λT: after all, the constant coefficients in (1.57) will cancel out, so we need not bother about them. So,

$$n_{eff} \propto \int_0^{\lambda''} \frac{d\lambda}{\lambda^4\left(\exp\frac{hc}{\kappa T\lambda} - 1\right)} \sim \int_{x''}^{\infty} \frac{x^2 dx}{\exp x - 1} = \mathrm{In}_1(2) . \qquad (1.58)$$

Now we find the derivative of this integral, which is known to be equal to the integral of the derivative of the integrand:

$$\frac{dn_{eff}}{dT} \propto \int_0^{\lambda''} \frac{d}{dT}\left(\frac{1}{\exp\frac{hc}{\kappa T\lambda} - 1}\right)\frac{d\lambda}{\lambda^4} = \int_0^{\lambda''} \frac{hc}{\kappa T^2 \lambda} \frac{\exp\frac{hc}{\kappa T\lambda}}{\left(\exp\frac{hc}{\kappa T\lambda} - 1\right)^2}\frac{d\lambda}{\lambda^4} . \qquad (1.59)$$

The differentiation of the exponent resulted in the additional factor $hc/\kappa T^2 = x/T$. Of course, it will be retained when we go over from variable λ to variable x: compare the following formula for dn_{eff}/dT with the expression (1.58) for n_{eff}:

$$\frac{dn_{eff}}{dT} \sim \frac{1}{T}\int_{x''}^{\infty} \frac{x^3 \exp x}{(\exp x - 1)^2} dx = \frac{\mathrm{In}_2(3)}{T} . \qquad (1.60)$$

This integral is an old acquaintance of ours; after all, the digression to this and other similar integrals was worth the effort. Now for calculating the temperature coefficient we shall take advantage of the analytical approximations (1.44), (1.42) for $\mathrm{In}_2(n)$ with

$n = 3$. In accordance with (1.57), (1.58), (1.60), we have

$$\xi = \frac{1}{\mathrm{T}} \frac{\mathrm{In}_2(3)}{\mathrm{In}_1(2)} = \frac{1}{\mathrm{T}} \frac{x''^3 + 3x''^2 + 6x'' + 6 \cdot 1.2}{x''^2 + 2x'' + 2 \cdot 1.2} . \tag{1.61}$$

This expression may be given some face-lifting. With this purpose, we expand the fraction:

$$\xi = \frac{1}{\mathrm{T}} \frac{\left(x''^3 + 2x''^2 + 2 \cdot 1.2x''\right) + \left(x''^2 + 2x'' + 2 \cdot 1.2\right) + (1.6x'' + 4 \cdot 1.2)}{x''^2 + 2x'' + 2 \cdot 1.2}$$

$$= \frac{1}{\mathrm{T}} (x'' + 1 + \zeta) , \tag{1.62}$$

where ζ is the following correction:

$$\zeta = \frac{1.6x'' + 4 \cdot 1.2}{x''^2 + 2x'' + 2 \cdot 1.2} = 0.1...0.48 . \tag{1.63}$$

The numerical estimate has been obtained for the long-wave limits of 3 μm ($x'' = 16.4$) and 14 μm ($x'' = 3.51$), respectively.

Now there are three possibilities for doing away with the correction ζ.

1. Since in the spectral range of interest we have $x'' + 1 >> \zeta$, the face-lifting may consist in just dropping the correction:

$$\xi \approx \frac{x''}{\mathrm{T}} \left(1 + \frac{1}{x''}\right) = \frac{\mathrm{h}c}{\kappa \mathrm{T}^2 \lambda''} \left(1 + \frac{1}{x''}\right) . \tag{1.64}$$

The error of this approximation for $\lambda \leq 14\mu\text{m}$ is less than 10%.

2. For a better approximation, the correction in (1.61) may be fixed at a certain mean value — say, $\bar{\zeta} = 0.4$. Then the error in the range of $\lambda < 14\mu\text{m}$ will be reduced down to 2...3%.

3. To ensure convergence in the entire range of long-wave limits, $\lambda'' = 0...\infty$, we must set $\zeta = 2$ (the value of the correction at $x'' = 0$). Then (1.61) is reduced to

$$\xi \approx \frac{1}{\mathrm{T}} (x'' + 3) = \frac{\mathrm{h}c}{\kappa \mathrm{T}^2 \lambda''} + \frac{3}{\mathrm{T}} . \tag{1.65}$$

The error of this formula in the neighborhood of $\lambda \sim 14\mu\text{m}$, however, increases to about 30%.

So, our tiresome calculations have brought us to a handsome and clear-cut formula (1.64) for the temperature coefficient of the background current in the spectral range of interest, up to 14 μm. This formula shows explicitly that the temperature coefficient depends on the long-wave sensitivity limit of the photodetector, and on the temperature itself: it is almost inversely proportional to the long-wave sensitivity limit (long-wave cut-off wavelength), and almost inversely proportional to the square of the temperature. We say "almost", because the dependence becomes weaker as $\mathrm{T}\lambda''$ increases, and assumes a different nature when $x'' = \mathrm{h}c/\kappa\mathrm{T}\lambda'' \ll 1$: such a broad-band photodetector registers practically the entire flow of photons $n_{\mathrm{eff}} = \sigma'\mathrm{T}^3$, so that further increase in λ'' has no effect on the background current, while the dependence of ξ on T is given by the Stefan – Boltzmann law,

$$x'' \to 0\ ,\quad \xi = \frac{1}{n_{\mathrm{eff}}}\frac{dn_{\mathrm{eff}}}{d\mathrm{T}} = \frac{1}{\sigma'\mathrm{T}^3}\frac{d\sigma'\mathrm{T}^3}{d\mathrm{T}} = \frac{3}{\mathrm{T}}\ . \tag{1.66}$$

Naturally, this expression follows also from the general formula (1.61) at $x'' = 0$, or from (1.65). Comparing (1.64) and (1.66), we see that the dependence $\xi \sim \mathrm{T}^{-2}\lambda^{-1}$, observed when $\lambda'' < 14$ μm, is replaced by $\xi \sim \mathrm{T}^{-1}$ when $\lambda'' \to \infty$. This "change of the scene" can also be seen from the approximation (1.65).

Expressions (1.64) – (1.65) allow making the following numerical estimate for the temperature coefficient of thermal background current:

$$\lambda'' = 3\ \ldots\ 14\ \mu\mathrm{m},\quad x'' = 16.4\ \ldots\ 3.51\ ,$$

$$\xi = \frac{x''+1}{\mathrm{T}} = \frac{(16.4\ \ldots\ 3.51)+1}{293} = 0.059\ \ldots\ 0.015\ , \tag{1.67}$$

$$\lambda'' \to \infty\ ,\quad x'' = 0\ ,\quad \xi = \frac{3}{\mathrm{T}} = \frac{3}{293} \approx 0.01\ .$$

As indicated, a short-wave IR photodetector registers a very small number of photons with the relatively high energy from the left-hand tail of the Planck distribution. The number of such photons increases sharply as the temperature increases (at a rate of about 6% per kelvin at 3 μm). If the photodetector is sensitive in the neighborhood of the maximum of the background blackbody radiation, it will register a considerable share of the photons, and the number of such photons increases with the temperature much slower (about 1...1.5% per kelvin).

The calculation of j_{db} for real photodetectors operating under real conditions must take due account of some additional factors.

Effects of the relative profile of spectral response and quantum yield. In real photodetectors, the photocurrent is caused by only a part of the falling quanta, which are referred to as "photoactive". Accordingly, in order to determine the number of such photoactive quanta n_{eff}, the spectral density of the equilibrium thermal radiation $n(\lambda)$ at each wavelength (1.30) we must multiply by the quantum yield at this wavelength $\eta(\lambda)$, and then carry out the integration. This technique has been used for calculating the values of n_{eff} for a certain typical spectral characteristic (curve 3 in Fig. 1.8, and condition $\max[\eta]=1$. How they differ from n_{eff} for an ideal photodetector (curve 2 in Fig. 1.8), one can see from the following juxtaposition:

Spectral range, μm	2...3	3...5	8...10	8...12	8...14
n_{eff} for spectral curve of ideal photodetector, $cm^{-2}s^{-1}$	$3.74 \cdot 10^{13}$	$1.01 \cdot 10^{16}$	$2.44 \cdot 10^{17}$	$5.42 \cdot 10^{17}$	$8.5 \cdot 10^{17}$
n_{eff} for spectral curve of typical photodetector, $cm^{-2}s^{-1}$	$4.71 \cdot 10^{13}$	$1.08 \cdot 10^{16}$	$2.47 \cdot 10^{17}$	$5.6 \cdot 10^{17}$	$9 \cdot 10^{17}$
Ratio	1.26	1.07	1.01	1.03	1.06

Due account for the real profile of $\eta(\lambda)$ results in an acceptable shift in the value of n_{eff}: 26% in the range 2...3 μm, and merely 1...7% in the range 5...14 μm. In real photodetector, the fall of n_{eff} in the range from λ' to λ'' because of the lower quantum yield is counterbalanced by the sensitivity of the photodetector outside of this range. In other words, the value of n_{eff} is determined primarily by the spectral range (by its long-wave limit λ''), and depends very little on the particular shape of the spectral curve (as long as the long-wave fall-off is sharp enough — at least such as in the example above).

In this calculation we have assumed that $\eta(\lambda_{max})=1$. If the quantum yield at peak sensitivity $\eta(\lambda_{max})$ is less than one, the effective number of quanta registered will be reduced proportionally.

Effects of field of view. So far we have been calculating the number of quanta falling on the photodetector from the hemisphere (solid angle 2π). In real photodetectors the field of view (the solid angle from which the radiation is received) is less than 2π steradian. If the field of view is a cone with the apex angle 2α (which is called the angular field of view), then the effective number of quanta is [5]

$$n_{eff}(2\alpha)=\sin^2\alpha\cdot n_{eff}(\pi)\ . \tag{1.68}$$

In the modern designs of IR photodetectors the heat flux $n_{eff}(2\alpha)$ is the external background thermal radiation, which is collected by the lens on to the photosensitive site, whereas the elements of the design give practically no contribution into the value of

$n_{eff}(2\alpha)$. The plane angle of view 2α is determined by the cooled aperture, the ratio of the lens diameter D to the focal length f_{lens} (the relative aperture). In a wide-angle system the relative aperture may be $D:f_{lens} = 1:1$; therefore,

$$\tan\alpha = \frac{D/2}{f_{lens}} = 0.5 \ , \quad \alpha = \arctan 0.5 = 27^\circ \ ,$$

and so

$$n_{eff}(2\cdot 27^\circ) = \sin^2 27^\circ \cdot n_{eff}(\pi) = 0.2\, n_{eff}(\pi) \ . \qquad (1.69)$$

This means that in well-designed IR photodetectors the heat flux falling on the site does not exceed 20% of the heat flux from the hemisphere. In narrow-angle systems the flux $n_{eff}(2\alpha)$ is even less than that. For example, when the plane angle of view is 5°...10° the heat flux is two orders of magnitude less than the flux from the hemisphere:

$$n_{eff}(10...20^\circ) = \sin^2(5...10^\circ)\cdot n_{eff}(\pi) = (0.76...3)\cdot 10^{-2} n_{eff}(\pi) \ . \qquad (1.70)$$

Coefficient of emission. Finally, the calculation of the background current must take care of the emissivity ε_e, which is the ratio of the radiation emitted by a surface to the radiation emitted by a perfect blackbody radiator at the same temperature. Although the value of ε_e varies broadly for different bodies, and may also depend on the wavelength, the natural backgrounds in our IR ranges may be regarded as gray: $\varepsilon_e \approx 0.7...0.9$ in the range 2...5 μm, and $\varepsilon_e > 0.9$ in the range 8...14 μm. The emissivity is very low (~ 0.05) for transparent and mirror-reflecting surfaces [46].

Regimes of high-level and low-level background. The above numerical estimates for the dark currents (thermal generation and thermal background components) are of fundamental importance: these currents determine the magnitude of the main noise (shot noise) in photodetectors (Sect. 2.3), and may also delimit the time of integration in integration devices by charging the capacitance (Sect. 7.1). The level of noise and the integration time are, in their turn, definitive for the design of the electronics. So, we had good grounds to argue that the spectral range is one of the main factors which defines the magnitude of j_{db}, and, therefore, the ideology of both the electronics and the PDA as a whole. This is also the reason why the development of high-quality PDAs depends not only on the high quantum yield, but also on the low dark current I_d. As we said, the quantum yield is now quite close to its theoretical limit. The magnitude of I_d has also been made as low as possible: the dark current of high-quality silicon and germanium PDs is so small that the noise caused thereby is entirely drowned in the noise of the most advanced amplifiers (Sect. 2.5); in case of IR photodetectors of ranges III – V the current I_d characterizes the optical system rather than the photodetector, being a function of the spectral range defined by the optics. Recall that the values of $j_d = j_{db} + j_{dg}$ in Table 1.1 are given for the background at normal temperature, for hemispheric angle of view (that is, for the solid angle of 2π steradian, or angular field of view ±90°, and for the specified spectral range from λ' to λ''. It follows that the struggle for the low value of I_d must be waged not by the designer of the photodetector (or, at least, not by him alone), but first of all by the optoelectronics engineer. All in all, there are just three ways of achieving this end: lowering the temperature of thermal radiation falling on the photodetector, reducing the field of view (aperture) of the photodetector, and narrowing the spectral window $\lambda' - \lambda'' \lambda' ... \lambda''$. Of course, the engineer has to comply with the requirements to the optoelectronic systems, with the operating conditions of the latter. In this context, we may distinguish two extreme cases, corresponding to strong and weak background. In the first case the system requires a wide enough field of view ($2\alpha \approx 54°$), the temperature of the background is close to normal (or even elevated), the spectrum of the source is broad. Then the values of j_d will be close to those quoted in Table 1.1. The other extreme corresponds to the situation when the field of view can be reduced considerably ($2\alpha \leq 10...20°$), or the temperature of thermal radiation can be brought down to minus 30...40°C or even lower, or the spectral window is narrow (monochromatic source). Then the current density component j_{db} may be reduced by several orders of magnitude, and the thermal generation component may come upstage. The electronics of such IR PDAs of ranges III – V is similar to the electronics of PDAs of ranges I – II, where the dark current of the photodetector is small (Sect. 2.5).

Apart from the dark current, the photodetector may carry another dc component: the stray light photocurrent I_{pb}. This component is caused by the light coming from all sources excluding the equilibrium thermal radiation at the temperature of the background. One example is the direct (or reflected, or scattered) solar radiation. The current I_{pb} in silicon photodetectors in the presence of stray light may become considerable, and the device will operate in the regime of strong background. We see that photodetectors of any of the five spectral ranges may operate under conditions of strong background or weak background. For the sake of clarity, however, we shall take into consideration the typical current values associated with particular spectral ranges, as presented in Table 1.1. This will not restrict the universality of our analysis, since the table embraces the entire spread of current densities, from $3 \cdot 10^{-10}$ to 10^{-1} $\mathrm{A/cm^2}$.

Equivalent diagram of photodetector for small variable signal. The process of generation of the primary photoactive carriers is reflected on the equivalent diagram by the signal photocurrent generator $I'_p = S_0 E$ (Fig. 1.9a). In a photodiode the current of this generator goes directly to the external circuit. If there is internal amplification, the diagram will include the amplifier stage with the gain K_p. If the inertiality of photodetector is not negligible, the transfer coefficient of the amplifier will have to be multiplied by $s(f)$. The diagram shown in Fig. 1.9a reflects what has been said in the previous section: the frequency transform in the photodetector may be simulated by the first stage of filter amplifier. This circuit, in which the generation of primary carriers is

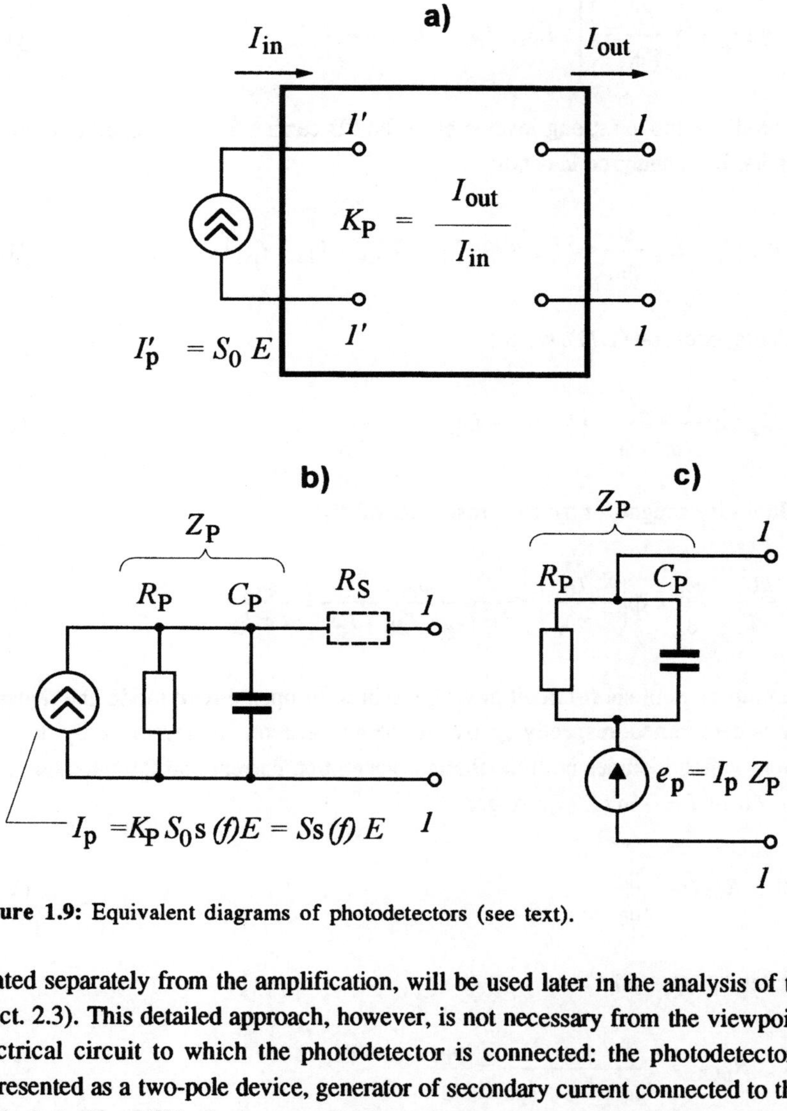

Figure 1.9: Equivalent diagrams of photodetectors (see text).

treated separately from the amplification, will be used later in the analysis of the noise (Sect. 2.3). This detailed approach, however, is not necessary from the viewpoint of the electrical circuit to which the photodetector is connected: the photodetector can be represented as a two-pole device, generator of secondary current connected to the output leads 1–1 (Fig. 1.9b). As in case of any real generator, one must take into account its internal resistance, which is generally a complex resistance connected in parallel to the current generator. Let us find this resistance for various types of photodetectors.

Photodiode. The dynamic active resistance of a PD is determined by its volt-ampere characteristic. Taking into account the direct ($I_{dg} \exp(U/\varphi_{th})$) and the inverse ($-I_{dg}$) components of thermal generation current, and the photocurrents I_{db}, I_{pb}, which in the first approximation do not depend on the bias, we obtain the following known expression for the volt-ampere characteristic [33, 34, 42]:

$$I(U)=I_{dg}\left[\exp\frac{U}{\varphi_{th}}-1\right]-I_{db}-I_{pb}\ ,\quad \varphi_{th}=\frac{\kappa T}{q}\ . \tag{1.71}$$

In the photodiode mode (strong inverse bias) the PD carries the back saturation current I_0, which has been analyzed in detail:

$$\frac{U}{\varphi_{th}}<<-1\ ,\quad \exp\frac{U}{\varphi_{th}}<<1\ ,\quad I\approx-I_0=-\left(I_{dg}+I_{db}+I_{pb}\right)\ . \tag{1.72}$$

Differentiating equation (1.71), we get

$$dI=I_{dg}\exp\frac{U}{\varphi_{th}}\frac{dU}{\varphi_{th}}=\left(I+I_{dg}+I_{db}+I_{pb}\right)\frac{dU}{\varphi_{th}}\ . \tag{1.73}$$

Hence follows the sought-for dynamic resistance of PD:

$$R_p=\frac{dU}{dI}=\frac{\varphi_{th}}{I_{dg}}\exp\left(-\frac{U}{\varphi_{th}}\right)=\frac{\varphi_{th}}{I+I_{dg}+I_{db}+I_{pb}}=\frac{\varphi_{th}}{I+I_0}\ . \tag{1.74}$$

In the extreme cases of short-circuit mode (zero bias) or open-circuit mode, the dynamic resistance is determined, respectively, by the thermal generation current, or by the total back current which includes both the thermal generation current and the photocurrents. Setting $U=0$ or $I=0$ in (1.74), we get

$$U=0\ ,\quad R_{pSC}=\frac{\varphi_{th}}{I_{dg}} \tag{1.75}$$

for short-circuit mode, and

$$I=0\ ,\quad R_{pOC}=\frac{\varphi_{th}}{I_{dg}+I_{db}+I_{pb}}=\frac{\varphi_{th}}{I_0} \tag{1.76}$$

for open-circuit mode.

Photodiodes operate either with zero bias (the regime of threshold-level PDs used to avoid excessive noise), or with negative bias (the regime of high-frequency PDs). The typical values of dynamic resistance with zero bias (short-circuit mode), calculated according to (1.75), are presented in Table 1.1. For spectral ranges I and II the diameter of the *pn* junction is assumed to be 0.5 mm, which is common with single-element photodiodes. For spectral ranges III and IV the dimensions of the site are 0.1 by 0.1 mm (sites of this size can be found in linear IR photodetectors with preamplifiers). Since the photodetectors for the first two ranges are not cooled, in our calculations we used the value of thermal potential 26 mV, which corresponds to normal temperature. The value of 7 mV, used for cooled photodetectors in ranges III and IV, corresponds to 77 K.

As follows from Table 1.1, R_{pSC} is a very large quantity (at least, in the ranges I – IV). In the photodiode regime the dynamic resistance is even larger, and, according to (1.74), it should tend to infinity. In practice, its value is, of course, limited, but it still is greater by several orders of magnitude than R_{pSC}, and depends on the leakage currents not accounted for in (1.74), and on I_0 being a function of the bias (as the bias increases, the SCR broadens, and therefore the generation in this region increases). The resistance R_p in the photodiode mode is an extremely unsteady quantity: it shows a strong dependence on the temperature, and varies from one specimen to another. Luckily, the load resistance is less than the resistance of PD in the photodiode mode (it does not exceed $(1...3)R_{pSC}$), and so, instead of calculating the resistance R_p in this mode, one may assume it to be infinite.

Although photodiodes in threshold-level PDAs do not, as a rule, operate in the short-circuit mode, it is the calculated resistance for this regime that determines the level of noise (Sect. 2.4). This is the reason why Table 1.1 includes the values of R_{pSC} calculated by (1.76) for site dimensions and temperatures indicated above, and the values of j_{dg}, j_{db} indicated in Table 1.1. In silicon and germanium PDs not sensitive to thermal radiation, the value of R_p is the same in the short-circuit and open-circuit modes: the bias across such PDs remains equal to zero, and there is no current. When IR PDs of ranges III – V are switched from short-circuit mode to open-circuit mode, the current of thermal radiation biases the junction in the forward direction; accordingly, the dynamic resistance decreases and is determined by the current I_{db} (the current caused by optical (thermal) radiation). In this case it is said that the PD operates in the regime of optical generation (OG mode). Since I_{db} depends on the quantum yield η and the plane angle of view 2α, we had to select the values of these parameters. To calculate the values shown in Table 1.1, we used typical values $\eta = 0.8$, $2\alpha = 2 \cdot 27^\circ$ (as indicated above, this angle corresponds to background thermal radiation collected by a lens with the relative aperture 1:1).

Important for the detection is also the capacitive component C_p of the complex resistance of PD (Fig. 1.9b). It determines not only the inertiality, but also, as will be shown in Sect. 2.4, the specific component of the high-frequency noise in a PDA. Structurally, a semiconductor diode is a capacitor, where the two conducting plates (n and p layers) are separated by the insulator (the space charge region) of thickness W (or W_i). Accordingly, the capacitance per unit area of the diode may be calculated by the formula for a parallel-plate capacitor:

$$\frac{C_p}{A_e} = \frac{\varepsilon\varepsilon_0}{W} = \frac{\varepsilon\varepsilon_0}{\sqrt{\dfrac{2\varepsilon\varepsilon_0 U'}{qN_d}}} = \sqrt{\frac{qN_d\varepsilon\varepsilon_0}{2U'}}$$

$$= \sqrt{\frac{1.6 \cdot 10^{-19} \cdot \left(10^{14} ... 10^{16}\right) \cdot (12...16) \cdot 8.8 \cdot 10^{-14}}{2 \cdot (1...10)}}$$

$$\approx (1...35) \cdot 10^{-9} \ \frac{\text{F}}{\text{cm}^2} \tag{1.77}$$

for *pn* structure, and

$$\frac{C_{\text{p}}}{A_{\text{e}}} = \frac{\varepsilon\varepsilon_0}{W_i} = \frac{16 \cdot 8.86 \cdot 10^{-14}}{(20...500) \cdot 10^{-4}} \approx (0.3...7) \cdot 10^{-10} \ \frac{\text{F}}{\text{cm}^2} \tag{1.78}$$

for silicon-based *pin* structure. In deriving (1.77) we have used the expression (1.5) for W.

In some applications important is the serial resistance of the photodiode R_{s} (Fig. 1.9b), caused by the contacts, the neutral regions of the base and the collector, through which the current flows.

Photodetector with internal gain. The capacitive nature of the complex resistance must be taken into account for all photodetectors with a *pn* junction, including those with internal amplification; because of the internal amplification, however, the importance of the active resistance R_{p} increases. If, for instance, the internal gain K_{p} does not depend on the bias (as is the case in the idealized phototransistor), the current I_{p} and the conductance $1/R_{\text{p}} = dI/dU$ increase K_{p}-fold. In the general case, however, the gain itself is a function of the bias. In an APD this dependence is definitive: assuming that the primary current I_0 is constant, and using the expression (1.7) for the multiplication, we get

$$I = I_0 \frac{U_{\text{B}}}{n(U_{\text{B}} - U)} \ , \quad dI = I_0 \frac{U_{\text{B}} dU}{n(U_{\text{B}} - U)^2} = M^2 n \frac{I_0 dU}{U_{\text{B}}} \ ,$$

$$R_{\text{pAPD}} = \frac{dU}{dI} = \frac{U_{\text{B}}}{nM^2} \frac{1}{I_0} = \frac{\varphi_{\text{eq}}}{I_0} \ , \quad \varphi_{\text{eq}} = \frac{U_{\text{B}}}{nM^2} \ . \tag{1.79}$$

The expression for the dynamic resistance of APD is on purpose presented here in the same form as that for the dynamic resistance of photodiode (compare (1.76) and (1.79)). Comparing the numerical values of the equivalent potential

$$M = 10...30 \ , \quad \varphi_{\text{eq}} = \frac{30...60}{(2...5)(10...30)^2} \approx 6...300 \text{ mV} \tag{1.80a}$$

(germanium-based APD), and

$$M = 50...200\ ,\quad \varphi_{\text{eq}} = \frac{30...60}{(2...5)(50...200)^2} \approx 0.15...12\ \text{mV} \tag{1.80b}$$

(silicon-based APD) with the thermal potential ($\varphi_{\text{th}} = 26$ mV), one can compare the resistances of PD and APD. Although the presence of multiplication reduces the resistance, it still remains noticeable: for a germanium APD the resistance is of the order of R_{pOC}, and one or two orders less than R_{pOC} for a silicon APD. In the latter case, however, the corresponding value of R_{pOC} itself is larger, since the current I_0 is small.

Photoresistor. As opposed to PD, the capacitance of photoresistor is negligibly small, and the active resistance has a concrete finite magnitude. It is especially easy to find the value of R_{p} in the regime of optical generation, when the photocurrent of thermal radiation I_{db} prevails. In a photoresistor this current is multiplied by the factor of K_{p}, and so, with due account for (1.15),

$$I = K_{\text{p}} I_{\text{db}} = \tau_{\text{life}}\left(\mu_n + \mu_p\right) U \frac{I_{\text{db}}}{l^2} = \tau_{\text{life}}\left(\mu_n + \mu_p\right) U j_{\text{db}}\ . \tag{1.81}$$

Hence, the dynamic resistance in the OG mode is

$$R_{\text{p}} = \frac{U}{I} = \frac{1}{\tau_{\text{life}}\left(\mu_n + \mu_p\right) j_{\text{db}}}\ . \tag{1.82}$$

This formula corresponds to a "square" photoresistor, whose length is equal to the width. According to (1.82), the resistance of MCT photoresistor for the range 8...14 μm is

$$R_{\text{p}} = \frac{1}{1.5\cdot 10^{-6}\cdot (1...2)\cdot 10^{5}\cdot 0.14} \approx 25...50\ \text{Ohm}\ . \tag{1.83}$$

The "normal" resistance of photoresistor (caused by thermally generated carriers) is

$$R_{\text{p}} = \frac{1}{q\left(n\mu_n + p\mu_p\right)W}$$

$$= \frac{1}{1.6\cdot 10^{-19}\left(5\cdot 10^{14}\cdot 2\cdot 10^{5} + 13\cdot 10^{12}\cdot 5\cdot 10^{2}\right)\cdot 10^{-3}} \approx 60\ \text{Ohm}\ . \tag{1.84}$$

Here the values of μ_n, μ_p, $p = n_i^2/n$ are taken from Table 1.1; the thickness of photoresistor is optimized and made equal to several lengths of absorption; this is a matter of compromise between higher resistance and higher absorption. We see that in this example the operating regime is close to the OG mode; because of the strong thermal background the resistance is low (several tens of ohms). Matching of such low-resistance sources with amplifiers is associated with considerable difficulties (Sect. 2.5). Fortunately, for the shorter-wavelength detectors the current density j_{db} is less, and the resistance (1.82) increases in proportion. In the second example (PbS-based photoresistor) the current density j_{db} is reduced by a factor of $\sim 2 \cdot 10^4$ (Table 1.1), and so one should expect that the resistance will be

$$R_p = (25...50) \cdot 2 \cdot 10^4 = 5 \cdot 10^5 ... 10^6 \text{ Ohm} . \tag{1.85}$$

The typical values of resistance for PbS-based photoresistors are close by the order of magnitude to this estimate (Table 1.1).

The equivalent diagram often used for photoresistors includes the voltage generator e_p and internal resistance R_p connected in series (Fig. 1.9c). Using the methods of theory of electrical circuits, it is possible to prove complete equivalence of the diagrams in Fig. 1.9b and c, provided that $e_p = R_p I_p$, and the internal resistances of the generators are the same. The equivalence of the two circuits is obvious: for example, in the open-circuit mode the generated voltage is e_p, and the current increment caused by the external bias δU is $\delta I = \delta U / R_p$. This equivalent diagram, however, is only reasonable in the case of photoresistor: in these photodetectors the resistance R_p is a concrete, stable and reproducible parameter. Further on we shall use the diagram with the current generator (Fig. 1.9b), which is suitable for all types of photodetectors under consideration.

We started this section with the description of operating principles of photodetector as counter of photons. Now we end by reflecting the result of such count as the current generator on the equivalent diagram, which supplies electrons into the external circuit in exact proportion to the number of incident quanta.

Summary. *Commonly used in PDAs are photodiodes (PD), avalanche photodiodes (APD), photoresistors (PR), and, more rarely, phototransistors (PT). These photodetectors are counters of photons. The number of electrons supplied by photodiode to the external circuit is almost equal to the number of photons falling on the photodetector: the so-called quantum yield in the modern advanced photodetectors is very high (50...95%). APDs, PRs and PTs exhibit internal current gain, which is useful because they allow "dragging out" the small signal and the noise of photodetector from the noise of subsequent amplifier. With the correct design (judicious use of* pin *and* npip *structures, optimal choice of lifetime and bias), the photodetector can, as a rule, be made inertialess, and represented on the equivalent diagram by the current generator whose electrical output signal is identical in shape to the incoming optical signal.*

1.3. Amplifier

In the traditional PDAs the detected signal is sent to the amplifier [2], where the second processing operation is performed, the reinforcement of the signal in the specified frequency band (in case of photodetectors with internal gain this is the second stage of amplification).

Serial load circuit. In the most simple situation (Fig. 10a), the load resistance R_L is connected in series with the photodetector, and the signal voltage drop across this resistance is the input of the amplifier. The permanent bias source must have low internal resistance for ac current (being shunted, for instance, with a large capacitance); then the photodetector will be connected in terms of the ac current in parallel with the load and the input of the amplifier (leads 1 – 0). As a matter of fact, the equivalent diagram of the input circuit with respect to a small signal is a parallel connection of all elements: the photocurrent generator I_p, the active resistances (dynamic resistance of photodetector R_p, load resistance R_L, input resistance of the amplifier R_a), and the capacitances (capacitance of photodetector C_p, stray capacitance C_c, input capacitance of the amplifier C_a). Accordingly, the voltage at the output of the amplifier is

$$U_{out} = K_{21} U_{in} = K_{21} Z_{in} I_p \ , \tag{1.86}$$

where

$$\frac{1}{Z_{in}} = \frac{1}{R_{in}} + \frac{1}{j\omega C_{in}} \ , \tag{1.87}$$

$$\frac{1}{R_{in}} = \frac{1}{R_p} + \frac{1}{R_L} + \frac{1}{R_a} \ , \quad C_{in} = C_p + C_c + C_a \ . \tag{1.88}$$

Here K_{21} is the voltage gain of the amplifier. Since the output is voltage, and the input is current, it will be convenient to define the transfer coefficient of amplifier as the ratio of the output U_{out} to the input I_p, which will have the dimension of resistance:

$$K = \frac{U_{out}}{I_p} = K_{21} Z_{in} \ . \tag{1.89}$$

The transfer coefficient is at maximum when the equivalent input resistance $Z_{in} \approx R_p$ is maximum; to make the value of $Z_{in} = R_p$ as large as possible one must choose a high-resistance amplifier ($R_a >> R_p$) and a large load resistance,

$$R_L \geq (3...10) R_p = (3...10)\left(3 \cdot 10^4 ... 4 \cdot 10^{10}\right) \approx 10^5 ... 4 \cdot 10^{11} \ \text{Ohm} \ . \tag{1.90}$$

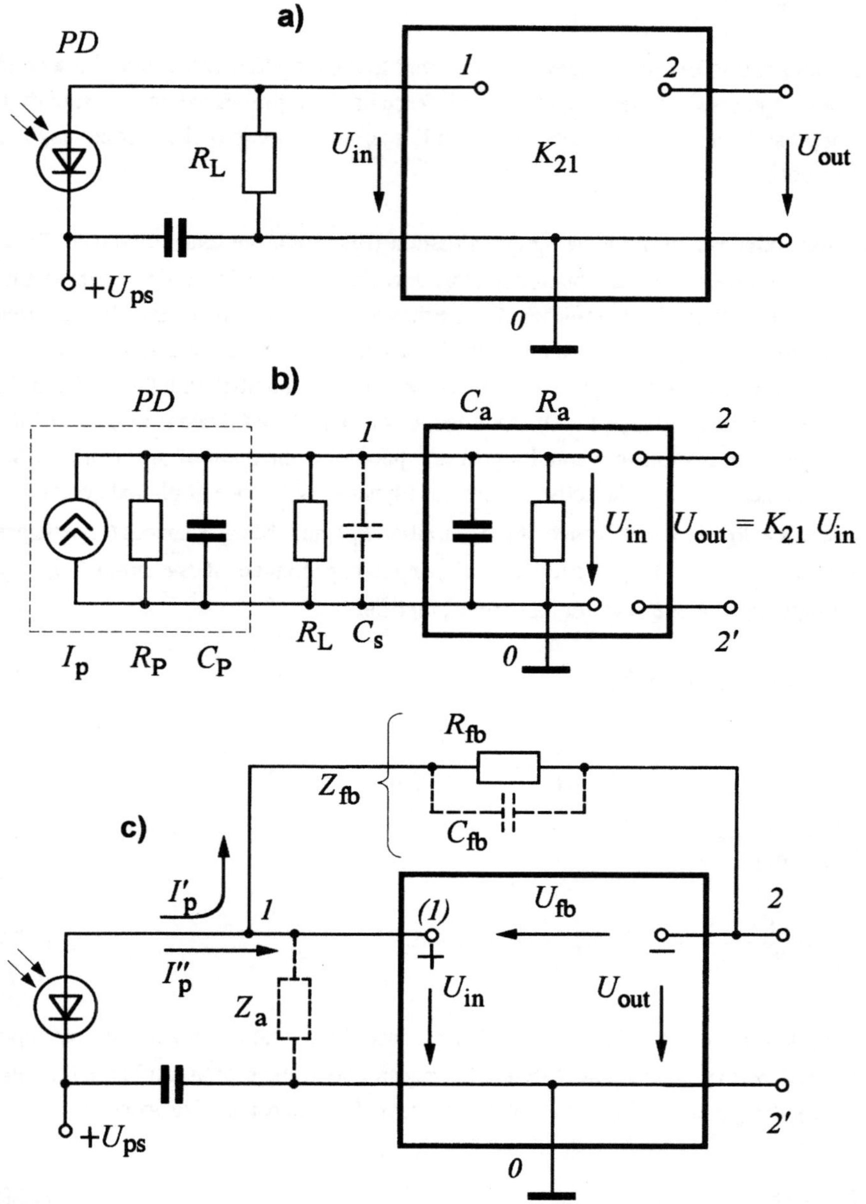

Figure 1.10: In traditional PDAs the signal of the photodetector is picked up: (a,b) from load resistor with the aid of amplifier; (c) by operational amplifier.

The values of the resistances are taken from Table 1.1 for PDs of ranges I – IV at zero bias, when condition $R_p = R_{pSC}$. In the photodiode mode the resistance R_p is larger by several orders of magnitude, and so in many cases the load resistance must be not just large, but extremely large. As it happens so often in the engineering tasks, this condition is antagonistic to other requirements imposed on the parameters of PDA.

Limitations on the supply voltage. The large value of the load resistance R_L (1.90) results in an unreasonably high supply voltage U_{ps}. For example, for a photoresistor with the bias voltage U, the supply voltage must be equal to $U_{ps} = U(R_L + R_p)/R_p$. In practice, the value of R_L is selected under assumption that the voltage U_{ps} applied to the photoresistor and its load resistance is fixed. Then the expression (1.86) at low frequencies, when the capacitance may be disregarded, with due account for (1.15), (1.88), and with the replacement $U = U_{ps}R_p/(R_L + R_p)$, can be transformed into

$$U_{out} = K_{21}R_{in}I_p = K_{21}\frac{R_L R_p}{R_L + R_p}\left[\frac{\tau_{life}}{l^2}(\mu_n + \mu_p)US_0E\right]$$

$$= K_{21}\frac{R_L R_p}{R_L + R_p}\left[\frac{\tau_{life}}{l^2}(\mu_n + \mu_p)\frac{U_{ps}R_p}{R_L + R_p}S_0E\right]. \tag{1.91}$$

Differentiating this expression with respect to R_L, we see that the maximum of U_{out} under the above condition $U_{ps} = \text{const}$ is achieved when the resistances are matched:

$$R_L = R_p ,$$

$$U_{out} = K_{21}\frac{\tau_{life}(\mu_p + \mu_n)}{l^2}\frac{U_{ps}R_p}{4}S_0E$$

$$= K_{21}\frac{\tau_{life}(\mu_p + \mu_n)}{l^2}\frac{UR_p}{2}S_0E . \tag{1.92}$$

If the power supply voltage U_{ps} is not limited, the load resistance can be made much larger than R_p, and then, according to (1.91), we have

$$R_L >> R_p , \quad U_{out} = K_{21}\frac{\tau_{life}(\mu_p + \mu_n)}{l^2}UR_pS_0E . \tag{1.93}$$

Comparison between (1.92), (1.93) reveals that, given that the bias voltage U across the photoresistor is the same, the gain in the regime of matching resistances (1.92) is one-half of the maximum possible gain (1.93), but the required supply voltage is much lower.

Stability of bias voltage. Variations of the temperature and the background irradiation cause variations in the current I_0 of the photodetector, and change the voltage drop I_0R_L across the load resistance. If R_L is large, the change in I_0R_L can be so large as to swing the bias across the photodetector from the full power supply voltage U_{ps}

down to zero, and even further to the forward-bias mode (in photodiodes with strong background). In such situation the photodetector may fail: because of breakdown and excess noise at strong back-bias when $U \approx U_{ps}$, or because of the sharp drop in amplification in APDs and PRs.

Inertiality. According to the equivalent diagram in Fig. 1.10b, the input circuit is a one-section RC filter, whose time constant is $\tau = R_{in}C_{in}$. Because of this, with the large value of the load resistance the PDA may be too sluggish to comply with the requirements.

Intercoupling. In multielement PDAs the illumination of the ith element may produce the signal in the neighboring kth element. One of the causes of such electrical intercoupling is the finite feed-through resistance Z_{ki} between these elements. In this case the load $Z_{in\,k}$ together with the feed-through resistance form a voltage divider; and the larger the load, the stronger the spurious signal from the non-illuminated element.

We see that the response time, the stability of bias, the decoupling of elements are all improved when the load resistance is made smaller, and only one parameter becomes worse — the signal voltage U_{out} (and, accordingly, the "voltage' sensitivity of the photodetector). It might seem wise to choose a small load resistance, and make up for the loss in U_{out} and the voltage sensitivity by increasing the gain K_{21}. In the threshold-level PDAs, however, this usually cannot be done. As will be demonstrated in Sect. 2.5, the component of thermal noise increases when the load resistance is small. In order to make this component smaller than the inherent (shot) noise of the photodetector, one has to use a very large load resistance, $R_L > R_{pOC}$. So, it is not possible to reconcile all the contradicting requirements, and the serial connection of photodetector and load often has to be renounced.

Photodetector with operational amplifier. If we disconnect one of the leads of the load on the diagram in Fig. 1.10b from the ground (point 0), and connect it to the output (point 2), we get a negatively feedbacked circuit (operational amplifier; Fig. 1.10c). Now the load is part of the feedback circuit; negative feedback is ensured by choosing an inverting amplifier, so that the sign of the output signal is opposite to the sign of the input signal. Using the theory of electrical circuits, one can easily demonstrate the following four features of such arrangement [49–52].

Feature one. *The output voltage is equal to the voltage drop across the feedback circuit,* $U_{out} = U_{fb}$. To be more precise, in tracing the circuit 2 – 1 – 0 we find that the input voltage U_{in} must be added to the output voltage U_{out}. Since, however, U_{out} is greater than U_{in} by a factor of K_{21} (usually, $K_{21} = 10^2 \ldots 10^5$), the input voltage may be neglected:

$$|U_{fb}| = |U_{out}| - |U_{in}| = |U_{out}| - \frac{|U_{out}|}{K_{21}} \approx |U_{out}| \,. \tag{1.94}$$

Feature two. *The input current goes into the feedback circuit.* More precisely, the input current I_p branches out at the input of the amplifier (point 1): part of this current I'_p goes into the feedback circuit, and part I''_p goes into the amplifier. Accordingly, the component I'_p causes the voltage drop $U_{fb} = I'_p Z_{fb} = U_{out}$ across the feedback circuit, and the component I''_p produces the input voltage $U_{in} = I''_p Z_{in}$. Knowing that $U_{out}/U_{in} = K_{21}$, we find the ratio of the two currents:

$$\left|\frac{I''_p}{I'_p}\right| = \left|\frac{U_{in}}{Z_{in}}\frac{Z_{fb}}{U_{out}}\right| = \left|\frac{1}{K_{21}}\frac{Z_{fb}}{Z_{in}}\right| << 1 \,. \tag{1.95}$$

The gain $|K_{21}|$ is always made greater than the ratio $|Z_{fb}/Z_{in}|$, which allows us to write the above inequality. Under this condition, we may neglect the leakage of the input current component I''_p into the amplifier, and assume that $I_p \approx I'_p$.

Feature three follows from the above characteristics of operational amplifier: *the output voltage equals to the voltage drop across the impedance of the feedback circuit*:

$$U_{out} \approx U_{fb} \approx I_p Z_{fb} \,, \quad K \approx \frac{U_{out}}{I_p} = Z_{fb} \,. \tag{1.96}$$

Feature four. *The equivalent input impedance Z_{eq} is by a factor of K_{21} smaller than the impedance of the feedback circuit.* This follows directly from the definition of the equivalent impedance as the ratio of the input voltage to the input current I_p:

$$Z_{eq} = \frac{U_{in}}{I_p} = \frac{U_{out}}{K_{21} I_p} = \frac{I_p Z_{fb}}{K_{21} I_p} = \frac{Z_{fb}}{K_{21}} \,. \tag{1.97}$$

In this way, all problems have been solved by simply rearranging the load connection. Comparing the diagrams in Fig. 1.10a and 1.10c under condition $R_L = R_{fb}$, we see that, owing to the K_{21}-fold reduction (that is, by a factor of one hundred to one hundred thousand!) of the equivalent input impedance of the operational amplifier, we have achieved both the stabilization of the photodetector bias voltage, and electrical decoupling in multielement photodetectors. The inertiality has decreased sharply: according to (1.96), the lag is now determined by the time constant of the feedback $R_{fb}C_{fb}$, and the capacitance of this circuit (0.1...0.5 pF) is one or two orders of magnitude less than the input capacitance C_{in}. At the same time, the resistance R_{fb} is large enough, which ensures that the contribution of its thermal noise to the total noise of the PDA is small

(Sect. 2.5). Yet another advantage of the operational amplifier consists in the highly stable transfer coefficient. This coefficient is equal to the total impedance Z_{fb}, and exhibits no dependence on the amplifier gain K_{21}, which is a highly changeful parameter. Of course, the negative feedback reduces the transfer coefficient (note that expressions (1.89) and (1.96) differ by the factor of K_{21}); after all, this is why the feedback is called negative. Firstly, however, in the ordinary amplifier the gain K_{21} is much less than the high gain ($K_{21} \sim 10^5$) typical of the operational amplifier with negative feedback. Secondly, the total impedance Z_{fb} of the operational amplifier has been increased (in comparison with the total impedance Z_{in} at $R_L = R_{fb}$) through reducing the capacitive component. Thirdly... thirdly, one has to put up with the reduced transfer coefficient of the operational amplifier, being prepared to add, if necessary, a second amplification stage.

Today the ordinary amplifiers are almost completely replaced by the operational amplifiers. Direct amplification circuits are still used in HF and UHF systems, where it is difficult to apply the feedback because of the risk of spurious generation at the high frequencies (where the phase distortion may revert the feedback from negative to positive). Modern integrated circuits, with a high input resistance and low noise, are often specially developed for PDAs. In multichannel PDAs it is possible to use multistage amplification, before and after switching. A detailed description of the circuitry of PDAs falls beyond the scope of this book; we only give here the necessary minimum of information about the amplifiers, as required for the analysis of signal processing in the traditional PDAs.

The readout of the photodetector signal in PDAs with integration is different in the respect that, before being converted into the output voltage, the photocurrent I_p is converted into the charge Q. Formally, this conversion can also be described by the transfer coefficient $K(f) = Q/I_p$ which has the dimension of time (the time of charge integration T_i. We postpone the detailed analysis of the current-charge conversion till Sect. 7.1, when we shall get acquainted with the requirements to the frequency response of such conversion.

Summary. *In the traditional PDAs the current of the photodetector is converted into the amplifier output voltage, with the transfer coefficient $K_{21}Z_{in}$ (in the absence of feedback), where K_{21} is the voltage gain of the amplifier, and Z_{in} is the total input impedance, which is defined by both ohmic and capacitive components of the photodetector, load, and input of the amplifier. The high load resistance, required for matching the high-resistance photodetector, causes instability of the bias voltage, stray electrical coupling in multielement PDAs), increases the time constant of the front-end RC circuit; the transfer coefficient is unstable because of the instability of K_{21}. These drawbacks are eliminated by the use of operational amplifiers, where the load resistance is connected to the negative feedback circuit (Z_{fb}). While retaining the high value of the load resistance (which keeps its thermal noise component to the minimum), it is possible to improve the stability of the bias voltage, to reduce the spurious coupling of* *elements (by virtue of the low equivalent input impedance of the operational amplifier), to stabilize the transfer coefficient, and to reduce the time constant of the RC circuit, since the transfer coefficient is now equal to Z_{fb}, and does not depend either on the gain of the amplifier, or on the amplifier input capacitance.*

Chapter 2

SIGNAL AND NOISE IN PHOTODETECTING ASSEMBLIES

Chapter 1 introduced us to the functional diagram of the PDA and its main elements which determine the lowest detectable optical power: the photodetector and the amplifier. Now we can embark upon the main topic of the present book. First we shall meet the protagonist — the pulsed optical signal, and then the antagonist — the noise which interferes with the detection of the signal. We shall follow the path of the signal and noise through the functional units of the PDA, and discuss the possible ways of minimizing the noise.

2.1. Pulsed Optical Signal

Typical shape of optical signal. The problem of signal detection is stated when defined are the distinguishing features of the signal which can be used for telling the signal from the noise. As pointed out in Sect. 1.1, in the majority of optoelectronic systems the optical signal is pulsed. Let us introduce the following notation for the optical signal:

$$E_s(t) = E_{sA} e_s(t) \ , \tag{2.1}$$

where E_{sA} is the signal amplitude (the pulse power of optical radiation falling on the photosensitive site), and $e_s(t)$ is the relative shape of the pulsed signal (a dimensionless quantity). At a certain time $t = t_0$ the signal reaches its maximum, $E_s(t_0) = E_{sA}$, $e_s(t_0) = 1$. It is the profile $e_s(t)$ that is the "fingerprint" of the signal.

In laser systems the profile e_s is determined by the design of the laser (LED), and by the nature of excitation; and in the scanning systems by the energy distribution across the spot and the relationship between the diameter of the spot and the size of the site, or the slit width of the mechanical modulator. The simplest profile is rectangular (Fig. 2.1a):

$$\begin{aligned} &-\frac{T_s}{2} < t < \frac{T_s}{2} \ , \quad e_s(t) = 1 \ ; \\ &t \le -\frac{T_s}{2} \ , \quad t \ge \frac{T_s}{2} \ , \quad e_s(t) = 0 \ . \end{aligned} \tag{2.2}$$

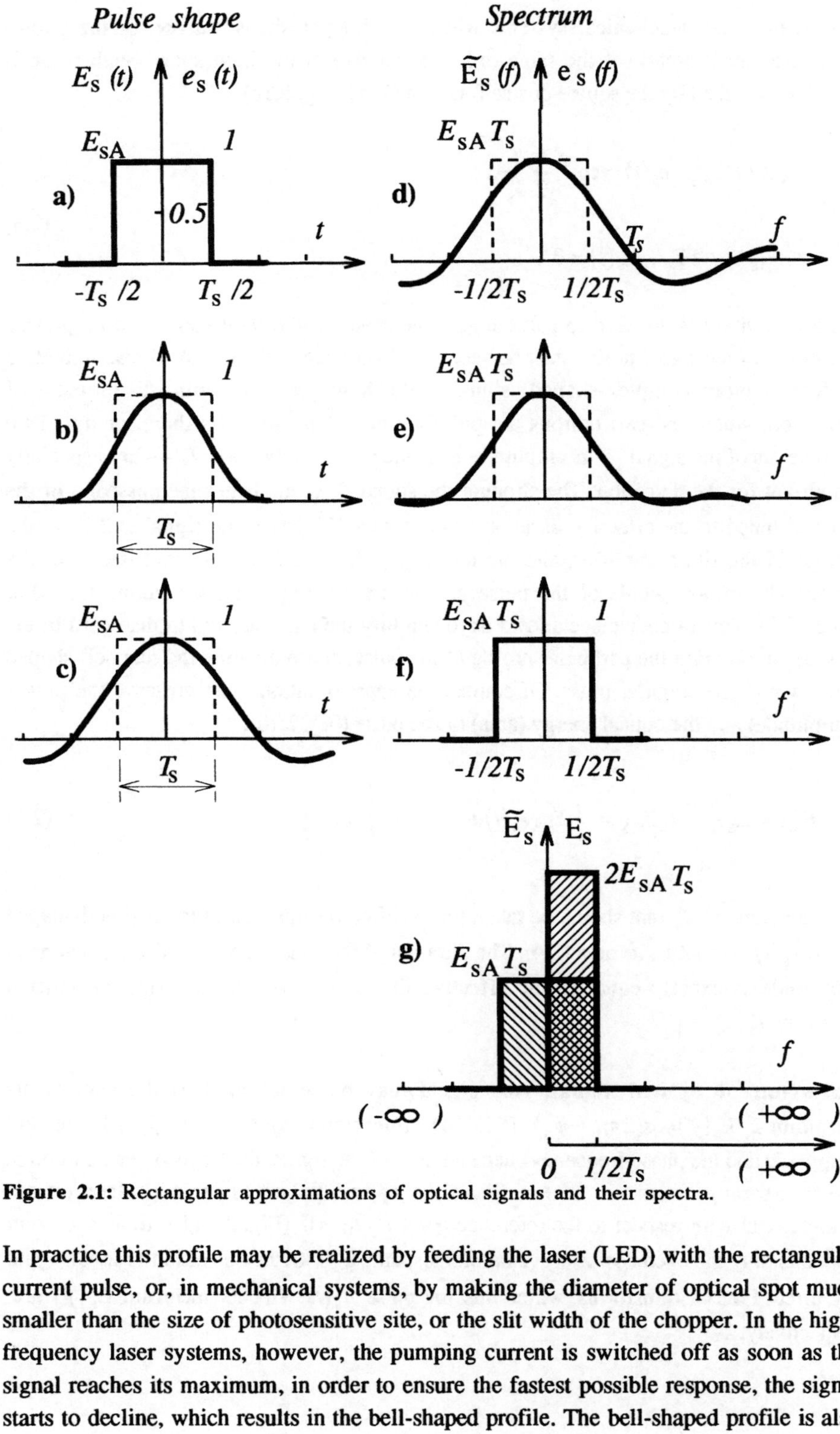

Figure 2.1: Rectangular approximations of optical signals and their spectra.

In practice this profile may be realized by feeding the laser (LED) with the rectangular current pulse, or, in mechanical systems, by making the diameter of optical spot much smaller than the size of photosensitive site, or the slit width of the chopper. In the high-frequency laser systems, however, the pumping current is switched off as soon as the signal reaches its maximum, in order to ensure the fastest possible response, the signal starts to decline, which results in the bell-shaped profile. The bell-shaped profile is also observed in the mechanical systems when the chopper slit is narrow, or the photosensitive site is small (of the same order as the size of the light spot). Such pulse is usually described by the square-cosine function [1, 4] (Fig. 2.1b):

$$-T_s \le t \le T_s \ , \quad e_s(t) = \cos^2 \frac{\pi}{2} \frac{t}{T_s} \ ;$$
$$t \le -T_s \ , \quad t \ge T_s \ , \quad e_s(t) = 0 \ . \tag{2.3}$$

As a rule, there is no need to particularize the shape of the light spot. A more precise definition of the signal profile may be necessary in dealing with the task of discriminating different (more complex-shaped) signals. We, however, start with the problem of detection, when just two options are possible: either the signal is there, or not. Two parameters of the signal — its amplitude E_{sA} and effective duration T_s — are especially important for the detection. The stronger the signal E_{sA}, the better the passband of the filter is tuned to the effective duration of the signal, the larger the signal will be at the output of the filter-amplifier, and the less likely the signal will be confused with the noise. The minor details of the pulse profile are not important for tuning the filter (Sect. 3.3). This circumstance allows us to simplify the analysis, and to disregard in the first approximation the particular profile of the pulse, approximating the real bell-shaped pulse with a rectangular pulse. Of course, the approximation must preserve the power (amplitude) and the optical energy (area) of the pulse (Fig. 2.1b):

$$E'_{sA} = E_{sA} \ , \quad T_s E_{sA} = \int_{-\infty}^{\infty} E_{sA} e_s(t) dt \ , \quad T_s = \int_{-\infty}^{\infty} e_s(t) dt \ . \tag{2.4}$$

It is the value of T_s that should be taken for the effective duration of the real bell-shaped pulse $e_s(t)$ — that is, its area (2.4). The duration of the square-cosine pulse (2.3) at half-amplitude is exactly equal to the effective duration (2.4); this justifies the unified notation T_s.

Spectrum of optical signal. Any signal may be represented as the sum of its harmonics, $\tilde{E}_s(f)\cos(2\pi ft + \tilde{\psi}_s)$ [53, 54]. Functions $\tilde{E}_s(f)$ and $\tilde{\psi}_s(f)$ are the amplitude and the phase-frequency characteristics of the signal; the frequency runs through the entire range from $-\infty$ to $+\infty$. Since the optical pulses under consideration are symmetrical with respect to the reference point $t = t_0 = 0$ (Fig. 2.1a,b), their spectrum contains only the even cosine-wave harmonics, and $\tilde{\psi}_s = 0$. In this case the inverse and the direct Fourier transforms, which link the pulse $E_s(t)$ with its spectrum $\tilde{E}_s(f)$ are, respectively,

$$E_S(t) = \int_{-\infty}^{\infty} \tilde{E}_S(f) \cos 2\pi f t \, df \ , \tag{2.5}$$

$$\tilde{E}_S(f) = \int_{-\infty}^{\infty} E_S(t) \cos 2\pi f t \, dt \ . \tag{2.6}$$

Since the cosine-waves at positive and negative frequencies are indistinguishable, $\cos 2\pi(-f)T_S = \cos 2\pi f T_S$, and, according to (2.6), the spectral densities for the symmetrical pulse at these frequencies are the same, $\tilde{E}_S(-f) \equiv \tilde{E}_S(f)$, the two cosine-waves add up:

$$\tilde{E}_S(-f)\cos 2\pi(-f)T_S + \tilde{E}_S(f)\cos 2\pi f T_S \equiv 2\tilde{E}_S(f)\cos 2\pi f T_S \ . \tag{2.7}$$

This is why only the positive frequencies are often taken into consideration; these frequencies are used for defining the transfer coefficient of the amplifier $K(f)$, and the spectrum of the signal:

$$f = [0, \infty] \ , \ E_S(f) = 2\tilde{E}_S(f) \ . \tag{2.8}$$

In the present book it will be convenient to use both the spectrum $\tilde{E}_S(f)$, corresponding to the expansion over the frequency band $f = [-\infty, \infty]$, and the double-power spectrum $E_S(f)$, corresponding to the expansion in the harmonics of only the positive frequencies $f = [0, \infty]$.

The direct and the inverse Fourier expansions (2.5), (2.6) have three general implications, which hold for the spectrum of symmetrical pulse of arbitrary shape.

Feature one. *The characteristic frequency of the spectrum of the pulse is the frequency* $1/2T_S$ *: the decline of the spectral density starts in the neighborhood of this frequency.*

This feature is especially easily traced for the example of rectangular pulse (Fig. 2.2a). According to (2.6), the spectral density $\tilde{E}_S$ in this case is equal to the area (integral) of the cosine wave on the interval T_S, whereas at the low frequencies $f << 1/2T_S$ both the cosine wave and its integral are constant (Fig. 2.2b). Only when the half-period $1/2f$ becomes equal (approximately equal) to the pulse length, the cosine wave area decreases (Fig. 2.2c). At the high frequencies $f >> 1/2T_S$, when several periods of the cosine fit into the interval T_S, the spectral density sharply decreases and may become equal to zero (Fig. 2.2d) or negative (Fig. 2.2e).

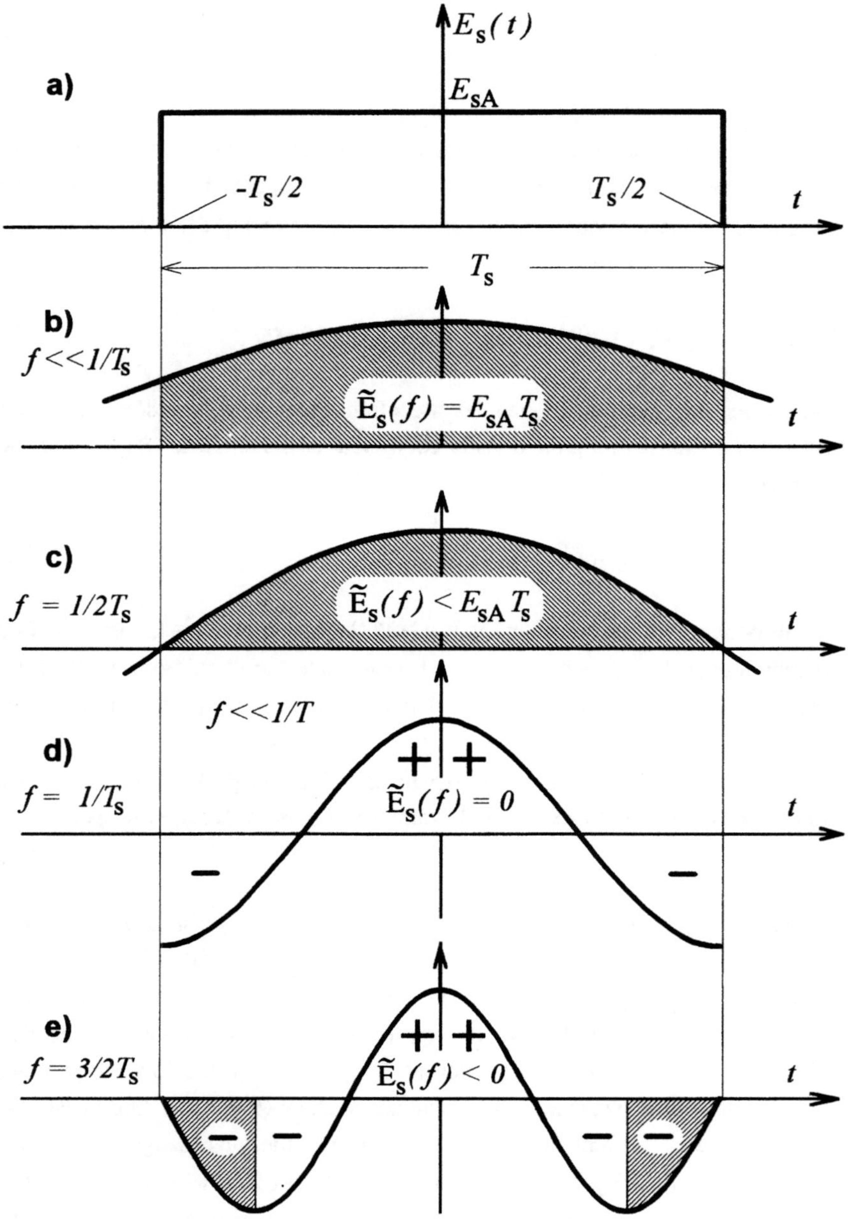

Figure 2.2: Spectral density E_s of rectangular pulse is determined by the area under cosine-wave on the interval T_s: (a) pulse shape; (b) low frequency; (c) limiting frequency; (d,e) high frequencies.

Feature two*. The spectral density* $\tilde{E}_s$ *at the low frequencies* $f << 1/2T_s$ *is equal to the energy (area) of the optical pulse.*

This follows directly from the Fourier transform (2.6) by setting $\cos 2\pi ft = 1$ (since $2\pi ft = 0$):

$$f << 1/2T_s \ , \ \tilde{E}_s(f) = \int_{-\infty}^{\infty} E_{sA}(t)dt = E_{sA}T_s \ , \tag{2.9}$$

$$E_s(f) = 2E_{sA}T_s \ . \tag{2.10}$$

Here we have used the definition of effective pulse duration (2.4). Since for the pulses under consideration the spectral density $\tilde{E}_s(f)$ is maximum at the low frequencies, and is always equal to $E_{sA}T_s$, it is convenient to normalize the spectral density (2.9) to this value (and relate the spectrum $E_s(f)$ to the quantity $2E_{sA}T_s$), and introduce the relative (dimensionless) spectrum $e_s(f)$:

$$\begin{aligned} \tilde{E}_s(f) &= E_{sA}T_s \, e_s(f) \ , \\ E_s(f) &= 2E_{sA}T_s \, e_s(f) \ , \end{aligned} \tag{2.11}$$

where $e_s(f) = \frac{1}{T_s} \int_{-\infty}^{\infty} e_s(t) \cos 2\pi ft \, dt$.

From the above definition of $e_s(f)$ it follows that at the zero frequency it is necessary to obtain $e_s(0) = 1$. This is confirmed also by expressions (2.4), (2.11).

Feature three*. The area of the spectrum is equal to the amplitude of the signal.* The pulse attains its amplitude value at $t = 0$; from the inverse Fourier transform for this instant it follows that

$$E_{sA} = E_s(0) = \int_{-\infty}^{\infty} \tilde{E}_s(f)df \ . \tag{2.12}$$

Physically, this is obvious: at $t = 0$ the harmonics of all frequencies reach their respective amplitude values and add up, totalling to the signal amplitude E_{sA}. The last two features are symmetrical, and have been proved by similar methods: by setting $\cos 2\pi ft = 1$ in (2.5), (2.6) (assuming, respectively, that $f \to 0$, or $t = 0$).

Now the spectra for the pulses considered above are:

$$\mathrm{e_s}(f)=\frac{1}{T_\mathrm{s}}\int\limits_{-\infty}^{\infty} e_\mathrm{s}(t)\cos 2\pi f\, t dt=\frac{2}{T_\mathrm{s}}\int\limits_{0}^{T_\mathrm{s}/2}\cos 2\pi f t\, dt=\frac{\sin \pi f T_\mathrm{s}}{\pi f T_\mathrm{s}} \tag{2.13}$$

for the rectangular pulse, and

$$\mathrm{e_s}(f)=\frac{2}{T_\mathrm{s}}\int\limits_{0}^{T_\mathrm{s}}\cos^2\left(\frac{\pi}{2}\frac{t}{T_\mathrm{s}}\right)\cos 2\pi f\, t dt=\frac{\sin 2\pi f T_\mathrm{s}}{2\pi f T_\mathrm{s}\left[1-\left(2 f T_\mathrm{s}\right)^2\right]} \tag{2.14}$$

for the square-cosine pulse.

These spectra are shown in Fig. 2.1e. For the square-cosine pulse $e_\mathrm{s}(t)$ we have introduced a simple (rectangular) approximation. Now a similar approximation suggests itself for the frequency response $\mathrm{e_s}(f)$. Let us replace the real spectrum $\mathrm{e_s}(f)$ with the rectangular spectrum with constant density $E_{\mathrm{sA}}T_\mathrm{s}$; $\mathrm{e_s}=1$ in the frequency range from 0 to $1/2T_\mathrm{s}$ (Fig. 2.1d,e). In this way we may introduce the effective band of the signal f_s:

$$f_\mathrm{s}=\frac{1}{2T_\mathrm{s}}=\int\limits_{0}^{\infty}\mathrm{e_s}(f)df\ . \tag{2.15}$$

This formula follows from (2.10), (2.12). All three obligatory conditions hold both for the exact spectrum and for the approximation: the spectral density at the low frequencies $\tilde{E}_\mathrm{s}$ is the area (energy) of the optical pulse $E_{\mathrm{sA}}T_\mathrm{s}$; the area of the spectrum is the amplitude $2(E_{\mathrm{sA}}T_\mathrm{s})(1/2T_\mathrm{s})=\tilde{E}_{\mathrm{sA}}$; the fall-off occurs in the neighborhood of the frequency $1/2T_\mathrm{s}$. Of course, the suggested rectangular approximation of the spectrum is not the spectrum of the rectangular pulse. The inverse Fourier transform can be used for finding the exact time profile of the pulse corresponding to this spectrum:

$$e_\mathrm{s}(t)=T_\mathrm{s}\int\limits_{-\infty}^{\infty}\mathrm{e_s}(f)\cos 2\pi f t\, df=T_\mathrm{s}\int\limits_{-1/2T_\mathrm{s}}^{1/2T_\mathrm{s}}\cos 2\pi f t\, df=\frac{\sin(\pi t/T_\mathrm{s})}{\pi t/T_\mathrm{s}}\ . \tag{2.16}$$

The mirror-like relationship of the direct and inverse Fourier transforms results in the mutual symmetry of functions $e_\mathrm{s}(t)$ and $\mathrm{e_s}(f)$: the rectangular pulse $e_\mathrm{s}(t)$ corresponds to the spectrum $\mathrm{e_s}(f)\propto \sin x/x$ (2.13) (Fig. 2.1a,d), and, conversely, the rectangular spectrum $\mathrm{e_s}(f)$ corresponds to the pulse profile $e_\mathrm{s}(t)\propto \sin x/x$ (2.16) (Fig. 2.1c,d).

In this way, we have approximated with rectangles both the bell-shaped signal $e_\mathrm{s}(t)$ and its spectrum $\mathrm{e_s}(f)$ (Fig. 2.1). Such "rectangular reasoning" is the basis of one of the methodological approaches used in this book. Carrying out all the necessary calculations for such approximations, we shall derive the main laws of the theory of detection in the rule-of-thumb manner, and get a clear-cut physical picture. Strictly speaking, these calculations will only apply to the rectangular pulse, and the pulse of the form $\sin x/x$. The power of our method, however, consists in that the results obtained for such pulses can be extended in the first approximation to the entire family of bell-shaped pulses. As a matter of fact, the main share of the energy of optical pulse is concentrated on the time interval T_S (in the frequency range from 0 to $1/2T_S$). The shape of the pulse edges (the rate of decline in the neighborhood of frequencies $f \propto 1/2T_S$) comes up in the effects of the second order (see Sect. 3.3, 5.1).

Asymmetrical pulses. For the sake of generality, let us consider three more complicated situations, when the symmetry of the pulse with respect to the zero-time point $t = 0$ is violated.

Assume that the profile of the pulse is symmetrical, but the pulse is displaced from the origin by the time t_0 (Fig. 2.3a). Then the expressions for the shape of the pulse $e_S(t)$ and its spectral components will be modified as follows:

$$\begin{gathered} e_S(t) \Rightarrow e_S(t - t_0) , \\ \mathrm{E}_S(f)\cos 2\pi f(t - t_0) = \mathrm{E}_S(f)\cos(2\pi ft - 2\pi ft_0) . \end{gathered} \tag{2.17}$$

In other words, the phase has become nonzero, $\psi_S = -2\pi ft_0$.

Next we assume that asymmetrical is not the position of the pulse, but its shape (Fig. 2.3b). Since the pulse is now not an even function, $E_S(t - t_0) \neq E_S(t_0 - t)$, the spectral expansion must include the odd terms (sine-waves), and the phase characteristic $\psi_S(f)$ will therefore also become nonzero. Accordingly, we must introduce the phase into (2.5):

$$E_S(t) = \int_{-\infty}^{\infty} \tilde{\mathrm{E}}_S(f)\cos(2\pi ft + \psi_S)\,df . \tag{2.18}$$

The first two features, as formulated for the symmetrical pulse profile, are still valid. As can be seen from Fig. 2.2, the proof of the decline of spectrum in the neighborhood of $f = 1/2T_S$ did not require symmetry of the pulse. In calculating the spectral density at zero frequency $\tilde{E}_S(0) = E_{SA}T_S$ we assumed that $\cos 2\pi ft = 1$. At zero frequency, however, $\psi_S = 0$ even when the pulse is not symmetrical, so $2\pi ft + \psi_S = 0$, $\cos(2\pi ft + \psi_S) = 1$, and expression (2.9) still holds. The third feature, however, is not preserved: the amplitude of the pulse is not equal to the area of the amplitude spectrum. According to (2.18), one has to take into account also the phase characteristic,

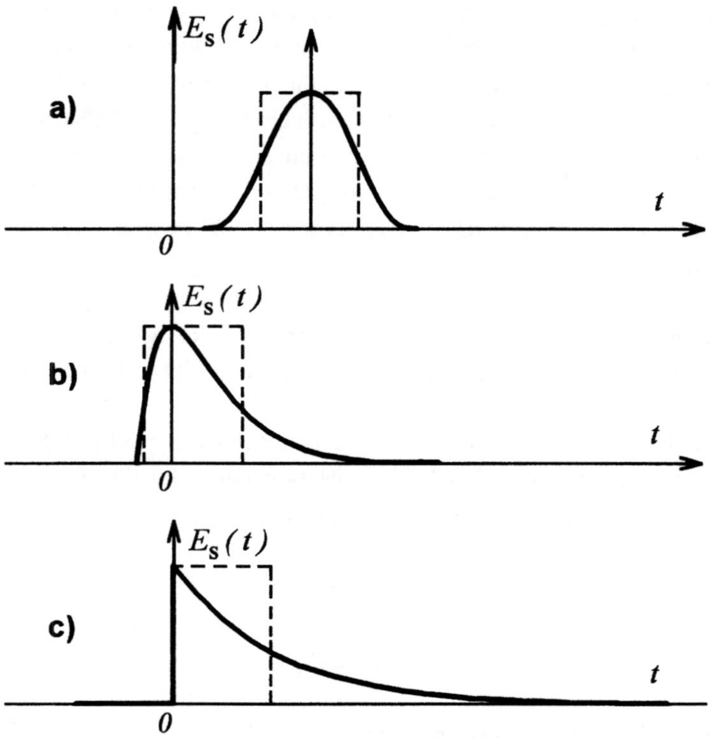

Figure 2.3: Examples of asymmetrical pulse positions with respect to the origin.

$$E_{sA} = E_s(0) = \int_{-\infty}^{\infty} \tilde{E}_s(f) \cos \psi_s(f)\, df \quad . \tag{2.19}$$

Even more complicated is the linkage between the amplitude of the pulse E_{sA} and its spectrum for the third and last example we are going to consider: the jump-like pulse (Fig. 2.3c). In this case there is ambiguity at point $t = 0$: the value of $E_s(t)$ changes suddenly from zero to E_{sA}, and the mean value, according to (2.19), is $E_{sA}/2$. If the spectrum is known, in this case the general formula (2.18) must be used for finding the amplitude E_{sA}; first of all, one should find the time function $E_s(t)$, and tend $t \to 0$ from the right.

Using the above-described method, all three non-symmetrical pulses can be approximated by rectangular pulses, equalizing the amplitudes and areas (Fig. 2.3). The spectrum of such rectangular pulse can, in turn, be approximated by the rectangular spectrum, having constant density $\tilde{E}_s = E_{sA} T_s$ in the frequency range $[-1/2T_s, 1/2T_s]$. As already indicated (see (2.19)), however, the area of such spectrum $2(E_{sA})(1/2T_s) = E_{sA}$ is not equal to the area of the real amplitude spectrum of the asymmetrical pulse.

Summary. *The typical optical signal, registered by a PDA, is a pulsed bell-shaped (square-cosine) signal. For further analysis it is sufficient to consider just two parameters of the pulse: its amplitude* E_{sA} *and effective duration* T_s. *In the first approximation the spectrum of optical pulse may be considered constant in the frequency range* $[-1/2T_s, 1/2T_s]$, *having constant density* $\tilde{E}_s = E_{sA}T_s$ *in this range (or* $E_s = 2E_{sA}T_s$ *in the range from* 0 *to* $1/2T_s$*). This simple approximation is an important methodological approach, and allows one to trace the physics associated with all stages of signal processing in the course of signal detection, without sacrificing the generality.*

2.2. Basic Notions of Noise Process

Noise as random function. Assume that only the photocurrent of the pulsed signal flows in the photodetector circuit (the input circuit of the amplifier) (Fig. 2.4a). Then, with sufficient amount of amplification, it is possible, in principle, to detect an arbitrarily small signal. Apart from the signal current, however, there also is a constant current I in the input circuit (Sect. 1.2). First, let us assume that this current is strictly permanent, $I = I_0$ (Fig. 2.4b). In this case, it is also theoretically possible to detect an arbitrarily small signal: one will only need to subtract the permanent current component I_0. Unfortunately, this is just a speculation, since the current always fluctuates because of the discrete nature of the charge (Fig. 2.4c). Fluctuations from other sources are superimposed on the fluctuations of the current of photodetector. Fluctuations interfere with the detection of the signal: a small signal pulse may be indistinguishable from the noise spike, as shown in Fig. 2.4c. If there were no noise, there would be no subject of the theory of detection.

The function $I(t)$ may be of arbitrary shape; such functions are called random functions [55, 56]. Their description is based on the concept of ensemble: a large number of entirely identical systems (in our case, PDAs), which occur under entirely identical conditions. If we monitor the current in the input circuit of these PDAs (photodetector current) during the observation time T, we get a certain function $I_1(t)$ for the first PDA, $I_2(t)$ for the second, and, generally, $I_n(t)$ for the nth (Fig. 2.5a). These functions $I_n(t)$, $n = 1,2,\ldots,N$ are called the realizations of the random function $I(t)$ over the observation time T. Let us select the value of random function at a certain time t_1; this value of $I(t_1)$ is a random quantity. The simplest statistical characteristics of a random quantity are its moments of order l — that is, its value to the power l, pertaining on the average to each sample:

$$N \to \infty\ , \quad \overline{I^l}(t_1) = \frac{1}{N}\sum_{n=1}^{N} I_n^l(t_1)\ . \tag{2.20}$$

Seeing that bar over any quantity in the future (like $\bar{I}$, or $\overline{U_n^2}$), we must remember that this seemingly simple mark indicates the cumbersome operation of adding up all the

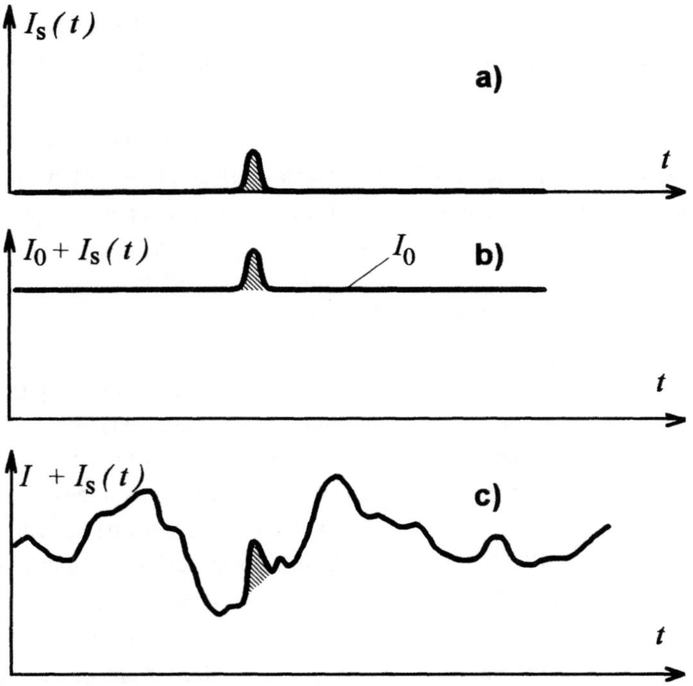

Figure 2.4: Noise prevents detection of the signal.

selected values (raised to the power $l = 1, 2, \ldots$), and dividing by the number of summands. The moment of the first order of the current $\overline{I(t)}$ is its mean value. Under steady conditions the ensemble-average value $\overline{I(t_1)}$ does not depend on the sampling time t_1, $\overline{I(t_1)} = \bar{I} = \text{const}$. Referring to "constant current" in Sect. 1.2, we meant that its mean value is constant.

The noise i_n is the deviation of the random quantity from its mean value,

$$i_n(t) = I(t) - \bar{I} \ . \tag{2.21}$$

Such functions are called centered functions: the value of the random quantity at any time is counted from the mean ("central") value $\bar{I}$. Since we have already deduced the mean value $\bar{I}$, the mean value of the centered function (the noise i_n) must be equal to zero:

$$\overline{i_n(t)} = \frac{1}{N}\sum_{n=1}^{N} i_{nn}(t) = \frac{1}{N}\sum_{n=1}^{N}\left[I_n(t) - \bar{I}\right]$$

$$= \frac{1}{N}\sum_{n=1}^{N} I_n(t) - \frac{1}{N}\sum_{n=1}^{N}\bar{I} = \bar{I} - \frac{1}{N}\left(N\bar{I}\right) = 0 \ . \tag{2.22}$$

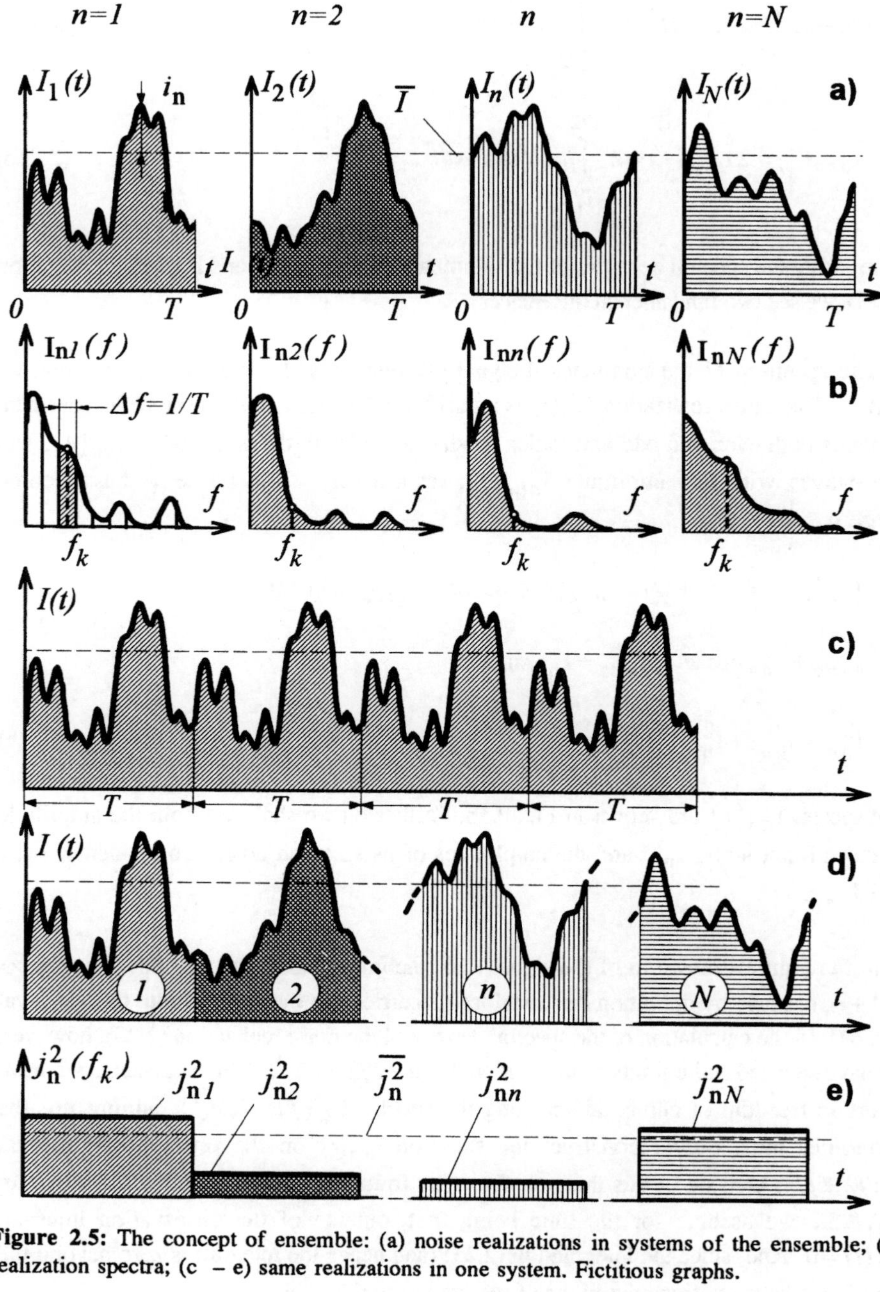

Figure 2.5: The concept of ensemble: (a) noise realizations in systems of the ensemble; (b) realization spectra; (c – e) same realizations in one system. Fictitious graphs.

Realization spectrum. Similarly to the signal pulse $E_s(t)$, each of the noise realizations $i_{nn}(t)$ on the interval $(-T/2, T/2)$ can be expanded in harmonics:

$$I_{nnc}(f) = 2\tilde{I}_{nnc}(f) = 2\int_{-T/2}^{T/2} i_{nn}(t)\cos\omega t dt \ , \tag{2.23a}$$

$$I_{nns}(f) = 2\tilde{I}_{nns}(f) = 2\int_{-T/2}^{T/2} i_{nn}(t)\sin\omega t dt \ . \tag{2.23b}$$

Comparing the spectral densities of the symmetrical signal (2.6), and the noise realization (2.23), we see two fundamental differences.

1. The spectrum of the symmetrical signal contains only the even harmonics (cosine-waves). The noise realization $i_{nn}(t)$ is an arbitrary function, and therefore its spectrum contains both even and odd harmonics (cosine-waves with the amplitude $I_{nnc}(f)$, and sine-waves with the amplitude $I_{nns}(f)$, respectively). The phase ψ has become nonzero:

$$I_{nn}(f)\cos(2\pi ft + \psi) = I_{nnc}(f)\cos 2\pi ft + I_{nns}(f)\sin 2\pi ft \ ,$$
$$I_{nnc} = I_{nn}\cos\psi \ , \ I_{nns} = I_{nn}\sin\psi \ ; \tag{2.24}$$

$$I_{nn}^2 = I_{nnc}^2 + I_{nns}^2 \ . \tag{2.25}$$

The spectra $I_{nn}(f)$ are shown in Fig. 2.5b. Further on we shall use both the amplitude of the full harmonic I_{nn}, and the amplitudes of its sine and cosine components, I_{nns} and I_{nnc}.

2. In the Fourier transform of the signal (2.6) the integration is carried out between $-\infty$ and $+\infty$, because by definition the transform is carried out over the infinite time interval [50, 51]. In the calculation of the spectral density of the noise realization (2.23), however, we have selected finite limits of integration, from $-T/2$ to $T/2$. This is because we have a certain freedom of choice in selecting the spectra $I_{nc}(f)$, $I_{ns}(f)$: adding up, the harmonics must only reconstruct the function $i_{nn}(t)$ on the observation interval $(-T/2, T/2)$. Outside of this interval, they may form any desired function; as shown in Fig. 2.5a, we assume for the time being that, outside of the observation interval, $i_{nn}(t) = 0$. And, since the integrand in (2.23) (and hence the integral) is zero beyond the time T, we have confined the limits of integration to this time interval.

In the model shown in Fig. 2.5a, the noise realization is actually considered as a single pulse of length T, and therefore its spectrum (2.23) is continuous. Further on, however, it will be convenient to use discrete harmonics in the calculation of the noise-related characteristics. The finite observation interval T offers such a possibility: the real continuous spectrum on this interval may be approximated by a discrete spectrum. With this purpose we split the continuous spectrum into narrow bands Δf, as shown in Fig. 2.5b (realization $n=1$). It is always possible to make the band Δf narrow enough, so that all the harmonics within this band exhibit little divergence in phase over the observation time T. Such harmonics, distributed over the frequency range from f to $f+\Delta f$, may be considered as a single train. On the time interval T these harmonics add up in unison, and the density of continuous spectrum $\mathrm{I}_{\mathrm{n}n}(f)$, spread out initially over the band Δf, pulls up and defines the amplitude of the resulting discrete harmonic (train) $\mathrm{j}_{\mathrm{n}n}(f)$:

$$\mathrm{j}_{\mathrm{n}n}(f)=\mathrm{I}_{\mathrm{n}n}(f)\Delta f=\frac{\mathrm{I}_{\mathrm{n}n}(f)}{T}\,;$$

$$\mathrm{j}_{\mathrm{n}nc}(f)=\frac{\mathrm{I}_{\mathrm{n}nc}(f)}{T}\,,\quad \mathrm{j}_{\mathrm{n}ns}(f)=\frac{\mathrm{I}_{\mathrm{n}ns}(f)}{T}\,. \qquad (2.26)$$

In these formulas we have anticipated the events by setting the band Δf equal to $1/T$. This amounts to the assumption that the harmonics in the train diverge over the entire time T at most by one period — that is, $2\pi\Delta f T=2\pi$, and $\Delta f=1/T$.

Of course, a discrete spectrum is a rather rough approximation of the model $i_{\mathrm{n}n}(t)$ in Fig. 2.5a: the choice of the limits of the train and its width Δf is arbitrary. It is possible, however, to obtain discrete spectrum in a perfectly rigorous way: such is the spectrum of periodic functions. Taking advantage of our freedom of choice, we can construct a periodic function. After all, nothing compels us to assume that $I\equiv 0$ outside of the interval T, like we did in the model just considered. Now we may construct a different model by replicating our realization an infinite number of times without change, as shown in Fig. 2.5c for the realization number one. This feat results in the time-dependent function which precisely repeats its values time after time with the period T: $i_{\mathrm{n}n}(t)=i_{\mathrm{n}n}(t-T)$. Because of this, its spectrum will contain only those harmonics which also repeat their values with the period T, $\cos 2\pi f t=\cos 2\pi f(t-T)$ [53, 54]:

$$2\pi f T=2\pi n\,,\quad f=\pm\frac{n}{T}\,,\quad f=\;\dots,-\frac{2}{T},-\frac{1}{T},0,\frac{1}{T},\frac{2}{T},\dots\,. \qquad (2.27)$$

We see that each frequency band $\Delta f=1/T$ in this model contains one harmonic. The only exception is the first interval with the zero frequency $f_0=0$. It is just half as wide as the rest, and is equal to $1/2T$ (the same frequency interval is occupied by the pulse of length T; recall Sect. 2.1). Figure 2.6 explains this peculiarity. In the space of frequencies

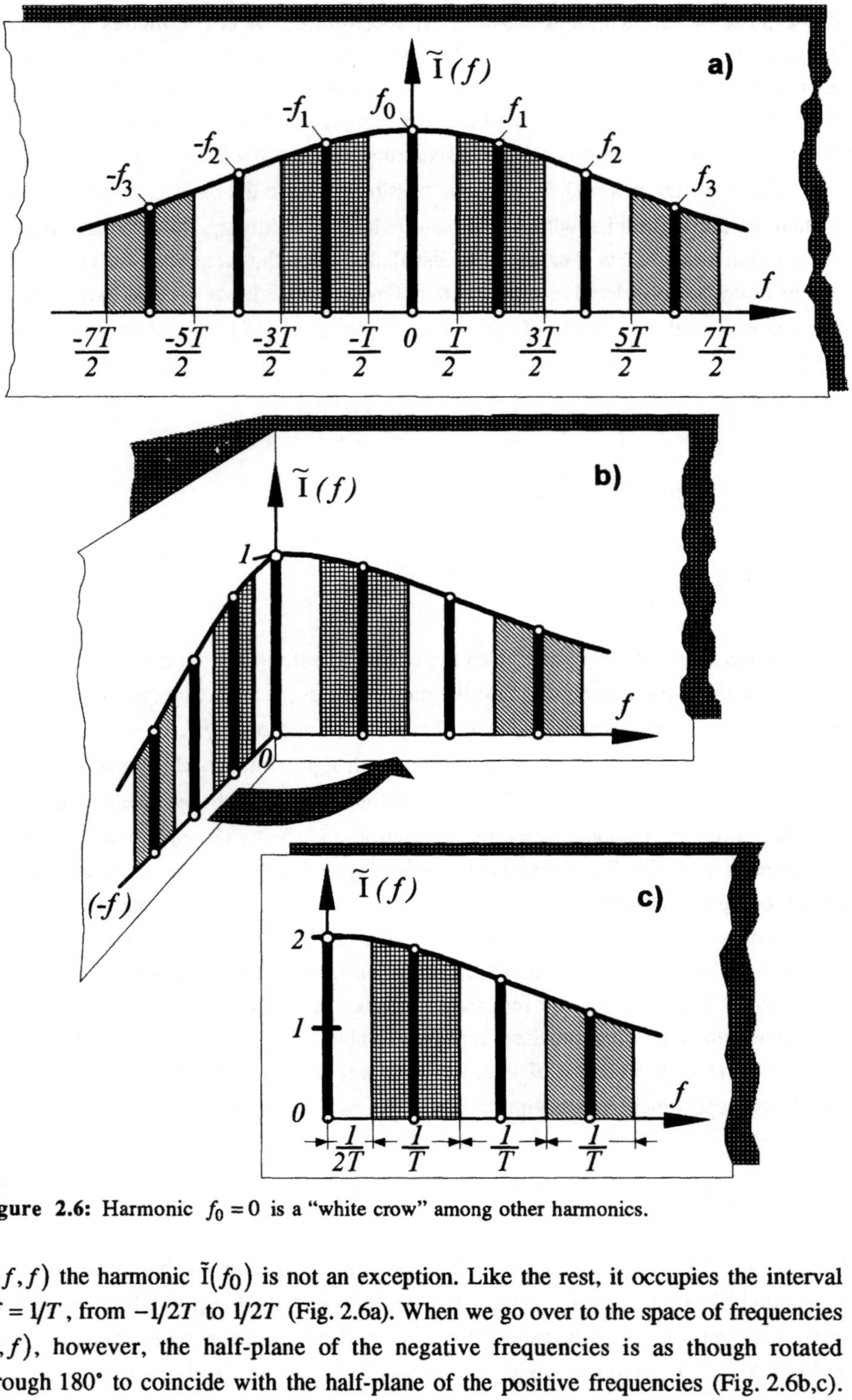

Figure 2.6: Harmonic $f_0 = 0$ is a "white crow" among other harmonics.

$(-f, f)$ the harmonic $\tilde{I}(f_0)$ is not an exception. Like the rest, it occupies the interval $\Delta f = 1/T$, from $-1/2T$ to $1/2T$ (Fig. 2.6a). When we go over to the space of frequencies $(0, f)$, however, the half-plane of the negative frequencies is as though rotated through 180° to coincide with the half-plane of the positive frequencies (Fig. 2.6b,c).

Since the axis of rotation passes through the frequency f_0, the band Δf occupied by this frequency may be regarded as being folded in two. This peculiarity comes up only in the analysis of the fluctuations of the permanent component (over different intervals T). It is only because we are going to discuss the permanent component in Sect. 2.3 below that we have paid some attention to this feature. Of course, in general for the spectrum we must accept $\Delta f = 1/T$.

Now let us return to the spectrum obtained. In both models under consideration the spectrum $\mathrm{j}_{\mathrm{n}n}(f)$ is discrete, with the envelope $\propto \mathrm{I}_{\mathrm{n}n}(f)$ and the frequency step $\Delta f = 1/T$. Of course, in our models there are some conventions (we have chosen an arbitrary and finite interval T, and an arbitrary function $i_{\mathrm{n}n}(t)$ outside of this interval). We need not be too unhappy about these conventions: later on we shall do away with them by letting $T \to \infty$. For now, however, we have at our disposal a clear-cut model which enables us to operate with random quantities, the amplitudes of the discrete harmonics $\mathrm{j}_{\mathrm{n}n}(f)$, $\mathrm{j}_{\mathrm{n}n\mathrm{c}}(f)$, $\mathrm{j}_{\mathrm{n}n\mathrm{s}}(f)$.

Spectral density of the noise power. Our next step must consist in calculating the statistical parameters (moments). We start with the moments of the first order: the mean values of amplitudes $\mathrm{j}_{\mathrm{n}n\mathrm{c}}(f)$. Averaging the spectrum (2.23), (2.26) over the ensemble, we get

$$N \to \infty ,$$

$$\overline{\mathrm{j}_{\mathrm{nc}}(f)} = \frac{1}{N}\sum_{n=1}^{N} \mathrm{j}_{\mathrm{n}n\mathrm{c}}(f) = \frac{1}{N}\sum_{n=1}^{N} \frac{2}{T}\int_{-T/2}^{T/2} i_{\mathrm{n}n}(t)\cos\omega t dt$$

$$= \frac{2}{T}\int_{-T/2}^{T/2}\left[\frac{1}{N}\sum_{n=1}^{N} i_{\mathrm{n}n}(t)\right]\cos\omega t dt = \frac{2}{T}\int_{-T/2}^{T/2}\overline{i_{\mathrm{n}}(t)\cos\omega t dt} . \tag{2.28}$$

It was only necessary to change the order of summation and integration to see that the mean value of the spectrum $\mathrm{j}_{\mathrm{nc}}(f)$ is the spectrum of the mean value of the noise realization $\overline{i_{\mathrm{n}}(t)}$. Since, however, the mean value of the noise $\overline{i_{\mathrm{n}}(t)}$ is zero, the mean value of the spectrum $\overline{\mathrm{j}_{\mathrm{nc}}(f)}$ is also zero. Similarly, $\overline{\mathrm{j}_{\mathrm{ns}}(f)} = 0$, and hence $\overline{\mathrm{j}_{\mathrm{n}}(f)} = 0$.

So, our attempt to calculate the mean value of the spectrum has failed: we wound up with the meaningless zero. Nonzero is the next (the second-order) moment, the mean square value, since the summation involves the positive quadratic terms $\mathrm{j}_{\mathrm{n}n}^2(f)$. It is the mean square value that is used as the characteristic of the noise spectrum (the mean square of the effective value rather than that of the amplitude):

$$N \to \infty , \quad \frac{\overline{\mathrm{j}_{\mathrm{n}}^2}}{2} = \frac{1}{N}\sum_{n=1}^{N}\left[\frac{\mathrm{j}_{\mathrm{n}n}(f)}{\sqrt{2}}\right]^2 . \tag{2.29}$$

The reason for the transition from the amplitudes to the effective values is clear: from the theory of electricity we know that the effective value of the harmonic defines the electric power — the quantity which is also expressed in terms of the squared current. Further on, in measuring the noise spectrum, and defining the mean square value of the noise sample at a given time $\overline{i_{\mathrm{n}}^2(t)}$ we shall also deal with the effective values.

Observe that in (2.29) we dealt with the amplitude of the full harmonic $\mathrm{j}_{\mathrm{n}n}\cos(\omega t+\psi)$. It will be interesting to find the linkage between the mean square amplitudes of its cosine and sine components, $\mathrm{j}_{\mathrm{n}nc}\cos\omega t$ and $\mathrm{j}_{\mathrm{n}ns}\sin\omega t$. Since the amplitudes are linked by the relations (2.24), $\mathrm{j}_{\mathrm{n}nc}=\mathrm{j}_{\mathrm{n}n}\cos\psi$ and $\mathrm{j}_{\mathrm{n}ns}=\mathrm{j}_{\mathrm{n}n}\sin\psi$, for the squared amplitudes we have

$$\overline{\mathrm{j}_{\mathrm{nc}}^2(f)}=\overline{\mathrm{j}_{\mathrm{n}}^2}\;\overline{\cos^2\psi}=\overline{\mathrm{j}_{\mathrm{n}}^2}\;\frac{1+\overline{\cos 2\psi}}{2}=\frac{\overline{\mathrm{j}_{\mathrm{n}}^2}}{2}\,; \tag{2.30}$$

$$\overline{\mathrm{j}_{\mathrm{ns}}^2(f)}=\overline{\mathrm{j}_{\mathrm{n}}^2}\;\overline{\sin^2\psi}=\overline{\mathrm{j}_{\mathrm{n}}^2}\;\frac{1-\overline{\cos 2\psi}}{2}=\frac{\overline{\mathrm{j}_{\mathrm{n}}^2}}{2}\,. \tag{2.31}$$

Here we have taken into account the fact that the phase ψ is equally likely to take on any value from 0 to 2π, and so, on the average, $\overline{\cos 2\psi}=0$. The result (2.30), (2.31) is rather obvious: because of the random nature of the noise process, the sine and the cosine components are equivalent, and their contributions to the mean square value of the full harmonic are equal.

Now we may go over to the definition of the spectral density of the noise power $\overline{J^2(f)}$. The frequency band Δf contains one harmonic $\mathrm{j}_{\mathrm{n}n}(f)$, and its power (that is, the effective value of the mean square) is $\mathrm{j}_{\mathrm{n}n}^2/2$. Since we are speaking of the density, the averaged power of the harmonic $\overline{\mathrm{j}_{\mathrm{n}}^2}/2$ must be related to the frequency band Δf occupied by the harmonic:

$$\overline{J^2(f)}=\frac{1}{\Delta f}\,\frac{\overline{\mathrm{j}_{\mathrm{n}}^2(f)}}{2}=\frac{1}{\Delta f}\,\overline{\mathrm{j}_{\mathrm{nc}}^2(f)}=\frac{1}{\Delta f}\overline{\left[\Delta f\,\mathrm{I}_{\mathrm{nc}}^2(f)\right]^2}$$
$$\equiv\Delta f\,\overline{\mathrm{I}_{\mathrm{nc}}^2(f)}\equiv\frac{\overline{\mathrm{I}_{\mathrm{nc}}^2(f)}}{T}=\frac{\overline{\mathrm{I}_{\mathrm{n}}^2(f)}}{2T}\,. \tag{2.32}$$

All members of this formula will be useful for future calculations. In the chain of transformations we made use of (2.30), (2.25), (2.26).

The spectral density of the noise power (or the noise spectrum for short) is one of the main characteristics of the noise process in radio engineering. As will be shown below, in most cases this characteristic is sufficient for calculating the other necessary parameters of the noise.

An aside is due concerning the term "power". This is, of course, a matter of convention: in general, we are dealing with the fluctuations not of the power, but of any random quantity (number of particles, flux of energy, resistance, etc.). In our particular case we are concerned with the spectral density of the power of the noise of the current (or the current noise). To go over from the spectral density of the noise power to the electrical power, the former must be multiplied by the resistance of the circuit which carries the noise current.

Ergodicity. Measurement of spectrum. The above definition of the noise spectrum may be deemed to imply that the measurement of the noise spectrum requires an ensemble comprising a large number of PDAs. Fortunately, this is not the case: in the real steady-state systems, including the PDAs, the noise is, as a rule, ergodic, and averaging over the ensemble may be replaced by averaging over the time. In case of ergodic processes, the same N realizations of the random function over the time T can be obtained not by watching the ensemble of N identical devices for the time T, but rather by observing just one device for the proportionally longer time $NT = T_N$.

This is illustrated by Fig. 2.5d. The entire interval T_N is divided into N equal intervals T; in this manner, from one "long" realization $I_n(t)$ we get N "short" realizations $I_{nn}(t)$. These N "short" realizations coincide with the realizations for the ensemble of N PDAs (compare Fig. 2.5a and 2.5d). This is a schematic demonstration of the property of ergodic process. Figure 2.5e shows the squared amplitude of one of the harmonics of the spectrum of these realizations, of frequency f_k. All N realizations are different, and so different are the squares of all N amplitudes. This schematic drawing helps to understand the technique used for measuring the noise spectrum $\overline{J^2(f)}$.

1. The noise is passed through the filter with the resonant frequency f_k and narrow band $\Delta f = 1/T$. Such filter lets through just one harmonic of frequency f_k. After equal time periods the realizations change, and the amplitude of the harmonic also changes, taking on consecutively the values $\mathrm{j_{n1}}(f_k)$, $\mathrm{j_{n2}}(f_k)$, ..., $\mathrm{j}_{\mathrm{n}n}(f_k)$.

2. From the output of the filter the noise is sent to the quadratic detector with the integration time T_N. Over the integration time, the detector will determine the mean of all N values of $\overline{\mathrm{j}^2_{\mathrm{n}n}}$ (Fig. 2.5e):

$$\overline{\mathrm{j}^2_\mathrm{n}} = \frac{1}{N}\sum_{n=1}^{N}\overline{\mathrm{j}^2_{\mathrm{n}n}(f_k)} \; . \tag{2.33}$$

The larger N (that is, the larger $T_N/T = T_N\Delta f$), the closer the measured value of $\overline{\mathrm{j}^2_\mathrm{n}}$ is to the true value.

3. Calibrating the detector in the effective values, one reads the value of $\sqrt{\overline{\mathrm{j}_{\mathrm{n}}^2}/2}$. Dividing this reading by $\sqrt{\Delta f}$, we find the spectral density of the noise at frequency f_k:

$$\sqrt{\overline{J^2}} = \sqrt{\frac{\overline{\mathrm{j}_{\mathrm{n}}^2}}{2\Delta f}} \ . \tag{2.34}$$

Of course, this is a schematic description; in any case, however, it gives an idea of the technique used for measuring the noise spectrum. We see that in case of ergodic processes we do not really need the ensemble — we can do with a single soloist, just one PDA.

Linear transformations of random function. In the linear systems (for instance, in the linear circuits of PDAs), two main operations (multiplication by a constant and addition) are carried out on the primary noise sources. Once again we single out the harmonic of frequency f_k in the nth realization of the input noise. This noise harmonic is transformed in exactly the same way as the deterministic harmonic of the signal. For example, the voltage drop across the load Z caused by the noise harmonic $\mathrm{I}_{\mathrm{n}n}(f_k)$ is $U_{\mathrm{nin}n}(f_k) = Z\,\mathrm{I}_{\mathrm{n}n\mathrm{c}}(f_k)$. Passing through the amplifier, the noise harmonic is multiplied by the transfer coefficient K, $U_{\mathrm{n}n}(f_k) = K\,\mathrm{I}_{\mathrm{n}n\mathrm{c}}(f_k)$. If the linear circuit carries components from several sources, these components add up.

The mean square product of random quantity X and constant K, and the mean square sum of two random quantities X, Y are found by the following known rules [56]:

1. The mean square product of random quantity X and constant K is equal to the mean square value of the random quantity times the squared constant:

$$\overline{(KX)^2} = K^2\overline{X^2} \ . \tag{2.35}$$

2. The mean square sum of two random independent centered quantities is equal to the sum of their mean square values:

$$\overline{(X+Y)^2} = \overline{X^2} + \overline{Y^2} \ . \tag{2.36}$$

These rules follow from the definition of the mean square value (2.20) with $l = 2$. Since K^2 is the same multiplier for all terms in the sum, it can be factored out of the sum in (2.20), which will result directly in the expression (2.35). Somewhat more complicated is the proof for the sum of random quantities. Assume that in the N realizations the random quantity consecutively takes on the values X_n, $n = 1,2,\ldots,N$ (the most probable values will occur more frequently). In the summation, any first term X_n with the given number n may be combined with any second term Y_m with arbitrary number m (recall that the value of Y_m does not depend on the value of X_n). Because of this, for each number n we get N possible sums $X_n + Y_m$, $m = 1,2,\ldots,N$. Since n varies from 1 to N, all told there are N^2 sums with different numbers nm. The square of each sum is

$$(X_n + Y_m)^2 = X_n^2 + Y_m^2 + 2X_nY_m \; . \tag{2.37}$$

First, we carry out averaging with respect to m with fixed n:

$$\frac{1}{N}\sum_{m=1}^{N}(X_n + Y_m)^2 = \frac{1}{N}\sum_{m=1}^{N}X_n^2 + \frac{1}{N}\sum_{m=1}^{N}Y_m^2 + 2X_n\underbrace{\frac{1}{N}\sum_{m=1}^{N}Y_m}_{=0}$$

$$= X_n^2 + \overline{Y_m^2} \; . \tag{2.38}$$

Here we have taken into account that the mean value of the centered random quantity Y is zero. Now we carry out averaging with respect to n, and arrive from (2.38) straight on to (2.36). The rule (2.36) applies to the sum of any number of independent quantities.

In the beginning of this subsection we gave an example of multiplication: the noise voltage is linked with the noise current by

$$U_{\mathrm{n\,in}\,n}(f_k) = Z_{\mathrm{in}}\,\mathrm{I}_{\mathrm{n}nc}(f_k) \; , \quad U_{\mathrm{n}n}(f_k) = K\,\mathrm{I}_{\mathrm{n}nc}(f_k) \; . \tag{2.39}$$

Now we may use (2.35) to find the linkage between the mean square values of these quantities:

$$\overline{U^2_{\mathrm{n\,in}\,n}(f_k)} = Z^2_{\mathrm{in}}\,\overline{\mathrm{I}^2_{\mathrm{nc}}(f_k)} \; , \quad \overline{U^2_{\mathrm{n}}(f_k)} = K^2\,\overline{\mathrm{I}^2_{\mathrm{nc}}(f_k)} \; . \tag{2.40}$$

The above two rules for the linear transformations of random quantities will be much used later on. As a matter of fact, we have used the multiplication rule in the derivation of (2.32), where $X = I(f)$, and $K = 1/T$.

Sampling of instant noise value. In the course of detection we observe the noise as a concrete function of the time, $i_{\mathrm{n}}(t)$. The value of the noise (current, voltage) at a given time $i_{\mathrm{n}}(t_1)$ is referred to as sample. Let us find the linkage between the sample and the noise spectrum $\overline{J^2(f)}$. For this purpose we return to the model constructed earlier in the subsection dealing with the noise spectrum. Then we were able to represent the time realization as the sum of discrete harmonics. Let us start with a particular case when the realization is comprised of just one harmonic:

$$i_{\mathrm{n}n}(t) = \mathrm{j}_{\mathrm{n}n}(f_k)\cos(2\pi f_k t + \psi_n) \; . \tag{2.41}$$

Such noise can be obtained by passing the realization through the above-described filter with the resonant frequency f_k and the passband $\Delta f = 1/T$. Averaging over the ensemble, and with due account for (2.32), we get

$$\overline{i_\mathrm{n}^2(t)} = \overline{i_\mathrm{n}^2} = \overline{\mathrm{j}_\mathrm{n}^2(f_k)}\ \overline{\cos^2(2\pi f_k t + \psi)}$$

$$= \overline{\mathrm{j}_\mathrm{n}^2(f_k)}\ \frac{1 + \overline{\cos 2(2\pi f_k t + \psi)}}{2} = \frac{\overline{\mathrm{j}_\mathrm{n}^2(f_k)}}{2} = \overline{J_\mathrm{n}^2(f_k)}\Delta f\ . \tag{2.42}$$

As indicated above, the phase $\varphi = 2\pi f_k t + \psi$ is equally likely to take on any value from 0 to 2π, which allows us to assume that $\cos 2\varphi = 0$. Two important results follow from expression (2.42):

1. The mean square value of the sample $\overline{i_\mathrm{n}^2(t)}$ does not depend on the time of sampling, just as should be the case for the steady ergodic process.

2. The mean square value of the sample $\overline{i_\mathrm{n}^2}$ is defined by the effective value of the harmonic (rather than by the squared amplitude). Because of this, the spectral density $\overline{J^2}$ was defined above in terms of the effective value.

It is easy to extend the solution (2.42) for the general case, when the noise is not passed through the narrow-band filter, and contains an arbitrary number of harmonics,

$$i_{\mathrm{n}n}(t) = \sum_k \mathrm{j}_{\mathrm{n}n}(f_k)\cos(2\pi f_k t + \psi_k)\ . \tag{2.43}$$

Since the harmonics are independent random functions, their mean square sum, by virtue of (2.36), is the sum of mean squares:

$$\overline{i_\mathrm{n}^2} = \sum_k \frac{\overline{\mathrm{j}_{\mathrm{n}k}^2}}{2} = \sum_k \frac{\overline{\mathrm{j}_{\mathrm{n}k}^2}}{2\Delta f}\Delta f = \sum_k \overline{J^2(f_k)}\Delta f \Rightarrow \int_0^\infty \overline{J^2(f)}df\ . \tag{2.44}$$

Here we have used the definition of the spectral density of noise power (2.32), and went over from summation to integration with $\Delta f \to 0$ ($T \to \infty$).

Thus, the mean square of the noise sample at any time is determined by the area confined by the curve $\overline{J^2(f)}$.

In statistics, there are three closely related concepts.

1. Mean square value of random quantity, like the mean square value of the noise sample $\overline{i_\mathrm{n}^2}$ under consideration.

2. Dispersion (variance) σ^2, which also is a squared value of the centered quantity (whose mean value is necessarily zero). Since the noise is a centered function by definition, $\overline{i_n(t)} = 0$, the meaning of the two terms is the same:

$$\sigma^2 \equiv \overline{i_n^2(t)} \equiv \overline{i_n^2} \; . \tag{2.45}$$

In statistics, dispersion is used for any random quantity. In radio engineering, however, the "dispersion of noise" is usually understood as the dispersion of noise sample in the entire working frequency range. It is in this sense (in full agreement with (2.45)) that we shall use this term.

3. Finally, the root mean square value (deviation) is defined as

$$i_n = \sqrt{\overline{i_n^2}} \equiv \sqrt{\sigma^2} \; . \tag{2.46}$$

The dimension of the root mean square (rms) value is the same as that of the random quantity itself — in this context, the current.

Two noise samples. Since the duration of the signal T_s is finite, one noise sample is certainly not sufficient: it is necessary to have information about the noise on the time interval of the order of several T_s. This means that we need to consider the values of noise at the nearby times t_1, t_2. As always, the first step consists in calculating the rms values. This will not, however, give us any new information, because we have just demonstrated that the rms values for steady ergodic processes do not depend on the time,

$$\sqrt{\overline{i_n^2(t_1)}} = \sqrt{\overline{i_n^2(t_2)}} \equiv i_n \; . \tag{2.47}$$

What is new then in the statement of the problem of two samples? Let us look at Fig. 2.7, which shows the realizations of two random processes. In the processes in Fig. 2.7a and 2.7b the two possible outcomes, +1 and –1, occur at the times t_1 and t_2 with the same likelihood, so that the mean square values for the two processes are the same at t_1 and t_2,

$$\sqrt{\overline{i_n^2(t_1)}} = \sqrt{\overline{i_n^2(t_2)}} = \sqrt{\frac{1}{6}\cdot\left[3\cdot 1^2 + 3\cdot(-1)^2\right]} = 1 \; . \tag{2.48}$$

But how different these processes are! In the first case (Fig. 2.7a), the value which occurs at the time t_1 is always repeated at the time t_2. Two events which are linked together by a deterministic relationship (in our present case, $i_n(t_2) = i_n(t_1)$) are called completely correlated.

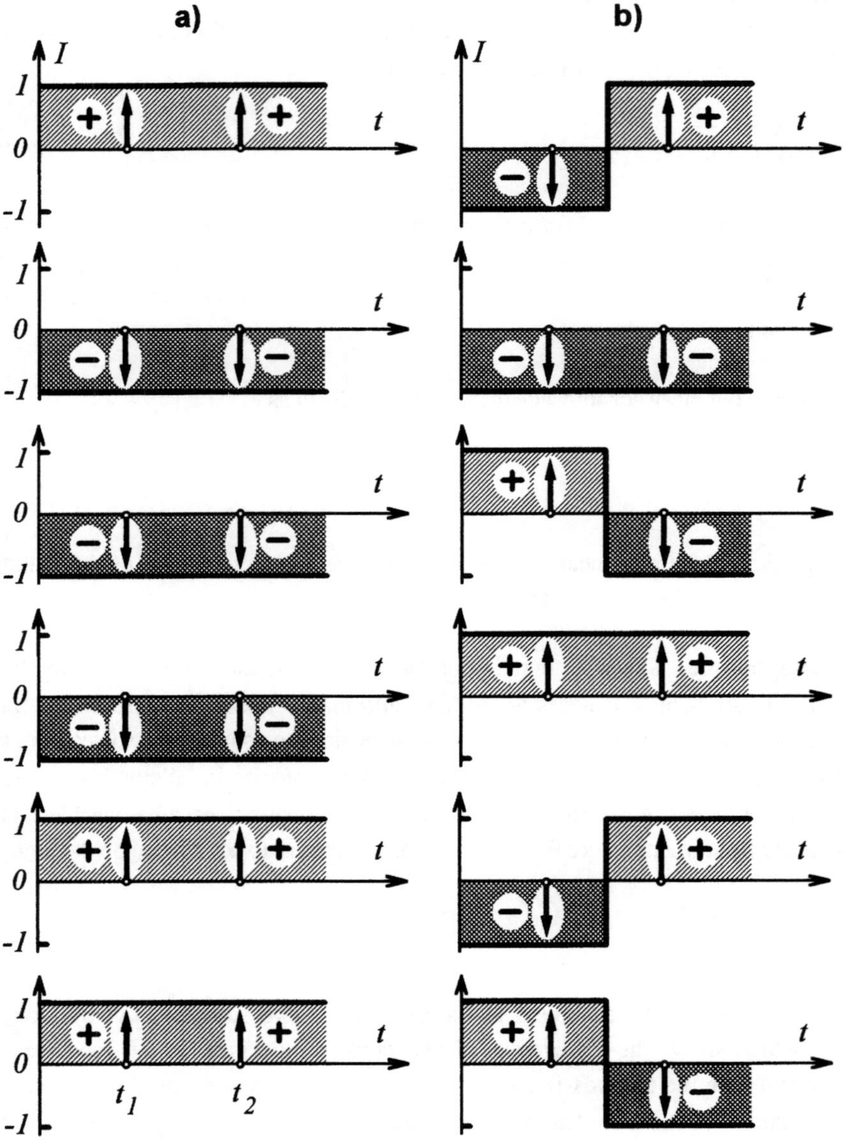

Figure 2.7: The processes may differ even though the distribution of fluctuations at t_1 and t_2 is the same: (a) correlated fluctuations; (b) independent fluctuations.

In the second case (Fig. 2.7b), the situation is quite different: the random events $i_n(t_1)$ and $i_n(t_2)$ are not correlated — in other words, there is no connection between the appearance of either +1 or –1 at the times t_1 and t_2. The statistical relationship between the results of two samples at the times t_1 and t_2 is described by the correlation function $\mathcal{K}(t_1,t_2)$.

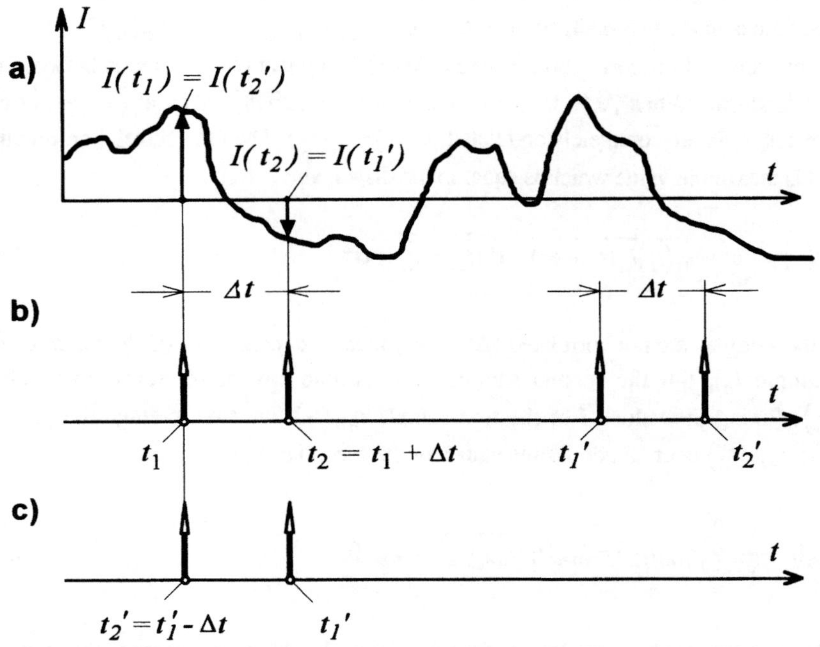

Figure 2.8: To the dependence of correlation function on the arrangement of samples.

Correlation function. The correlation function is constructed similarly to the mean square: in case of one sample (one-dimensional random quantity) we took one value $i_n(t_1)$, and multiplied it by itself: $i_n(t_1)i_n(t_1)$. Now we have two samples $i_n(t_1)$ and $i_n(t_2)$ (two-dimensional random quantity), and multiply one times the other. Then follows the standard averaging procedure:

$$\mathcal{K}(t_1,t_2) = \mathcal{K}(\Delta t) = \overline{i_n(t_1)i_n(t_2)} = \overline{i_n(t_1)i_n(t_1+\Delta t)} \; . \tag{2.49}$$

In the case of steady noise process (in the narrow sense) the statistical characteristics do not depend on the time (in the present context, on t_1). This allowed us to write the correlation function as a function of one argument Δt. Placing the sampling times t_1, t_2 arbitrarily on the time axis, we get one and the same value of the correlation function $\mathcal{K}$ as long as the time between the samples Δt is the same (Fig. 2.8a,b). Because the process is steady, the correlation function has the property of being even. Making the replacement $t_1' = t_1 - \Delta t$, we get

$$\mathcal{K}(-\Delta t) = \overline{i_n(t_1)i_n(t_1-\Delta t)} = \overline{i_n(t_1'+\Delta t)i_n(t_1')} = \mathcal{K}(\Delta t) \; . \tag{2.50}$$

The evenness of $\mathcal{K}(\Delta t)$ is also seen from the comparison between Fig. 2.8b and 2.8c. The samples in Fig. 2.8c are taken at $t_2' < t_1'$, $\Delta t = t_2' - t_1' < 0$. Since the correlation function $\mathcal{K}$ does not depend on the absolute position of the samples t_2', t_1' on the time axis, we have aligned them with the samples t_1, t_2. Accordingly, the product of the selected values in the two realizations is the same, $i_n(t_1)i_n(t_2) \equiv i_n(t_2')i_n(t_1')$.

Large values of the correlation function $\mathcal{K}(\Delta t)$ imply that the correlation between the samples is strong. When $\Delta t \to 0$, the two samples are actually taken at the same time, and they naturally are completely correlated, $i_n(t_1) \approx i_n(t_2)$. Then the correlation function reaches is maximum value which is equal to the dispersion:

$$\mathcal{K}(0) = \lim_{\Delta t \to 0} \overline{i_n(t_1)i_n(t_1 + \Delta t)} = \overline{i_n^2(t_1)} = \overline{i_n^2} = \sigma^2 . \tag{2.51}$$

When the samples are not correlated ($\Delta t \to \infty$), then, independently of the value of the first sample $i_{nn}(t_1)$, the second sample may assume any of its admissible values $i_{nm}(t_2)$. Fixing the value of of the first sample $i_{nn}(t_1)$, we first average the product $i_{nn}(t_1) \times i_{nm}(t_2)$ over all admissible values of the second sample:

$$i_{nn}(t_1)\frac{1}{N}\sum_{m=1}^{N} i_{nm}(t_2) = i_{nn}(t_1)\overline{i_{nm}(t_2)} = i_{nn} \cdot 0 . \tag{2.52}$$

In order to obtain the correlation function, we should now average the product $i_{nn}(t_1)\overline{i_{nm}(t_2)}$ with respect to the first sample. This is not necessary, since this product is equal to zero anyway, and so

$$\Delta t \to \infty , \quad \mathcal{K}(\Delta t) = \overline{i_n(t_1)i_n(t_1 + \Delta t)} = 0 . \tag{2.53}$$

We see that the correlation function is indeed a measure of the interdependence (correlation) of random quantities: its value is the largest when the samples are completely correlated ($\Delta t \to 0$), and falls to zero when the samples are statistically independent ($\Delta t \to \infty$)*. With due account for the above properties, Fig. 2.9 shows a typical example of the correlation function, which may be regarded as a symmetrical pulse. The effective length of this pulse may be called the correlation time, which, using our customary technique, can be defined through approximating the function $\mathcal{K}(\Delta t)$ by a rectangle (Fig. 2.9). The amplitude and the area of the rectangle are made equal to the amplitude and the area of the pulsed function. The length of the rectangle is taken for the correlation time:

$$\mathcal{K}(0)\tau_{cor} = \overline{i_n^2}\tau_{cor} = \int_{-\infty}^{\infty} \mathcal{K}(\Delta t)d(\Delta t) . \tag{2.54}$$

The meaning of the term "correlation time" is clear: the values of the samples taken

---

* We have already used the fact that the correlation function of two independent quantities is zero when we assumed that $\overline{X_n Y_m}=0$ in (2.36).

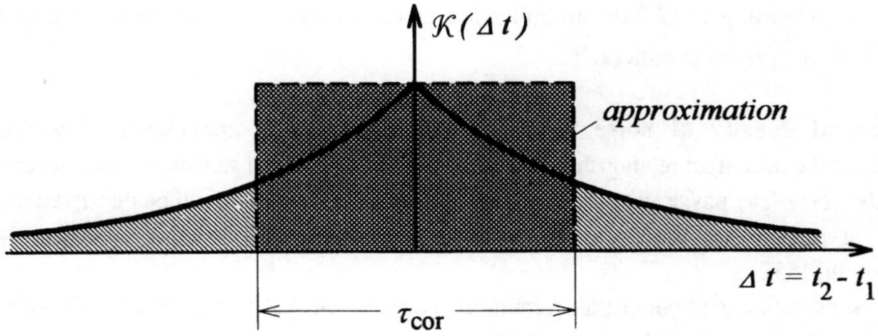

Figure 2.9: Correlation time is the effective duration of the correlation function.

within the correlation time τ_{cor} (from $-\tau_{cor}/2$ to $\tau_{cor}/2$) are correlated, because $\mathcal{K} = \sigma^2$; otherwise, when the time between the samples is $|t_2 - t_1| = |\Delta t| > \tau_{cor}$, the selected values are independent, since $\mathcal{K} = 0$.

The correlation between the noise samples $i_n(t_1)$ and $i_n(t_2)$ must always be taken into account when the signal is processed in the PDA. If the samples are added up (as, for example, in the integrator), for the dispersion of the sum we have

$$\overline{[i_n(t_1) + i_n(t_2)]^2} = \overline{i_n^2(t_1)} + \overline{i_n^2(t_2)} + 2\overline{i_n(t_1)i_n(t_2)}$$
$$= 2\overline{i_n^2} + 2\overline{i_n(t_1)i_n(t_2)} \ . \tag{2.55}$$

Hence, for the "nearby" samples we get

$$|t_2 - t_1| << \tau_{cor} \ , \quad \overline{i_n(t_1)i_n(t_2)} = \overline{i_n^2} \ ,$$
$$\overline{[i_n(t_1) + i_n(t_2)]^2} = 4\overline{i_n^2} \ ; \tag{2.56}$$

and for "faraway" samples

$$|t_2 - t_1| >> \tau_{cor} \ , \quad \overline{i_n(t_1)i_n(t_2)} = 0 \ ,$$
$$\overline{[i_n(t_1) + i_n(t_2)]^2} = 2\overline{i_n^2} \ . \tag{2.57}$$

We see that the dispersion of the sum may differ by a factor of two. The same conclusion follows from the above-formulated rules of linear transformations of random quantities. According to (2.36), the dispersion of the sum of independent quantities is the sum of their dispersions, the same as in (2.57). Summation of two completely correlated and equal terms $i_n(t_1)+i_n(t_1)=2i_n(t_1)$ amounts to multiplying the term by the coefficient $K=2$; according to (2.35), multiplication increases the dispersion by the factor of $K^2=4$, in agreement with (2.56).

Spectral density of noise power is the spectrum of correlation function. Such is the important relationship that exists between two main statistical characteristics of the noise. The physical implications of this remarkable property will be discussed later on; at present we are going to prove this statement (the Wiener – Khintchine theorem) [55].

Substituting the Fourier transform of the noise current (2.23) into the formula for the spectral density of the noise power (2.32), we get

$$\overline{J^2(f)}=\frac{1}{T}\overline{I^2_{nc}(f)}=\frac{1}{T}\overline{\left[2\int_{-T/2}^{T/2} i_n(t)\cos\omega t dt\cdot 2\int_{-T/2}^{T/2} i_n(t)\cos\omega t dt\right]}\ . \tag{2.58}$$

Since in each of the two integrals the integration is carried out independently, we can make the replacement $t\Rightarrow t_1,\ t\Rightarrow t_2$. Then, going over to the double integral and substituting $t_2=t_1+\Delta t$, we find the correlation function:

$$\begin{aligned}\overline{J^2}&=\frac{4}{T}\overline{\left[\int_{-T/2}^{T/2} i_n(t_1)\cos\omega t_1 dt_1\cdot\int_{-T/2}^{T/2} i_n(t_2)\cos\omega t_2 dt_2\right]}\\ &=\frac{4}{T}\int_{\Delta t_{\text{low}}}^{\Delta t_{\text{high}}}\int_{-T/2}^{T/2}\overline{i_n(t_1)i_n(t_1+\Delta t)}\cos\omega t_1\cos\omega(t_1+\Delta t)dt_1 d\Delta t\\ &=\frac{4}{T}\int_{\Delta t_{\text{low}}}^{\Delta t_{\text{high}}}\mathcal{K}(\Delta t)\left[\int_{-T/2}^{T/2}\cos\omega t_1\cos\omega(t_1+\Delta t)dt_1\right]d\Delta t\ .\end{aligned} \tag{2.59}$$

Now we have to work out the integral in brackets. Replacing the product of cosines with their difference, we reduce the integrand to

$$\cos\omega t_1\cos\omega(t_1+\Delta t)=\frac{1}{2}\cos\omega\Delta t-\frac{1}{2}\cos(2\omega t_1+\Delta t)\ . \tag{2.60}$$

The first term is a constant which does not depend on the time t_1. Because of this, the integral of this term grows linearly with the time t_1, and becomes arbitrarily large when the observation time T (integration from $-T/2$ to $T/2$) is arbitrarily large. The second term in (2.60) is the alternating-sign cosine wave, and its integral cannot be greater than the half-period area. In other words, it is negligibly small compared to the unlimited integral of the first term.

So, the value of the integral in brackets is determined by the first term in the integrand, $(\cos\omega\Delta t)/2$. Retaining this term in (2.59), we get

$$\overline{J^2} = \frac{4}{T}\int_{\Delta t_{\text{low}}}^{\Delta t_{\text{high}}} \mathcal{K}(\Delta t)\left[\int_{-T/2}^{T/2} \frac{1}{2}\cos\omega\Delta t\, dt_1\right] d\Delta t$$

$$= \frac{4}{T}\int_{\Delta t_{\text{low}}}^{\Delta t_{\text{high}}} \mathcal{K}(\Delta t)\frac{T}{2}\cos\omega\Delta t\, d\Delta t \ . \tag{2.61}$$

With the large observation time, the limits of integration

$$\Delta t_{\text{high}} = (t_2)_{\text{high}} - t_1 = \frac{T}{2} - t_1 \ ,$$

$$\Delta t_{\text{low}} = (t_2)_{\text{low}} - t_1 = -\frac{T}{2} - t_1 \tag{2.62}$$

are much greater (in the absolute value) than the damping constant of the integrand (with the exception of small regions at the limits of the range, $|t_1| > |T/2 - \tau_{\text{cor}}|$, whose contribution to the integral is negligible. On these grounds we may go over to infinite integration limits in (2.61):

$$\overline{J^2} = 2\int_{-\infty}^{\infty} \mathcal{K}(\Delta t)\cos\omega\Delta t\, d\Delta t = 4\int_{0}^{\infty} \mathcal{K}(\Delta t)\cos\omega\Delta t\, d\Delta t \ , \tag{2.63}$$

where we have taken into account the evenness of $\mathcal{K}(\Delta t)$. In this way, we have found the relationship between $\overline{J^2(f)}$ and $\mathcal{K}(\Delta t)$.

Observe that the inverse Fourier transform follows uniquely from (2.63):

$$\mathcal{K}(\Delta t) = \int_{0}^{\infty} \overline{J^2(f)}\cos\omega\Delta t\, df \ . \tag{2.64}$$

The earlier expressions (2.44), (2.45), which define the dispersion $\sigma^2 = \mathcal{K}(0)$ as the area confined by the noise spectrum curve $\overline{J^2(f)}$, is but a special case of the general formula (2.64) with $\Delta t = 0$.

Distribution function. The statistical characteristics considered above (dispersion, noise spectrum, correlation function) are not sufficient for the detection of signals. The random function (the noise $i_n(t)$) at the time of sampling t_1 may take on any arbitrary value, so we need to know the probability of occurrence of a particular value. When the number of systems N in the ensemble is large ($N \to \infty$), the values of the samples $I_1, I_2, \ldots, I_n, \ldots, I_N$ will occur more than once, so that we shall get only M different values $I_{\mathrm{I}}, I_{\mathrm{II}}, \ldots, I_m, \ldots, I_M$. Each of the values I_m will occur n_m times (if the quantity $I(t_1)$ is continuous, the value of n_m is related to the range from I_m to $I_m + \Delta I$), whereas $\Sigma n_m = N$. By definition, the distribution function is the probability ρ (frequency) of occurrence of the given value of I_m (or the probability that the observed value falls within the range from I_m to $I_m + \Delta I$):

$$\rho(I_m)\Delta I = \frac{n_m}{N} , \tag{2.65}$$

$$M \to \infty , \quad \sum_{m=1}^{M} \frac{n_m}{N} = \sum_{m=1}^{M} \rho(I_m)\Delta I \Rightarrow \int_{-\infty}^{\infty} \rho(I) dI = 1 . \tag{2.66}$$

The distribution function (the probability density $\rho(I)$) is a comprehensive and sufficient statistical characteristic of a random quantity (but not of a random function). It can be used for calculating all other characteristics, such as the moments of all orders, including the dispersion. For example, combining in (2.20) the terms with the same value of the current, we get

$$\begin{aligned} \overline{I^l} &= \frac{1}{N}\left(I_1^l + I_2^l + \ldots + I_n^l + \ldots + I_N^l\right) \\ &= \frac{1}{N}\left(n_{\mathrm{I}} I_{\mathrm{I}}^l + n_{\mathrm{II}} I_{\mathrm{II}}^l + \ldots + n_m I_m^l + \ldots + n_M I_M^l\right) \\ &= \sum_{m=\mathrm{I}}^{M} \frac{n_m}{N} I_m^l = \sum_{m=\mathrm{I}}^{M} \rho(I_m) I_m^l \Delta I \Rightarrow \int_{-\infty}^{\infty} \rho(I) I^l dI . \end{aligned} \tag{2.67}$$

The solution of the inverse problem, however, (finding the probability of occurrence of the specified value of I given the dispersion σ^2) is only possible when the distribution function is known, and includes the dispersion σ^2 as the only parameter. It is exactly this situation that is usually encountered in radio engineering, where the noise obeys the normal distribution law [56]

$$\rho(i_n) = \frac{1}{\sqrt{2\pi\sigma^2}} \exp\left(-\frac{i_n^2}{2\sigma^2}\right) . \tag{2.68}$$

The normal (Gaussian) distribution is often referred to as the law of large numbers: as shown by the theory, when a large number of independent random quantities distributed according to one and the same law, the distribution of the sum is normal whatever the distribution law of the individual terms of the sum (the elementary fluctuations). Because of this, the normal distribution law is quite universal. The noise process has long come to be associated in our mind with the normal distribution law: in many problems this distribution is assumed a priori to hold; accordingly, speaking about the noise, one usually refers to its spectrum $\overline{J^2(f)}$, or the area of the spectrum (the dispersion σ^2), or the mean square value $i_n = \sqrt{\overline{i_n^2}} = \sqrt{\sigma^2}$, which is quite sufficient for describing the normal distribution. PDA is not an exception: the theory of detection of optical signals is based on the normal law. One should not forget, however, about the possible deviations from the normal distribution, when individual fluctuations become discernible — they no longer mix, and the law of large numbers does not hold. For example, explosive noise (noise bursts) may be observed in photodetectors [57], and in APDs there also is the possible occurrence of microplasmas, due to the bistability of local regions in the pn junction [59–60]. Random "on-off" flips result in pulses of random length and, as a rule, fixed amplitude. The task of the engineer is to eliminate such noise. Deviations from the normal distribution are also observed when the current is so small that it becomes possible to register individual electrons. We shall discuss this in detail in Sect. 6.2, 6.3.

For the purposes of the present book the expression for probability distribution of one sample $I(t_1)$ is sufficient. For the sake of generality, however, we should point out the following. In case of two samples $I(t_1)$, $I(t_2)$ we shall deal with a two-dimensional random quantity, so the distribution law will also become two-dimensional, and will involve not only the dispersion, but also the correlation function $\mathcal{K}(t_2 - t_1)$. A continuous random quantity, however, is not one or two values, but rather an infinite number of random values (samples) over the observation interval T. In principle, therefore, its description will require an infinite-dimensional distribution function (with the infinite number of arguments $I(t)$). The correlation time τ_{cor} of the real noise processes, however, is finite; the value of the random function on this interval may be considered finite, and so the dimensionality of the random function (the number of variables, which is the number of independent readings) is finite, being of the order of T/τ_{cor}.

Main ideas on sources of noise. The main and inherent sources of noise (not only in PDAs, but in radio engineering in general) are the shot noise, the generation-recombination noise, and the thermal noise.

The shot noise occurs in a device with space charge region because of the discrete nature of charge carriers. The carrier ("pellet"), coming into the space charge region, induces a current in the external circuit during the entire time τ_{tr} it takes to traverse the SCR (Sect. 1.2). In this way, a pulse of length τ_{tr} is observed in the external circuit.

The generation-recombination noise is observed in photoresistors (and in semiconductors in general). Essentially, this also is a kind of shot-noise process (or even two independent shot-noise processes). The generation is due to the fact that the "pellets" (photons or phonons) being absorbed generate electron-hole pairs in the bulk of semiconductor. The recombination is the inverse process: after a while each of the pairs recombines to emit a phonon (or a photon). As long as the pair exists (from the time of generation till the time of recombination), it increases the concentration and hence the current. In this way, generation and recombination of a pair results in a pulsed contribution to the current. The length of this pulse is random, and its mean value is the mean lifetime of the pair in the semiconductor τ_{life}.

The physical nature of the thermal noise is also reflected in its name: thermal motion of a carrier in the semiconductor may violate the neutral balance, giving rise to local field and local emf. Let one electron in the resistor of constant cross section area A be displaced to the distance l. The capacitance of the layer of such length is C. The thermal displacement of the electron may be regarded as the transfer of charge q from one plate of the capacitor to the other — that is, as charging of the capacitor. The discharge time depends on the capacitance and resistance of the layer:

$$R = \frac{\rho l}{A}\ , \quad C = \frac{\varepsilon\varepsilon_0 A}{L}\ , \quad RC = \rho\varepsilon\varepsilon_0 = \tau_{\text{M}}\ . \tag{2.69}$$

We see that the relaxation does not depend on the geometry, being determined by a physical constant of the material, the Maxwell time constant τ_{M} (the resistivity ρ times the permittivity $\varepsilon\varepsilon_0$).

All these sources of noise produce instantaneous statistically independent elementary fluctuations; the system responds to such brief disturbance with the elementary pulse which relaxes with the characteristic time constant of the system τ_{n}. This time constant is equal to τ_{tr} for the shot noise in photodiodes, to τ_{life} for the generation-recombination noise in photoresistors, to τ_{M} for the thermal noise. When the noise passes through electrical circuits and amplifiers, the time constant of the path will affect the value of τ_{n}.

The elementary noise pulses, mixing together, form a random function $i_{\text{n}}(t)$ (Fig. 2.10). If the times of arrival of a large number of elementary pulses are close enough, a net pulse of a large amplitude will stand out in the noise process. Its effective length cannot be less than that of its constituent elementary noise pulse τ_{n}. Approximating, as before, such resultant pulses with rectangles, we may represent the noise as a sequence of pulses of average length τ_{n} and random amplitude (Fig. 2.10). Of course, this approximation is rather arbitrary, and some characteristics of the noise fall out of consideration: the fluctuations of pulse length and shape are disregarded, the superposition of pulses is not taken into account. Nevertheless, this approximation is acceptable for large fluctuations, which are especially dangerous: such fluctuations are rare and stand alone; the interval between the large pulses in the approximation is random (Fig. 2.10). We should not be too unhappy about this "rectangular reasoning", however: even such a simple rectangular approximation of the noise will allow us to endow the above statistical characteristics with straightforward physical meaning, and will be useful later on for deriving some of the principles of the theory of detection.

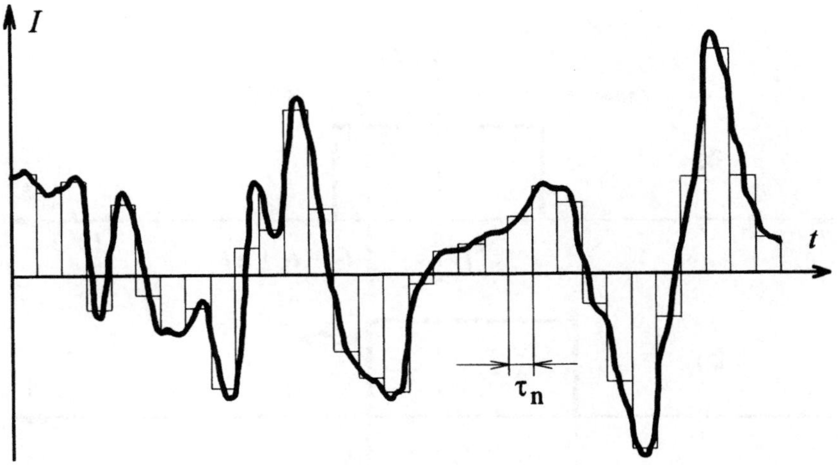

Figure 2.10: Approximation of the noise with a sequence of pulses of average length τ_n and random amplitude.

First of all, we can now give a simple physical interpretation of correlation. On the observation time interval we select one noise pulse of amplitude I_A and length τ_n (Fig. 2.11a). With due account for the ergodicity of the process, we first find the dispersion by averaging with respect to the time:

$$\sigma^2 = \overline{i_n^2} = \frac{1}{T} \int_{-T/2}^{T/2} I^2(t)dt = \frac{\tau_n}{T} I_A^2 . \quad (2.70)$$

Next we find the correlation function. If the pulse had been deterministic, and its location on the time axis had been known, we could have measured $I(t_1 + \Delta t)$ by aligning the first sample with the signal maximum (the origin of coordinates in Fig. 2.11a), and varying the displacement Δt of the second sample. This is very important: under these conditions the correlations function (the product $I(t_1)I(t_1 + \Delta t)$) is simply the pulsed signal (up to the factor of $I(t_1)$). The noise pulse is random; in our example, as shown in Fig. 2.11, random is its position on the time axis. Accordingly, the product $I(t_1)I(t_1 + \Delta t)$ must be averaged with respect to all possible positions of the pulse. In this case the correlation function may also be treated as a pulse (up to the same factor); this time, however, as a certain typical noise pulse, averaged over all possible random pulses in the noise process.

Let us perform the described procedure of averaging. Since the noise pulse may occur with equal probability at any point of the time interval T, averaging will consist in uniformly moving the pulse along the time axis, for example, from left to right. Figure 2.11b shows the instant when both readings have become nonzero, and Fig. 2.11c depicts the last such instant. From these two diagrams we see that the product $I(t_1)I(t_1 + \Delta t)$ is nonzero and is equal to I_A^2 on the interval $\tau_n - \Delta t$. Therefore,

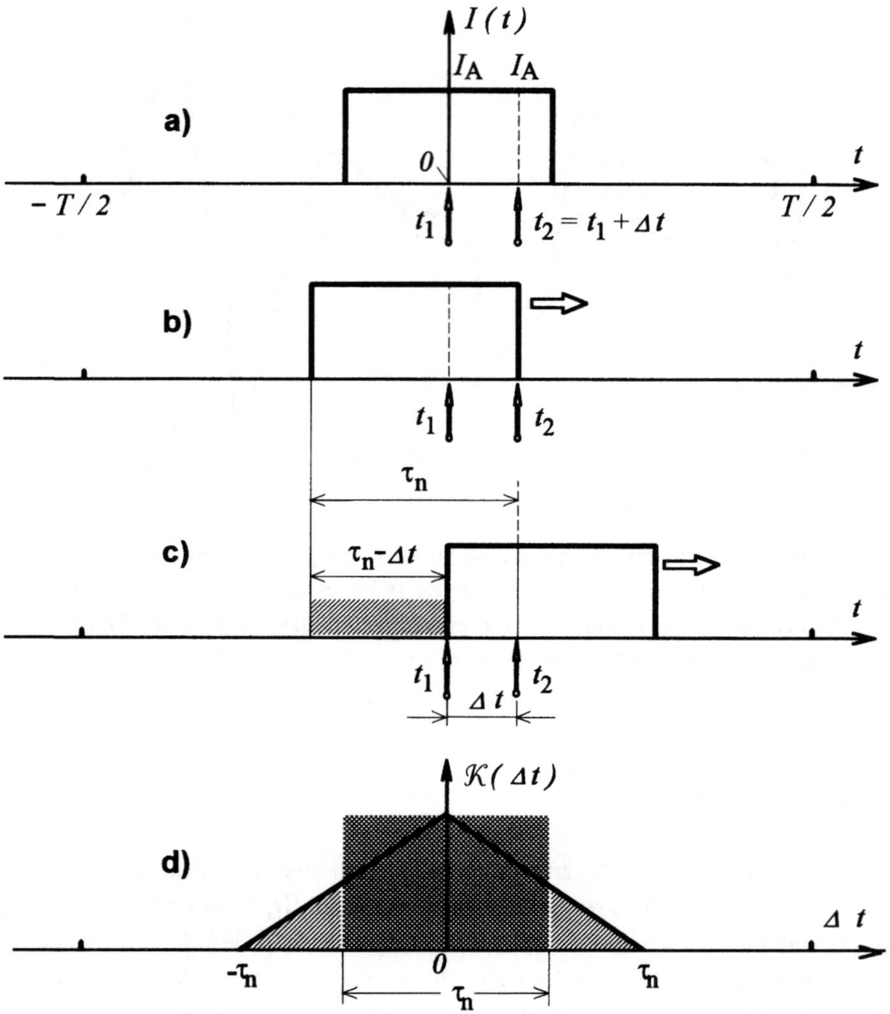

Figure 2.11: Correlation function may be regarded as noise pulse averaged over all its equiprobable positions on the observation interval T.

$|\Delta t| < \tau_n$,

$$\mathcal{K}(\Delta t) = \overline{I(t_1)I(t_1 + \Delta t)} = \frac{\tau_n - \Delta t}{T} I_A^2 = \frac{\tau_n}{T} I_A^2 \left(1 - \frac{\Delta t}{\tau_n}\right) = \sigma^2 \left(1 - \frac{\Delta t}{\tau_n}\right). \tag{2.71}$$

Here we have used expression (2.70) for the dispersion. From the above-described procedure of calculation of $\mathcal{K}(\Delta t)$ it follows that this function depends not on the location of the samples (the time t_1), but only on Δt (as should be the case), and that the function is even (as it should be). The resulting correlation function is shown in

Fig. 2.11d; the shape of the pulse averaged according to the rule described above (the function $\mathcal{K}(\Delta t)$) is triangular. Averaging resulted in the spread-out of the pulse edges, but the effective duration (the correlation time τ_{cor}) for the rectangular pulse in question is equal to the duration of the random noise pulse τ_n (in the general case of a noise pulse of arbitrary shape, $\tau_n \leq \tau_{cor} < 2\tau_n$). Amazingly simple is the nature of correlation: the system stores ("remembers") the instantaneous fluctuation for the time $\tau_n = \tau_{cor}$, and therefore the values of all samples on this time interval, which falls on one noise pulse, are the same. After the time $\tau_n = \tau_{cor}$ the pulse is completely wipes out, and the system memorizes the next fluctuation for the time τ_n, and we shall get new independent readings, and so on.

It also becomes clear why the spectrum of the noise power is obtained through the Fourier expansion of the correlation function (2.63). As a matter of fact, the function $\mathcal{K}(\Delta t)$ is a kind of the average noise pulse, and so its spectrum is the spectrum of such noise pulses (the factor of $I(t_1)$ in the function $\mathcal{K}(\Delta t)$ ensures that the function has the dimension of the squared quantity). All these arguments remain in force also when N independent pulses arrive within the observation time T. If all these pulses are similar in the form, and random is only the time of their arrival, the dispersion is multiplied by the factor of N, and in place of (2.70) we get $\sigma^2 = N\tau_n I_A^2 / T$. If the pulses differ in length, the correlation function (2.71) and the dispersion (2.70) will be expressed in terms of their mean duration; if they differ in amplitude, the expression for the dispersion will involve the mean square amplitude I_A^2.

Noise and signal: e pluribus unum. Despite the antagonism between the signal we are eager to detect, and the noise which tries to prevent us from doing that, they have much in common.

1. The signal is a pulse; the noise, if not a pulse, is a sequence of pulses (random pulse sequence).

2. The signal is described by the deterministic function $E_s(t)$; its main parameters are the amplitude E_{sA} and the effective length (duration) T_s. The random noise pulses are described in terms of the correlation function $\mathcal{K}(\Delta t)$, which may be regarded as the statistical-average noise pulse: the amplitude $\mathcal{K}(0)$ is nothing else but the mean square amplitude of the noise pulse, and the effective duration of the function $\mathcal{K}(\Delta t)$ (correlation time τ_{cor}) is nothing else but the typical length of the statistical-average noise pulse τ_n.

3. In the space of frequencies the pulsed signal is represented by its spectrum; the noise, in its turn, is represented by the spectrum of its statistical-average noise pulse — that is, by the spectrum of the function $\mathcal{K}(\Delta t)$.

4. The area of the pulses gives the maximum density of spectra:

$$\mathrm{E_s}(f=0)=2\int_{-\infty}^{\infty}E_s(t)dt\ , \tag{2.72}$$

$$\overline{J^2(0)}=2\int_{-\infty}^{\infty}\mathcal{K}(\Delta t)dt\ . \tag{2.73}$$

Conversely, the area of spectra gives the maximum value of the pulse (the signal amplitude and the noise dispersion):

$$E_{sA}=\int_{0}^{\infty}\mathrm{E_s}(f)df\ , \tag{2.74}$$

$$\sigma^2=\mathcal{K}(0)=\int_{0}^{\infty}\overline{J^2(f)}df\ . \tag{2.75}$$

All these expressions have been obtained earlier (see (2.10, (2.12), (2.44), (2.51), (2.63)).

We see that the noise tries to disguise itself as the signal: without stopping for a microsecond, it sends its pulses to the PDA, trying to sneak through the circuits all the way down to the decision-making device, making it take the wrong decision.

The interpretation of noise as a random pulse sequence may meet with difficulties arising from the divergence of the integral (2.75). First of all, this may be associated with the excessive low-frequency noise $\overline{J^2} \propto 1/f$, when the integral will diverge in the neighborhood of zero frequency. Alternatively, the integral may diverge in the high-frequency range ($f \to \infty$): the high-frequency components of the noise may be amplified as the primary noise pulses pass through the electronic circuits, resulting in $\overline{J^2} \propto f^2$. Of course, in the real devices the power is limited, so that the dependences like $\overline{J^2} \propto 1/f$ or $\overline{J^2} \propto f^2$ will only pertain to a limited frequency range. When such dependences arise, however, it is no longer possible to disregard the actual shape of the noise pulses, approximating them with rectangles. One of the tasks of the theory of low-frequency noise is the study of the possible shapes of noise pulses and the mechanism of their formation; the definition of the lower frequency limit of the low-frequency noise.

Noise sources on the equivalent diagram. As indicated above, each separate harmonic in the noise realization $i_{nn}(t)$ passes through the electrical circuits in exactly the same way as the harmonic of the deterministic signal. This is the reason why the source of the noise harmonic, similarly to the source of the signal harmonic, is represented on the equivalent diagram by a current or voltage generator, and further analysis is based on the standard theory of electrical circuits. Two features distinguish it from the harmonic of the deterministic signal. Quantitatively, the noise harmonic is characterized not by the amplitude (amplitude density), but rather by the mean square value (or the density of the mean square value) of the amplitude $\overline{J^2}$. When several harmonics are added together in one element of the circuit, the rule of summation is geometrical rather than algebraic (recall that the dispersion of the sum is equal to the sum of dispersions). This rule holds both for the harmonics from the same source of noise (since they are statistically independent), and for the harmonics from different sources (in radio engineering, the equal-frequency harmonics from different sources sometimes may be correlated; in our present context such situations are not going to arise).

A consistent presentation of the theory of noise in radio engineering can be found in the extensive literature on statistical physics, radio engineering, mathematical statistics [55–57, 61, 62]. In this section we had to sacrifice the mathematical precision for the sake of straightforward interpretation of the main statistical characteristics of the noise. The model of the noise as a superposition of statistical-average pulses, developed in this section, will work for us throughout the remainder of this book.

Summary. *The main inherent (basically irremovable) sources of noise in PDAs, like in many other radio electronic and semiconductor devices, are the shot noise, the generation-recombination noise, and the thermal noise. Shot noise arises in the devices having a* pn *junction, whereas the generation-recombination noise is observed in homogeneous semiconductor (photoresistor) because of the discrete nature of the charge. The carrier entering the SCR of photodiode, or the pair generated in the photoresistor, produce a current increment in the external circuit. The length of this increment (the pulse length) is determined, respectively, by the time of transit across the SCR, or by the lifetime of the pair in semiconductor. Thermal current in resistors is produced by the thermal motion of mobile charge carriers: when displaced, they create local field and the emf, which then relaxes with the Maxwell time constant. Passing through electrical circuits, these pulses may be prolonged. In this way, random pulses arise in the system, whose length* τ_n *is determined by the characteristic time constant of the system. A random superposition of these random pulses is the noise. In time, the noise is described by the correlation function, which may be treated as the statistical average noise pulse: its amplitude* $\mathcal{K}(0)$ *is the dispersion of the amplitude of the noise pulse, and its characteristic time constant* τ_{cor} *is the length of the pulse* τ_n*. The time constant* $\tau_{cor} = \tau_n$ *defines the time interval on which two adjacent noise samples are correlated. The spectrum of the statistical-average pulse (the spectrum of correlation function* $\mathcal{K}(\Delta t)$*) is the noise spectrum (the spectral density of the noise power)* $\overline{J^2(f)}$*, which is the main characteristic of noise in radio engineering. The area of the spectrum* $\int_0^\infty \overline{J^2(f)}\,df$ *defines the dispersion, the mean square value of the noise at any time. In the overwhelming majority of cases the noise complies with the normal distribution law; so, knowing the dispersion, one may use the normal law to calculate the probability of occurrence of given value of the fluctuating quantity.*

2.3. Photodetector Noise

In the preceding section we got a general idea of the nature of noise in photodetectors, and its statistical characteristics. Now we shall derive formulas which can be used for calculating the spectral density of the photodetector noise knowing the photoelectric parameters of the device.

Photodiode noise is a chaotic superposition of pulses induced by each separate charge carrier (Sect. 2.2). Assume that one electron arrives during the observation time T. This electron induces a current pulse of length τ_{tr} in the photodiode circuit. The area of this stand-alone current pulse is equal to the charge of one electron q. Since the induced pulses are very short ($\tau_{tr} \approx 10^{-11}...10^{-8}$ s; Sect. 1.2), the decline of the spectrum starts at very high frequencies (Sect. 2.1):

$$f \approx \frac{1}{2\tau_{tr}} = \frac{1}{2\left(10^{-11}...10^{-8}\right)} = 5\cdot\left(10^{7}...10^{10}\right)\text{Hz} . \qquad (2.76)$$

In physics, the so-called delta-pulse $\delta(t-t_0)$ is defined as infinitesimally short pulse of infinitely large amplitude [55]. In other words, the magnitude of delta-pulse is zero at all times except $t = t_0$, when the magnitude tends to infinity. In practice, "infinitesimally short" means shorter than any of the characteristic times of the system (observation time, the time constants of the electronic processing sections). Accordingly, the decline in the spectrum of the pulse which may be regarded as a delta-pulse must start above the working frequencies (as is the case with the PD-based PDA). The delta-pulse is quantitatively characterized by its area (rather than by its duration and amplitude), which by definition is assumed to be equal to one:

$$\int_{-\infty}^{\infty} \delta(t-t_0)dt = 1 . \qquad (2.77)$$

Then the current pulse may be written as

$$I(t) = q\delta(t-t_0) . \qquad (2.78)$$

We see that the noise pulses of the photodiode current may be regarded as delta-pulses, characterized by their area (the charge q). The spectrum of such solitary pulse in the working frequency band does not depend on the frequency; the spectrum of this kind is called "white noise". According to Sect. 2.1, its density $\mathrm{I_n}(f)$ is equal to twice the area of the pulse:

$$\mathrm{I_n}(f) = 2q . \qquad (2.79)$$

The amplitude spectrum of the one-electron pulse turns out to be deterministic; but the phase, which corresponds to the random position of the pulse t_0 on the observation interval, is random. As shown in the preceding section, in this case we must introduce the spectral density of the noise power, which by definition is linked with the spectrum of realization (2.79) by (2.32):

$$\overline{J^2(f)} = \frac{\overline{\mathrm{I}_{\mathrm{n}}^2(f)}}{2T} = \frac{(2q)^2}{2T} = \frac{2q^2}{T} . \tag{2.80}$$

Assume now that N_m charge carriers have arrived in the given mth realization within the observation time T. The total current is the sum of current pulses induced by each carrier, and the mean current over the time T in the mth realization is

$$I_m = \frac{qN_m}{T} = qn_m , \tag{2.81}$$

where n_m is the flow of electrons per unit time in the mth realization. Since the dispersion of the sum is equal to the sum of dispersions of N_m independent terms (and the pulses are independent because independent are the times of their arrival t_0), the spectral density in the given realization is multiplied by N_m:

$$\overline{J^2(f)} = N_m \frac{2q^2}{T} = 2q\frac{qN_m}{T} = 2q \cdot qn_m = 2qI_m \Rightarrow 2qI_0 . \tag{2.82}$$

Here we have replaced the current in the mth realization I_m with the mean current I_0, which is justified when $T \to \infty$.

This last expression is the known Shottky formula for the shot noise, one of the fundamental equations in the statistical radio engineering [57, 62]. This expression holds for the photodiode regime, when the current I_0 is the back current of the photodiode $I = I_0$, caused by thermal and photo generation. In case of small inverse bias, and especially in case of forward bias, the forward current $I_{\mathrm{dg}} \exp(U/\varphi_{\mathrm{th}})$ is also important (see (1.71)). The forward and the back (1.72) currents are statistically independent, and the powers of these noises are added [62]:

$$\overline{J^2(f)} = 2qI_0 + 2qI_{\mathrm{dg}} \exp\frac{U}{\varphi_{\mathrm{th}}} . \tag{2.83}$$

When the forward and the back currents balance out, this expression is simplified:

$$2qI_0 = 2qI_{\mathrm{dg}} \exp\frac{U}{\varphi_{\mathrm{th}}} ,$$

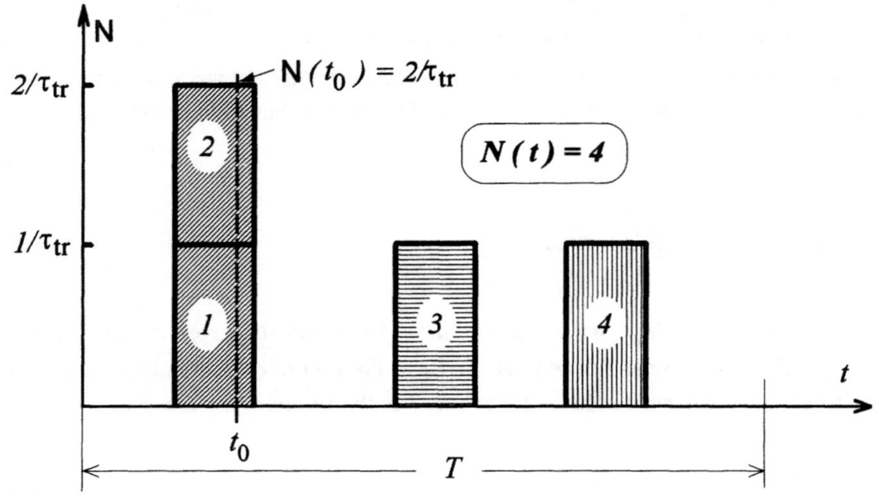

Figure 2.12: Do not confuse the instantaneous flux of particles $\mathcal{N}(t)$ and the flux of particles N on the observation interval T.

$$\overline{J^2(f)} = 4qI_{dg}\exp\frac{U}{\varphi_{th}} = 4\kappa T\frac{qI_{dg}}{\kappa T}\exp\frac{U}{\varphi_{th}} = \frac{4\kappa T}{R_p} . \tag{2.84}$$

Here we have used the expression for the dynamic resistance of photodiode (1.74). Observe that in the presence of background generation ($I_{db} + I_{pb} \neq 0$) this regime is realized in the open-circuit mode, $R_p = R_{pOC}$ (1.76). When there is no background generation ($I_0 = I_{dg}$, $U = 0$), the forward and the back components of thermal generation current balance out also in the short-circuit mode, and $R_{pOC} = R_{pSC} = R_p$ (1.74)–(1.76).

As will be shown below, (2.84) is the formula for the thermal noise of the resistance R_p. So, in this particular case, two basic mechanisms of noise (shot noise and thermal noise) result in the same expression for the spectral density of the noise power.

Fluctuations of flux of particles. In photoelectronics, the concept of electric current is used along with the concept of the flux of particles (electrons, photons). Earlier we have already used the flux of photons for evaluating the intensity of thermal background (Sect. 1.2). In Sect. 6.2 we shall characterize the noise of APD-based PDA in terms of the number of noise electrons. As a matter of fact, the electric current $I(t)$ and the flux of particles (charge carriers) $\mathcal{N}(t)$ are the same random function; the difference is only in the dimension:

$$I(t) = q\mathcal{N}(t) . \tag{2.85}$$

In this representation the flux of particles $\mathcal{N}(t)$, like the electric current, is a random pulse sequence. As indicated above, each elementary pulse in the PD circuit is induced by the passage of one charge carrier across the SCR. The length of such pulse is τ_{tr} — the pulse is as though "spread out" over this time interval, and therefore its amplitude is $1/\tau_{tr}$. The area of the pulse is equal to the number of particles involved:

$$\frac{1}{\tau_{tr}} \cdot \tau_{tr} = 1 \ . \tag{2.86}$$

Of course, the area is equal to dimensionless unity: carried over is one particle (drawing the analogy with the electric current, we may note that the area of elementary current pulse is equal to the charge of one electron, $(q/\tau_{tr}) \cdot \tau_{tr} = q$). This is illustrated in Fig. 2.12, which shows one of the possible realizations $I(t)$, $\mathcal{N}(t) \equiv I(t)/q$.

Elementary particle, elementary pulse... even the most elementary things may cause confusion. Confused are the concepts of the instantaneous flux of particles $\mathcal{N}$ and the flux of particles over the observation interval T.

1. The instantaneous flux of particles $\mathcal{N}(t)$: we have just introduced this random function. Its value (sample) at the time t is defined by the number of elementary pulses induced in the given section of the circuit at this moment.

2. The flow of particles n over the observation time T is defined as the ratio of the total number of particles N to the observation time, $n = N/T$.

The difference between $\mathcal{N}(t)$ and n is clear from Fig. 2.12. Two pulses arrive at $t = t_0$, and so $\mathcal{N}(t_0) = 2/\tau_{tr}$. Altogether there are four pulses over the time T, $N = 4$, and so $n = 4/T$.

Averaging $\mathcal{N}(t)$ over the ensemble of the systems, we get $\overline{\mathcal{N}}$. Averaging n over different time intervals T, we get $\bar{n}$ (the intervals T may be selected within the same system, which amounts to averaging with respect to time). We have already indicated that the ensemble-average and time-average values in the ergodic system coincide, and therefore $\overline{\mathcal{N}(t)} = \overline{\mathcal{N}} = \bar{n}$. This is the reason why the two concepts are confused. There is, however, an important difference between the random functions $\mathcal{N}(t)$ and n, because their dispersions are different. To find the dispersion $\overline{\mathcal{N}^2}$, one must add up all the acting spectral harmonics $\overline{J^2_{\mathcal{N}}(f)}$ in the effective frequency band Δf_{eff} (2.44). In the model under consideration this band is determined by the length of the elementary pulse, $\Delta f_{eff} = 1/2\tau_{tr}$; therefore,

$$\overline{\mathcal{N}^2} = \overline{J^2_{\mathcal{N}}}\, \Delta f_{eff} = \overline{J^2}\, \frac{1}{2\tau_{tr}} \ . \tag{2.87}$$

To find the dispersion $\overline{n^2}$, we must first find the flux of particles n for the interval T, which requires counting all the N pulses arriving within this time. The counting procedure is performed by the integrator with the integration time T. In greater detail this counting procedure will be discussed in Sect. 6.2, 6.3. In particular, we shall demonstrate that the effective frequency band of such integrator is $1/2T$, which agrees with the arguments developed above in Sect. 2.2. Then we noted that the fluctuations of the permanent component, measured over different time intervals T, is determined by the fluctuations of harmonics in the range from zero to $1/2T$ (this is the band occupied by the frequency $f_0 = 0$, Fig. 2.6). Passing the noise $\overline{J^2_{\mathcal{N}}}$ through such integrator, we get at the output the noise in the integrator's band:

$$\overline{n^2} = \overline{J^2_{\mathcal{N}}} \frac{1}{2T} \; . \tag{2.88}$$

From comparison between (2.87), (2.88) we see that the dispersion $\overline{\mathcal{N}^2}$ is much greater than the dispersion $\overline{n^2}$: by the factor of T/τ_{tr}.

The foregoing was actually a long prelude, which has been performed with the sole purpose of enabling us to give another derivation of the Shottky formula, from only the Poisson distribution. If the probability of arrival of N particles over the given time T obeys the Poisson distribution, the dispersion of the number of particles $\overline{N^2}$ is equal to the mean number, $\overline{N^2} = \overline{N}$ [55]. Then, using some straightforward algebra, we come from (2.88) directly to the Shottky formula for the spectral density $\overline{J^2_{\mathcal{N}}(f)}$ of the flux of particles $\mathcal{N}(t)$:

$$\overline{J^2_{\mathcal{N}}(f)} = 2T\overline{n^2} = 2\frac{\overline{(Tn)^2}}{T} = 2\frac{\overline{N^2}}{T} = 2\frac{\overline{N}}{T} = 2\overline{n} \; . \tag{2.89}$$

Since the current $I(t)$ and the flux of particles $\mathcal{N}(t)$ are linked through the constant q (see (2.85)), the moments of the second order (the spectral densities of the current $\overline{J^2(f)}$ and the flux of particles $\overline{J^2_{\mathcal{N}}(f)}$) are linked through the square of this constant,

$$\overline{J^2(f)} = q^2 \overline{J^2_{\mathcal{N}}(f)} = q^2 \cdot 2\overline{n} = 2q(q\overline{n}) = 2qI \; . \tag{2.90}$$

Is the Shottky formula a dogma? The immediate reaction to this question is yes, it is — the Shottky theorem $\overline{i^2_{\mathrm{n}}} = (2q)I_0$ that we have just derived is a dogma at least for the photodiode noise.

The factor of $(2q)$ is the spectrum of elementary one-electron current pulse (twice the area of such pulse). The second multiplier I_0 is proportional to the number of such pulses. The Shottky dogma could possibly be violated in other worlds, where the electrons would have the charge $q' \neq q$. Then the formula for the noise will have to be modified to include the noise factor $\mathcal{F}$:

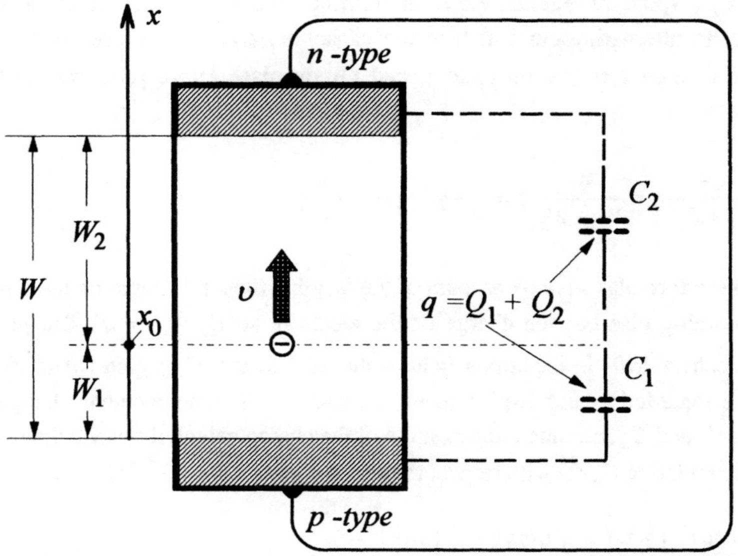

Figure 2.13: Drift of charge carriers in SCR induces current pulses in the external circuit.

$$\overline{j^2} = 2q'I_0 = 2qI_0\left(\frac{q'}{q}\right) = 2qI_0\mathcal{F} \ , \quad \mathcal{F} = \frac{q'}{q} \ , \tag{2.91}$$

or the noise factor $\mathcal{F}_n$:

$$\overline{j^2} = 2(q')^2 n = 2q^2 n\left(\frac{q'}{q}\right)^2 = 2q^2 n\mathcal{F}_n \ , \quad \mathcal{F}_n = \left(\frac{q'}{q}\right)^2 \ , \tag{2.92}$$

As it turns out, however, there is no need to appeal to other worlds: in our own world with its ordinary PDAs and ordinary electrons, the elementary current pulses may carry charge q' which is different from the charge of electron q, and this q' may be smaller or larger than q. Let us consider both of these theoretically feasible situations.

Generation in SCR $(q' < q)$ [63]. So far we have assumed that an electron traversing the SCR induces in the PD circuit an elementary current pulse whose area is equal to the electron charge q. Let us consider in greater detail the mechanism of induction by carriers generated in the SCR.

Assume that the generation-recombination centers are located at the arbitrary cross section $x = x_0$ of the SCR, and at the time $t = 0$ a phonon generates one electron from these centers (Fig. 2.13). The plane $x_0 = W_1$, where the mobile electron is located, divides the SCR into two parts. These parts may be regarded as capacitors with the capacitances $C_1 = \varepsilon\varepsilon_0 A/W_1$ and $C_2 = \varepsilon\varepsilon_0 A/W_2$. The plates of these capacitors are represented by the plane $x_0 = W_1$ where our electron is located, and by the conducting n^+ (plate 1) and p^+ (plate 2) regions. We assume further that the external circuit is shorted with respect to alternating current; then the capacitors C_1 and C_2 are connected in parallel, and their charges (Q_1 on plate 1, and Q_2 on plate 2) are proportional to the capacitances:

$$Q_1 = \frac{C_1}{C_1 + C_2} q = \frac{W_2}{W_1 + W_2} q = \frac{W_2}{W} q \, , \quad Q_2 = \frac{W_1}{W} q \, . \tag{2.93}$$

In writing these formulas we took account of the fact that the total charge on the common plate x is nothing else but the charge of the electron, so $Q_1 + Q_2 = q$. The phonon-generated electron drifts in the strong field of the SCR to the n^+ region (plate 2). This drift may be regarded as the displacement of plate x. This displacement changes the capacitances C_1 and C_2, causing redistribution of the charge, which flows via the external circuit. Differentiating (2.93) with respect to time, we get

$$I(t) = \left|\frac{dQ_1}{dt}\right| = \left|\frac{dQ_2}{dt}\right| = \frac{q}{W}\left|\frac{dW_1}{dt}\right| = \frac{q}{W}\left|\frac{dW_2}{dt}\right| = \frac{q}{W} v \, . \tag{2.94}$$

We see that the current induced in the external circuit is proportional to the magnitude of the charge and the velocity of the charge, and inversely proportional to the thickness of the SCR (the distance between plates 1 and 2). Hence it follows that the transferred charge (the area of the current pulse) is proportional to the displacement — that is, to the path length of the carrier l:

$$Q = \int_0^t I(t) dt = \frac{q}{W} \int_0^t v \, dt = \frac{q}{W} l \, . \tag{2.95}$$

Expression (2.94) is the known Rameau theorem [64], and (2.95) is its modification. The Rameau theorem is commonly used for calculating the kinetics and the current fluctuations in photodiodes in the most general case. For our qualitative analysis it will suffice to restrict ourselves to the most simple case of the electron and the hole having constant and equal velocities.

If the electron traverses the entire SCR, $l = W$, then, according to (2.95), it will induce a current pulse in the external circuit which carries the entire charge q, Fig. 2.14a. This situation occurs when the generation-recombination center is located in the SCR near its boundary with the p^+ layer (as, for example, in asymmetrical junctions), or when the carriers are generated in the p^+ layer and leave it by diffusion (the same applies to the holes generated in the n^+ layer). It was such pulses that were considered above in connection with the derivation of the Shottky formula. If, however, the sites are located in the SCR far enough from the p^+ layer, then $l = W_1 < W$, and,

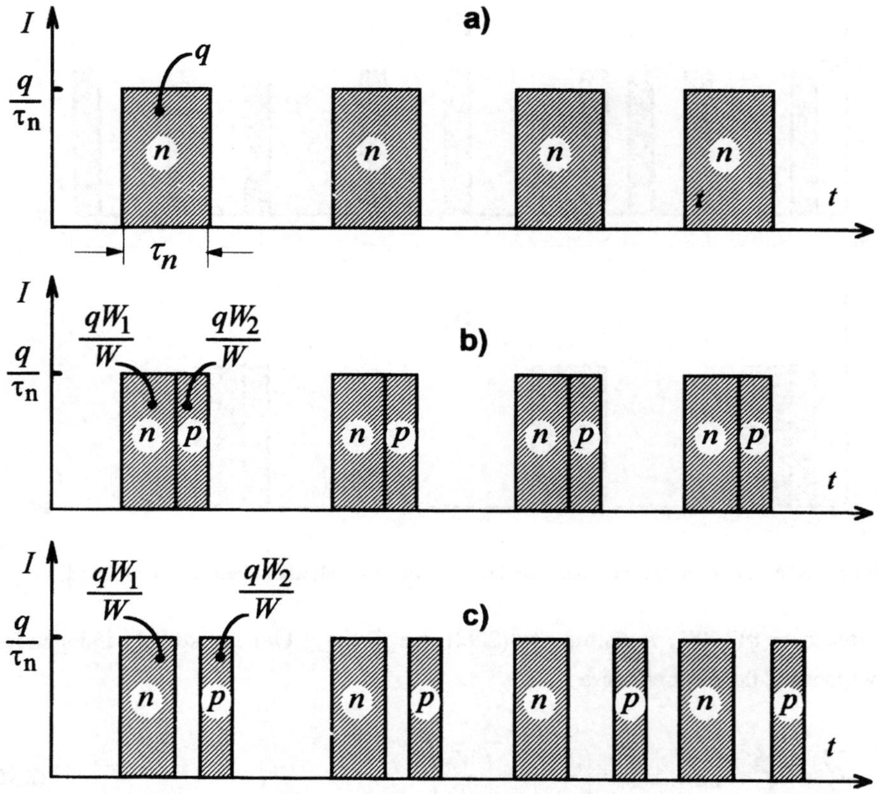

Figure 2.14: When charge carriers are generated in SCR, the elementary current pulse may carry a charge less than the charge of electron, and Shottky theorem is violated.

according to (2.95), the electron passing through the SCR carries the charge $q' = qW_1/W < q$. This is not a miracle: the electron-hole pair carries the same integral charge in the external circuit, but now the electron and the hole "share the work": the electron drifts towards the p^+ layer and covers the distance $l = W_2$, and the hole drifts towards the n^+ layer and covers the distance $l = W_1$. According to (2.95), each of them carries, respectively, W_2/W-part and W_1/W-part of the charge. The entire current pulse is thus split up into the electron and the hole components (Fig. 2.14b). If the ejection of the electron into the conduction band is almost immediately followed by the ejection of the hole into the valence band (as shown in Fig. 2.14b), there would be nothing new for us: the area of the resulting pulse would be equal to q, as before, and the possible variation of the pulse shape would only affect the high-frequency decline of the spectrum. A similar situation is observed in the case of photogeneration: the electron and the hole are simultaneously ejected into their respective bands.

If, however, the statistics of generation of electrons and holes is similar (the ejections occur, on the average, in comparable time intervals), the sequence of pulses is as though split into two sequences (Fig. 2.14c). Each of the n pulses induced by the electrons carries the charge $q'_n = qW_1/W$, and each of the n pulses induced by the holes carries the

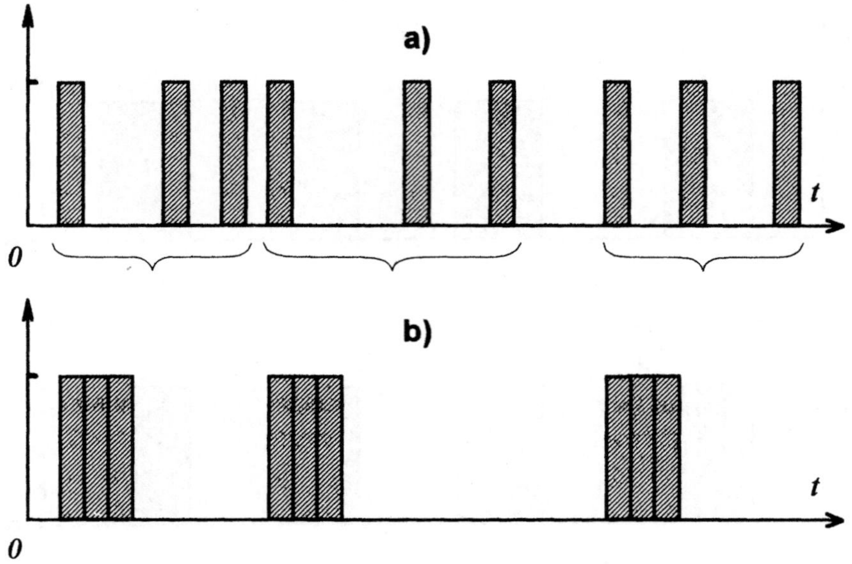

Figure 2.15: Low-energy photons bundle together, and Shottky theorem is violated.

charge $q'_p = qW_2/W$. According to (2.92), we then get two statistically independent components: the electron noise

$$\overline{J_n^2} = 2q^2 n \mathfrak{F}_{nn} , \quad \mathfrak{F}_{nn} = \left(\frac{q'_n}{q}\right)^2 = \left(\frac{W_1}{W}\right)^2 , \tag{2.96}$$

and the hole noise

$$\overline{J_p^2} = 2q^2 n \mathfrak{F}_{np} , \quad \mathfrak{F}_{np} = \left(\frac{q'_p}{q}\right)^2 = \left(\frac{W_2}{W}\right)^2 . \tag{2.97}$$

Accordingly, the total noise is

$$\overline{J^2} = \overline{J_n^2} + \overline{J_p^2} = 2q^2 n \mathfrak{F}_n = 2qI_0 \mathfrak{F}_n , \tag{2.98}$$

where

$$\mathfrak{F}_n = \mathfrak{F}_{nn} + \mathfrak{F}_{np} = \left(\frac{W_1}{W}\right)^2 + \left(\frac{W_2}{W}\right)^2 . \tag{2.99}$$

The largest deviation from the Shottky formula is observed when the site is located halfway across the SCR. Then $W_1 = W_2 = W/2$, and the noise factor is

$$\mathcal{F}_n = \left(\frac{1}{2}\right)^2 + \left(\frac{1}{2}\right)^2 = \frac{1}{2}. \tag{2.100}$$

The physics of this noise reduction is clear: each elementary current pulse carries half the electron charge, $q'_n = q'_p = q' = q/2$, and the noise power, according to (2.91), is proportional to the charge (the area) of the elementary pulse.

It is rather unlikely, however, that all the generation-recombination sites happen to concentrate in the middle of the SCR. More natural is the situation when the sites are uniformly distributed within the SCR. Then, averaging $\mathcal{F}_n$ (2.99) with respect to coordinate $x = W_1$, we get:

$$\overline{\mathcal{F}_n} = \overline{\mathcal{F}_{nn}} + \overline{\mathcal{F}_{np}} = 2\overline{\mathcal{F}_{nn}} = 2\overline{\mathcal{F}_{np}} = \frac{2}{W}\int_0^W \left(\frac{W_1}{W}\right)^2 dW_1 = \frac{2}{3}\left(\frac{W_1}{W}\right)^3\Bigg|_0^W = \frac{2}{3}. \tag{2.101}$$

This result may be formulated as follows: many "ifs" are required to observe deviations from the Shottky formula. If

— thermal generation prevails over photogeneration,

— this thermal generation occurs in the SCR,

— the statistics of generation of electrons and holes is the same,

then the root mean square noise will be reduced by the factor of $\sqrt{\mathcal{F}_n} = \sqrt{2/3} = 0.82$. So even with all these "ifs and ans" the reduction is not considerable, and is primarily of academic interest (for example, precision noise spectrum measurements allow one to diagnose the parameters of the sites). Let us add that the theoretically feasible case of $\mathcal{F}_n = 2/3$ (when the generation in SCR prevails) corresponds to a situation when the noise of PD is small, and can hardly be discriminated from the general noise of the PDA (Sect. 2.5).

Background noise. We have just considered the case when the elementary current pulse carried the charge $q' < q$. Now let us consider the opposite situation, when the elementary current pulse carries the charge $q' > q$. It is easy to imagine such model: independent elementary pulses (Fig. 2.15a) will group into packages (Fig. 2.15b), with random intervals between the packages. This is not a speculation, but a quite realistic assumption. Such grouping is observed in the fluctuations of the background current, when ideally they reproduce the fluctuation of the long-wave component of blackbody radiation, the emission of low-energy photons with $h\nu << \kappa T$. One such photon cannot carry the entire energy of the phonon κT: this energy can be used up for the emission of a package comprised, on the average, of $m = \kappa T/h\nu$ quanta, as shown in Fig. 2.15b. Absorption of the package of m photons results in the elementary current pulse of m electrons, which now carries the charge mq. According to (2.91), the noise factor is also increased m-fold:

$$\mathcal{F} = \frac{q'}{q} = m = \frac{\kappa \mathrm{T}}{\mathrm{h}\nu} . \tag{2.102}$$

The emission of high-energy photons is, however, not associated with their grouping: to cause emission of one quantum, $\mathrm{h}\,\nu/\kappa\,\mathrm{T}$ phonons have to be absorbed. Accordingly, emitted are statistically independent "solitary" quanta, which cause the appearance of "solitary" electrons (Fig. 2.15a). Then the Poisson distribution is preserved, and the values $m = 1$, $q' = q$, $\mathcal{F} = 1$.

Of course, this is just a qualitative description of fluctuations of photons. From Bose-Einstein statistics one can derive the general expression for the noise factor, or, as they call it, the coefficient of degeneracy of photon fluctuations [65]:

$$\mathcal{F} = \frac{\exp\dfrac{\mathrm{h}\nu}{\kappa \mathrm{T}}}{\exp\dfrac{\mathrm{h}\nu}{\kappa \mathrm{T}} - 1} . \tag{2.103}$$

It is easy to prove that in the limits of low and high photon energies this formula yields those values of $\mathcal{F}$ that we have just derived:

$$x = \frac{\mathrm{h}\nu}{\kappa \mathrm{T}} << 1 , \quad \exp\frac{\mathrm{h}\nu}{\kappa \mathrm{T}} \approx 1 + \frac{\mathrm{h}\nu}{\kappa \mathrm{T}} , \quad \mathcal{F} = \frac{1 + \overset{\to 0}{\dfrac{\mathrm{h}\nu}{\kappa \mathrm{T}}}}{1 + \dfrac{\mathrm{h}\nu}{\kappa \mathrm{T}} - 1} \approx \frac{\kappa \mathrm{T}}{\mathrm{h}\nu} >> 1 , \tag{2.104}$$

$$x = \frac{\mathrm{h}\nu}{\kappa \mathrm{T}} >> 1 , \quad \exp\frac{\mathrm{h}\nu}{\kappa \mathrm{T}} >> 1 , \quad \mathcal{F} = \frac{\exp\dfrac{\mathrm{h}\nu}{\kappa \mathrm{T}}}{\exp\dfrac{\mathrm{h}\nu}{\kappa \mathrm{T}} - 1} \approx 1 . \tag{2.105}$$

In Sect. 1.2 we calculated the effective number of quanta of thermal radiation n_{eff} registered by the photodetector, and the background current $I_0 = I_{\mathrm{db}} = qn_{\mathrm{eff}}$ due to that radiation. We showed that in the spectral ranges we are concerned with ($\lambda \le 14\mu\mathrm{m}$) we have $x'' >> 1$ (1.38), so that Wien's approximation holds ($\exp x''$ is much greater than unity). Now in the calculation of the noise factor we have reworked through this approximation, and obtained $\mathcal{F} = 1$. Accordingly, at $\lambda \le 14\mu\mathrm{m}$ there is no grouping of photons, the Poisson distribution holds for the electrons generated both by the phonons and by the photons of thermal radiation, $\overline{n^2} = 2\bar{n} = 2n_{\mathrm{eff}}$ (2.89). Because of this, the Shottky formula (2.82) holds not only for the noise of thermal generation current, but also for the noise of background current.

Grouping of photons and the associated noise increase become noticeable only when $\lambda'' \ge 50\mu\mathrm{m}$ ($x'' \le 1$; see (2.95)). For the sake of generality, let us consider this case also. It will be convenient to perform side-by-side calculations for $\mathcal{F} = 1$ and $\mathcal{F} \ne 1$. In the first case, calculating the spectral density of the effective photons $\overline{n^2_{\text{eff}}}$ with $\mathcal{F} = 1$, we have to calculate, like in Sect. 1.2, the total number of such photons, $\overline{n^2_{\text{eff}}} = 2n_{\text{eff}}$. This amounts to integrating the Poisson distribution $n_\lambda(\lambda)$ over the spectral sensitivity range of the photodetector, from 0 to λ'' (recall (1.31)–(1.33)):

$$\overline{n^2} = 2n_{\text{eff}} = 2\int_0^{\lambda''} n_\lambda(\lambda)d\lambda = 2\left(2\pi\frac{\kappa^3 \mathrm{T}^3}{\mathrm{h}^3 c^2}\right)\int_{x''}^{\infty}\frac{x^2 dx}{\exp x - 1} \propto \mathrm{In}_1(2) . \tag{2.106}$$

In the second case ($\mathcal{F} \neq 1$) we must, in accordance with (2.91), find the integral of $n_\lambda(\lambda)$, this time with the weight function $\mathcal{F}(\lambda)$ (2.103). So the integrand will now be multiplied by $\mathcal{F} = \exp x/(\exp x - 1)$:

$$\overline{n^2} = 2\int_0^{\lambda''} n_\lambda(\lambda)\mathcal{F}d\lambda = 2\left(2\pi\frac{\kappa^3 \mathrm{T}^3}{\mathrm{h}^3 c^2}\right)\int_{x''}^{\infty}\frac{x^2 \exp x}{(\exp x - 1)^2}dx \propto \mathrm{In}_2(2) . \tag{2.107}$$

This actually ends the calculation: the grouping of photons has resulted in the "regrouping" of integrals: the integral $\mathrm{In}_1(2)$ is replaced by $\mathrm{In}_2(2)$. Accordingly, the noise factor is equal to the ratio $\mathrm{In}_2(2)/\mathrm{In}_1(2)$. In Sect. 1.2 we have studied these integrals in detail. Now, taking advantage of the approximations (1.43), (1.44), we can write the expression for the noise factor:

$$\overline{\mathcal{F}} = \frac{\overline{n^2}}{2n_{\text{eff}}} = \frac{\mathrm{In}_2(2)}{\mathrm{In}_1(2)} \approx \frac{x''^2 + 2x'' + 2\sum_{m=1}^{\infty} m^{-3}}{x''^2 + 2x'' + 2\sum_{m=1}^{\infty} m^{-2}}$$

$$= \frac{x''^2 + 2x'' + 2\cdot 1.6}{x''^2 + 2x'' + 2\cdot 1.2} = 1 + \frac{2\cdot 0.4}{x''^2 + 2x'' + 2\cdot 1.2} . \tag{2.108}$$

The maximum noise increase because of the grouping of photons is achieved theoretically at $\lambda'' \to \infty$, and is

$$\mathcal{F} = \frac{\sum_{m=1}^{\infty} m^{-3}}{\sum_{m=1}^{\infty} m^{-2}} = \frac{1.6}{1.2} = 1.33 . \tag{2.109}$$

Accordingly, the root mean square noise gain is merely $\sqrt{\mathfrak{F}} = \sqrt{1.33} = 1.155$. This modest value is explained by the fact that the maximum of blackbody radiation (at $\lambda_{max} \approx 10\,\mu\text{m}$) corresponds to $x'' \approx 5$, so that $\mathfrak{F} \approx 1$. The excessive noise is cause by the long-wave photons, whose number is small.

Two expressions for the noise of the background current will be useful for future calculations. One follows from (2.106), (2.108),

$$\overline{J^2} = q^2\overline{n^2} = 2qI_0\mathfrak{F} \ , \tag{2.110}$$

where $\mathfrak{F}$ is given by (2.108), and the other follows directly from (2.107), (1.44), (1.42), (1.46):

$$\begin{aligned}
\overline{J^2} &= 2q^2\left(2\pi\frac{\kappa^3 \mathrm{T}^3}{\mathrm{h}^3 c^2}\right)\mathrm{In}_2(2) \\
&\approx 2q^2\left(2\pi\frac{\kappa^3 \mathrm{T}^3}{\mathrm{h}^3 c^2}\right)\exp(-x'')\left(x''^2 + 2x'' + 2\cdot 1.6\right) \\
&= 2q^2\left(2\pi\frac{\kappa^3 \mathrm{T}^3}{\mathrm{h}^3 c^2}\cdot 2\cdot 1.2\right)\exp(-x'')\left(\frac{x''^2}{2\cdot 1.2} + \frac{x''}{1.2} + \frac{1.6}{1.2}\right) \\
&= 2q^2\sigma'\mathrm{T}^3\exp(-x'')\left(0.416x''^2 + 0.83x'' + 1.33\right) \ .
\end{aligned} \tag{2.111}$$

We have considered a second case of possible deviation from the Shottky formula, and the conclusion remains the same: it is only under a very hypothetical condition (when the photodetector register very far-infrared radiation) that a deviation from the Shottky formula is observed, and this deviation will still be quite modest: at $\lambda'' \sim 20...50\,\mu\text{m}$ the mean square noise will increase just 3 to 7%; the increase close to the theoretical limit of 13 to 15% is observed when λ'' is as large as 200...500μm.

Now we can give a final answer to the query stated at the beginning of this subsection: the Shottky formula is more than a dogma, it is a guide to action. This formula must be used for calculating the noise of the primary current of all photodetectors used in PDAs considered in this book. Of course, we are speaking of the white noise; the noise with $1/f$ must be taken into account additionally.

"The noise of the primary current of *all* photodetectors" must be taken literally. Not only in ordinary photodiodes, but also in photodetectors with internal gain, also considered in this book, the source of the primary noise is the generation of charge carriers by phonons and photons. The rate of this generation determines the primary current I_0', and the primary noise $2qI_0'$. Of course, the noise of the secondary (output) current in photodetectors with internal gain is different from the noise of the primary current, but this is the topic of the forthcoming subsection.

Noise of APD. Assume that each incoming (or generated) carrier in the SCR is multiplied exactly by the factor of M. The transferred charge (the area of the elementary pulse) is multiplied by the same factor, $q' = Mq$. So, it is not only the grouping of photons, but also the internal gain, that can increase the elementary charge q'. In this case the spectral density of the noise, according to (2.92), is

$$\overline{J^2} = 2(q')^2 n = 2M^2 q^2 n = 2qI_0 M^2 \ . \tag{2.112}$$

Here and below by $I_0 = qn$ we denote the density of the primary (non-multiplied) current of the carriers which enter the SCR and start the avalanche, not to be confused with the secondary (multiplied) current $I_0 M$.

It follows that the multiplication of each carrier exactly M-fold increases the spectral density of the noise power by the factor of M^2. Multiplication, however, is a random process, and for N primary carriers which arrive over the time T we have N random values of the multiplication factor $M_1, M_2, \ldots, M_i, \ldots, M_N$. When we speak of the multiplication in APD, we imply the mean value $M = \overline{M}$.

It turns out, however, that one cannot use in place of M^2 in (2.112) the square of the mean value $\overline{M}^2$. According to (2.92), the spectral density of the noise $\overline{J_i^2}$ due to the ith elementary pulse is

$$\overline{J_i^2} = 2(q')^2 n = 2M_i^2 q_i \frac{1}{T} \ .$$

Here we have taken into account that, for a solitary pulse, $N = 1$, and $n = N/T = 1/T$. Adding up these partial spectra of the solitary elementary pulses, we get the overall spectrum:

$$\overline{J^2} = \sum_{i=1}^{N} \overline{J_i^2} = \sum_{i=1}^{N} \frac{2q^2}{T} M_i^2 = 2q\frac{qN}{T}\left(\frac{1}{N}\sum_{i=1}^{N} M_i^2\right) = 2qI_0 \overline{M^2} \ . \tag{2.113}$$

Comparing this with (2.112), we see that the formula for the noise (2.113) involves *not the squared mean* $\overline{M}^2$, *but the mean square* $\overline{M^2}$.

Fluctuation of the multiplication factor results in additional noise (the mean square $\overline{M^2}$ is greater than the squared mean $\overline{M}^2$). This noise is called the excess avalanche noise (not to be confused with the excess flicker noise with $1/f$ spectrum; see Sect. 2.4). Equation (2.113) is represented in the following form [66–68]:

$$\overline{J^2(f)} = 2qI_0 \overline{M}^2 \mathcal{F} \ , \quad \mathcal{F} = \frac{\overline{M^2}}{\overline{M}^2} \ , \tag{2.114}$$

where $\mathcal{F}$ is the noise factor equal to the ratio of the noise power of the real APD to the noise power of a hypothetical perfect multiplier with the constant (noise-free) multiplication factor $M_i = \overline{M} = M$.

For the calculation of the noise factor one needs to know the distribution function of the multiplication factor M_i. This function, and the avalanche noise, exhibit strong dependence on the ratio of the coefficients of impact ionization k. When multiplied is just one type of carriers — for example, the electrons ($k \to 0$) — then the avalanche goes in one direction (in the direction of movement of electrons). The noise of such avalanche is the smallest. In this case the distribution function of the multiplication factor is exponential [67]:

$$\rho(M) = \frac{1}{\overline{M}} \exp\left(-\frac{M}{\overline{M}}\right) . \tag{2.115}$$

Then for the mean square we have

$$\overline{M^2} = \int_0^\infty M^2 \rho(M) dM = \int_0^\infty M^2 \frac{1}{\overline{M}} \exp\left(-\frac{M}{\overline{M}}\right) dM$$

$$= \overline{M}^2 \int_0^\infty x^2 e^{-x} dx = \overline{M}^2 \left(x^2 + 2x + 2\right) e^{-x} \Big|_\infty^0 = 2\overline{M}^2 , \tag{2.116}$$

and, according to (2.114), the noise factor is

$$\mathcal{F} = \frac{\overline{M^2}}{\overline{M}^2} = 2 . \tag{2.117}$$

Generally, the distribution function is very sophisticated. Appropriate calculations result in the following expressions for the noise factor [68]:

$$\mathcal{F} = k\overline{M} + \left(2 - \frac{1}{\overline{M}}\right)(1-k) ; \tag{2.118}$$

$$\overline{M} >> 1 , \quad \mathcal{F} = k\overline{M} + 2(1-k) ; \tag{2.119}$$

$$\overline{M} >> 1 , \quad k \to 0 \quad (k\overline{M} << 1) , \quad \mathcal{F} = 2 ; \tag{2.120}$$

$$\overline{M} >> 1 , \quad k \to 1 , \quad \mathcal{F} = \overline{M} . \tag{2.121}$$

We see that when both types of carriers are multiplied ($k \to 1$), the noise factor may ultimately reach the value of $\overline{M}$. In this case the carrier traversing the SCR experiences, on the average, one collision. The thus generated carrier of the opposite sign moves in the opposite direction, and also experiences one collision, etc. In this way a ramified avalanche of $\overline{M}$ appears; each branch supplies a secondary pair into the external circuit ($\overline{M}$ pairs altogether). The noise powers of all $\overline{M}$ branches add up, which results in the noise increase by the factor of $\mathcal{F} = \overline{M}$.

Equation (2.118) was used for plotting the noise factor $\mathcal{F}$ versus the mean multiplication coefficient $\overline{M}$ on the log-log scale (Fig. 2.16a). In the typical range of $\overline{M} = 10\ldots200$ we made an attempt to approximate this function with a "straight line"

$$\log\mathcal{F} \approx \text{æ}\log\overline{M} \ , \quad \mathcal{F} = \overline{M}^{\text{æ}} \ . \tag{2.122}$$

Figure 2.16b shows the values of such noise coefficient æ as a function of the ratio of coefficients of impact ionization k. The value of æ(k) can be adjusted to within approximately 15% (on the finer scale this value depends also on $\overline{M}$ itself). With the aid of approximation (2.122), the formula (2.114) for the noise of APD can be rewritten in the following most commonly known form:

$$\overline{J^2} = 2q\overline{I_0}\overline{M}^{2+\text{æ}} \ . \tag{2.123}$$

Silicon, the staple material of semiconductor electronics, is unrivaled also in this respect. The ratio k for silicon is small, of the order of $(1\ldots5)\cdot10^{-2}$ [39]; accordingly, the noise coefficient æ is small, æ $\sim 0.2\ldots0.3$ (Fig. 2.16b). This feature, combined with the low dark currents (see Table 1.1), permits using silicon for creating the most advanced low-noise APDs. The structures are designed in such a way (Fig. 1.7b,c) as to facilitate the multiplication of electrons, whose coefficient of impact ionization is higher than that of the holes.

Silicon-based APDs, however, are sensitive up to 1...1.1 μm. For the spectral range II (wavelengths 1.3 and 1.5 μm, used in the lightwave communication technologies), APDs have been designed on the basis of germanium, and triple compounds GaInAs – InP (the last-named material ensures a lower density of the dark current of PD: 10^{-8} A/mm^2 as compared with 10^{-6} A/mm^2 for germanium-based PDs). The disadvantage of these materials lies in the close values of the coefficients of impact ionization of electrons α and holes β ($k = 0.5$). Because of this, one of the major directions of development of photoelectronics in the past decade has been associated with the search for semiconductor compounds and structures having a low value of k. The search is aimed at synthesizing the materials with the desired band structure (it has been demonstrated that $\beta >> \alpha$ in case of the so-called valence-band resonance); at the development of channel structures, where one of the carrier types is withdrawn from the multiplication region; at the creation of variable band structure and structures exhibiting a potential jump in the bands, so that the energy needed for impact ionization is acquired selectively by one type of carriers [39].

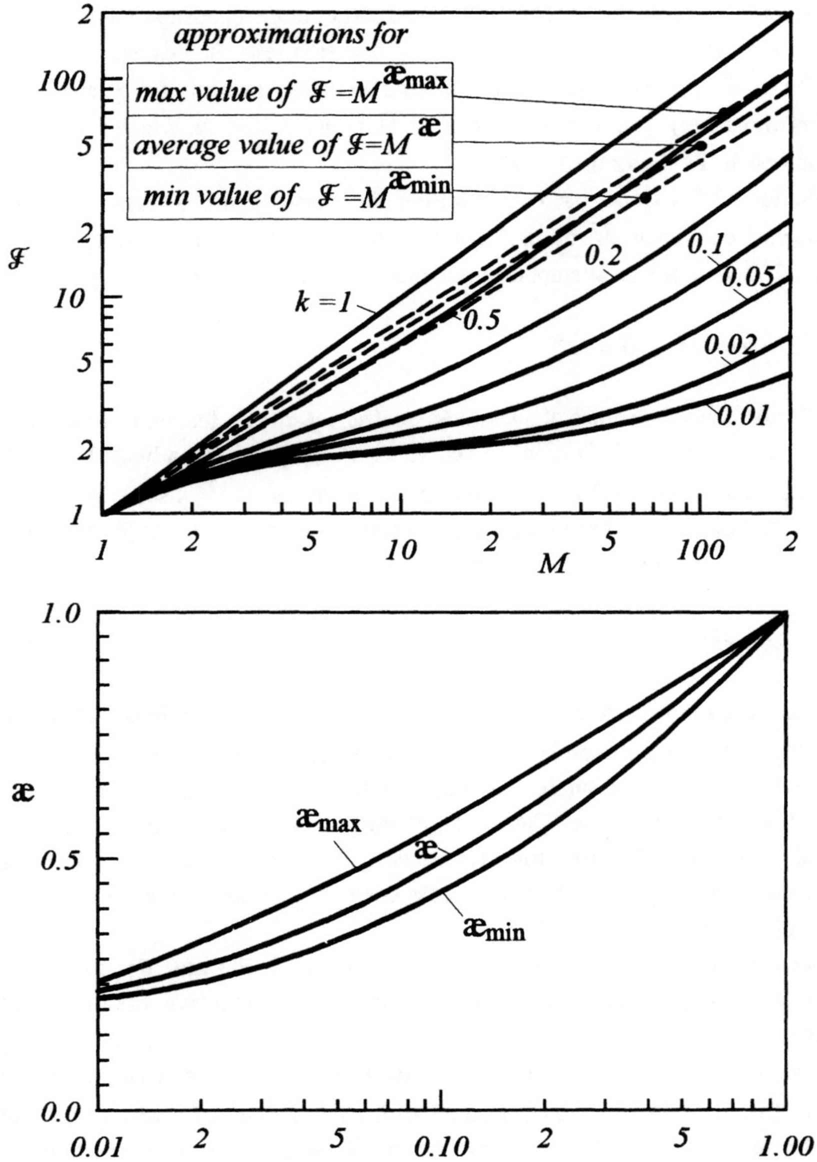

Figure 2.16: (a) Approximation of APD noise factor with a power-law function (dashed line); (b) noise coefficient $æ$ as function of the ratio k.

Photoresistor noise is assessed in a similar way. In APDs, the amplitude of the elementary pulse is increased M-fold; in photoresistors the length of the pulse is multiplied by the factor of τ_{life}/τ_{tr} [Sect. 1.2]. The most important, however, is that in both cases the charge q' carried by the elementary pulse (the pulse area) is increased: it is equal to Mq in APDs, and to $(\tau_{life}/\tau_{tr})q$ in PRs. This is the reason why the expressions for the noise spectrum are similar: we only have to replace in (2.113) the mean square gain of APD $\overline{M^2}$ by the mean square gain of photoresistor $\overline{(\tau_{\text{life}}/\tau_{\text{tr}})^2}$:

$$\overline{J^2(f)} = 2qI_0\overline{\left(\frac{\tau_{\text{life}}}{\tau_{\text{tr}}}\right)^2} = 2qI_0\mathcal{F}\frac{\bar{\tau}^2_{\text{life}}}{\tau^2_{\text{tr}}} = 4qI_0\frac{\bar{\tau}^2_{\text{life}}}{\tau^2_{\text{tr}}} \; . \tag{2.124}$$

Here, as in case of APD, we introduce the noise factor $\mathcal{F} = \overline{\tau^2_{\text{life}}}/\bar{\tau}^2_{\text{life}}$, and use the value $\mathcal{F} = 2$. This value is exactly the same as in case of APD in which ionization is performed by one carrier type ($k = 0$; see (2.120)). This is not by accident: both the multiplication factor of such APD, and the lifetime in PR obey the same exponential distribution law.

The additional amount of gain has to be paid for: because of the extra noise both APDs and PRs are inferior to PDs in terms of the intrinsic SNR (by the factor of $\sqrt{\mathcal{F}}$). The internal gain becomes an advantage when the noise of the amplifier is considerable (see Sect. 1.2, 2.5).

The expression for the noise of photoresistor can also be found in a different way, using the equivalent diagram in Fig. 1.9a (following the guidelines developed for calculating the noise of APD). It was not without reason that we emphasized the existence of two independent shot noise processes in photoresistors, generation and recombination. Each of these processes is represented on the equivalent diagram by a separate generator $2qI_0$ at the input, so that the total power is doubled. At the output the noise power is reinforced by the factor of $K^2_{\text{p}}(f)$, where $K_{\text{p}}(f)$ is the gain of the photoresistor (1.14), (1.27):

$$\overline{J^2(f)} = 4qI_0K^2_{\text{p}}(f) = 4qI_0\frac{\tau^2_{\text{life}}}{\tau^2_{\text{tr}}\left(1+\omega^2\tau^2_{\text{tr}}\right)} \; . \tag{2.125}$$

Equation (2.124) was obtained for the noise density at low frequencies, where it coincides with (2.125), which holds for the entire spectral range. Since the noise pulses in PRs are much longer than in PDs and APDs $\tau_{\text{life}} >> \tau_{\text{tr}}$, one has to account for the fact that the noise is not white: the decline starts at the frequency

$$f = \frac{1}{2\pi\tau_{\text{life}}} = \frac{1}{2\pi\left(10^{-6}...10^{-2}\right)} \approx 15 \text{ Hz}...150 \text{ kHz} \; .$$

CCD-cell noise. As shown elsewhere (see Sect. 1.1, 1.2), the physics of operation of CCD cell is similar to that of a PD, with the difference that the carriers do not escape into the external circuit, but are instead stored by a capacitance. Because of this, the noise of CCD cell are of the same nature as those in a PD (shot noise), and are represented by generator $2qI_0$ connected to the integration capacitance.

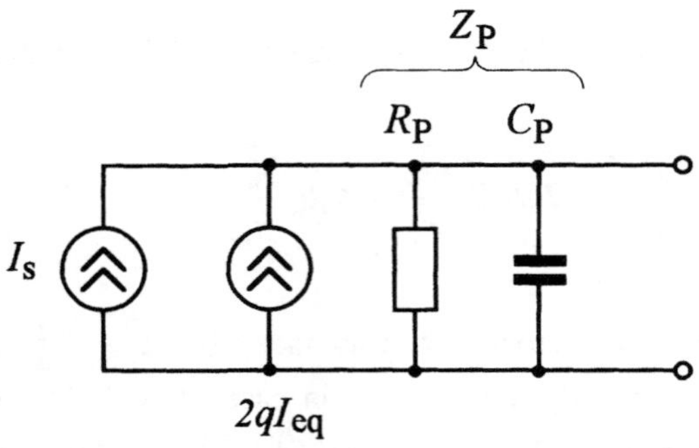

Figure 2.17: On the equivalent diagram of photodetector the noise is represented by generator $2qI_{eq}$.

Noise equivalent diagram of photodetector. The above analysis allows us to supplement the generalized equivalent diagram of photodetector by the noise current generator (cf. Fig. 1.9b and 2.17), representing this generator as

$$\overline{J^2(f)} = 2qI_{eq} \; . \tag{2.126}$$

Comparing the last formula with our earlier expressions (2.82), (2.84), (2.114), (2.123), (2.125), we find the following expressions for the equivalent current I_{eq}:

PD in photodiode mode: $I_{eq} = I_0$; (2.127)

PD in open-circuit mode: $I_{eq} = \dfrac{4\kappa T}{qR_P}$; (2.128)

APD: $I_{eq} = I_0 \overline{M^2 \mathcal{F}} = I_0 \overline{M}^{2+æ}$; (2.129)

PR: $I_{eq} = 2I_0 \dfrac{\tau_{life}^2}{\tau_{tr}^2 \left(1+\omega^2 \tau_{life}^2\right)}$. (2.130)

The current I_{eq} is called the current of equivalent noise diode. The reason is obvious: any source of noise $\overline{J^2}$ can be replaced by the equivalent diode with the back current I_{eq} calculated according to (2.126); the shot noise generated by such diode will be equivalent to that of the replaced noise source.

Summary. *The current pulses produced by individual carriers in PD may be regarded as infinitesimally short, and their spectrum as white. Because of this, the spectrum of PD noise is white; the density of noise power under back bias is given by the known Shottky formula* $2qI_0$, *where* I_0 *is the back current of the PD. In photodetectors with* *internal gain K_p, the noise power increases by a factor greater than K_p^2, because of the fluctuations of the gain itself. The noise factor due to photoresistor gain fluctuations is two; the noise factor due to avalanche multiplication is either two (when the like-sign carriers are multiplied), or equal to the multiplication coefficient (this extreme value of the noise factor is observed when the coefficients of impact ionization of electrons and holes are the same).*

2.4. Noises of PDA

Our next task consists in giving the expressions for the noises of active and passive elements of the amplifier (resistors and transistors), build the noise diagram of the amplifier as a whole, and finally compile a complete inventory of PDA noises. We shall need some patience: compilation of any inventory is a job which is dull and meticulous, but this job is necessary and must be done with care. We must take every component of noise into account, because any of them may turn out to be definitive for the threshold-level characteristics of the PDA. Our burden may be lightened by a consistent and systematic approach (like the rectangular approximation assisted us in the analysis of signal and noise pulses). We shall use the following two simple principles.

1. Different components of the PDA noise will be referred to the PDA input.

2. We shall use parallel equivalent diagrams, which represent the circuit as a parallel connection of passive and active elements (resistors, capacitors, current generators), like we did in case of photodetectors. The reason is clear: the conductivities and the currents in such circuit are simply added together.

Thermal noise of resistor. This noise, along with the shot noise, is the main inherent source of noise not only in PDAs, but in any other electronic device. Our general technical culture obliges us to have a notion about the origins of these two kinds of noise. The nature of the shot noise and thermal noise is different, but the opposites unite: following the guidelines developed in [57], we shall demonstrate the similarities between the mechanisms of thermal and shot noise.

Figure 2.18a shows a resistor of length L, cross section A, resistivity ρ, total resistance $R = \rho L/A$. By analogy with the diode, we short the terminals 1–2 of the resistor across the external circuit with infinitesimally low equivalent resistance, and calculate the noise of the current in the external circuit. The thermal noise of resistor, like the shot noise, is caused by the discrete nature of the charge. The corpuscules (electrons) perform an unceasing thermal motion. On the observation interval T we single out one elementary act: the displacement of one electron to the distance l_t, which is the free path length between two consecutive collisions (in order to separate this elementary displacement from the general chaos, we "freeze" for a moment all the other electrons).

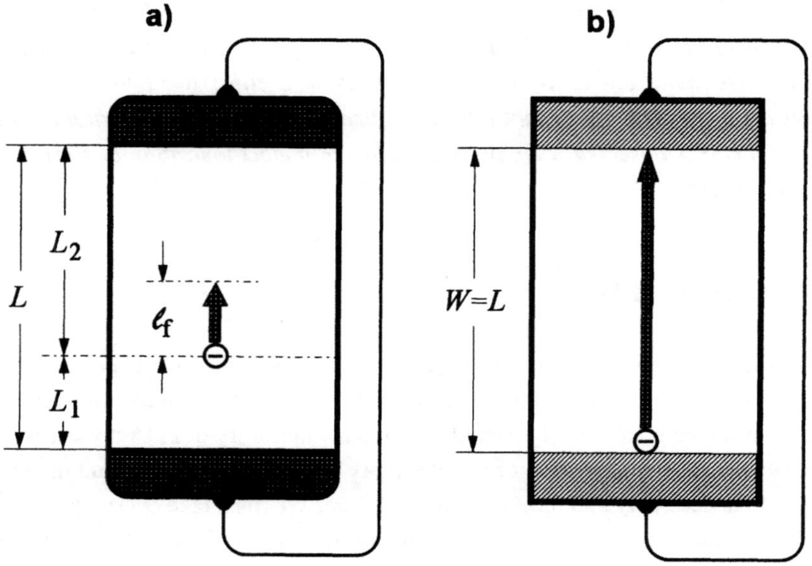

Figure 2.18: Thermal motion of electron in the resistor induces a noise current pulse.

The displacement of the charge q between the interconnected terminals 1–2 produces the current $I(t)$ in the external circuit, and, according to the modified Rameau theorem (2.86), the charge Q carried by this current is proportional to the displacement $l = l_t$:

$$Q = \int_0^\infty I(t)dt = \frac{ql_t}{L} \ . \tag{2.131}$$

We see that the nature of the random pulses in a diode and in a resistor is the same: they are caused by the motion of electron (charge carrier). To facilitate the comparison, Fig. 2.18b shows the passage of electron across the diode, whose space charge region length W is made equal to the length of photoresistor L. The difference consists primarily in the path the electron covers per one elementary act. In a diode, as in vacuum, the electron traverses the entire space charge layer from contact to contact, and $l_t = L$. In a resistor such feat is prevented by the presence of other mobile carriers, so that the free path length (between two consecutive collisions) is a random quantity, and is much smaller than the distance between the contacts L. Of course, the nature of forces which drive the carriers, is different. In the bulk of resistor (Fig. 2.18a) the mobile carriers are at equilibrium, and perform chaotic thermal motion. In the bulk of diode (in the SCR; Fig. 2.18b), there are no equilibrium carriers: each generated carrier is immediately withdrawn by the field into the external circuit. In the induced elementary noise pulse, however, these differences are only revealed in the free path length (l_t or $L = W$).

Another distinction concerns the number of elementary pulses (elementary "flight legs"). In a diode it was determined by the magnitude of the current. In a resistor this number is defined in a different way: the mean time between collisions is $\overline{\tau_t}$, and so, during the observation interval T, the electron is displaced on the average $\overline{N} = T/\overline{\tau_t}$ times, and induces $\overline{N}$ pulses. All told, there are nAL electrons in the resistor (where n is their bulk concentration), and all of them induce pulses. The total number of pulses $\overline{N}(nAL)$ is similar to the flux of electrons (carriers) I_0T/q across the diode over the observation time T. Now, if the mechanism which causes the noise pulses is the same in a diode and in a resistor, the method of calculation of the spectral density of the noise should also be the same. We follow the procedure of Sect. 2.3, where we calculated the shot noise. First of all, we state that pulses induced in the resistor are short: the free path time of the electron is $10^{-12}...10^{-15}$ s [69]. The spectrum of such a short pulse is white, and its density is twice the pulse area (Sect. 2.1):

$$\mathrm{I_n}(f) = 2q\frac{l_t}{L} . \tag{2.132}$$

Next we find the spectral density of the noise from one elementary pulse on the observation interval; by definition (2.32) it is

$$\overline{j^2(f)} = \frac{\overline{\mathrm{I_n^2}(f)}}{2T} = \frac{2q^2}{T}\frac{\overline{l_t^2}}{L^2} . \tag{2.133}$$

For all $\overline{N}(nAL)$ pulses the density is $\overline{N}(nAL)$ times greater (since the pulses are statistically independent):

$$\overline{j^2(f)} = \overline{N}(nAL)\frac{2q^2}{T}\frac{\overline{l_t^2}}{L^2} = \frac{T}{\overline{\tau_t}}(nAL)\frac{2q^2\overline{l_t^2}}{TL^2} = \left(\frac{2q^2An}{L}\right)\frac{\overline{l_t^2}}{\overline{\tau_t}} . \tag{2.134}$$

We have almost derived the desired formula for thermal noise. So far it involves the microscopic parameters $\overline{l_t^2}$, $\overline{\tau_t}$, which must obviously be replaced by the macroscopic ones. For this we shall need the following four relations from the molecular theory of gases (as applied to the electrons) [70, 71]:

$$\overline{v_t^2} = \frac{\overline{l_t^2}}{\overline{\tau_t^2}} , \tag{2.135}$$

$$\overline{v_t^2} = \frac{\kappa \mathrm{T}}{m} , \tag{2.136}$$

$$\overline{\tau_t} = \frac{m\mu}{q} , \tag{2.137}$$

$$\frac{\overline{\tau_t^2}}{\overline{\tau}_t^2} = 2 \ , \tag{2.138}$$

where m is the mass of electron (charge carrier), and v_t is its thermal velocity. It is not difficult to recall how these relations are derived.

The first follows from the ordinary equation of motion, which links the path (the free path length) with the thermal velocity v_t and the free path time τ_t: $l_t = v_t \tau_t$; hence, $l_t^2 = v_t^2 \tau_t^2$. Since v_t and τ_t are statistically independent, their averaging is independent too, $\overline{l_t^2} = \overline{v_t^2}\, \overline{\tau_t^2}$. The second equation (2.136) follows directly from the definition of the temperature via the kinetic energy of the particle: $m\overline{v_t^2}/2 = \kappa T/2$. The third equation (2.137) follows from the definition of mobility μ. In the field $\mathcal{E}$ the electron is acted upon by the force $q\mathcal{E}$. Its momentum over the mean free path time $q\mathcal{E}\overline{\tau_t}$ increases the momentum of the particle on the average by $m\Delta\bar{v}$. The mobility is the ratio of the increment of the mean velocity $\Delta\bar{v}$ along the field to the strength of the field:

$$q\mathcal{E}\overline{\tau_t} = m\Delta\bar{v} \ , \quad \mu = \frac{\Delta\bar{v}}{\mathcal{E}} = \frac{q\overline{\tau_t}}{m} \ . \tag{2.139}$$

The last relation (2.138) is a consequence of the exponential distribution $\rho(\tau_t)$ of the free path time [71]. For the third time in this book we encounter such distribution, and for the third time state that $\overline{x^2}/\bar{x}^2 = 2$ (2.138).

Now, using (2.135) – (2.138), we replace the microscopic parameters with the macroscopic ones:

$$\frac{\overline{l_t^2}}{\overline{\tau_t}} = \frac{\overline{l_t^2}}{\overline{\tau_t^2}}\overline{\tau_t}\frac{\overline{\tau_t^2}}{\overline{\tau}_t^2} = \overline{v_t^2}\,\overline{\tau_t}\frac{\overline{\tau_t^2}}{\overline{\tau}_t^2} = \frac{\kappa T}{m}\frac{m\mu}{q}\cdot 2 = 2\kappa T\frac{\mu}{q} \ . \tag{2.140}$$

Substituting this into the expression for the spectral density (2.134), we arrive at the famous Nyquist formula for thermal noise:

$$\overline{J^2} = \left(\frac{2q^2 nA}{L}\right) 2\kappa T\frac{\mu}{q} = 4\kappa T\cdot qn\mu\frac{A}{L} = 4\kappa T\frac{A}{\rho L} = \frac{4\kappa T}{R} \ . \tag{2.141}$$

Now, how does the spectral density of thermal noise depend on the resistance? The designer of PDA will reply that it is inversely proportional to R, whereas the radio engineer will argue that the spectral density of thermal noise is directly proportional to R. Both are right: all depends on how the thermal noise is represented on the equivalent diagram. Equation (2.141) describes the noise current generator $\overline{J^2} \propto 1/R$,

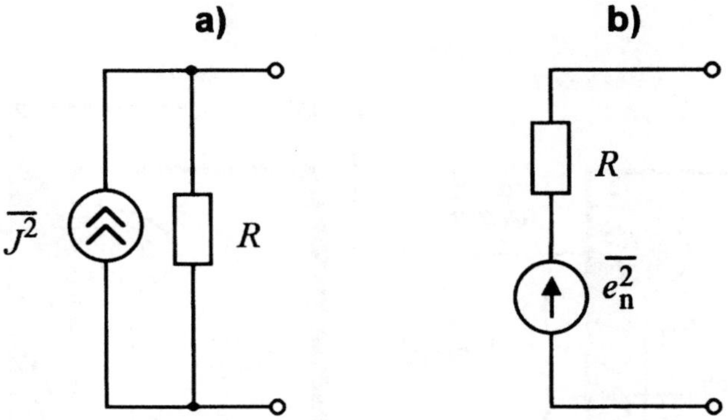

Figure 2.19: Equivalent noise diagrams of a resistor.

which on the equivalent diagram is connected in parallel to the resistor (Fig. 2.19a). In case of short circuit, all the noise current flows into the external circuit. This current generator $\overline{J^2}$ with the internal resistance R can be replaced by the noise emf generator $\sqrt{\overline{e_n^2}} = R\sqrt{\overline{J^2}}$ with the serial resistance of the same magnitude (Fig. 2.19b):

$$\overline{e_n^2} = \overline{J^2}R^2 = \left(\frac{4\kappa T}{R}\right)R^2 = 4\kappa T R \ . \tag{2.142}$$

This representation of the Nyquist formula is more familiar; here $\overline{e_n^2} \propto R$.

One final remark. The thermal noise and the shot noise are related not only by the mechanism of induction of random one-electron pulses. There is an interesting situation when they become simply indistinguishable. This happens when the diode is operated in the open-circuit mode. Then the spectral density of the noise is the same by both the Shottky formula (2.84), where the forward and the back currents are taken into account ($4qI_0$), and the Nyquist formula (2.141) ($4\kappa T/R_{pOC}$), since $R_{pOC} = \kappa T/qI_0$ (1.76).

FET noise. The simplest structure of a field-effect transistor (FET) is shown in Fig. 2.20a: the n^+ regions of the source and the drain are located on the p-type substrate. The source and the drain are connected electrically through the resistance of the channel, a thin high-resistance n-layer. The common electrode is the source; the input electrode is the gate, and the output electrode is the drain. In the initial state the potentials of electrodes are zero (or a small back bias is applied to the gate). As the voltage U_D is applied to the drain, the drain current $I_D = U_D/R_C$ starts to flow, where R_C is the channel resistance in the initial state (Fig. 2.20a). The voltage U_D is distributed along the channel, back-biasing the latter with respect to the bulk (substrate); the SCR of the bulk-channel junction broadens and penetrates the channel. The bias is the strongest ($\sim U_D$) near the drain, and at a certain cut-off voltage $U_D = U_{co}$ the SCR extends to the entire cross-section of the channel (Fig. 2.20b). This is a characteristic point of the

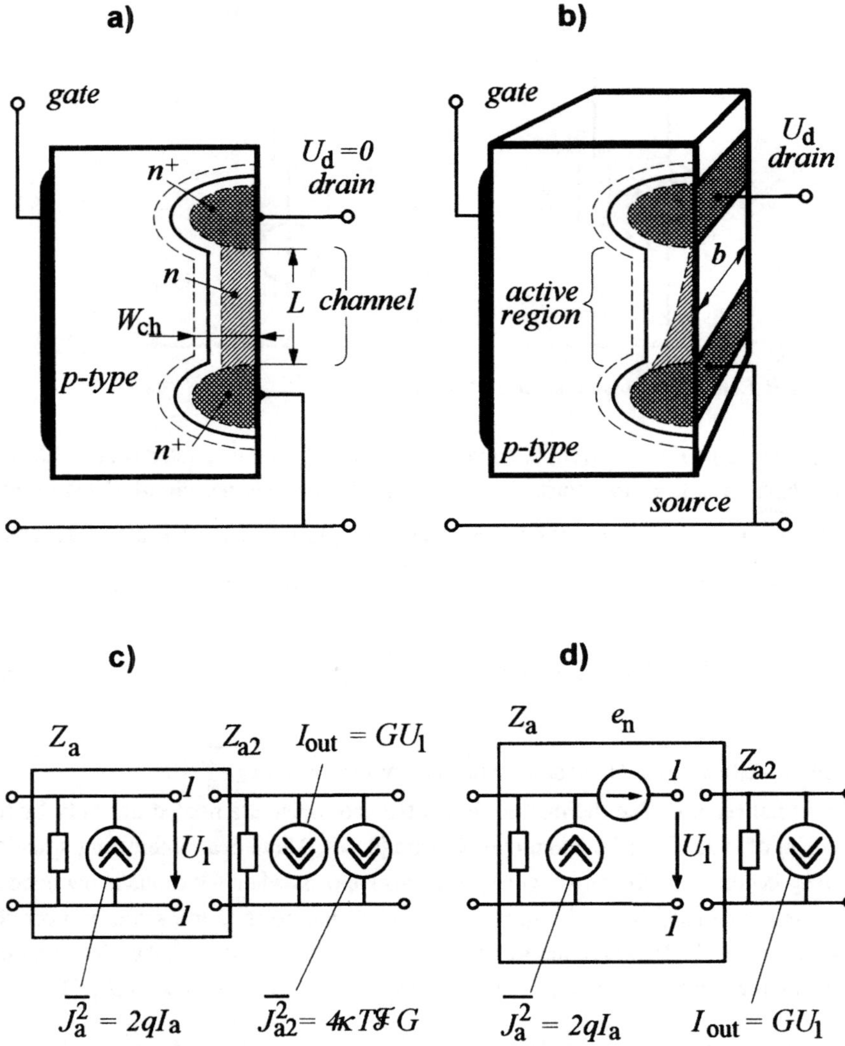

Figure 2.20: The noise of FET is the shot noise of junction and the thermal noise of the channel. Conducting regions are hatched (a,b).

device: the channel is cut off from the drain, and the voltage applied to the channel cannot be higher than U_{co} even though U_D may be further increased. The difference $(U_D - U_{co})$ drops across the narrow depleted SCR at the drain-channel boundary. Since the voltage across the channel is stabilized at the level of U_{co}, the drain current is also stabilized at the level of $\sim U_{co}/R_C$.

FET amplification property is based on the modulation of channel resistance by the voltage applied to the gate. By applying the bias $U_G = U_{co}$ to the gate, we deplete the channel over its entire length, and disconnect the source from the drain (the off state of the transistor), so that the drain current falls to zero. The increment of the input voltage

$U_1 = |\Delta U_G| = U_{co}$ causes the increment of the output current, the drain current $|\Delta I_D| = I_D$, and so the transconductance is

$$G = \frac{|\Delta I_D|}{|\Delta U_G|} = \frac{U_{co}}{R_C}\frac{1}{U_{co}} = \frac{1}{R_C} . \quad (2.143)$$

In other words, the transconductance is determined by the conductance of the channel. Figure 2.20c shows the equivalent diagram of FET with respect to alternating current. The increment of the output current $I_{out} = \Delta I_D$ is represented by the current generator GU_1; Z_a and Z_{a2} are the input and the output impedances, respectively (the effects of the input-output impedance will be discussed later).

We had a good reason for describing the structure of FET and recalling the definition of transconductance: this will help us to find the expression for the noise of this device. From the structure of FET (Fig. 2.20b) we see that it contains two sources of noise. The p-type gate forms a back-biased junction with the n-regions of drain, channel, and source. Accordingly, through the gate flows the back saturation current of this junction, the input leakage current I_a, which causes shot noise $\overline{J_a^2}$ at the input. At the output, the source-drain path includes the channel resistance, which is the source of thermal noise. So,

$$\overline{J_a^2} = 2qI_a , \quad (2.144)$$

$$\overline{J_{a2}^2} = 4\kappa T\frac{\mathcal{F}}{R_C} = 4\kappa T\mathcal{F}G . \quad (2.145)$$

In the last expression we have taken due account of the linkage between the channel resistance and the transconductance (2.143). Coefficient $\mathcal{F}$ takes care of the nonuniform channel thickness. In Fig. 2.20b it varies from W to zero, so its mean value may be taken to be $W/2$. Accordingly, the conductivity of the channel in this mode is one-half of that in the initial regime (cf. Fig. 2.20a), and equals $1/2R_C$ (that is, $\mathcal{F} = 1/2$). Precise calculations, taking the nonuniform thickness of the channel into account, confirm the expressions for transconductance (2.143) and thermal noise (2.145), obtained here from general considerations.

Coefficient $\mathcal{F}$ may be regarded as the noise factor of FET. It increases somewhat when the gate is back-biased; its maximum value is 2/3. Formula (2.145) also holds for other FET structures. For a FET with MIS gate the coefficient $\mathcal{F}$ for thermal noise of the channel does not depend on the bias applied to the gate, and is equal to 2/3 [62, 72].

Unfortunately, it is far not always possible to attain such a low value of noise factor with FETs. At low frequencies, the generation-recombination noise may come into play along with the thermal noise, like in case of photoresistor. This time its mechanism is somewhat different: the concentration of carriers may be changed because of capture of carriers by traps in the SCR, which alters the thickness of the space charge region. This is equivalent to random variations of voltage applied to the gate [73]. In some structures and in some modes (narrow channel, lowered temperatures, strong bias) the mobility may come to saturation, also resulting in the increased noise factor, which may become greater than one [74].

Now we have to stick to the principles proclaimed in the beginning of this section, and refer the thermal noise of the channel $\overline{J_{a2}^2}$ to the input — that is, replace this current generator by the emf generator $\overline{e_n^2}$ at the input, which means going over from the diagram in Fig. 2.20c to the diagram in Fig. 2.20d. The emf generator produces additional voltage e_n at the input, which causes additional drain current Ge_n. Equating this current to the current of the replaced noise generator $\overline{J_{a2}^2}$ (2.145), we get

$$Ge_n = J_{a2}\ , \quad \overline{e_n^2} = \frac{\overline{J_{a2}^2}}{G^2} = \frac{4\kappa T \mathcal{G} G}{G^2} = 4\kappa T \frac{\mathcal{G}}{G}\ . \tag{2.146}$$

Numerical evaluation of the noises of FETs used in PDAs will be given in the next section.

Noises of bipolar transistor. The structure of bipolar transistor (BT) is shown schematically in Fig. 2.21a. The electrons injected by the forward-biased emitter diffuse across the thin base layer, and almost all of them reach the collector. Only a small part of the electrons recombine in the base, causing the base current I_b. The collector current is about equal to the emitter current, $I_c \approx I_e >> I_b$. The equivalent low-signal common-emitter diagram is shown in Fig. 2.21b. We are going to use the following notation: input current $I_1 = dI_b$, output current $I_2 = dI_c$, transfer coefficient (current gain) $h_{21e} = I_2/I_1 = dI_c/dI_b$. We may point out three features which distinguish the equivalent diagram of BT from that of FET.

1. The input resistance R_a of a FET is an uncontrollable resistance of the back-biased gate junction, whereas the input resistance R_a of a BT is known exactly, and is the dynamic resistance of the forward-biased emitter junction (more precisely, that part of the junction from which the current flows into the base electrode). The expression for the dynamic resistance of the forward-biased junction has been derived earlier (see (1.76)). Setting in (1.76) $I = I_b >> I_{dg}$, $I_{db} = 0$, $I_{pb} = 0$, we get

$$R_a = \frac{U_1}{I_1} = \frac{dU}{dI_b} = \frac{\varphi_t}{I_b}\ , \quad \varphi_t = \frac{\kappa T}{q}\ . \tag{2.147}$$

2. FET is a voltage-current converter, whereas a BT is a current amplifier. Since the input resistance of the latter is precisely known, it is easy to calculate the variation of the output current I_2 caused by the given variation of the input voltage $U_1 = R_a I_1$:

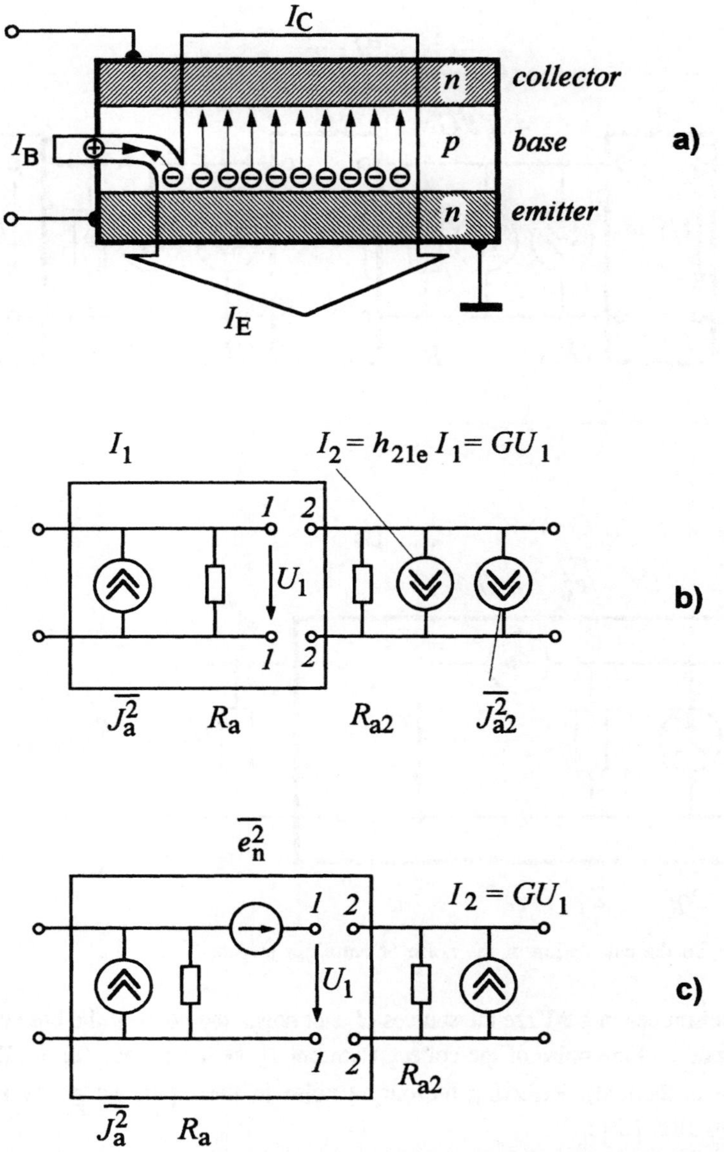

Figure 2.21: The noise of BT is the shot noise of base current I_B and collector current I_C.

$$I_2 = h_{21e} I_1 = \frac{h_{21e}}{R_a} R_a I_1 = GU_1 \ , \tag{2.148}$$

$$G = \frac{h_{21e}}{R_a} = \frac{h_{21e} I_b}{\varphi_t} = \frac{I_c}{\varphi_t} = \frac{qI_c}{\kappa T} \ . \tag{2.149}$$

The transconductance of BT turns out to be h_{21e} times its input conductivity $1/R_a$; recall that the transconductance of FET is equal to its output conductivity (before the cut-off).

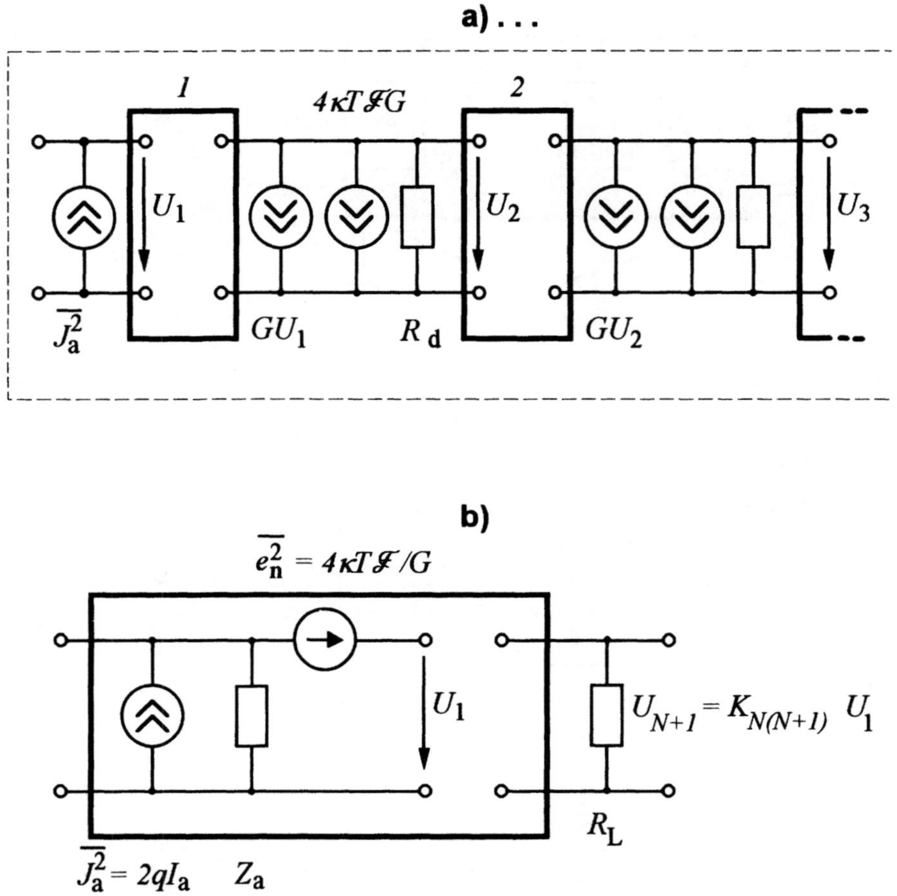

Figure 2.22: To the calculation of the noise of amplifier stages.

3. Both noise sources in a BT are the sources of shot noise: the noise of the base current I_b at the input, and the noise of the collector current I_c at the output (in a FET the output noise is thermal). Referring the output noise to the input, we go over from Fig. 2.21b to Fig. 2.21c:

$$\overline{J_a^2} = 2qI_b ,$$

$$\overline{e_n^2} = \frac{\overline{J_{a2}^2}}{G^2} = \frac{2qI_c}{G^2} = 4\kappa T \cdot \frac{qI_c}{\kappa T} \cdot \frac{1}{2} \cdot \frac{1}{G^2} = 4\kappa T \frac{\mathcal{F}}{G} , \tag{2.150}$$

$$\overline{e_n^2} = 4\kappa T \frac{\mathcal{F}\varphi_t}{h_{21e} I_b} , \quad \mathcal{F} = \frac{1}{2} . \tag{2.151}$$

. . . a)

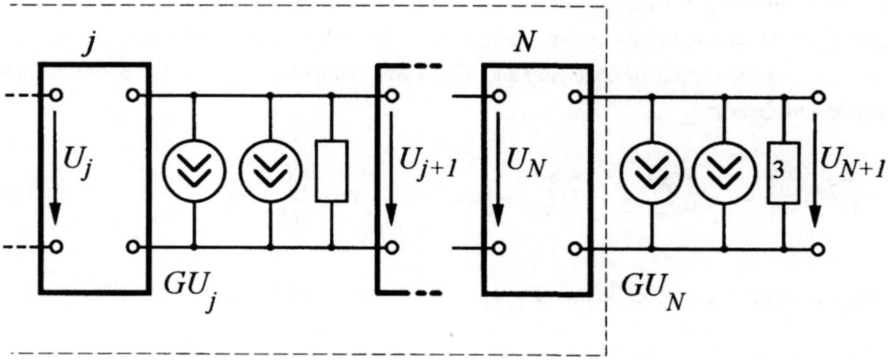

In the transition from (2.150) to (2.151) we have used expression (2.149) for the transconductance. We see that the FET and the BT can be represented by the same equivalent diagram: a four-pole with the transconductance G (conductance-dimensional) of the voltage-to-current conversion, and two noise generators: the shot noise of the input current, and the emf of the transconductance noise. Ideally, the noise factor is $\mathfrak{F} \approx 0.5...0.66$. This equivalent diagrams also applies to other active amplifying elements which may be represented as four-poles, such as radio valves. Although we have managed to represent FET and BT by a unified equivalent diagram, in a BT both noise generators depend on the same parameter, the base current I_b, which is a nuisance when it comes to minimizing the PDA noise. This factor will play a definitive role in the comparison of the noise characteristics of the two types of transistors in the threshold-level PDAs.

Intrinsic noise of amplifier is determined by noise of input transistor. Knowing the noises of active and passive elements of the amplifier (transistors and resistors), one can calculate the noise of amplifier as a whole. This calculation, however, is usually not necessary: in a properly designed amplifier the intrinsic input noise is practically the same as the noise of the first-stage transistor. For the sake of simplicity, let us consider an amplifier built up of N identical sections (Fig. 2.22). On the diagram, $Z_{a2} = R_{a2} = R_D$ is the active load at the output of the amplifier (in the drain circuit), with $R_D << R_a$. The voltage gain is the same for all stages; so, for one, j, and N stages the gain is

$$j = 1...N \ ,$$

$$K_{21} = \frac{U_2}{U_1} = GR_D \ , \ K_{(j+1)1} = \frac{U_{j+1}}{U_1} = (GR_D)^j \ , \ K_{(N+1)1} = (GR_D)^N \ . \qquad (2.152)$$

Two noise sources operate at the output of each stage: the thermal noise of the load

$4\kappa T/R_D$, and the output noise of the transistor $4\kappa T \mathcal{G} G$ (here it is more convenient to the equivalent diagram of Fig. 2.20c, where this noise is not referred to the input). The input shot noise $\overline{J_a^2} = 2qI_a$ is only taken into account for the first stage, because in the subsequent stages it is negligible compared to the other two noise components (see Sect. 2.5). At the output of each jth stage these two components cause the same voltage drop (in unit band):

$$\overline{U_2^2} = \overline{U_3^2} = \ldots = \overline{U_{j+1}^2} = \ldots = \overline{U_{N+1}^2} = \left(4\kappa T \mathcal{G} G + 4\kappa T \frac{1}{R_D} \right) R_D^2 . \tag{2.153}$$

Using the voltage gain coefficients $K_{(j+1)j}$, we can refer all these noises to the input:

$$\overline{e_n^2} = \frac{\overline{U_2^2}}{K_{21}^2} + \frac{\overline{U_3^2}}{K_{31}^2} + \ldots + \frac{\overline{U_{N+1}^2}}{K_{(N+1)1}^2}$$

$$= \overline{U_2^2} \left[\frac{1}{(GR_D)^2} + \frac{1}{(GR_D)^4} + \ldots + \frac{1}{(GR_D)^{2N}} \right] . \tag{2.154}$$

The contribution of each subsequent stage to the noise power at the input is by the factor of $1/K_{21}^2 = (1/GR_D)^2$ less than that of the preceding one, so the sum of the contributions of all stages is the sum of geometrical progression. Setting $N \to \infty$, we get the following expression for the noise emf at the input:

$$\overline{e_n^2} = \frac{\overline{U_2^2}}{(GR_D)^2} \frac{1}{1 - \dfrac{1}{(GR_D)^2}} \approx \frac{\overline{U_2^2}}{(GR_D)^2} = \frac{4\kappa T \mathcal{G} G + \dfrac{4\kappa T}{R_D}}{G^2}$$

$$= 4\kappa T \left(\mathcal{G} + \frac{1}{GR_D} \right) \frac{1}{G} \approx 4\kappa T \frac{\mathcal{G}}{G} . \tag{2.155}$$

To ensure the existence of voltage gain, one always chooses $GR_D >> 1$ (usually, $GR_D = 5 \ldots 20$). In this case, as follows from (2.154), (2.155), the second and all subsequent stages (even if their number is infinite) contribute negligibly little to the noise as compared to the noise of the first stage (as long as $\mathcal{G} >> 1/GR_D$), which was the thing to be proved. By virtue of arguments developed above, at the input of the four-pole on the equivalent diagram of the amplifier we may retain only the noise sources of the first-stage (input) transistor (Fig. 2.22b). In future it will be convenient to calculate the output voltage using the voltage gain coefficient for the amplifier as a whole, denoting $K_{(N+1)1} \Rightarrow K_{21}$, $U_{N+1} \Rightarrow U_2$.

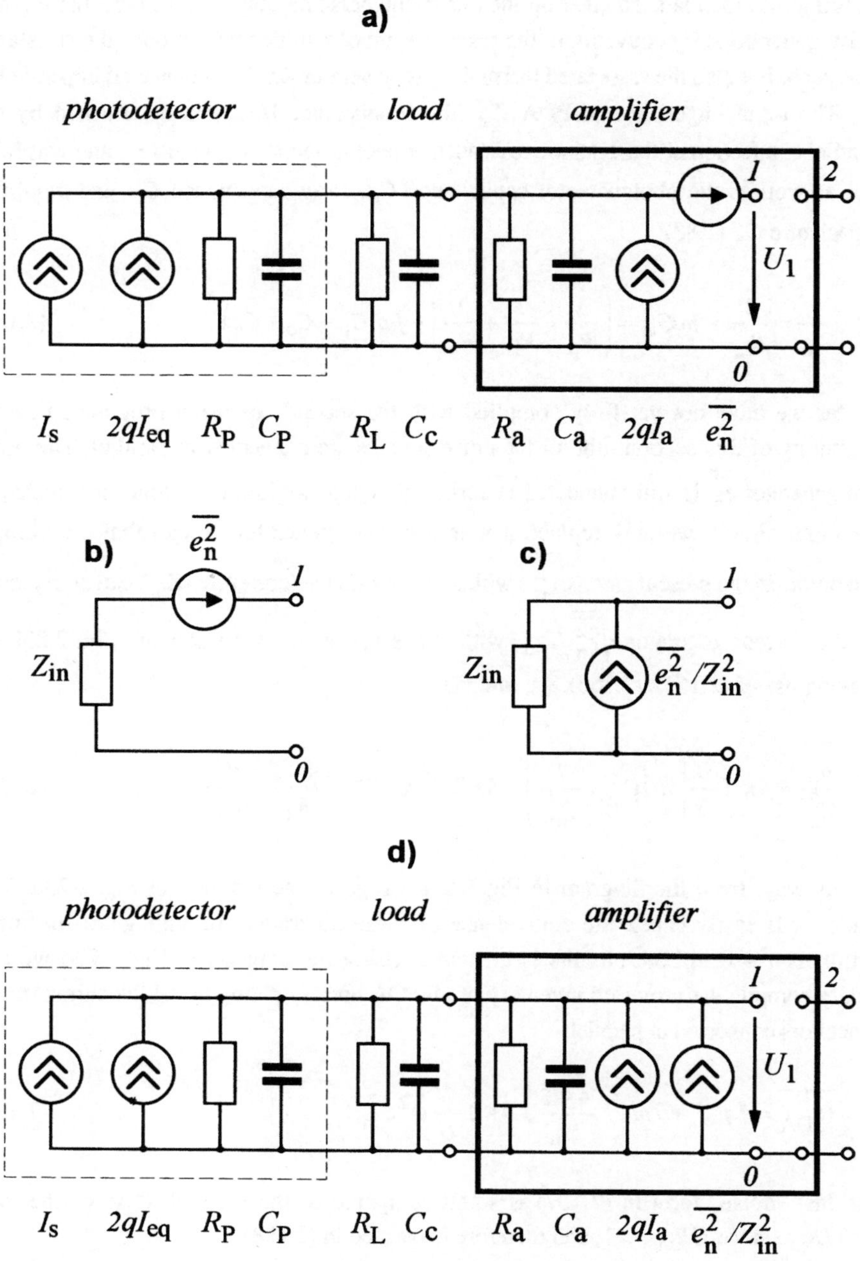

Figure 2.23: The last two steps in the construction of equivalent diagram of PDA: putting the two halves (photodetector and amplifier) together (a), and replacing the serial emf generator with the current generator (b,c), we come to the desired circuit (d).

PDA noises. Now is the time to put together the two "halves", the equivalent diagrams of photodetector (Fig. 2.17) and amplifier (Fig. 2.22b), and get the equivalent diagram of PDA as a whole (Fig. 2.23a). This diagram is presented for the case of photodetector working into the load. To clear up the rather cumbersome diagram, we omit the thermal noise generators: by convention, the resistor symbol will denote not only the resistance R as such, but also the associated thermal noise generator $4\kappa\mathrm{T}/R$ connected in parallel.

The input impedance of PDA Z_{in} (the conductance $1/Z_{\mathrm{in}}$) is determined by the parallel connection of the dynamic resistance of photodetector R_{p}, load R_{L}, and amplifier R_{a}, as well as the photodetector capacitance C_{p}, stray capacitance C_{c}, and amplifier capacitance C_{a} (1.88):

$$\frac{1}{Z_{\mathrm{in}}}=\frac{1}{R_{\mathrm{in}}}+j\omega C_{\mathrm{in}}=\left(\frac{1}{R_{\mathrm{p}}}+\frac{1}{R_{\mathrm{L}}}+\frac{1}{R_{\mathrm{a}}}\right)+j\omega\left(C_{\mathrm{p}}+C_{\mathrm{c}}+C_{\mathrm{a}}\right) . \tag{2.156}$$

So far we have not yet fully complied with the second condition proclaimed in the beginning of this section: not all the noise sources are connected in parallel. The noise emf generator $\overline{e_{\mathrm{n}}^2}$ is still connected in series. Now that we know the total impedance of the input circuit, we may replace it with a current generator. Recall that the voltage generator (in the present case, $\sqrt{\overline{e_{\mathrm{n}}^2}}$) with serial resistance (currently, Z_{in}) can be replaced by the current generator $\sqrt{\overline{e_{\mathrm{n}}^2}}/Z_{\mathrm{in}}$ with the same internal resistance (Fig. 2.23b,c). Making use of (2.155), (2.156), we find that

$$\frac{\overline{e_{\mathrm{n}}^2}}{Z_{\mathrm{in}}^2}\approx 4\kappa\mathrm{T}\frac{\mathcal{F}}{G}\left(\omega^2C_{\mathrm{in}}^2+\frac{1}{R_{\mathrm{in}}^2}\right)=4\kappa\mathrm{T}\frac{\mathcal{F}}{G}\omega^2C_{\mathrm{in}}^2+\frac{4\kappa\mathrm{T}}{R_{\mathrm{in}}}\frac{\mathcal{F}}{GR_{\mathrm{in}}} . \tag{2.157}$$

In this way, from the diagram in Fig. 2.23a we go to the diagram in Fig. 2.23d. The reader will easily check the equivalence of these diagrams: the voltage at the input terminals 1 – 0 in both circuits is the same. Using the diagram in Fig. 2.23d we can finally compile the promised inventory of all PDA noises, adding up all the current noise generators connected in parallel:

$$\overline{J_{\mathrm{PDA}}^2}=2qI_{\mathrm{eq}}+2qI_{\mathrm{a}}+\frac{4\kappa\mathrm{T}}{R_{\mathrm{in}}}+4\kappa\mathrm{T}\frac{\mathcal{F}}{G}\omega^2C_{\mathrm{in}}^2 . \tag{2.158}$$

The last "noise" term in (2.157) is small compared to the thermal noise of the load $4\kappa\mathrm{T}/R_{\mathrm{in}}$, since $GR_{\mathrm{in}}>>1$, and therefore is omitted in (2.158).

The inventory of noises runs as follows:

(1) $2qI_{\mathrm{eq}}$, the noise of photodetector (as a rule, it is the shot noise, as well as the excess noise of internal gain, if any);

(2) $2qI_a$, the noise of leakage current of the amplifier (shot noise);

(3) $4\kappa T/R_{in}$, thermal noise of the total impedance of the input circuit;

(4) $\overline{e_n^2}\omega^2 C_{in}^2 = (4\kappa T\mathcal{F}/G)\omega^2 C_{in}^2$, the noise component from the emf of amplifier noise.

Further transformations in the amplifier are linear transformations which are the same for the signal and for all components of the input noise, and so the signal-to-input-noise ratio cannot be changed as a result of these transformations. This applies also to the more sophisticated transformations in feedbacked circuits. This brings us to the following important conclusion:

Feedback cannot change the signal-to-input-noise ratio. Let us consider this statement in greater detail, using the exemplary current-feedbacked circuit (Fig. 2.24a). The feedback four-pole, like the amplifier four-pole, is a voltage-to-current converter. The feedback four-pole, however, is connected "the other way round": it converts the voltage at the output terminals of the amplifier U_2 into the current generator I_{fb}, connected to the input terminals of the amplifier 0 – 1:

$$I_{fb} = G_{fb}U_2 \ , \tag{2.159}$$

where G_{fb} is the transconductance of the feedback circuit.

Let us select in the signal and in the noise realization, which acted at the input of the amplifier before the application of feedback, a certain harmonic $I_s(f) + I_{nnc}(f)$. The input voltage is found by multiplying these components by the constant Z_{in}, the output voltage is this product times the constant K_{21}, and the current generator of the feedback four-pole is found by multiplying the result by the constant G_{fb}:

$$I_{fb} = G_{fb}U_2 = G_{fb}K_{21}U_1 = G_{fb}K_{21}Z_{in}\left(I_s + I_{nnc}\right) . \tag{2.160}$$

In other words, the feedback brings additional amounts of signal and noise to the input, which are mixed, however, in the same proportion as existed before the application of feedback, so that the signal-to-noise ratio remains unchanged. Of course, the absolute values of signal and noise at the input and output of the amplifier will be modified, but modified by one and the same factor:

$$U_1 = Z_{in}I_{fb} + Z_{in}\left(I_s + I_{nnc}\right) = Z_{in}K_{21}G_{fb}U_1 + Z_{in}\left(I_s + I_{nnc}\right) .$$

Hence, after some simple algebra, we get

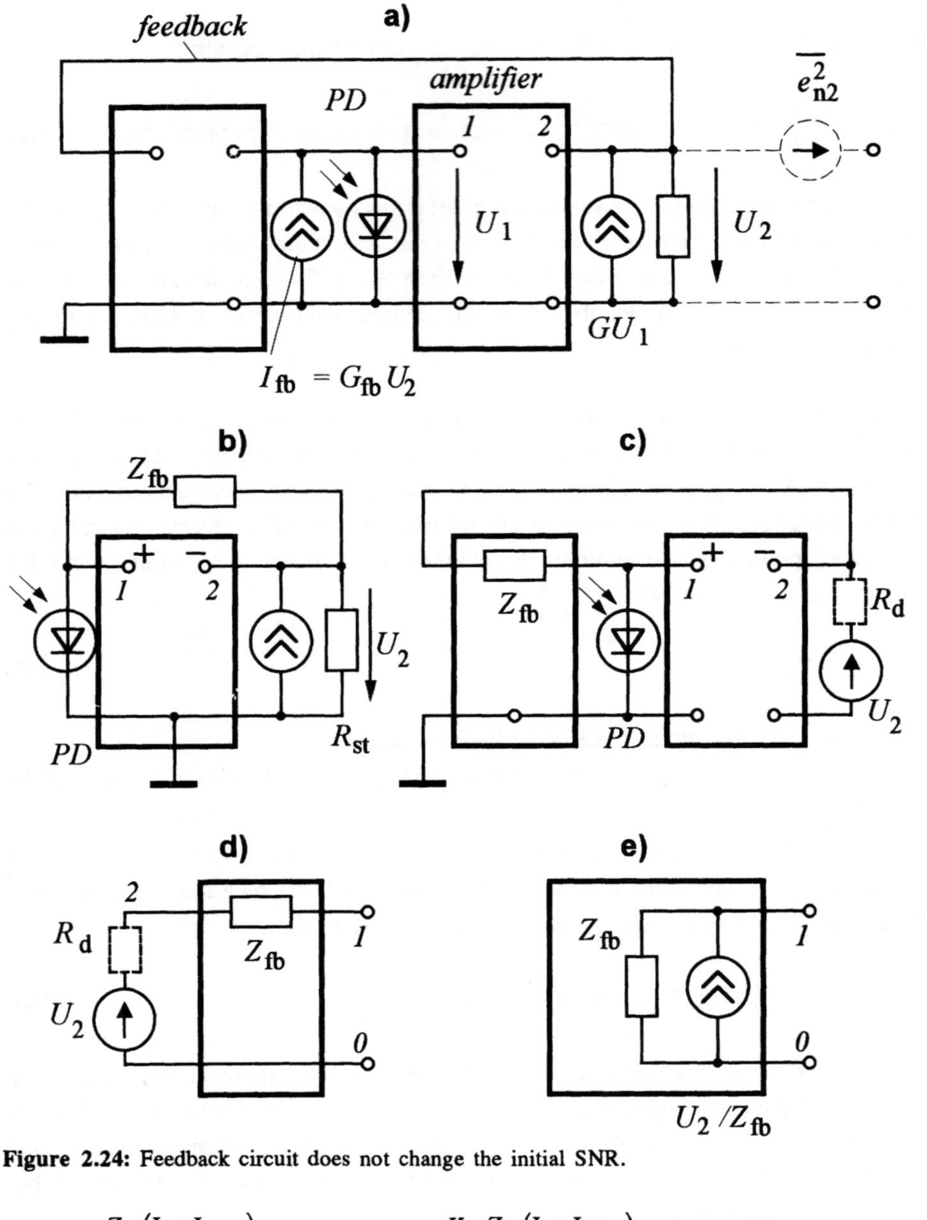

Figure 2.24: Feedback circuit does not change the initial SNR.

$$U_1 = \frac{Z_{in}(I_s + I_{nnc})}{1 - Z_{in}K_{21}G_{fb}}, \quad U_2 = K_{21}U_1 = \frac{K_{21}Z_{in}(I_s + I_{nnc})}{1 - Z_{in}K_{21}G_{fb}},$$

$$U_2 = K_{eff}(I_s + I_{nnc}), \quad K_{eff} = \frac{K_{21}Z_{in}}{1 - Z_{in}K_{21}G_{fb}}, \tag{2.161}$$

where K_{eff} is the transfer coefficient of the feedbacked amplifier.

In Sect. 1.3 we considered the combination of photodetector with operational amplifier; let us reproduce this diagram once again (Fig. 2.24b). Now transform it in accordance with Fig. 2.24c. This representation shows clearly that the feedback four-pole in our case consists simply of a resistor connected between the input (point 1) and the output (point 2). But where in this four-pole is the feedback current generator I_{fb}? A generator is indeed connected to the input of the amplifier, but it is a generator of voltage U_2. The impedance Z_{fb} may be considered as the high internal resistance of this source (Fig. 2.24d) (more precisely, the internal resistance is $R_D + Z_{fb}$, but $R_D << |Z_{fb}|$). We know that the voltage generator U_2 with internal impedance Z_{fb} can be replaced by the current generator $I_{fb} = U_2/Z_{fb} = -|U_{fb}|/Z_{fb}$ with the same internal impedance Z_{fb} (Fig. 2.24e). The minus sign indicates that the feedback is negative (phase-inverting amplifier). In this way, in case of the operational amplifier there is the current generator $I_{fb} = -G_{fb}U_2$ and the impedance Z_{fb} at the input, and the transconductance is equal to the conductance of the feedback circuit, $G_{fb} = -1/Z_{fb}$. Substituting this value into (2.161), for the operational amplifier with negative feedback we get:

$$K_{eff} = \frac{K_{21}Z_{in}}{1 - G_{fb}K_{21}Z_{in}} = \frac{K_{21}Z_{in}}{1 - \left(-\frac{1}{Z_{fb}}\right)K_{21}Z_{in}} = Z_{fb}\frac{K_{21}Z_{in}}{Z_{fb} + K_{21}Z_{in}} \approx Z_{fb} . \quad (2.162)$$

Here we have taken into account the condition of sufficient amplification, $|K_{21}Z_{in}| >> |Z_{fb}|$ (1.95). We obtained the result which is already known to us from Sect. 1.3: the input current – output voltage transfer coefficient is equal to the impedance of the feedback circuit. Now we may add that this applies to any input current, whether signal or noise.

Two amendments must be made to the above statement that the signal-to-input-noise ratio does not change upon application of feedback. As a matter of fact, this statement concerns not the noise in general, but rather the generators which acted at the input before the feedback was applied. Amendment one: a strong enough negative feedback may suppress the gain to such an extent that the noises of the second and subsequent stages will become noticeable (generator $\overline{e_{n2}^2}$ in Fig. 2.24a), spoiling the overall SNR. Amendment two: the feedback four-pole itself is not immaterial, and will introduce its own noise (if only it is not a pure capacitance). In the operational amplifier, for instance, the thermal noise of the feedback resistance R_{fb} will be supplied to the input.

From these arguments it follows that equation (2.158), which defines the relationship between different noise components, remains in force, with the correction that the input resistance and the input capacitance must now be calculated with due account for the feedback resistance R_{fb} and capacitance C_{fb}:

$$\frac{1}{R_{in}} = \frac{1}{R_p} + \frac{1}{R_L} + \frac{1}{R_a} + \frac{1}{R_{fb}} , \quad C_{in} = C_p + C_c + C_a + C_{fb} . \quad (2.163)$$

These amendments are not too restrictive. Usually it is technically feasible to make the gain large enough so as to eliminate the noise of the subsequent stages $\overline{e_{n2}^2}$. Also, the feedback resistance, as shown in Sect. 1.3, is equivalent to the load, and the existence of Z_{fb} makes the load R_L unnecessary (the latter is replaced by the equal resistance R_{fb}).

"Triad" of resistances of operational amplifier. The simplest combination of photodetector – load – amplifier is characterized by the single total impedance of the input circuit: the impedances of photodetector, load and amplifier connected in parallel (2.156). This impedance determines both the input signal voltage U_1, and the noise intensity (the active component R_{in} being responsible for the thermal noise, and the capacitive component C_{in} for the share of the noise emf (2.158)). Our analysis reveals, however, that the circuit with operational amplifier calls for the introduction of three characteristic impedances.

1. First of all, it is the equivalent total impedance Z_{eff}, which determines the input voltage U_1. As shown in Sect. 1.3, the main advantage of the operational amplifier is that Z_{eff} is small: for the operational amplifier itself this impedance is smaller than the feedback impedance Z_{fb} (1.73) by the factor of K_{21}. In the general case, with all the other resistances connected in parallel, the expression for Z_{eff} is

$$\frac{1}{Z_{eff}} = \left(\frac{1}{R_p} + \frac{1}{R_L} + \frac{1}{R_a} + \frac{K_{21}}{R_{fb}}\right) + j\omega\left(C_p + C_c + C_a + K_{21}C_{fb}\right) . \tag{2.164}$$

This formula also holds for the more complicated kinds of feedback. In an amplifier, for instance, there is at least one other negative feedback, due to the feedthrough capacitance C_{ft} of the first stage. Similarly to the feedback impedance Z_{fb}, this capacitance increases the equivalent conductance $1/Z_{eff}$ by the amount of $K'_{21}C_{ft}$, where $K'_{21} = GR_D$ is the gain of the first stage (since this feedback only embraces the first stage of the amplifier).

2. It would be a serious mistake to calculate the noise intensities on the basis of the same values of Z_{eff} and its components R_{eff}, C_{eff} as those used for calculating the input voltage, like we did in case of amplifier without feedback. As shown above, the feedback does not change the ratio between the signal and the noise component. Because of this, the formula for calculating the noise intensity (2.158) involves the total (net) input impedance Z_{in} (2.163) — that is, the input impedance with the disconnected feedback (or, more precisely, with the feedback impedance Z_{fb} connected to the ground bus).

3. The third characteristic impedance of operational amplifier is the feedback impedance Z_{fb}. This is the "pure" load for calculating the output voltage U_2: the impedance Z_{fb} is not shunted by either the photodetector impedance, or the impedance of the amplifier.

These three characteristic impedances may be much different in magnitude, and stand in the following order: $|Z_{eff}| < |Z_{in}| < |Z_{fb}|$.

Excess flicker noise. Our calculations of the noise were based on the following clear-cut guidelines:

1. Encountering a resistance (load, channel) in the PDA, we associated it with the thermal noise $4\kappa T/R$;

2. Encountering a *pn* junction (photodiode, transistor), a process of generation-recombination of carriers in semiconductor, we associated them with the shot noise $2qI_0$;

3. Encountering the existence of internal gain, we used the noise factor $\mathcal{F}$ to take care of its fluctuations.

However, our inventory will be incomplete without the excess flicker noise, the companion of all electronic devices. The spectral density of flicker noise increases as the frequency goes down:

$$\overline{J^2(f)} = \frac{A}{f^{\alpha}} , \tag{2.165}$$

where $\alpha \approx 0.8...1.2$ [57, 62, 75]. Because of this peculiar spectrum, flicker noise is often referred to as the low-frequency noise, or $1/f$ noise. There is no general model of flicker noise, and so it is not possible to relate its intensity (constant A) to other electrical parameters of the device, like this was done in case of shot noise and thermal noise. The constant A may vary broadly from one specimen to another, even though all the other parameters of the specimens are the same. If flicker noise occurs in the photodetector, generator A/f is connected in parallel to generator $2qI_{eq}$:

$$\overline{J^2} = \frac{A}{f} + \left(2qI_{eq} + 2qI_a + \frac{4\kappa T}{R_{in}} \right) + \omega^2 C_{in}^2 \overline{e_n^2} . \tag{2.166}$$

For the sake of definiteness, we set $\alpha = 1$. Flicker noise is especially often encountered in polycrystalline photoresistors (for example, PbS PR), in photoresistors based on triple compounds (MCT); it is also likely to occur in forward-biased or strongly back-biased photodiodes. A FET may also be a source of flicker noise. Equation (2.166) reflects also the case when flicker noise occurs in the FET junction; then the generator A/f is connected in parallel with the generator $2qI_a$.

Let us now analyze expression (2.166) for the noise of PDA.

Noise spectrum of PDA splits into three regions: low-frequency noise ($1/f$), white noise, and high-frequency noise ($\sim f^2$). As a matter of fact, we hardly need to go any further: the result is summarized in the heading of this subsection.

For describing the noise intensity in the plateau (white-noise) region, we introduce the current of the equivalent noise diode I_n; then

$$2qI_n = 2q\left(I_{eq} + I_a\right) + \frac{4\kappa T}{R_{in}} . \tag{2.167}$$

Normalizing the noise (2.166) to its intensity at the plateau (2.136), we find the relative noise spectrum of PDA:

$$\Theta(f) = \frac{\overline{J^2(f)}}{2qI_n} = \frac{A}{2qI_n}\frac{1}{f} + 1 + \frac{(2\pi C_{in})^2 \overline{e_n^2}}{2qI_n} f^2 = \frac{f_{low}}{f} + 1 + \frac{f^2}{f_{high}^2} , \tag{2.168}$$

where

$$f_{low} = \frac{A}{2qI_n} , \quad f_{high} = \frac{\omega_{high}}{2\pi} = \frac{\sqrt{2qI_n}}{2\pi e_n C_{in}} . \tag{2.169}$$

Strange as it is, despite the diversity of PDAs, all these devices exhibit, as a rule, the typical noise spectrum (2.168) with three characteristic regions, as indicated in the heading of this section. The relative spectrum is described by just two parameters, the frequencies f_{low} and f_{high} which mark off, respectively, the limits of low-frequency and high-frequency regions of the spectrum. At these frequency limits the power of the low-frequency noise and the high-frequency noise is equal to the power of the white noise. For the absolute spectrum we must also know the third parameter I_n.

Let us emphasize once again that the spectra of the low-frequency and white noises are defined by the noise spectra of the primary sources. The high-frequency noise $\propto f^2$, however, is associated not with the spectrum of primary source (the noise emf spectrum e_n^2 is white), but with the transformations in the linear circuit. At high frequencies, the input impedance Z_{in} becomes capacitive ($\left|Z_{in}^2\right| = 1/\omega^2 C_{in}^2$), and decreases with the frequency. Therefore, decreases the voltage drop across this impedance, caused by the signal and by the current noise. As a result, the importance of the serially connected noise emf generator increases (in proportion with $\omega^2 C_{in}^2$). Capacitance is usually regarded as a parameter of inertiality. As follows from equation (2.166), it also belongs to the list of photoelectric parameters which defined the noise of PDA (the so-called noise parameters): I_0, R_{in}, I_a, G, C_{in}.

The high-frequency noise comes from all serially connected resistances of the input circuit — for example, the serial resistance of the PD R_s, the resistance of the transistor

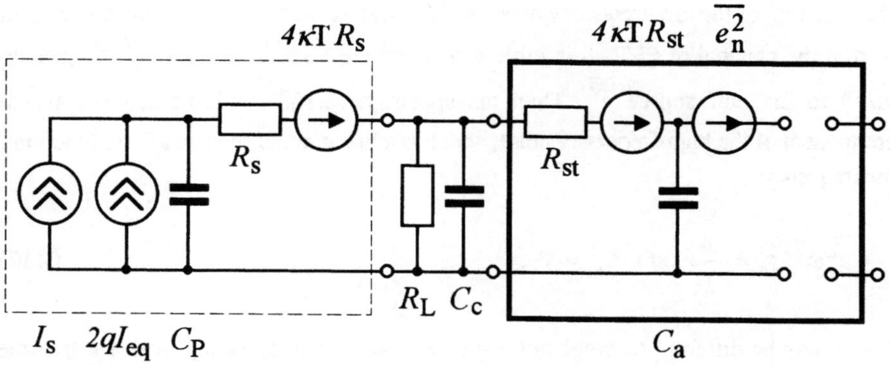

Figure 2.25: Serial emf generators give rise to the f^2 high-frequency noise component.

base R_{st} (Fig. 2.25). At working frequencies, these resistances are much smaller than the capacitive resistances ($R_s << 1/\omega C_p$, $R_{st} << 1/\omega(C_p + C_c)$), and we can, using the already familiar method, replace the emf noise generators $4\kappa T R_s$, $4\kappa T R_{st}$ with the current generators, and add the noise currents of the latter to the total noise current (2.167).

Peculiar is the nature of high-frequency noise in PR-based PDA. In the expression for the input noise of such PDAs we retain two main terms: the noise of photoresistor, and the noise of the load (2.125), (2.166):

$$\overline{J^2(f)} = 2qI_{eq} + \frac{4\kappa T}{R_{in}} = 4qI_0 \frac{\tau_{life}^2}{\tau_{tr}^2\left(1+\omega^2\tau_{life}^2\right)} + \frac{4\kappa T}{R_{in}} \; . \tag{2.170}$$

In this case, the share of the PR noise also decreases at high frequencies, not because of the reduced total input impedance, but rather because of the reduced internal gain. As follows from the last expression, the share of thermal noise increases according to the known ω^2 law. It is possible to obtain the relative spectrum of the form (2.168) by referring this noise to the input of photoresistor — the input leads $1'-1'$ of the resistor amplifier, to which the sources of the signal and generation-recombination noise are connected (see Fig. 1.9a):

$$\left[\overline{J^2(f)}\right]' = \frac{\overline{J^2(f)}}{K_p^2} = \overline{J^2(f)}\ \frac{\tau_{tr}^2\left(1+\omega^2\tau_{life}^2\right)}{\tau_{life}^2}$$

$$= 4qI_o + \frac{4\kappa T}{R}\frac{\tau_{tr}^2}{\tau_{life}^2} + \frac{4\kappa T}{R}\tau_{tr}^2\omega^2 \; . \tag{2.171}$$

And the last. It was not by accident that we said: "PDAs exhibit, *as a rule*, the typical noise spectrum (2.168)". There are exceptions to the rule. In general, for example, the spectrum of the low-frequency noise is $1/f^{\alpha}$ (rather than $1/f$). This noise may occur also in the channel of FET — in other words, we must add the source A/f^{α} (usually, A/f) to the emf source $\overline{e_{\mathrm{n}}^2}$. Then the spectrum (2.168) will include yet another component of the high-frequency noise, which is a linear rather than quadratic function of the frequency:

$$\overline{e_{\mathrm{n}}^2}\omega^2 C_{\mathrm{in}}^2 \Rightarrow \frac{A}{f}(2\pi f)^2 C_{\mathrm{in}}^2 \propto f \ . \tag{2.172}$$

There may be different recombination sites in photoresistors, having different lifetimes τ_{life}. On the basis of the above discussion, the formulas of the detection theory in the forthcoming sections will be given in the general form, valid for arbitrary relative noise spectrum $\Theta(f)$. Exceptions, however, prove the rule: the spectrum of the form $\Theta(f) = f_{\mathrm{low}}/f + 1 + f^2/f_{\mathrm{high}}^2$ remains typical for the majority of situations encountered with PDAs. So, having derived a general formula, we shall also analyse it for all typical cases, when one of the three main noise components prevails (low-frequency f^{-1}, white f^0, or high-frequency f^2).

Summary. *In the general case, the noise properties of PDA are described by the noise generator $2qI_{\mathrm{n}}\Theta(f)$, which connected at the PDA input in parallel with the signal generator I_{s}. As a rule, the relative noise spectrum $\Theta(f)$ splits into three regions: the excess low-frequency noise $\Theta(f) = f_{\mathrm{low}}/f$, the white noise $\Theta(f) = 1$, the high-frequency noise $\Theta(f) = f^2/f_{\mathrm{high}}^2$. Accordingly, the noise is described by three parameters: the frequency f_{low} (which characterizes the frequency limit and the intensity of the low-frequency noise), the current of equivalent noise diode I_{n} (which characterizes the intensity of the white noise), and the frequency f_{high} (which characterizes the frequency limit and the intensity of the high-frequency noise).*

2.5. Minimization of PDA Noises

The task of the PDA is to realize the potential of the photodetector, to "be aware" of its noise. Accordingly, the amplifier must be designed in such a way as to make its noise as small as possible compared to the noise of the photodetector. As follows from the formulas derived in the preceding sections, almost all components of the noise are expressed in terms of the electrical parameters ("the noise parameters") of the circuit elements. Accordingly, it is possible to "design" the noise by adjusting these parameters. The exception is the flicker noise: the spectral density of this noise is determined primarily by the technology rather than by the electrical parameters of the circuit, and almost defies theoretical assessment.

PD-based PDA. As follows from (2.166), the input shot noise of the amplifier (the noise of the leakage current I_a of the transistor), and the thermal noise of the total input resistance R_{in} become equal to the noise of the photodiode at the following values of these parameters (in brackets):

$$[I_a] = I_{eq} = I_0 \; ; \tag{2.173}$$

$$\frac{4\kappa T}{[R_{in}]} = 2qI_0 \, , \; [R_{in}] = \frac{2\kappa T}{qI_0} = \frac{2\varphi_t}{I_0} \, . \tag{2.174}$$

The emf component of the amplifier noise can be characterized by the frequency $[f_{high}]$, at which this component becomes equal to the noise of the photodiode (2.169):

$$\left(2\pi[f_{high}]e_n C_{in}\right)^2 = 2qI_0 \, , \; [f_{high}] = \frac{\sqrt{2qI_0}}{2\pi e_n C_{in}} \, , \; [\omega_{high}] = \frac{\sqrt{2qI_0}}{e_n C_{in}} \, . \tag{2.175}$$

To make all the noises of the amplifier negligibly small compared to the noise of the photodiode, the input load of the amplifier R_{in} must be large, and its current I_a small; small must also be the product $e_n C_{in}$, to ensure that the frequency $[f_{high}]$ is much above the band of the signal:

$$I_a \le \frac{[I_a]}{2...5} \, , \; R_{in} \ge (2...5)[R_{in}] \, , \; [f_{high}] \ge \frac{1}{T_s} \, . \tag{2.176}$$

Under these conditions the noise of the PDA will exceed the noise of the photodiode by no more than 20...40%; they may be considered as the conditions of matching of photodiode and photodetector in terms of noise. It is interesting that these conditions involve the resistance of the PD: as seen from comparison between (2.174) and (2.176), the resistance $[R_{in}]$ is twice the dynamic resistance of photodiode in the open-circuit mode. From the theory of radio engineering we know that the best match between the voltage generator and the amplifier in the sense of the lowest possible value of the noise factor is achieved when the input resistance of the amplifier and the internal resistance of the voltage generator are equal. In our case, the input resistance of the amplifier must be greater than the internal resistance of the signal generator (photodiode). It should be borne in mind that the calculation is based on the resistance φ_t / I_0, which coincides with the actual value only in the open-circuit mode, whereas in the photodiode mode the actual dynamic resistance may be greater than φ_t / I_0 by many orders of magnitude.

Now is it possible to satisfy conditions (2.176). To answer this question, we have to carry out numerical evaluation of $[I_a]$, $[R_{in}]$, and $[f_{high}]$. Let us do this for typical parameters of photodetectors, for all five spectral ranges (Table 2.1). The dark current densities will be taken from Table 1.1. For estimating the values of R_{pSC}, R_{pOC} in

Table 2.1. Requirements to noise parameters of amplifier depend on the spectral range of photodiode

Spectral Range	I	II	III	IV	V	
Wavelength, μm	0.4...1.1	1...2	2...3	3...5	8...10	8...12
Material	Si	Ge	InAs	InSb	MCT	MCT
T, K (design value)	293	293	77	77	77	77
PSE size, mm	0.5dia	0.5dia	0.1×0.1	0.1×0.1	0.1×0.1	0.1×0.1
$[I_a]$, A	$6\cdot10^{-13}...2\cdot10^{-10}$	$(2...6)\cdot10^{-7}$	10^{-10}	$2\cdot10^{-8}$	$(0.6...1)\cdot10^{-6}$	$2.4\cdot10^{-6}$
$[R_{in}]$, Ω	$3\cdot10^{8}...10^{11}$	$(0.8...3)\cdot10^{5}$	$\sim10^{7}$	$5\cdot10^{5}$	$(1...2)\cdot10^{4}$	$6\cdot10^{3}$
$[f_{high}]$, Hz	$2\cdot10^{3}...4\cdot10^{4}$	$(1...2)\cdot10^{6}$	$3\cdot10^{4}$	$5\cdot10^{5}$	$(2...3)\cdot10^{6}$	$5\cdot10^{6}$

Table 1.1 we have already selected the design parameters of photodetector: the site 0.5 mm in diameter (spectral ranges I and II), or 0.1 by 0.1 mm (spectral ranges III – V); quantum yield $\eta = 0.8$, plane angle of view $\alpha = 27^\circ$. Now we use the same values for calculating $[I_a]$, $[R_{in}]$, and $[f_{high}]$. We also assume that the values of input capacitance and noise emf are: $C_{in} = 10\,\text{pF}$, $e_n = 3 \cdot 10^{-9}\ \text{V/Hz}^{1/2}$. Of course, in practice all these values vary broadly, and so the results presented in Table 2.1 should be regarded as order-of-magnitude estimates. This accuracy, however, is good enough to enable us to distinguish three possible situations — three types of PDA which may occur in practice.

Low-frequency PDA with high-resistance PD. Such PDAs are distinguished by the highly restrictive requirements to R_{in} and I_a of the amplifier, because one has to realize the noise of photodiode with ultimately low value of I_d. In our table this case is represented by the silicon-PD-based PDA (spectral range I). It is very hard to make R_{in} as large as required: the value of $R_{in} = (2...5)[R_{in}] = 2 \cdot 10^8 ... 10^{11}$ Ohm is very difficult to achieve, especially in microminiature devices, because the resistance of insulation may become comparable with such values of R_{in}. Also, it is hard to achieve the desired low value of the leakage current $I_a = [I_a]/(2...5) \approx 10^{-13} ... 10^{-10}$ A with junction-gate FETs (the use of MIS FETs, whose leakage current is as low as $I_a < 10^{-13}$ A, is prevented by their large noise emf, $\sim 10^{-7}\ \text{V/Hz}^{1/2}$).

It might seem that obtaining the desired values of I_a with junction-gate silicon FETs is not much of a problem. After all, the nature of the current of PD and that of FET is the same: it is thermal generation in the same region (SCR), in the same material (silicon), while the area of the *pn* junction of FET is even smaller than that of the *pn* junction of PD, whose minimum size is limited by the optical system. Since, however, the voltage across the FET is higher than the voltage across the PD (the latter may be zero), the currents I_0 and I_a may be of the same order. Because of this, it is rather hard to ensure that $I_a < I_0$ with a junction-gate FET. From these arguments it follows that in the low-frequency PDA with high-resistance PD under consideration the white noise is equally likely to be determined by any of the three comparable components: $2qI_0$, $4\kappa T/R_{in}$, $2qI_a$. The high-frequency noise component in such PDAs also becomes considerable at the sonic (ultrasonic) frequencies as low as 2...40 kHz (see Table 2.1); in the presence of the excess noise emf, when $e_n \approx 10...100\,\text{nV/Hz}^{1/2}$, this noise may start to prevail even at 100...200 Hz.

Background limited PDAs. As a rule, such PDAs are based on photodetectors pertaining to spectral ranges III – V, with the relatively large currents ($\sim 10^{-9} ... 10^{-4}$ A), and are primarily intended for systems with mechanical modulation or scanning (IR range finders, IR imagers). The signal spectrum in this case corresponds to the low or medium frequencies, from 10 to 10^5 Hz. The contemporary semiconductor technology is able to ensure that

$$R_{in} = (2...5)[R_{in}] = 10^4...10^9 \text{ Ohm} \ , \ I_a < 10^{-9} \text{ A} \ , \tag{2.177}$$

and the high-frequency noise does not yet descend into this frequency range, because $[f_{high}] \approx 10^5...3 \cdot 10^7$ Hz (Table 2.1). In other words, this is the luckiest situation for the PDA: the PDA fully complies with its mission, and secures the threshold-level properties of the photodetector; all the noise components of amplifier and input circuit are negligible compared to photodetector's own noise $2qI_0$. If, on top of that, the thermal generation current I_{dg} is small, so that the current I_0 is determined by the current due to thermal radiation, as is the case for the spectral ranges III – V in Table 2.1, we get an ideal solution. The noise of PDA is determined by the fundamental ineradicable noise, the fluctuations of the number of thermal background photons falling on the photodetector ("photon noise"). This is the so-called background limited (BLIP) mode, the radiation limit. Neither the photodetector not the PDA can be blamed for the existence of the background noise: they faithfully reproduce the fluctuations of the background radiation coming from the optical system.

High-frequency PDAs. As follows from Table 2.1, the high-frequency noise component $[f_{high}] \lesssim 10^6$ Hz prevails when the signal modulation frequency is high, of the order of 10^7 Hz and more, which is typical for the laser-based systems of spectral ranges I and II. The main methods of reducing the noise consist in choosing the transistor with the noise emf e_n as low as possible (the lowest attainable values will be discussed later), and minimizing the total input capacitance C_{in}. All conceivable ways are good for reducing the input capacitance.

First of all, it is the reduction of capacitance of the photodiode itself by using the *pin* structure, miniaturizing the photosensitive element (reducing the diameter down to 0.1...0.5 mm, further decrease being prevented by the optical system).

It is the selection of the transistor with not only low noise, but also with the smallest possible input and feedthrough capacitances.

It is the reduction of stray ("circuit") capacitance of the input circuit. Recall that small stray capacitance was cited among the advantages of the hybrid-film and solid-state PDA technologies over the devices assembled from separate elements. In Sect. 1.2 we mentioned the PD–FET tandem, designed to reduce the wiring capacitance and the elimination of the casing capacitance.

It would seem that one of the possible ways to reduce the high-frequency noise consists in reducing the effective input capacitance through application of feedbacks. But this is a fallacy. Of course, a positive feedback can reduce the effective input capacitance; in Sect. 2.4, however, we made a point of the existence of the "trinity of impedances": calculation of the high-frequency noise must be based not on the equivalent capacitance, but on the total input capacitance C_{in}, which does not depend on the feedbacks, and builds up of the PD capacitance, stray capacitance of the circuit, input and feedthrough (transfer) capacitance of the amplifier (2.163). Combining the above described techniques, it is possible to make the input capacitance of modern PDAs as low as 5...10 pF, and even 0.5...1 pF in a *pin* PD–FET tandem. With C_{in} as low as that, further struggle goes for the prize of 0.05...0.1 pF — even such a tiny victory may reduce the noise by a considerable amount: $\Delta C_{in}/C_{in} = (0.05...0.1)/0.5 \approx 10...20\%$.

But still, even though the emf-noise and the capacitance may be small, the high-frequency noise component at the high frequencies remains so strong that there is no need to employ high-resistance loads and demand that the currents of photodiode and transistor should be low. Equating the thermal noise of the resistance to the high-frequency noise, and using (2.146) for the emf noise, we find:

$$\frac{4\kappa T}{[R_{in}]} = \omega^2 C_{in}^2 \overline{e_n^2} = \omega^2 C_{in}^2 \frac{4\kappa T \mathcal{F}}{G} \; , \quad R_{in} > [R_{in}] = \frac{G}{\mathcal{F}\omega^2 C_{in}^2} \; . \tag{2.178}$$

Hence follows the estimate for the minimum allowable value of $[R_{in}]$, given the typical values of $f = 10^7$ Hz, $C_{in} = 10^{-11}$ F, $G = 10^{-3}$ A/V, $\mathcal{F} = 0.5$:

$$[R_{in}] = \frac{G}{\mathcal{F}(\omega C_{in})^2} = \frac{10^{-3}}{0.5\left(2\pi \cdot 10^7 \cdot 10^{-11}\right)^2} \approx 5 \cdot 10^3 \text{ Ohm} \; . \tag{2.179}$$

With such a low-resistance load, rather large currents are admissible. Even the shot noise of germanium-based PD with the relatively large current (Table 2.1) will be buried in the high-frequency noise:

$$[I_d] = \frac{2\varphi_t}{[R_{in}]} = \frac{2 \cdot 0.026}{5 \cdot 10^3} = 10^{-5} \text{ A} \; . \tag{2.180}$$

Condition (2.178) restricts the value of the input load from below: if the load is smaller, its thermal noise becomes considerable. In radio engineering the load is often restricted from above, to make the time constant of the RC-circuit correspond to the working frequency band:

$$[R_{in}] C_{in} \omega = 1 \; , \quad R_{in} < [R_{in}] = \frac{1}{\omega C_{in}} \; . \tag{2.181}$$

Both requirements (low thermal noise and the desired time constant) can only be met simultaneously, as follows from (2.178), (2.181), above the following frequency:

$$\frac{G}{\mathcal{F}(\omega C_{in})^2} \le R_{in} \le \frac{1}{\omega C_{in}} \; , \tag{2.182a}$$

$$f > \frac{G}{2\pi\mathcal{F}C_{\text{in}}} = \frac{10^{-3}}{2\pi \cdot 0.5 \cdot 10^{-11}} \approx 3 \cdot 10^7 \text{ Hz} , \tag{2.182b}$$

$$R_{\text{in}} = \frac{1}{\omega C_{\text{in}}} = \frac{1}{2\pi \cdot 3 \cdot 10^7 \cdot 10^{-11}} \approx 500 \text{ Ohm} . \tag{2.182c}$$

We see that it is only for very high signal frequencies (30 MHz, with $C_{\text{in}} = 10^{-11}$ F, $G = 10^{-3}$ A/V) that one can employ direct broad-band amplifier with low-resistance load, which does not delimit the frequency and does not introduce thermal noise. At lower frequencies the two requirements are incompatible. In the threshold-level PDAs one has to sacrifice the frequency band: the load, in accordance with (2.178), is chosen large enough, so that the time constant of the input RC-circuit cuts the high-frequency response. The High-frequency response is corrected by special equalizing stages, or feedbacks, like in operational amplifier.

From the above analysis it is clear that the distinguishing feature we used in the classification of PDAs is the input resistance of the amplifier, required for noise-matching with the photodiode. The first class comprises the high-resistance PDAs — that is, PDAs using a high-input-resistance amplifier ($R_{\text{in}} \sim 10^9 \ldots 10^{11}$ Ohm) (spectral range I). The second class includes PDAs with the medium-valued input resistance ($R_{\text{in}} \sim 10^4 \ldots 10^8$ Ohm) (spectral ranges II – IV). The third class includes the high-frequency low-resistance PDAs ($R_{\text{in}} \sim 10^2 \ldots 10^5$ Ohm) (spectral ranges I – II). This is the reason why in Sect. 1.2 we said that the spectral range and the frequency of modulation define the entire technical design of the PDA, including the design of signal processing electronics. At the same time, in Sect. 1.2 we pointed out that the current $I_{\text{dg}} + I_{\text{db}}$ cannot be unambiguously related to the spectral range: this current may be high in the spectral ranges I – II (for example, because of the solar radiation), and low in the spectral ranges III – IV (with the low background temperature, small angle of view). Accordingly, the linkage between the type of PDA and the spectral range in Table 2.1 is mostly statistical rather than obligatory.

APD-based PDA. Our analysis indicates that PDAs are far from being perfect when the PD current is small, and especially when the modulation frequency is high: the noise of PD is buried in the noise of the amplifier. The only possible way to correct the situation is to take advantage of the internal gain of photodetector. Because of the multiplication in APD, the power of its own noise sharply increases (by the factor of $M^{2+\text{æ}}$; see (2.123)). Accordingly, the share of the other components of the total noise of PDA decreases, and the restrictions may be loosened; then equations (2.173) – (2.175) become

$$[I_{\text{a}}] = I_0 M^{2+\text{æ}} , \quad [R_{\text{in}}] = \frac{2\varphi_{\text{t}}}{M^{2+\text{æ}} I_0} , \quad [f_{\text{high}}] = \frac{M^{(2+\text{æ})/2}\sqrt{2qI_0}}{2\pi e_{\text{n}} C_{\text{in}}} . \tag{2.183}$$

The low level of intrinsic (non-multiplied) noise combined with the high multiplication M and fast response make APDs unrivalled among other single or few-element photodetectors in the registration of weak and short optical signals.

Now, how large must the multiplication coefficient be? On the one hand, the larger it is, the stronger the signal, and the better the suppression of the amplifier noise (2.183). On the other hand, as the multiplication coefficient increases, the noise of APD becomes stronger. This brings us to the problem of optimization of the operating regime — a typical task encountered by the designer of PDA, and, of course, not only the APD-based PDA.

Optimization, optimization, optimization... The noble desire to reduce the total noise of the PDA may drive one into the situation when "what is good for the back is bad for the head". Many parameters of the system are interlocked, and, while reducing one component of the noise, one may unwillingly increase another. In such case there is some optimal value of the parameter, corresponding to the lowest possible total noise. The designer of PDA every now and then has to deal with the problem of optimization by this criterion: choosing the type of active element, its features, operating conditions. Working at reducing the noise, one may run to the limit of allowable operating values; then it will be these values that will define the attainable noise minimum.

Let us consider the main problems of optimization of PDA with respect to the minimum noise (maximum signal-to-noise ratio).

Optimal multiplication coefficient. Let us continue the analysis of APD-based PDA. In Sect. 1.2, devoted to photodetectors, we formulated the main task of multiplication as "withdrawal" of the noise of photodetector above the noise of amplifier (while also reinforcing the signal). Our qualitative analysis revealed the existence of a certain optimal value of the multiplication coefficient M_{opt}, with which the noise of APD becomes equal to the noise of the circuit, so that further increase ($M > M_{\text{opt}}$) is not worthwhile because APD itself becomes too noisy. Now that we have exact formulas for signal and noise of APD and PDA as a whole, we can obtain quantitative definition for the optimal multiplication coefficient M_{opt}.

From the total noise of PDA (2.166) we separate the APD noise component $2qI_{\text{d}}M^{2+æ}$; all the other components pass as the amplifier noise:

$$\overline{J^2(f)} = 2qI_{\text{d}}M^{2+æ} + \overline{J_{\text{a}}^2(f)} \,,$$

$$\overline{J_{\text{a}}^2(f)} = \frac{4\kappa T}{R_{\text{in}}} + 2qI_{\text{a}} + \omega^2 C_{\text{in}}^2 \overline{e_{\text{n}}^2} \,, \quad I_0' = I_{\text{d}} \,. \tag{2.184}$$

Since the signal photocurrent of APD I_{s} is M times its primary (non-multiplied) photocurrent I_{s}' (Sect. 1.2), we can write the signal-to-noise ratio (SNR) in unit band as

$$\mathrm{SNR} = \frac{I_s}{\sqrt{\overline{J^2}}} = \frac{MI'_s}{\sqrt{2qI_d M^{2+æ} + \overline{J_a^2}}} . \tag{2.185}$$

Now we divide the numerator (the signal) and the denominator (the noise) into the multiplication coefficient:

$$\mathrm{SNR} = \frac{I'_s}{\sqrt{2qI_d M^{æ} + \dfrac{\overline{J_a^2}}{M^2}}} . \tag{2.186}$$

The noise in the denominator

$$\overline{J^2} = 2qI_d M^{æ} + \frac{\overline{J_a^2}}{M^2} \tag{2.187}$$

can be given a straightforward physical interpretation: this is the sum of different components of the PDA noise referred to the input of the equivalent APD circuit (before the multiplication stage): to points $1'-1'$ whereto the primary (non-multiplied) photocurrent I'_s is applied (Fig. 1.9a). By definition, the optimal value of multiplication coefficient $M = M_{opt}$ corresponds to the minimum noise $\overline{J^2}$. Before calculating the value of M_{opt} we are going to make a brief digression. As a matter of fact, equation (2.187) is a binomial of the form

$$A(x) = Bx^n + \frac{C}{x^m} , \tag{2.188}$$

and we shall be dealing with such model (and the calculation of its minimum) more than once in this section and in the forthcoming chapters. Because of this, it will be helpful to find the minimum of $A(x)$ in the general form. Setting the derivative equal to zero,

$$\left.\frac{dA}{dx}\right|_{x=x_{min}} = nBx_{min}^{n-1} - m\frac{C}{x_{min}^{m+1}} = 0 , \tag{2.189}$$

and carrying out some straightforward manipulations, we see that at $x = x_{min}$, corresponding to the minimum value of A, the two terms relate as $m : n$:

$$nBx_{min}^{n} \equiv m\frac{C}{x_{min}^{m}} . \tag{2.190}$$

Given this, the minimum value of the binomial is

$$A(x_{\min}) = Bx_{\min}^n + \frac{C}{x_{\min}^m} = \frac{n+m}{m} Bx_{\min}^n = \frac{n+m}{n}\frac{C}{x_{\min}^m}, \tag{2.191}$$

$$x_{\min} = \sqrt[(m+n)]{\frac{m}{n}\frac{C}{B}} . \tag{2.192}$$

The calculation is simplified when $m = n = 1$. Then at $x = x_{\min}$ the two terms of the binomial are equal, and the minimum value of A is twice the value of any term:

$$\frac{dA}{dx} = \frac{d}{dx}\left(Bx + \frac{C}{x}\right)\Bigg|_{x=x_{\min}} = B - \frac{C}{x_{\min}^2},$$

$$Bx_{\min} = \frac{C}{x_{\min}}, \quad x_{\min} = \sqrt{\frac{C}{B}},$$

$$A(x_{\min}) = Bx_{\min} + \frac{C}{x_{\min}} = 2Bx_{\min} = \frac{2C}{x_{\min}} = 2\sqrt{BC} . \tag{2.193}$$

These convenient formulas will be useful for calculating the extremes of expressions involving binomials of the form $Bx^n + C/x^m$.

Now back to the calculation of the optimal multiplication coefficient. According to (2.190), M_{opt} is found from the following condition:

$$n = æ, \quad m = 2, \quad 2qI_d M_{\text{opt}}^{æ} = \frac{2}{æ}\frac{\overline{J_a^2}}{M_{\text{opt}}^2} . \tag{2.194}$$

In other words, the noise is minimal when the noise of APD exceeds the noise of amplifier by a factor of $2/æ$. For "noisy" APDs with $æ = 1$ this factor is equal to two; for the low-noise APDs this factor increases, and is as high as twenty when $æ = 0.1$. For a perfect APD ($æ \to 0$) we have $2/æ \to \infty$, which is obvious: when the noise of APD does not increase, the noise of the circuit is suppressed the better, the higher the value of M.

To calculate the optimal multiplication coefficient (2.194) one needs to know $æ$, the measurement whereof is a painstaking procedure. It is easier to establish the optimal operating mode of the APD directly by going in for the best SNR value. In such case, the signal and the noise are measured while the bias is varied, and then the signal-to-noise ratio is calculated. Often the procedure is simplified even further: monitored is only the noise, and the operating voltage is increased until the noise power of APD exceeds the initial noise power (the amplifier noise) by a preset factor n. Generally, this mode is quasi-optimal, and the corresponding multiplication coefficient M_{qua} is found from the above relationship

$$2qI_{\mathrm{d}}M_{\mathrm{qua}}^{2+\text{æ}} = n\overline{J_{\mathrm{a}}^2} \ . \tag{2.195}$$

The recommended value is $n = 2$. This is the optimal regime for APD with $\text{æ} = 1$ (then $M_{\mathrm{qua}} = M_{\mathrm{opt}}$); in the other extreme ($\text{æ} = 0$) the deterioration of SNR as compared with the optimal regime can be easily calculated by (2.185):

$$\mathrm{SNR} = \frac{M_{\mathrm{qua}}I_{\mathrm{s}}'}{\sqrt{2qI_{\mathrm{d}}M_{\mathrm{qua}}^2 + \overline{J_{\mathrm{a}}^2}}} = \frac{M_{\mathrm{qua}}I_{\mathrm{s}}'}{\sqrt{2qI_{\mathrm{d}}M_{\mathrm{qua}}^2 + \dfrac{2qI_{\mathrm{d}}M_{\mathrm{qua}}^2}{n}}} = \frac{I_{\mathrm{s}}'}{\sqrt{2qI_{\mathrm{d}}}}\frac{1}{\sqrt{1+\dfrac{1}{n}}} \ . \tag{2.196}$$

In the optimal regime, we have $M_{\mathrm{opt}} \to \infty$, $n \to \infty$, and hence $\mathrm{SNR} = I_{\mathrm{s}}'/\sqrt{2qI_{\mathrm{d}}}$. Accordingly, in the quasi-optimal regime, with a finite n ($n = 2$), the deterioration factor is

$$\Pi = \frac{1}{\sqrt{1+\dfrac{1}{n}}} = \frac{1}{\sqrt{1+\dfrac{1}{2}}} = 0.816 \ . \tag{2.197}$$

As implied by this estimate, the simplified choice of quasi-optimal M_{opt} (2.195) entails a relatively small forfeit in terms of the noise and the signal-to-noise ratio. The rigorous analysis of the maximum $\mathrm{SNR}(M)$ in the neighborhood of M_{opt} shows that this peak is not sharp, which allows us to replace M_{opt} with M_{qua}. As a matter of fact, it is not even necessary to maintain a rigidly prescribed value of n (for example, $n = 3$): the ratio of the APD noise to the circuit noise may fall within a relatively broad margin, $n = 1...3$. In this case the deterioration of SNR at $\text{æ} = 0.2$ will not exceed 4...18%, and will be as small as 3% when $\text{æ} = 1$ [76].

At high frequencies, when the high-frequency noise increases dramatically, the coefficient M_{opt} (or M_{qua}) must be very large (see (2.194), (2.195)). The value of M, however, is limited from above (for silicon APDs, for example, it does not exceed $50...200$; see Sect. 1.2). Because of this, a situation is possible when the multiplication factor is not large enough to "withdraw" the noise of APD above the noise of the amplifier. Then the advantage of APD over a PD in terms of SNR is due to its increased sensitivity because of the avalanche multiplication, and the gain (the above-mentioned advantage) is equal to M.

Optimal PD bias. As the bias voltage across the PD is increased, its capacitance (and therefore the high-frequency noise) decreases, but at the same time the dark current increases (for instance, because of broadening of the SCR), and so does the shot noise. Let us single out these two components from the total noise of PDA (2.166). We set

$C_{\rm in} = C_{\rm p}$, and assume that the dark current is only caused by generation in the SCR. Using expressions (1.77), (1.29), (1.5) for $C_{\rm p}(U)$, $I_{\rm d}(W)$, $W(U)$, we find the explicit dependence of the total noise of PDA on the bias voltage:

$$\overline{J^2} = 2qI_{\rm d} + e_{\rm n}^2\omega^2C_{\rm p}^2 = 2q\left(\frac{qn_i}{\tau_i}WA_{\rm e}\right) + e_{\rm n}^2\omega^2\frac{(\varepsilon\varepsilon_0)^2}{W^2}A_{\rm e}^2$$

$$= 2q^2\frac{n_i}{\tau_i}A_{\rm e}\sqrt{\frac{2\varepsilon\varepsilon_0 U}{qN_{\rm d}}} + (e_{\rm n}\omega\varepsilon\varepsilon_0 A_{\rm e})^2\frac{qN_{\rm d}}{2\varepsilon\varepsilon_0 U}$$

$$= B\sqrt{U} + \frac{C}{U} . \tag{2.198}$$

$$B = 2q^2\frac{n_i}{\tau_i}A_{\rm e}\sqrt{\frac{2\varepsilon\varepsilon_0}{qN_{\rm d}}} , \quad C = \frac{1}{2}(e_{\rm n}\omega A_{\rm e})^2\varepsilon\varepsilon_0 qN_{\rm d} . \tag{2.199}$$

This expression for the noise is the familiar binomial, in which the variable now is the voltage. The optimal regime in terms of minimum noise is one in which the noise power of the photodetector is twice the high-frequency noise of the amplifier (similarly to the quasi-optimal regime of the APD):

$$\frac{d\overline{J^2}}{dU} = \frac{1}{2}\frac{B}{\sqrt{U_{\rm opt}}} - \frac{C}{U_{\rm opt}^2} = 0 , \quad B\sqrt{U_{\rm opt}} = 2\frac{C}{U_{\rm opt}} ,$$

$$U_{\rm opt} = \left(2\frac{C}{B}\right)^{2/3} . \tag{2.200}$$

With this bias voltage the noise attains its minimum value of

$$\overline{J^2} = 2qI_{\rm d} + C_{\rm p}^2e_{\rm n}^2\omega^2 = 3qI_{\rm d}(U_{\rm opt}) . \tag{2.201}$$

Optimal regime of BT. Does PDA need BT? An important distinction between FET and BT is that the two noise generators in the latter (the input noise current generator $\overline{i_{\rm na}^2}$, and the noise emf generator $\overline{e_{\rm n}^2}$) depend on one and the same parameter, the base current $I_{\rm b}$ (Sect. 2.4). So, we have another exhibit for our collection of "what's good for the back is bad for the head": driving for the best, and trying to make the shot noise of the input base current $I_{\rm b}$ as small as possible by reducing this current, one can make things worse, because the emf noise component may increase owing to the drop in transconductance at low base currents. Separating from the total noise the two components for which the BT is responsible, using (2.150), (2.151), and noting that the noise-factor for BT is $\mathcal{F} = 0.5$, we get (with $I_{\rm a} = I_{\rm B}$):

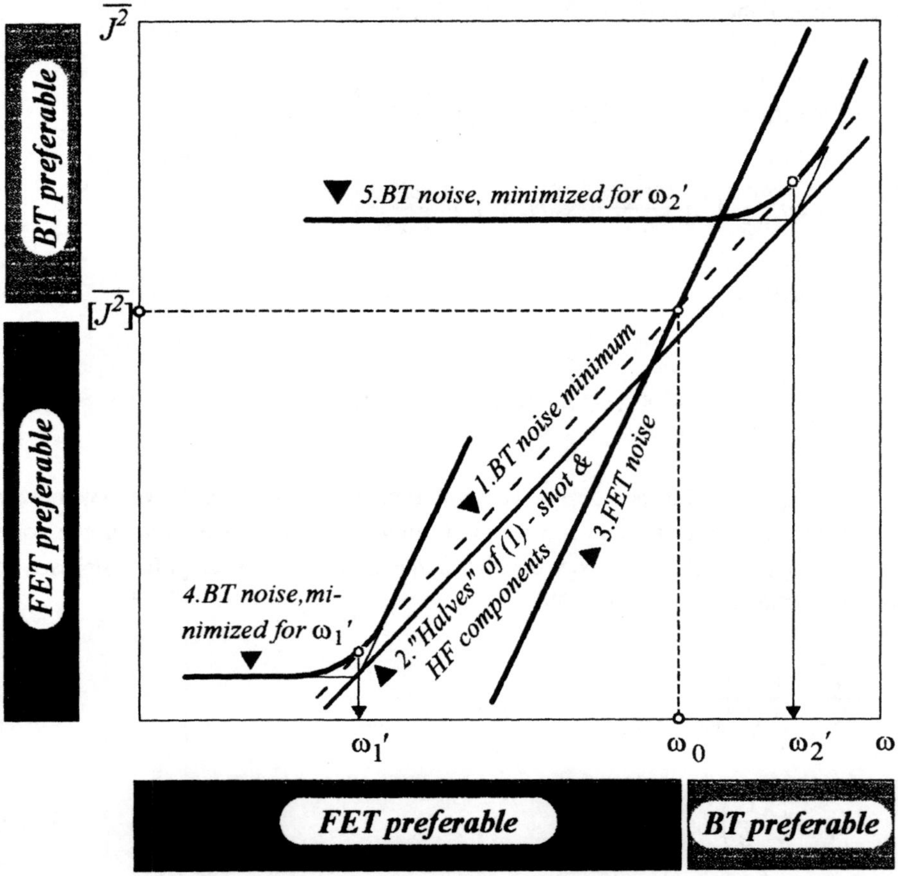

Figure 2.26: The noise of FET is, as a rule, less than the noise of BT.

$$\overline{J_{\mathrm{a}}^2} = 2qI_{\mathrm{a}} + \omega^2 C_{\mathrm{in}}^2 \overline{e_{\mathrm{n}}^2} = 2qI_{\mathrm{B}} + \omega^2 C_{\mathrm{in}}^2 \frac{2(\kappa \mathrm{T})^2}{h_{21\mathrm{e}} q I_{\mathrm{B}}} . \tag{2.202}$$

Now let us minimize this noise at some fixed frequency ω' by adjusting the optimal current I_{Bopt}. According to (2.193), the minimum of this expression occurs when the two terms are equal. Hence for the current I_{Bopt} we have

$$2qI_{\mathrm{Bopt}} = (\omega' C_{\mathrm{in}})^2 \frac{2(\kappa \mathrm{T})^2}{h_{21\mathrm{e}} q I_{\mathrm{Bopt}}} , \tag{2.203}$$

$$I_{\mathrm{Bopt}} = \frac{\kappa \mathrm{T}}{q} \frac{\omega' C_{\mathrm{in}}}{\sqrt{h_{21\mathrm{e}}}} . \tag{2.204}$$

It is convenient to rewrite the expression for the spectral density (2.202) as

$$\overline{J_a^2} = 2qI_{B\,opt} + \omega^2 C_{in}^2 \frac{2(\kappa T)^2}{h_{21e} q I_{B\,opt}} = 2qI_{B\,opt}\left[1+\left(\frac{\omega}{\omega'}\right)^2\right]. \tag{2.205}$$

Equation (2.205) was used for plotting the noise of bipolar transistor against frequency in Fig. 2.26, curve 4. Below the frequency ω' for which the optimization has been performed, the noise spectrum is white, and is determined by the shot noise of the base current. The high-frequency component ($\propto \omega^2$) dominates above the frequency ω'. It is worth noting that at the frequency ω' itself, the noise power is proportional to the frequency:

$$\overline{J_a^2(\omega')} = 4qI_{B\,opt} = 4\kappa T \frac{\omega' C_{in}}{\sqrt{h_{21e}}} . \tag{2.206}$$

This function (curve 1 in Fig. 2.26) is different from the sharper quadratic dependence of the noise power of FET, when the high-frequency component dominates in the frequency range in question:

$$\overline{J_a^2} = \omega^2 C_{in}^2 \overline{e_n^2} = \omega^2 C_{in}^2 \frac{4\kappa T \mathfrak{F}}{G} . \tag{2.207}$$

This dependence is also shown in Fig. 2.26 (curve 3); it intersects with curve 1 (the curve of minimum BT noise). Equating (2.206) and (2.207), we find the intercept of these two curves at frequency ω_0:

$$\omega_0^2 C_{FET}^2 4\kappa T \frac{\mathfrak{F}}{G} = 4\kappa T \frac{\omega_0 C_{BT}}{\sqrt{h_{21e}}}, \quad \omega_0 = \frac{G}{\mathfrak{F} C_{FET}\sqrt{h_{21e}}} \frac{C_{BT}}{C_{FET}} , \tag{2.208}$$

where C_{BT} and C_{FET} are the input capacitances of PDAs based on the respective transistors. FET is advantageous below ω_0 (Fig. 2.26, curve 3), whereas BT is less noisy at the high frequencies ($\omega > \omega_0$; curve 1). For the typical values of $h_{21e} = 100$, $G = 5 \cdot 10^{-3}$ A/V, $C_{in} = C_{FET} = C_{BT} = 2...5$ pF, $\mathfrak{F} = 0.5$, we get:

$$f_0 = \frac{G}{2\pi \mathfrak{F} C_{in}\sqrt{h_{21e}}} = \frac{5 \cdot 10^{-3}}{2\pi \cdot 0.5 \cdot (2...5) \cdot 10^{-12} \cdot \sqrt{100}} \approx 30...80\,\text{MHz}. \tag{2.209}$$

Most PDAs operate at lower frequencies, which justifies the use of FETs. In the high-frequency range, however, as follows from the above estimate, it may be expedient to use a BT: by choosing a large value of I_b, the noise emf is reduced to a minimum (curve 5 in Fig. 2.26).

The intercept of the minimum noise curves for BT and FET (intersection of curves 1 and 3) also marks another "watershed", the noise current limit $[I_n]$. If the photodiode current falls below this limit, then it is more advantageous to use a FET; otherwise a BT (if $I_d > [I_n]$). This is very clear from comparison between curves 4 and 5 in Fig. 2.26. So that the shot noise of BT should not exceed the shot noise of PD, the base current must not exceed the PD current. When the photodiode current is small ($I_b < I_d < [I_n]$), the noise spectrum of BT is described by curve 4. From the diagram we see that in such case a FET registers the noise ($2qI_d$) in a broader band than a BT. On the contrary, when the current is large, the band of BT in the above sense is broader (curve 5). The value of $[I_n]$ can be estimated by calculating the base current (2.204) at $\omega' = \omega_0$ (2.208):

$$[I_n] = I_B\big|_{\omega'=\omega_0} = \frac{\varphi_t C_{BT}}{\sqrt{h_{21e}}} \frac{G C_{BT}}{\mathcal{G} C_{FET}^2 \sqrt{h_{21e}}} = \frac{\varphi_t G C_{BT}^2}{\mathcal{G} h_{21e} C_{FET}^2}$$

$$= \frac{0.026 \cdot 5 \cdot 10^{-3}}{0.5 \cdot 100} = 2.6 \cdot 10^{-6}\ \text{A}\,. \qquad (2.210)$$

The values of parameters are the same as before. This estimate, together with the values presented in Table 2.1, indicate that it is better to use FETs with photodetectors of practically all spectral ranges (with the exception of low-resistance photodetectors of range V). From condition $I_B < I_d$ it also follows that the BT must operate in the low-current mode: according to Table 2.1, the photodetector current in spectral ranges I – IV is as low as $5 \cdot 10^{-13} \ldots 3 \cdot 10^{-6}$ A. At the same time, it is very difficult to operate a BT in this regime (and simply impossible at the lower end of the range) because of the loss (or even disappearance) of gain, especially at low temperatures.

The results of this subsection agree with the general rule: the high-resistance signal sources (low-current low-capacitance PDs) must be matched with high-resistance amplifiers (based on low-current FETs), and the low-resistance sources (high-frequency PDs with high capacitive conductance, large-current photodetectors of range V) should be matched with low-resistance amplifiers (based on BTs with large base currents).

Optimal design of low-noise FET. Let us consider the component of the high-frequency noise for which the FET is responsible. Separating this component from the total noise (2.166), and using (2.146) for the noise emf of FET, we get

$$\overline{J_h^2} = \omega^2 \overline{e_n^2} C_{in}^2 = \omega^2 \overline{e_n^2} \left(C_{FET} + \tilde{C}_p\right)^2 = \omega^2 4\kappa T \mathcal{G} \frac{\left(C_{FET} + \tilde{C}_p\right)^2}{G}\,. \qquad (2.211)$$

Here we have singled out the capacitance of the FET from the total capacitance, whereas all the other contributions are lumped together as $\tilde{C}_p = C_{in} - C_{FET} = C_p + C_c + C_{fb}$. The capacitance of photodiode usually prevails, so that $\tilde{C}_p \approx C_p$. From (2.211) we see that the noise parameters of the field-effect transistor are C_{FET}, G, and, if $C_p << C_{FET}$, the ratio C_{FET}^2/G. To reduce this ratio, one has to increase the transconductance G; however, the capacitance C_{FET} tends to increase in sympathy with G. This can be understood from the following arguments. Connecting in parallel n transistors, each of which has the transconductance G, we increase the transconductance of the composite transistor n-fold. Its capacitance, however, also increases n-fold, and the ratio C_{FET}^2/G gets even worse. So, in order to define the optimal design of FET, we must first of all find the linkage between the transconductance G and the capacitance C_{FET}. For this purpose we return to Fig. 2.20a, which shows the construction of the device. This diagram will help us first to express the electrical parameters of FET (G, C_{FET}, U_{co}) in terms of its structural features (resistivity of the channel ρ, its thickness W_{ch}, width b, and length L). Since the SCR at the cut-off voltage spreads to the entire thickness of the channel, we use equation (1.5) for the thickness of SCR and get

$$W_{ch} = W = \sqrt{\frac{2\varepsilon\varepsilon_0 U'_{co}\mu}{qN_d\mu}} = \sqrt{2\varepsilon\varepsilon_0\rho\mu U'_{co}} , \quad U'_{co} = \frac{W_{ch}^2}{2\varepsilon\varepsilon_0\rho\mu} . \tag{2.212}$$

Now we idealize our transistor model by taking into account only the ineradicable component of the input capacitance which is caused by the active region of the pn junction between the channel and the gate which controls the depth of the channel and is responsible for amplification (Fig. 2.20b). The role of the remaining stray components of the capacitance will be clarified later. As seen from Fig. 2.20b, the thickness of SCR of the active region of the junction in the operating mode varies from 0 to W_{ch}, so, as indicated in Sect. 2.4, the expected mean value is about $W_{ch}/2$. A more precise calculation, with due account for the actual channel profile, yields a refined value of $W_{ch}/3$ [72]. Since the area of the channel-substrate junction is bL, the sought-for capacitance is

$$C_{FET} = \frac{3\varepsilon\varepsilon_0 bL}{W_{ch}} . \tag{2.213}$$

Finally, the transconductance is determined by the conductivity of the channel (2.143) (see Fig. 2.20a),

$$G = \frac{1}{R_{ch}} = \frac{W_{ch}b}{\rho L} . \tag{2.214}$$

Now that we have expressed G, C_{FET}, U'_{co} in terms of the common structural parameters, we see that

$$G = \frac{2\mu}{3L^2} C_{FET} U'_{co} . \tag{2.215}$$

This formula is easily checked: substituting the values of C_{FET}, U'_{co} (2.213), (2.212), we get the expression for the transconductance (2.214):

$$G = \frac{2\mu}{3L^2} C_{FET} U'_{co} = \frac{2\mu}{3L^2} \frac{3\varepsilon\varepsilon_0 bL}{W_{ch}} \frac{W_{ch}^2}{2\varepsilon\varepsilon_0 \mu\rho} = \frac{bW_{ch}}{L\rho} .$$

By virtue of (2.215), the component (2.211) of the high-frequency noise can be rewritten as

$$\overline{J_h^2} = \omega^2 4\kappa T \mathcal{G} \frac{\left(C_{FET} + \tilde{C}_p\right)^2}{\frac{2\mu}{3L^2} C_{FET} U'_{co}} = \frac{\omega^2 4\kappa T \mathcal{G}}{\frac{2\mu}{3L^2} U'_{co}} \left[\sqrt{C_{FET}} + \frac{\tilde{C}_p}{\sqrt{C_{FET}}} \right]^2 . \tag{2.216}$$

The term in brackets is once again the binomial of the form $Bx^n + C/x^m$. According to (2.190), its minimum occurs at

$$n = m = \frac{1}{2} , \quad \frac{1}{2}\sqrt{C_{FET}} = \frac{1}{2} \frac{\tilde{C}_p}{\sqrt{C_{FET}}} , \quad C_{FET} = \tilde{C}_p . \tag{2.217}$$

This is a very instructive result: the amplifier must be matched with the PD not only in terms of ohmic resistance (recall (2.174), (2.176)), but also in terms of the capacitive resistance. As follows from (2.216), the lowest high-frequency noise value under the condition of matching capacitances (2.217) is

$$\sqrt{\overline{J_h^2}} = \omega \sqrt{16\kappa T \frac{\mathcal{G}}{U'_{co}} \left(\frac{3L^2}{2\mu} \right) \tilde{C}_p} , \quad (C_{FET} = \tilde{C}_p) . \tag{2.218}$$

This formula lays forth the guidelines for the designer of low-noise FETs for PDAs: reduce the channel length, increase the cut-off voltage, choose a material with high mobility for the channel (if silicon, then the electron-type silicon with $\mu_n = 1300\,\mathrm{cm^2/s\cdot V}$; GaAs is even better, $\mu_n \approx 7000...8000\,\mathrm{cm^2/s\cdot V}$). Let us first evaluate the noise parameters of a silicon FET, matched with a low-capacitance PD ($C_p = 0.3\,\mathrm{pF}$). Let the channel length be $L = 5\,\mu\mathrm{m}$, the cut-off voltage $U_{co} \approx U'_{co} = 3\,\mathrm{V}$ (as will be shown below, further increase of the cut-off voltage increases the power dissipation). Using (2.215), we get the following values of transconductance and noise emf for the optimal structure of FET:

$$C_{FET} = 0.3\,\mathrm{pF} ,$$

$$G = \frac{2\mu}{3L^2} C_{\text{FET}} U_{\text{co}} = \frac{2 \cdot 1300}{3 \cdot \left(5 \cdot 10^{-4}\right)^2} \cdot 3 \cdot 10^{-13} \cdot 3 = 3 \cdot 10^{-3} \text{ A/V} , \tag{2.219}$$

$$e_{\text{n}} = \sqrt{4\kappa T \frac{\mathcal{F}}{G}} = \sqrt{1.6 \cdot 10^{-20} \frac{0.5}{3 \cdot 10^{-3}}} \approx 2 \cdot 10^{-9} \text{ V/Hz}^{1/2} . \tag{2.220}$$

This estimate is too optimistic: in a real device it is hardly possible to obtain such a low noise emf (about $2\,\text{nV/Hz}^{1/2}$) together with such a low input capacitance ($\sim 0.3\,\text{pF}$). As a matter of fact, so far we have been considering only the basically inherent component of the capacitance of the active region of *pn* junction between the channel and the gate, which is responsible for amplification. In real structures there are other contributions to the total capacitance — such as the capacitances of the *pn* junctions between the substrate (the gate) and the source and drain regions (see Fig. 2.20a). These stray capacitances may be as large as units of picofarads, being definitive for the input capacitance of the FET. Special technologies and designs are used for reducing these undesirable capacitances.

The channel length in (2.220) is assumed to be rather large. Today it is possible to manufacture transistors with channel length as small as 1...2μm. As indicated in the previous subsection, the noise-factor increases in short-channel FETs, which prevents any substantial improvement on the noise-related characteristics of the device (in relation to the estimate of (2.220)). The choice of materials is also limited. Of course, the transition from silicon to GaAs, in which the mobility is higher by a factor of five or six, improves the noise parameters e_{n}, C_{FET}. However, the high flicker-noise, which extends into the megacycle frequency range, has so far permitted using this material only in the high-frequency range.

Such are the estimates for the parameters of FETs matched with the low-capacitance PDs. It would seem possible to design a FET with a lower noise emf for large-capacitance PDs (say, with $C_{\text{p}} = 30\,\text{pF}$). Namely, while increasing the capacitance of FET by two orders of magnitude (to $C_{\text{FET}} = C_{\text{p}} = 30\,\text{pF}$), we increase the transconductance by the same two orders of magnitude (to $300\,\text{mA/V}$), and reduce the noise ten-fold (to $0.2\,\text{nV/Hz}^{1/2}$). Such a high transconductance, however, entails yet another contradiction.

Power dissipation and FET noise emf. High transconductance implies high drain current I_{D}, and hence high power dissipation P:

$$P = I_{\text{D}} U = \frac{1}{3} G U_{\text{co}} U . \tag{2.221}$$

Here we have used the linkage between the drain current I_{D}, the cut-off voltage and the transconductance. The conductivity of the channel in the initial state is equal to the transconductance (see (2.143) and Fig. 2.20a). When a small potential is applied to the channel (between the source and the gate), the current $I = GU$ starts to flow through the channel. In the operating mode, $U = U_{co}$, and the channel depth, as indicated above, is reduced by about one half, so the conductivity also falls to half its initial value. More precise calculations, with due account for the actual profile of the channel, reveal a three-fold reduction, and hence $I_D = GU_{co}/3$ [72]. Now, substituting into (2.221) $U_{co} = 3\,\text{V}$, a typical value of working drain voltage $U = 10\,\text{V}$ (which ensures stable operation irrespective of the possible spread of U_{co} values, and linearity with the signal amplitude up to $U - U_{co} = 7\,\text{V}$), we get:

$$P = \frac{G}{3} \cdot 3 \cdot 10 = 10G \ , \tag{2.222}$$

where P is measured in watts, and G in ampere/volts.

When the transconductance is as high as $G = 0.3\,\text{A/V}$ (which, as shown above, is the target value in case of high-capacitance PDs, $C_p = 30\,\text{pF}$), the power dissipation becomes quite considerable:

$$P = 10G = 10 \cdot 0.3 = 3 \ \text{W} \ . \tag{2.223}$$

In some cases such a high dissipation power is not acceptable. The greater the number of channels, the less is the allowable power per channel. In accordance with (2.221), (2.222), the limitations on the power delimit also the transconductance. If, for example, the maximum allowable power per channel is $[P] = 10 \ldots 100\,\text{mW}$, then

$$G < [G] = \frac{3[P]}{U_{co}U} = \frac{[P]}{10} = \frac{10^{-2} \ldots 10^{-1}}{10} = 10^{-3} \ldots 10^{-2}\ \text{A/V} \ . \tag{2.224}$$

Hence, the FET noise emf will inevitably be not less than

$$e_n = \sqrt{4\kappa T \frac{\mathcal{F}}{G}} = \sqrt{1.6 \cdot 10^{-20} \frac{0.5}{10^{-3} \ldots 10^{-2}}} \approx 1 \ldots 3 \cdot 10^{-9}\ \text{V/Hz}^{1/2} \ . \tag{2.225}$$

This value does not depend either on the structure or on the material of FET, and improving it (given the power limitation) is beyond the power of the designer.

Transistor needed by PDA. The choice of the input transistor is the choice between a number of alternatives. The above analysis allows one to make the right decision.

Alternative number one. What kind of transistor, bipolar or field-effect, should serve as the first stage? A typical silicon FET is better than a BT in terms of noise-related properties in the frequency range from infrasound to 30 ... 80 MHz. As follows from (2.209), in order to expand this range up to 300 ... 1000 MHz, the transconductance of

FET must be made as high as 10 ... 30 mA/V (rather than the value of 5 mA/V for which we have obtained the above frequency limit $f_0 \sim 30$ ... 80 MHz). The high frequency also calls for reducing the capacitance of FET down to 0.1...0.2 pF. From (2.215) it follows that such a high transconductance with such a low capacitance can only be obtained using a material with a very high mobility,

$$\mu \geq G\frac{3L^2}{2G_{\text{FET}}U_{\text{co}}} = (1...3)\cdot 10^{-2}\frac{3\left(5\cdot 10^{-4}\right)^2}{2(1...2)\cdot 10^{-13}\cdot 3} \approx (6...40)\cdot 10^3\ \frac{\text{cm}^2}{\text{s}\cdot\text{V}}\ . \quad (2.226)$$

This condition is satisfied by GaAs, whose mobility is $7000...8000\,\text{cm}^2/\text{s}\cdot\text{V}$. So, a FET will also outdo a BT at frequencies above 30 MHz, if we make it from GaAs and achieve the transconductance of 10 ... 30 mA/V. According to (2.222), the power consumption will be 100 to 300 mW. One can put up with this fact, since the high-frequency PDAs have one channel (or a few channels).

There is a case, however, when a BT should be preferred to a FET. This is the case of PDA based on low-resistance PR, which will be discussed later on.

Alternative number two. What must be the structure of FET: gate junction or MIS gate? The theory points to the MIS gate, since the input current of such transistor is extremely small ($10^{-15}...10^{-13}$ A); as a consequence, the shot noise of this current is small too. Practice, however, dictates the use of the junction gate FET, especially for the low-frequency PDAs, since the excess channel noise in the MIS transistor is very high ($50...500\,\text{nV}/\text{Hz}^{1/2}$ at 1 kHz), which reduces to nothing its advantage with respect to the input current even at low frequencies. To wit, we have

$$\begin{aligned}\sqrt{\overline{J^2}} &= \sqrt{2qI_{\text{a}} + \left(e_{\text{n}}\omega C_{\text{in}}\right)^2} \\ &= \sqrt{2\cdot 1.6\cdot 10^{-19}\cdot 10^{-15} + \left(10^{-7}\cdot 2\pi\cdot 10^3\cdot 10^{-11}\right)^2} \\ &= 6\cdot 10^{-15}\ \text{A}/\text{Hz}^{1/2} \qquad (2.227\text{a})\end{aligned}$$

for MIS FET, and

$$\begin{aligned}\sqrt{\overline{J^2}} &= \sqrt{2\cdot 1.6\cdot 10^{-19}\cdot 10^{-12} + \left(3\cdot 10^{-9}\cdot 2\pi\cdot 10^3\cdot 10^{-11}\right)^2} \\ &= 0.6\cdot 10^{-15}\ \text{A}/\text{Hz}^{1/2} \qquad (2.227\text{b})\end{aligned}$$

for junction gate FET.

The accepted values of parameters are clear from the calculation. Of course, the excess noise of FET at higher frequencies ($\approx 100...1000$ kHz) may be small, and contribute

Table 2.2. **Typical parameters of input-stage transistors of PDAs**

Item	Power consumption	Type	Material	Name	e_n, nV/Hz$^{1/2}$	G, mA/V	C_a, pF	U_{co}, V	I_{dr}, mA	I_a, A
1.	Micro 0.2...0.5mW	J FET[1]	Si	2П202	20...30	0.2...0.3 (0.65)	6	0.4...1	0.05 (0.8)	$10^{-12}...3 \cdot 10^{-10}$
2.	Low 30...50mW	J FET	Si	2П312 (2П307)	2 (>1kHz)	4...6	2.5 (4...6)	~3	3...5	$10^{-10}...10^{-9}$
3.	Low 10...20mW	MIS FET[2]	Si	Incorporated in CCD	50...200 (1kHz)	0.2...0.6	0.03...0.3	—	~1	$<10^{-13}$
4.	Medium 500mW	J FET	Si	2П314	~1...1.5	15	10	~3	100	10^{-6}
5.	Medium 500...800mW	J FET	GaAs	3П331 (3П339)	~1...1.5	10...25	0.3	2.5...5	90...150	10^{-6}
6.	Low 10...20mW	BT[3]		Incorporated in IC 140УД3	1.5...2	—	—	—	Incorporated in IC 1...2	$5 \cdot 10^{-6}$

Notes: [1] For typical operating regimes [72].
[2] FET in the output stage of CCD.
[3] Input transistor of IC.

nothing to the total noise. At such frequencies, however, the HF-component of the noise prevails (Table 2.1), and so the advantage of the MIS FET in terms of the input current is unimportant, whereas in all other noise-related aspects the two types of transistors are identical. At frequencies above 100 kHz the traditional PDAs use junction gate FETs, and technological reasons dictate the use of MIS FETs for charge detection in charge-integrating matrices. Such frequencies (≈ 100...1000 kHz) are typical for scanning in these devices.

Other alternatives. Must the transistor be high-power or low-power; must its channel be broad or narrow; must it be made of silicon or of some other material? These points have already been discussed earlier.

Hit parade of transistors for PDA. The above analysis has given us a clear-cut idea of "who is who". The list of the most popular transistors is topped by a low-power silicon junction gate FET with the transconductance about 1 ... 5 mA/V. This general-purpose transistor is the optimal choice for the most diverse PDAs; an example can be found in Table 2.2, item 2. Of course, there is room for variations on the theme of "power – transconductance – channel noise". By applying the back-bias voltage to the gate, the transistor can be brought into the low-current mode (Table 2.2, item 1). This mode is useful when the number of channels is very large (several hundreds or thousands). By contrast, the power consumption can be increased in high-frequency PDAs having one channel or a few channels (Table 2.2, item 4); GaAs-based FETs can be used (Table 2.2, item 5). The third line in the table is occupied by a typical representative of CCD output transistors, a MIS FET. The last one is a BT, which is useful for PDAs with low-resistance photodetector, as will be shown in the forthcoming subsection.

This table gives us a comprehensive notion of the achievable level of noise for not only the transistors themselves, but also for PDA amplifier as a whole, since the total noise is determined by the input transistor as long as the design is correct. Observe that a differential (balance) first stage contains two transistors, and so the power of noise emf is doubled and becomes $2\overline{e_n^2}$. Doubled is also the electric power dissipation (the power consumed by the subsequent stages can be made less than that of the first stage).

Bipolar transistor is required for low-resistance-PR-based PDA. We still have to analyze matching of amplifier with photoresistor. It would seem that, owing to the presence of internal gain, it is much easier to match amplifier to a PR than to a PD. This is really so, but only for sufficiently high-resistance PRs. Like in APDs, they amplify signal and noise K_p-fold, thus slackening the requirements to the noise level of the amplifier. We set $\kappa = 0$ in equation (2.183), obtained for APD-based PDAs, and replace M by the photoresistor gain K_p (1.15).

It is much more difficult, however, to match amplifier with low-resistance PRs, despite their internal gain. Let us compare the input voltages supplied to the amplifier by a PD and a PR:

$$U_{\mathrm{nPD}} = \sqrt{2qI_{\mathrm{d}}} R_{\mathrm{LPD}} \ , \tag{2.228}$$

$$U_{\mathrm{nPR}} = \sqrt{4qI_{\mathrm{d}}} K_{\mathrm{p}} R_{\mathrm{LPR}} \ , \tag{2.229}$$

where the subscripts PD and PR mark the respective values for photodiode and photoresistor. As an example, we select a MCT-based photoresistor (range V in Table 1.1). The relatively high dynamic resistance of the photodiode in short-circuit mode or under a small inverse bias (0.1...0.5 V) allows us to choose $R_{\mathrm{LPD}} \approx R_{\mathrm{pSC}} = 10...100\,\mathrm{k\Omega}$ (subrange 8...10 µm, PSE size 0.1×0.1mm). The equivalent load R_{LPR} is limited by a very low proper resistance of the PR, $R_{\mathrm{LPR}} = R_{\mathrm{p}}/2 = 12...25\Omega$ (subrange 8...14 µm). Even the highest gain of $K_{\mathrm{p}} = 100$ (see Sect. 1.2) does not compensate for such a low load resistance, and the noise voltage of a low-resistance PR turns out to be very small — as a rule, much less than that of a PD:

$$\frac{U_{\mathrm{nPR}}}{U_{\mathrm{nPD}}} = \frac{\sqrt{2} K_{\mathrm{p}} R_{\mathrm{LPR}}}{R_{\mathrm{LPD}}} = \frac{\sqrt{2} \cdot 100 \cdot (12...25)}{10^4 ... 10^5} = 0.02...0.4 \ , \tag{2.230}$$

$$U_{\mathrm{nPR}} = \sqrt{4qI_{\mathrm{pb}}} K_{\mathrm{p}} R_{\mathrm{LPR}} = \sqrt{4 \cdot 1.6 \cdot 10^{-19} \cdot 1.4 \cdot 10^{-5}} \cdot 10^2 \cdot (12...25)$$
$$\approx (4...8) \cdot 10^{-9}\ \mathrm{V/Hz^{1/2}} \ . \tag{2.231}$$

With the loads as low as that, the condition $GR_{\mathrm{in}} > 1$ used in (2.158) may be violated; it is more convenient therefore to deal with the input noise voltage rather than with the current. From (2.231) we see that the noise emf of the input transistor must be of the order of $1...3\,\mathrm{nV/Hz^{1/2}}$ (and even less than that in case of reduced background current, $\sim 0.5\,\mathrm{nV/Hz^{1/2}}$). Theoretically, this low noise value in a FET can be achieved at the cost of a very large power dissipation, 100 mW per channel (Table 2.2, item 4). This amount of dissipation is not permissible in multielement PDAs. Because of this, we once again take a look at the BT: with a little bit of luck, its power dissipation may turn out to be small. Using expression (2.142), which links the collector current I_{c} with the transconductance, we find that

$$P = I_{\mathrm{c}} U = G\varphi_{\mathrm{t}} U \ . \tag{2.232}$$

Comparing (2.232) with the similar expression (2.221) for a FET, we see that we are lucky indeed: G and U being equal, the power consumed by a BT is less by a factor of $U_{\mathrm{co}}/3\varphi_{\mathrm{t}} = (1...3)/3 \cdot 0.026 = 13...40$. The low values of noise emf $0.5...1.5\,\mathrm{nV/Hz^{1/2}}$ can be attained, according to (2.143), with the following values of transconductance and power dissipation:

$$G = \frac{4\kappa T \mho}{e_n^2} = \frac{1.6 \cdot 10^{-20} \cdot 0.5}{\left[(0.5 \ldots 1.5) \cdot 10^{-9}\right]^2} \approx 3 \ldots 30\,\text{mA/V} \,, \tag{2.233}$$

$$P = G\varphi_t U = (3 \ldots 30) \cdot 10^{-3} \cdot 0.026 \cdot 10 = 0.8 \ldots 8\,\text{mW} \,. \tag{2.234}$$

Of course, in order to achieve this high transconductance the base current must be sufficiently large (see (2.149)):

$$G = \frac{h_{21e} I_b}{\varphi_t} \,, \quad I_b = \frac{G\varphi_t}{h_{21e}} = \frac{(3 \ldots 30)10^{-3} \cdot 0.026}{100} = (0.8 \ldots 8) \cdot 10^{-6}\,\text{A} \,. \tag{2.235}$$

Although the current I_b is not small, this is currently not a nuisance: after all, the thermal background current of photodetectors of range V is just as large (see Table 1.1), and since the background noise is amplified by the photoresistor, the shot noise of the base current is negligibly small compared to the background noise.

In this way, to the list of transistors needed by PDAs we add another one, this time a BT (Table 2.2, item 6), which is a better match for a low-resistance photodetector than a FET.

So, we have considered the typical situations encountered in connection with the attempts to minimize the noise of PDA. Our analysis was carried out "in the pure form": we simplified the models, and considered only two competing mechanisms. The reality may be more diversified and more complicated. For example, the choice of the optimal PD bias may be influenced by its leakage current and its flicker noise; its capacitance-voltage characteristic may be different from the one assumed above ($C_p \propto U^{-1/2}$), and its sensitivity may be a function of the bias. Optimization of the transistor design is also not that simple: it is possible to take better account of the stray capacitance of the substrate, of the diffusion source and drain regions, to pay attention to the excess noise of the FET channel. The analysis of the noise model of BT may include the thermal noise of the base, the $h_{21e}(I_b)$ dependence, etc. Many other circumstances may be involved in the choice of the transistor: cost, availability, feasibility of producing special topologies. At the same time, the above estimates are universal (see equation (2.227) and Table 2.2), and characterize the lowest noise level achievable in PDAs [78].

And one final comment. The high-frequency noise component in this section was evaluated at a fixed frequency (in unit band). The PDAs for pulsed signal operate in a broad band, and so the relationship between the white noise and the high-frequency components must be finally established after averaging over a broad frequency band. The guidelines for choosing the operating frequency band of PDA will be set forth below, and then we will be able to return to the problem of optimization of regimes (Sect. 5.3). The noise in the integration devices also depend on the frequency band (integration time), and will be considered in Sect. 7.2, when we shall be aware of the frequency band of these devices.

Summary. *Perfect PDAs can be based on PDs with* $I_d = 10^{-8}...10^{-4}$ A, *for the frequency range* 10...10^5 Hz*: then it is possible to eliminate the effects of the amplifier noise. If the current is determined by thermal background radiation (as in IR photodetectors of ranges III – V), the theoretical limit is achieved — the background limited (BLIP) operating mode.*

In case of very high-resistance PDs (with the currents as low as $10^{-13}...10^{-10}$ A*), and especially at high frequencies, the amplifier noise dominates, and is ineradicable in principle. Accordingly, APDs are unrivalled at the high frequencies, because the existence of internal gain (multiplication) effectively suppresses the effects of the high-frequency amplifier noise.*

To minimize the noise of PDA, one must follow a number of mandatory rules which concern the choice of circuit elements and their operating parameters. First of all, one must use a FET (junction gate, as a rule); a BT is to be used only with the low-resistance photodetectors — for example, with IR photoresistors of range V. The leakage current of the FET must be lower than the PD current, and its capacitance must be equal to that of the PD (to the circuit capacitance). If the power dissipation in PDA is restricted to $[P]$, *the transconductance of FET must be not greater than* $3[P]/UU_{co}$, *where* U *and* U_{co} *are, respectively, the power-supply and the cut-off voltages of the FET. The input circuit load must be two or more times the design dynamic resistance of the PD, which is equal to* $\kappa T/qI_d$. *The working voltage across the PD or APD should be chosen so as to make their noise power equal to that of the amplifier (or two or three times that value).*

Chapter 3

FOUNDATIONS OF THEORY OF OPTIMAL FILTERING AS APPLIED TO PHOTODETECTING ASSEMBLIES

We have analyzed the passage of signal through the linear circuits of the PDA, and the noises associated with these circuits. We also derived appropriate formulas for calculating the voltage at the output of the amplifier: the noise voltage, or the noise plus signal voltage, if the signal arrives at the input of PDA.

Knowing the voltage at the output of each channel, we may now embark upon the problem of detection itself, and answer the main question — the existential question for the PDA — whether or not the optical signal has arrived at the input of PDA. In this chapter we shall consider the rule which governs the decision concerning the presence or absence of the signal, the errors which may be associated with such decision, the ways to minimize the occurrence of errors (optimal filtering).

3.1. Signal – No Signal Decision Making

Correct decisions and errors in signal detection. In optoelectronic systems the choice is made between just two possible decisions: "there is a signal", and "there is no signal". Also, there are just two possible events: either the signal has arrived at the input of PDA, or not. Logically, this leads to the following four possible situations:

1. The signal has not arrived. The system decides that there is no signal. Correct decision as to the absence of signal.

2. The signal has not arrived. The system decides that there is a signal. This wrong decision is known as false alarm (error of the first kind)

3. The signal has arrived. The system decides that there is no signal. This wrong decision is known as target miss (error of the second kind)

4. The signal has arrived. The system decides that there is a signal. Correct decision as to the presence of signal.

The corresponding quantitative characteristics are the probabilities of errors: the false alarm probability P_{fa} (the probability that the "positive" decision will be taken, if it is known a priori that the signal is not there), and the target miss probability P_{tm} (the probability that the "negative" decision will be taken, if it is known a priori that the signal is there) [4]. Then the probabilities of correct decisions as to the absence or presence of signal are, respectively, $(1-P_{fa})$ and $(1-P_{tm})$. The quantities P_{fa} and P_{tm} are the most important characteristics of the detection system: the lower they are, the better the system, the more reliable its decisions.

Rule of decision-making. The decision in the detection system is taken by a threshold device (discriminator). The particular channel signal-noise mix is sent to the input of the discriminator. Assume that the time when the optical pulse may arrive at PDA is known in advance; then we also know the instant t_m when the signal at the output of amplifier (at the input of the discriminator) achieves its maximum. This opportune moment is used for taking the decision as to the presence or absence of the signal. The discriminator works by a very simple algorithm: if its input voltage $U(t_m)$ exceeds the preset threshold (discrimination threshold) U_t, the decision is "yes"; if $U(t_m)$ falls below the threshold U_t, the decision is "no":

$$\begin{aligned} &U(t_m) < U_t : \text{decision "no signal"} ; \\ &U(t_m) \geq U_t : \text{decision "signal"} ; \end{aligned} \tag{3.1}$$

False alarm probability. By definition, the probability P_{fa} is calculated when the signal is absent, and the noise realization is sent to the input of the discriminator (Fig. 3.1a). The cause of the wrong decision in this case is the noise pulse, if at the decision-taking time t_m the noise $U(t_m)$ exceeds the discrimination threshold U_t. It is easy to calculate the probability of this event. The probability that at the time t_m the noise voltage falls between U and $U+dU$ we denote by $\rho(U)dU$; this probability obeys the normal distribution law (2.68) (Fig. 3.1b). In our case, $\sigma^2 = \overline{U_n^2}$, $U_n = \sqrt{\overline{U_n^2}}$ (the possible deviations from the normal law in PDAs are discussed in Sect. 6.2, 6.3). Since the decision that the signal is there is taken when $U \geq U_t$, we must find the probability that U belongs to the interval $[U_t, \infty]$, which means that we must integrate the function ρdU between these limits:

$$P_{fa} = \int_{U_t}^{\infty} \rho(U)dU = \frac{1}{\sqrt{2\pi}} \int_{U_t}^{\infty} \frac{1}{U_n} \exp\left(-\frac{U^2}{2\overline{U_n^2}}\right) dU =$$

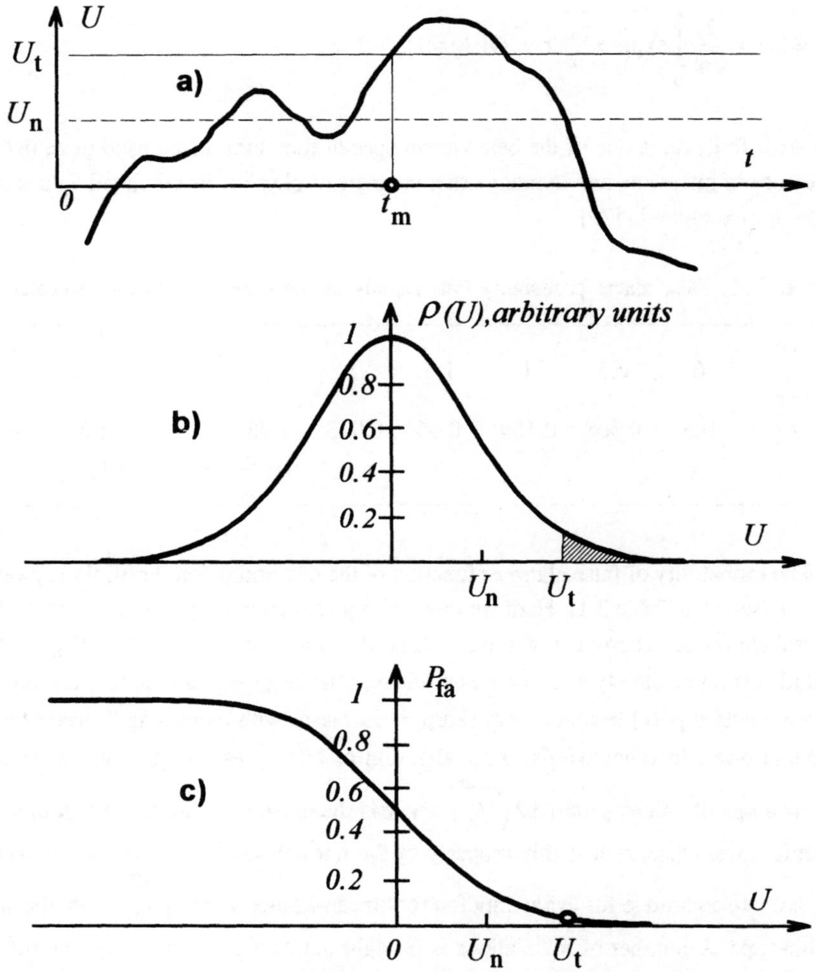

Figure 3.1: (a) False alarm caused by strong positive noise spike; and (b,c) calculation of false alarm probability.

$$= \frac{1}{\sqrt{\pi}} \int_{U_t}^{\infty} \exp\left[\left(-\frac{U}{\sqrt{2}U_n}\right)^2\right] d\frac{U}{\sqrt{2}U_n} = \frac{1}{\sqrt{\pi}} \int_{U_t/\sqrt{2}U_n}^{\infty} \exp(-x^2) dx \ . \tag{3.2}$$

This integral is expressed in terms of the error function $\Phi(z)$:

$$P_{fa} = \frac{1}{2}\left[\frac{2}{\sqrt{\pi}} \int_0^{\infty} \exp(-x^2) dx - \frac{2}{\sqrt{\pi}} \int_0^{U_t/\sqrt{2}U_n} \exp(-x^2) dx\right]$$

$$= \frac{1}{2}\left[1 - \Phi\left(\frac{U_t}{\sqrt{2}U_n}\right)\right] , \tag{3.3}$$

$$\Phi(z)=\frac{2}{\sqrt{\pi}}\int_0^z \exp\left(-x^2\right)dx \ , \quad \Phi(\infty)=1 \ .$$

The error function is one of the best known special functions, often used in statistics, in the theory of diffusion, and in many other chapters of physics. Its tabulated values can be found, for example, in [79].

Table 3.1. **False alarm probability falls rapidly as the threshold voltage increases**

$\frac{U_t}{U_n}$	0	0.5	1	1.5	2	3	4	5	6
P_{fa}	0.5	0.308	0.159	0.06	0.023	1.35 $\times10^{-3}$	3.17 $\times10^{-5}$	2.9 $\times10^{-7}$	10^{-9}

The probability of false alarm as function of the discrimination threshold is plotted in Fig. 3.1 (see also Table 3.1). Here we meet with an amazing property of the normal law: the probability of occurrence of small values of U (not greater than $(1...2)U_n$) is large, and falls off rather slowly as U increases, whereas the large values $U > 3U_n$ are unlikely, and the function $\rho(U)$ becomes very sharp: it decreases with increasing U faster than the exponent does. In other words, the realization of $U(t_m)$ as though concentrates in a narrow range of values within $\pm 2\sqrt{\overline{U_n^2}}$, whereas the noise overshoots beyond this range are quite rare. Observe that this property of the normal law is used in the well-known approximate technique for evaluating the root-mean-square noise $\sqrt{\overline{U_n^2}}$ with the aid of oscilloscope. A number of realizations is brought out on the display, with the probable overshoots within $\pm(2...3)\sqrt{\overline{U_n^2}}$, and a "noise trail" is observed, whose width (two-way swing) is $U \approx (4...6)\sqrt{\overline{U_n^2}}$. The width of the trail is then used for calculating the noise, $\sqrt{\overline{U_n^2}} \approx U/(4...6)$. In this context, the noise voltage $\sqrt{\overline{U_n^2}}$ is not simply a statistical characteristic (rms value): the term acquires a straightforward meaning. Indeed, for a deterministic harmonic its rms (with respect to the time) value $\sqrt{\overline{U_n^2}}$ characterizes the swing of this quantity (within $\pm\sqrt{2\overline{U_n^2}}$). For a random quantity (the normal noise) we have found complete analogy with a deterministic one: the probable values of the noise tend to concentrate between almost the same limits, $\pm 2\sqrt{\overline{U_n^2}}$.

This property of the normal law (the weak dependence of $\rho(U)$ on small values of U, and the strong decline of $\rho(U)$ as U becomes large) is emphasized even more dramatically in the error function, and hence in the above-derived function $P_{fa}(U_t)$. When the threshold is set at $U_t = 0$ the probability is $P_{fa} = 0.5$, since the positive and negative noise peaks are equiprobable. As the ratio U_t/U_n increases from 0 to 1, the probability P_{fa} falls off relatively slowly, from 0.5 to 0.159. When, however, $U_t/U_n = 2...3$, the probability P_{fa} becomes small enough, $P_{fa} = 2\cdot\left(10^{-2}...10^{-3}\right)$, and falls off very quickly as U_t further increases: as U_t/U_n increases by one, the probability P_{fa} falls by two orders of magnitude. This can be easily demonstrated with the aid of approximation for P_{fa} which is valid for the high discrimination thresholds. We rewrite (3.2) as

$$P_{fa} = \frac{1}{\sqrt{2\pi U_n^2}} \int_{U_t}^{\infty} \exp\left(-\frac{U^2}{2U_n^2}\right) dU$$

$$= \frac{\exp\left(-\frac{U_t^2}{2U_n^2}\right)}{\sqrt{2\pi U_n^2}} \int_{U_t}^{\infty} \exp\left(-\frac{U^2 - U_t^2}{2U_n^2}\right) dU , \tag{3.4}$$

and transform the exponent in the integrand:

$$-\frac{U^2 - U_t^2}{2U_n^2} = -\frac{(U - U_t)(U + U_t)}{2U_n^2} = -\frac{\Delta U(2U_t + \Delta U)}{2U_n^2} \approx -\frac{U_t}{U_n^2}\Delta U , \tag{3.5}$$

$\Delta U = U - U_t$.

Here we introduce the overrun above the threshold ΔU, and take advantage of the fact that in case of large thresholds $U_t > 3U_n$ the exponent falls off quickly even with small increments $\Delta U < U_t$. Then (3.4) becomes

$$P_{fa} \approx \frac{\exp\left(-\frac{U_t^2}{U_n^2}\right)}{\sqrt{2\pi U_n^2}} \int_0^{\infty} \exp\left(-\frac{U_t}{U_n^2}\Delta U\right) d\,\Delta U$$

$$= \frac{\exp\left(-\frac{U_t^2}{U_n^2}\right)}{\sqrt{2\pi U_n^2}} \frac{U_n^2}{U_t} = \frac{1}{\sqrt{2\pi}} \frac{U_n}{U_t} \exp\left(-\frac{U_t^2}{2U_n^2}\right) . \tag{3.6}$$

Expanding the error function in (3.3), we arrive at the same result. This approximation is very convenient for practical calculations of $P_{fa}(U_t)$. For example, this formula implies that, owing to the strong dependence of P_{fa} on the discrimination threshold, the threshold may be selected within a relatively narrow interval ($U_t = (3...6)U_n$). As the threshold is changed within a mere factor of two, the false alarm probability goes down as much as by six orders of magnitude, which spans the entire range of this parameter usually required by optoelectronic systems:

$$P_{fa} \approx \frac{1}{\sqrt{2\pi}} \frac{U_n}{U_t} \exp\left(-\frac{U_t^2}{2U_n^2}\right)$$

$$= \left(\frac{1}{\sqrt{2\pi}} \frac{1}{3} e^{-9/2}\right) ... \left(\frac{1}{\sqrt{2\pi}} \frac{1}{6} e^{-36/2}\right) = 1.5 \cdot 10^{-3} ... 10^{-9} . \tag{3.7}$$

Target miss probability. Assume now that the PDA receives an optical signal. The corresponding signal at the output of the amplifier $U_s(t)$ is shown in Fig. 3.2a. At the time t_m the amplitude of the output signal $U_{sA} = U_s(t_m)$ exceeds the threshold, and the logical device decides that the signal is there. However, a negative noise burst at t_m may happen to be so large as to make the signal-cum-noise voltage fall short of U_t. Then the logical device will take the wrong "no signal" decision (target miss; Fig. 3.2b).

The target miss probability is calculated similarly to our previous case. Since we know a priori that the signal at t_m assumes its amplitude value U_{sA}, the probability of occurrence of signal-cum-noise voltage value $U = U_{sA} + U_n(t_m)$ is equal to the probability of obtaining the noise value $U_n(t_m) = U - U_{sA}$:

$$\rho(U)dU = \frac{1}{\sqrt{2\pi U_n^2}} \exp\left[-\frac{(U - U_{sA})^2}{2U_n^2}\right] . \tag{3.8}$$

At all values $U < U_t$ the decision-making device will make a mistake (miss the target), so

$$P_{tm} = \int_{-\infty}^{U_t} \rho(U)dU = \int_{-\infty}^{U_t} \frac{1}{\sqrt{2\pi U_n^2}} \exp\left[-\frac{(U - U_{sA})^2}{2U_n^2}\right] dU$$

$$= \frac{1}{\sqrt{\pi}} \int_{U_t}^{-\infty} \exp\left[-\frac{(U_{sA} - U)^2}{2U_n^2}\right] d\frac{U_{sA} - U}{\sqrt{2}U_n} = \frac{1}{\sqrt{\pi}} \int_{(U_{sA} - U_t)/\sqrt{2}U_n}^{\infty} \exp(-x^2) dx . \tag{3.9}$$

We see that the calculations of P_{fa} and P_{tm} bring us to the same integrals, which only differ in the lower integration limit (cf. (3.2)). Accordingly, the formula for P_{tm} follows from that for P_{fa} simply by replacing $U_t \Rightarrow U_{sA} - U_t$:

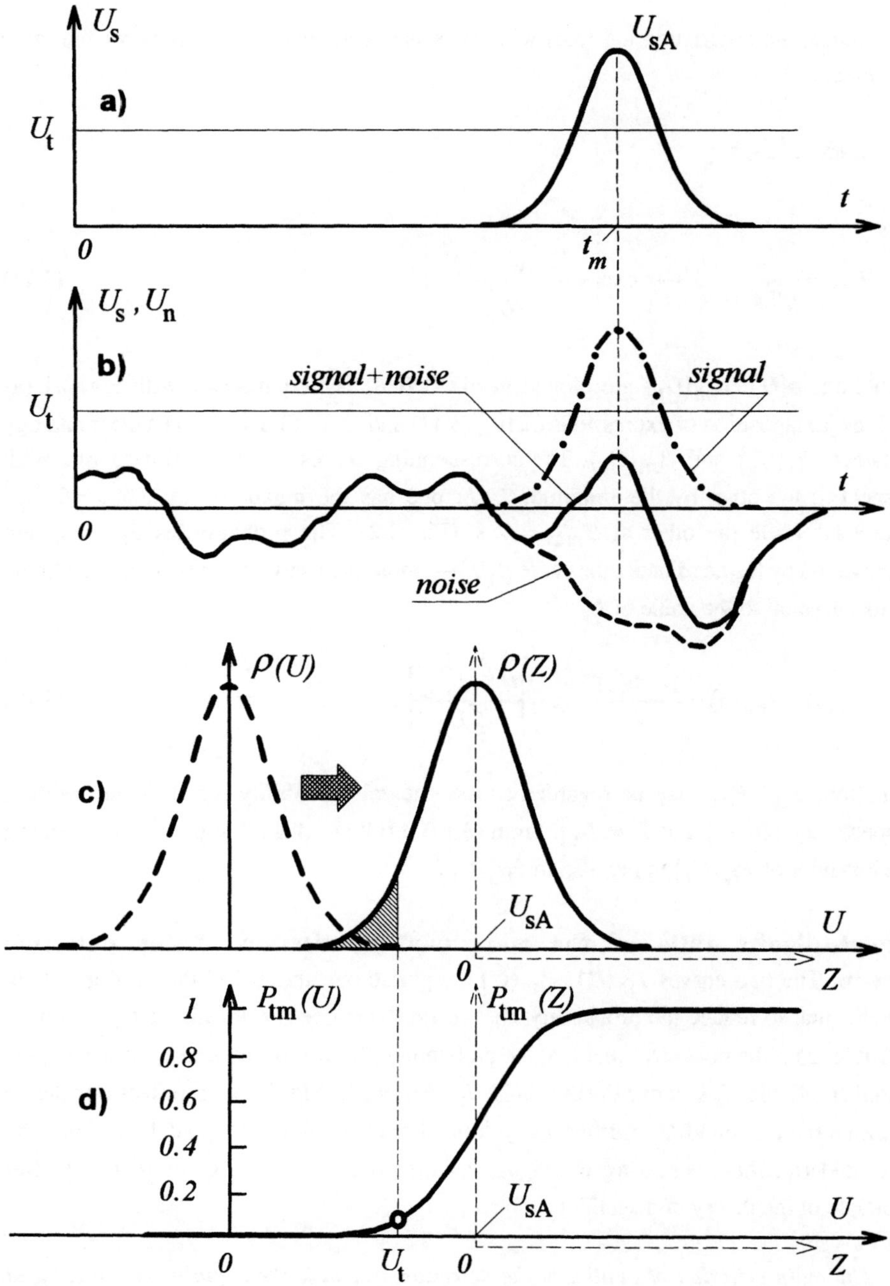

Figure 3.2: Target miss caused by strong negative noise spike superimposed on the signal: (a) signal); (b) signal cum noise mix; (c,d) calculation of target miss probability.

$$P_{tm} = \frac{1}{2}\left[1 - \Phi\left(\frac{U_{sA} - U_t}{\sqrt{2}\ U_n}\right)\right]. \tag{3.10}$$

Of course, the approximation (3.6) with the same replacement also holds for the large arguments:

$$\frac{U_{sA}-U_t}{U_n}>3\ ,$$

$$P_{tm}\approx\frac{1}{\sqrt{2\pi}}\frac{U_n}{U_{sA}-U_t}\exp\left[-\frac{(U_{sA}-U_t)^2}{2U_n^2}\right]. \tag{3.11}$$

Functions $\rho(U)$, $P_{tm}(U_t)$ are plotted in Fig. 3.2c,d. The comparison with Fig. 3.1 (as well as juxtaposition of expressions (3.10), (3.11) and (3.3), (3.6)) reveals a clear analogy between $P_{fa}(U_t)$ and $P_{tm}(U_t)$. The corresponding curves $\rho(U)$ are shifted one with respect to the other by the amplitude U_{sA}: one has the maximum at $U_{max}=0$ (see Fig. 3.1), while the other at $U_{max}=U_{sA}$ (Fig. 3.2). The probabilities P_{fa}, P_{tm} are represented by the areas under the curve $\rho(U)$ — more precisely, the areas under the "tail" of its curve up to the value of U_t:

$$P_{fa}(z)=P_{tm}(z)=\frac{1-\Phi(z)}{2}\ ,\quad z=\left|\frac{U-U_{max}}{\sqrt{2}\,U_n}\right|. \tag{3.12}$$

Functions P_{fa}, P_{tm} may be regarded as the integral probability for all values within, respectively, $[U_t,\infty]$, and $[-\infty,U_t]$. From (3.12) it follows that all arguments concerning the behavior of $P_{fa}(U_t)$ apply also to $P_{tm}(U_t)$.

Signal-to-noise ratio as the main quality criterion of the detection system. The two curves $P_{fa}(U)$, $P_{tm}(U)$ are placed one above the other in Fig. 3.3a,b. We see that to reduce the probability P_{fa} we must remove the threshold U_t as far as possible from the noise U_n, and that the probability P_{tm} can be reduced by increasing the signal amplitude U_{sA} above the threshold U_t. Arrows in Fig. 3.3a,b,c reflect our desire to move the threshold U_t further away from the noise, and the signal U_{sA} from the threshold U_t, thus increasing the signal-to-noise ratio. This brings us to the central principle of the theory of detection:

The main criterion of quality of the detection system is the signal-to-noise ratio at the input of the logical device, SNR $=U_{sA}/U_n$.

This statement hardly calls for any comment. Every engineer, concerned with the development of signal detectors (and not only those for optical signals) knows that the signal-to-noise ratio must be made as high as possible. It seems self-evident that the detection is the more reliable, the better the signal stands out against the noise which

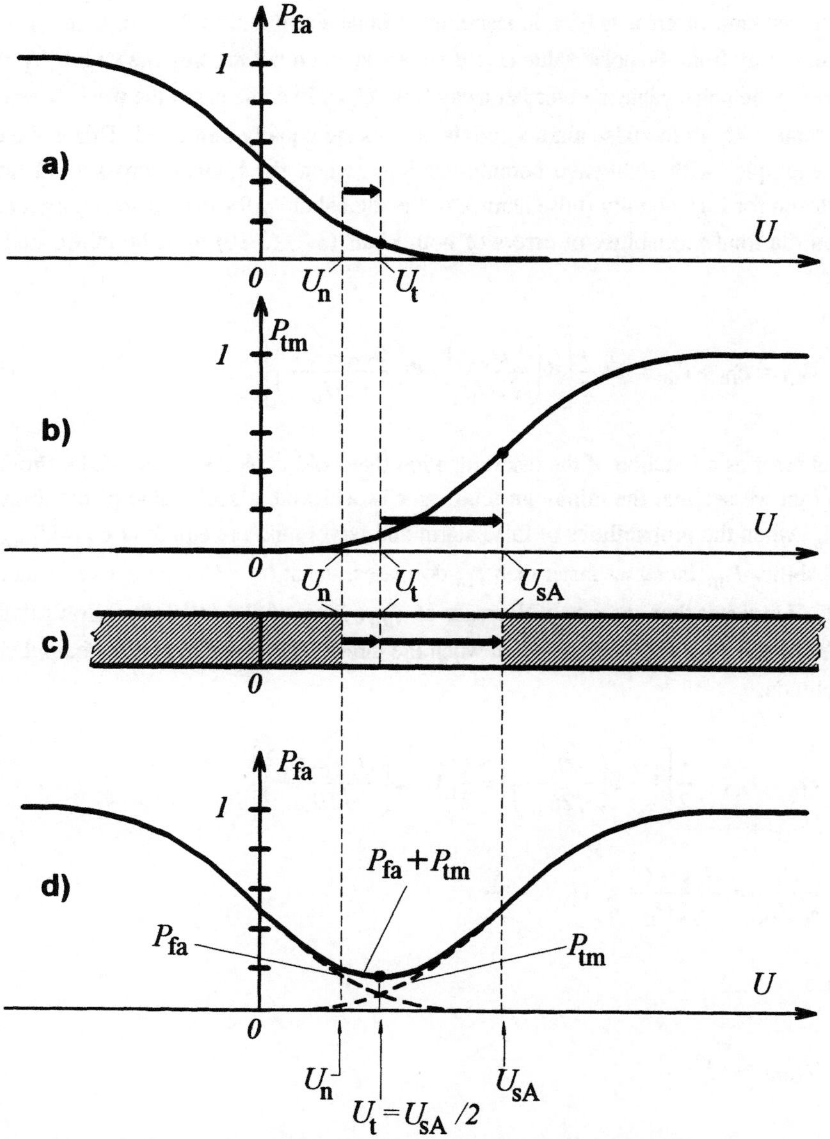

Figure 3.3: To reduce the probabilities of error P_{fa}, P_{tm} one must increase the signal-to-noise ratio.

interferes with the detection. We have already used the concept of SNR in connection with the internal gain of photodetector (Sect. 1.2). Now we have obtained the SNR as criterion of quality not from general considerations: now its value is uniquely related through (3.3), (3.10) with detection errors: the probabilities of false alarm and target miss.

The choice of discrimination threshold. Desirable values of SNR. As shown above, the discrimination threshold is usually chosen as $U_t = (3...6)U_n$ (see

(3.7)). Now we may add that the location of U_t on the interval $[U_n, U_{sA}]$ depends on whatever kind of error is less undesirable: if false alarm is unwelcome, then U_t is set further away from the noise value U_n; if we are keen on not missing the target, U_t is set closer to the noise value, and further away from U_{sA}. In some cases the price of errors is the same — both the false alarms and the misses are equally unwanted. This is the case, for example, with lightwave communications, when the logical zero should not be mistaken for logical unity (false alarm), or the logical unity for logical zero (target miss). Then the total probability of errors of both kinds (3.3) (3.10) must be minimized [10, 39]:

$$P_{err} = P_{fa} + P_{tm} = 1 - \frac{1}{2}\left[\Phi\left(\frac{U_t}{\sqrt{2}U_n}\right) + \Phi\left(\frac{U_{sA} - U_t}{\sqrt{2}U_n}\right)\right]. \tag{3.13}$$

Total error as a function of the discrimination threshold is plotted in Fig. 3.3d. From this diagram we see that the minimum total error is achieved at such value of the threshold $U_{t\,opt}$ when the probabilities of false alarm and target miss are equal: at $U_t > U_{t\,opt}$ the probability P_{tm} increases faster than P_{fa} decreases, and at $U_t < U_{t\,opt}$ the rate of increase of P_{fa} is greater than the rate of decrease of P_{tm}. From (3.3), (3.10) it follows that the probabilities P_{fa}, P_{tm} become equal when the threshold is equal to half the output signal amplitude:

$$P_{fa} = P_{tm} = \frac{1}{2}\left[1 - \Phi\left(\frac{U_t}{\sqrt{2}U_n}\right)\right] = \frac{1}{2}\left[1 - \Phi\left(\frac{U_{sA} - U_t}{\sqrt{2}U_n}\right)\right],$$

$$\frac{U_t}{\sqrt{2}U_n} = \frac{U_{sA} - U_t}{\sqrt{2}U_n}, \quad U_t = \frac{U_{sA}}{2}. \tag{3.14}$$

Hence

$$U_{t\,opt} = \frac{U_{sA}}{2},$$

$$\min\{P_{err}\} = 2P_{tm}(U_{t\,opt}) = 2P_{fa}(U_{t\,opt}) = 1 - \Phi\left(\frac{U_{sA}}{2\sqrt{2}U_n}\right). \tag{3.15}$$

The same formulas can be obtained by setting the derivative of the error (3.13) equal to zero.

In the lightwave communication systems any distortion of the transmitted information cannot be tolerated, and the requirements to the probabilities of error are especially severe, $P_{err} = 10^{-9}$, $P_{tm} = P_{fa} = 5 \cdot 10^{-10}$ [10, 39]. Then, according to (3.3), (3.14), we have

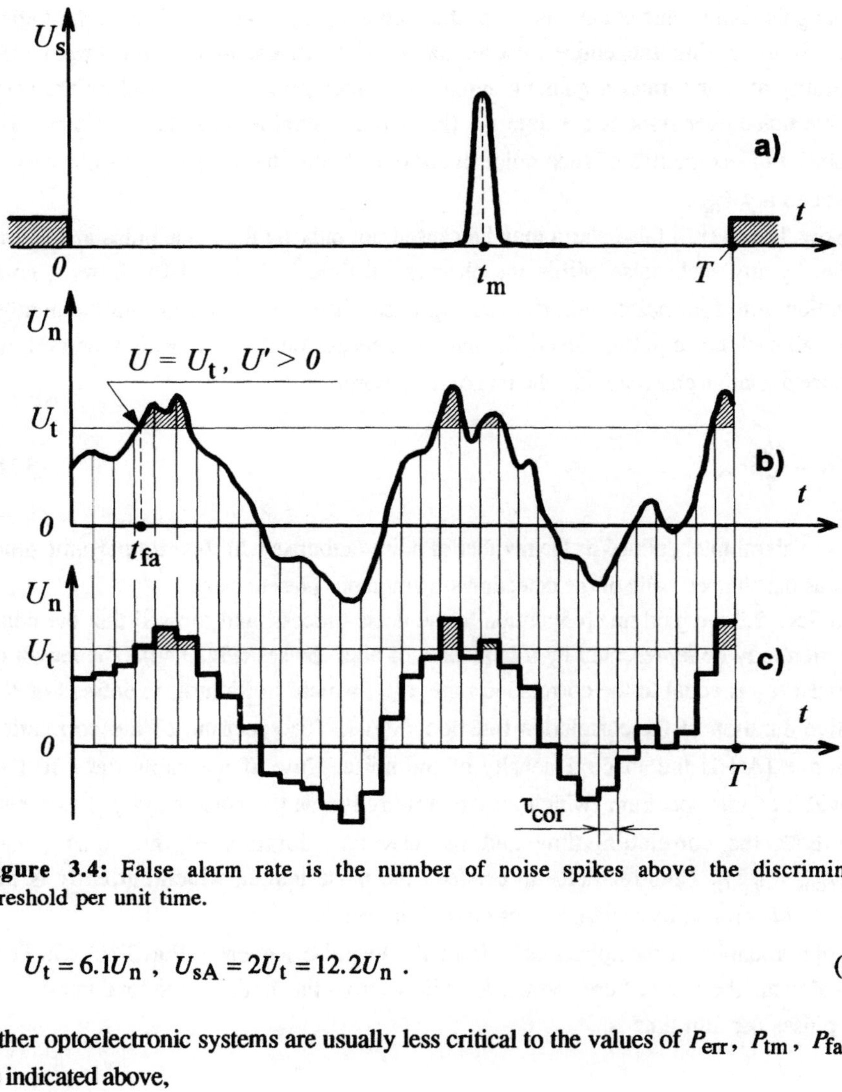

Figure 3.4: False alarm rate is the number of noise spikes above the discrimination threshold per unit time.

$$U_t = 6.1U_n\ ,\quad U_{sA} = 2U_t = 12.2U_n\ . \tag{3.16}$$

Other optoelectronic systems are usually less critical to the values of P_{err}, P_{tm}, P_{fa}; so, as indicated above,

$$U_t \approx (3...6)U_n\ ,\quad U_{sA} \approx 2U_t = (6...12)U_n\ . \tag{3.17}$$

These values of the discrimination threshold and the desirable SNR may be regarded as typical of the optoelectronic systems.

False alarm rate. So far we have been concerned with the case when the time of arrival of optical pulse t_0 is known, so that known is also the time when the pulse at the output of the amplifier reaches its maximum, and the discriminator can be switched on at exactly this instant. The time of arrival of the optical pulse, however, is far not always known; instead, we only know a rather lengthy time interval (the observation time T), known; instead, we only know a rather lengthy time interval (the observation time T), and the pulse may come at any instant of this interval (Fig. 3.4a). In this case the logical device is on during the entire observation time T. In calculating the target miss probability all our former arguments remain in force: the miss is caused by the same negative noise overshoot at the time t_m the output signal attains its maximum. The probability of occurrence of such noise overshoot of amplitude $|U| > |U_{sA} - U_t|$ defines the probability P_{tm}.

Now, however, a false alarm may be caused not only by the noise pulse at the time t_m, but by any such pulse within the observation time T. Figure 3.4b shows a noise realization with four noise outbursts causing false alarms. We need to find the number N_{fa} of all such noise pulses. Since this number depends on the observation interval, we introduce a relative characteristic, the rate of false alarms:

$$f_{fa} = \frac{N_{fa}}{T} \ . \tag{3.18}$$

The false alarm rate, defined as the number of noise outbursts with $U > U_t$ per unit time, serves as the characteristic of the detection system in our present case.

In Sect. 2.2 we paid much attention to the noise process, and showed that the noise realization may be represented by a sequence of noise pulses (Fig. 3.4c). The length of each pulse τ_n is equal to the correlation time τ_{cor}, which, in its turn, is defined as the effective duration of the correlation function $\mathcal{K}(\Delta t)$. The spectrum of the correlation function $\mathcal{K}(\Delta t)$ is the spectral density of the noise. Now, if we know the effective bandwidth of this spectrum (which may be referred to as the noise band f_n), we also know both the correlation time and the effective duration of the noise pulse, $\tau_n = \tau_{cor} = 1/2f_n$ (this relationship between the pulse length, which currently is the function $\mathcal{K}(\Delta t)$, and its spectrum, was proved in Sect. 2.1).

Approximation of the noise realization with the pulse sequence (Fig. 3.3c) will help us to calculate the rate of false alarms. As follows from the diagram, the total number of noise pulses per unit time is

$$f_{fa\,tot} = \frac{1}{\tau_n} = 2f_n \ . \tag{3.19}$$

It is only some of these pulses that exceed the discrimination threshold U_t; this part has already been calculated in (3.6). Accordingly, for the frequency of false alarms we get the following estimate:

$$U \geq U_t \geq 3U_n \ , \quad f_{fa} = \frac{1}{\tau_n}\left[\frac{1}{\sqrt{2\pi}}\frac{U_n}{U_t}\exp\left(-\frac{U_t^2}{2U_n^2}\right)\right] , \tag{3.20}$$

or, with due account for (3.17),

$$f_{\mathrm{fa}} = \left[\sqrt{\frac{2}{\pi}}\frac{U_{\mathrm{n}}}{U_{\mathrm{t}}}f_{\mathrm{n}}\right]\exp\left(-\frac{U_{\mathrm{t}}^2}{2U_{\mathrm{n}}^2}\right). \tag{3.21}$$

The term in brackets is called the characteristic rate,

$$f_{\mathrm{char}} = \sqrt{\frac{2}{\pi}}\frac{U_{\mathrm{n}}}{U_{\mathrm{t}}}f_{\mathrm{n}} \approx \sqrt{\frac{2}{\pi}}\frac{f_{\mathrm{n}}}{3...6} \approx (0.13...0.26)f_{\mathrm{n}} , \tag{3.22}$$

where the numerical estimate is based on the typical values of $U_{\mathrm{t}}/U_{\mathrm{n}}$ according to (3.17). Now we may rewrite (3.21) as

$$f_{\mathrm{fa}} = f_{\mathrm{char}}\exp\left(-\frac{U_{\mathrm{t}}^2}{2U_{\mathrm{n}}^2}\right). \tag{3.23}$$

The four false alarms are shown in Fig. 3.4b,c just by way of an illustration. Real optoelectronic systems cannot afford such a luxury: the average number of false alarms N_{fa} must be kept well below unity, and should not be greater than the above-quoted values of $10^{-3}...10^{-9}$. Then N_{fa} actually coincides with the probability of occurrence of one false alarm within the time T, and, with due account for (3.18), (3.20), we get the following expression for P_{fa}:

$$P_{\mathrm{fa}} = N_{\mathrm{fa}} = f_{\mathrm{fa}}T = \frac{T}{\tau_{\mathrm{n}}}\left[\frac{1}{\sqrt{2\pi}}\frac{U_{\mathrm{n}}}{U_{\mathrm{t}}}\exp\left(-\frac{U_{\mathrm{t}}^2}{2U_{\mathrm{n}}^2}\right)\right]. \tag{3.24}$$

Let us now compare (3.24) with the false alarm probability (3.6) calculated under assumption that the time of arrival of the signal is known, and therefore one sample was sufficient. The noise retains it value for the time $\tau_{\mathrm{n}} = \tau_{\mathrm{cor}}$; accordingly, we have T/τ_{cor} different noise values over the observation time T, when the discriminator is on. This is the reason why the probability (the average number) of false alarms has increased by the factor of T/τ_{cor} (compare (3.5) and (3.24)).

The exact calculation of f_{fa} in statistical radio engineering is based on the two-dimensional probability distribution for two random quantities: the noise U_{n}, and its derivative U_{n}'. A false alarm at the time t_{m} occurs when $U_{\mathrm{n}}(t_{\mathrm{m}}) = U_{\mathrm{t}}$, $U_{\mathrm{n}}'(t_{\mathrm{m}}) > 0$ (Fig. 3.4b). This approach leads to the following expression [55]:

$$f_{\mathrm{fa}} = f_{\mathrm{char}}\exp\left(-\frac{U_{\mathrm{t}}^2}{2U_{\mathrm{n}}^2}\right),$$

$$f_{\text{char}} = \sqrt{\frac{\int_0^\infty f^2 \Theta_{\text{out}}(f)df}{f_\text{n}}} , \tag{3.25}$$

$$f_\text{n} = \int_0^\infty \Theta_{\text{out}}(f)df ,$$

where $\Theta_{\text{out}}(f)$ is the relative noise spectrum at the output. When the input spectrum is white ($\Theta = 1$), the noise spectrum at the output Θ_{out} is determined by the frequency response of the amplifier, as follows from (2.40). Let us calculate f_{char} for the simplest case of the amplifier with the rectangular passband:

$$f_\text{a} = f_\text{n} , \quad k(f) = \begin{cases} 1, & f \le f_\text{a}; \\ 0, & f > f_\text{a}; \end{cases}$$

$$f_{\text{char}} = \sqrt{\frac{\int_0^{f_\text{n}} f^2 df}{\int_0^{f_\text{n}} df}} = \sqrt{\frac{\frac{1}{3} f_\text{n}^3}{f_\text{n}}} = \frac{f_\text{n}}{\sqrt{3}} = 0.58 f_\text{n} . \tag{3.26}$$

In Sect. 4.1 we shall consider the frequency response of real PDA amplifiers, shaped by RC circuits. The characteristic rate of false alarms with such amplifiers is $f_{\text{char}} \approx f_\text{n} \pm 30\%$.

The comparison between the approximate expression (3.22) and exact ones (3.25), (3.26) reveals that they coincide up to a factor of $0.58/(0.13 \ldots 0.26) = 2.2 \ldots 4.4$. Observe that the factor of such magnitude does not play any decisive role: the rate of false alarms exhibits a very acute dependence on the threshold U_t, and so a minor increase of the threshold (by 6...23%) will reduce f_{fa} by an order of magnitude:

$$\frac{f_{\text{fa}1}}{f_{\text{fa}2}} \approx \frac{\exp\left(-\frac{U_{\text{t}1}^2}{2U_\text{n}^2}\right)}{\exp\left(-\frac{U_{\text{t}2}^2}{2U_\text{n}^2}\right)} = \frac{\exp\left(-\frac{3.7^2}{2}\right)}{\exp\left(-\frac{3^2}{2}\right)} \approx 0.1 ;$$

$$\frac{\exp\left(-\frac{6.4^2}{2}\right)}{\exp\left(-\frac{6^2}{2}\right)} \approx 0.1\,. \qquad (3.27)$$

Hence follows another important conclusion: the false alarm probability over the observation time T, like that in case of the unit sample, is controlled primarily by the selected ratio U_t/U_n, and to a much smaller extent by the passband of the amplifier (the frequency f_{char}).

Summary. *To decide whether or not there is a signal, the mixture of signal with noise from the linear circuit must be sent to the decision-making device (discriminator). If the applied voltage exceeds the preset discrimination threshold* U_t*, the device decides that the signal has come. Noise outbursts lead to errors. If the noise overshoots the threshold* U_t *in the absence of signal, the device mistakes it for the signal. This wrong decision is called false alarm. A large negative noise pulse, superimposed on the signal, may reduce the voltage below the discrimination threshold, and then the wrong decision is that there is no signal (target miss). The probabilities of false alarm* P_{fa} *and target miss* P_{tm} *belong to the main criteria of quality of the detection system. The higher the ratio* U_t/U_n*, the lower the probability* P_{fa}*; the higher the ratio* U_{sA}/U_t*, the lower the probability* P_{tm}*. This brings us to the main issue in the design of PDAs: the need to maximize the signal-to-noise ratio* U_{sA}/U_n.

As a rule, the probabilities of error P_{fa}, P_{tm} *in optoelectronic systems should not exceed* $10^{-3}...10^{-9}$*, which corresponds to* $U_{sA}/U_n \approx 6...12$ *and* $U_t = (3...6)U_n$*. We see that the margin of the desirable values of* U_{sA}/U_n *and* U_t/U_n *in PDAs is relatively narrow. This is explained by the fact that the probabilities* P_{fa}, P_{tm} *exhibit a very strong dependence on these ratios.*

When the time of expected arrival of the signal is known precisely, it is sufficient to take one reading at the instant the signal attains its maximum. If, conversely, the signal may arrive at any moment within the observation time T, it is necessary to take T/τ_{cor} *readings over the time T (where* τ_{cor} *is the noise correlation time), and the probability of error is then increased by the same factor of* T/τ_{cor}*. To eliminate the dependence on the observation time T, the rate of false alarms is defined as the number of noise outbursts per unit time,* $f_{fa} = N_{fa}/T = P_{fa}/T$.

3.2. Optimal Filter (Frequency Analysis)

Now the target of the designer of threshold-level PDAs is clear: first of all, it consists in maximizing the signal-to-noise ratio at the output of the channel (amplifier). Let us write the expression for SNR at the output of the amplifier, and seek its maximum with respect to all controlling parameters.

Calculation of SNR at the output of the amplifier. The signal and the noise at the input of the amplifier were described by current generators (2.10), (2.11), (2.168):

$$I_s(f) = S\,\mathrm{E}_s(f) = SE_{sA}\,2T_s e_s(f)\ , \tag{3.28}$$

$$\overline{J^2(f)} = 2qI_n\Theta(f)\ . \tag{3.29}$$

Knowing these input spectra, and the frequency dependence of the transfer coefficient of the amplifier $K_{21}(f)$ (Sect. 1.3), we may calculate the spectra of signal and noise at the output of the amplifier:

$$U_s(f) = K_0 SE_{sA}\,2T_s k(f) e_s(f)\ , \tag{3.30}$$

$$\overline{U_n^2(f)} = K_0^2\,2qI_n k^2(f)\Theta(f)\ . \tag{3.31}$$

Here we have introduced the relative frequency response of the amplifier $k(f)$ as the ratio of the coefficient $K_{21}(f)$ to its maximum value K_0 at the plateau, so that $K_{21}(f) = K_0 k(f)$. Note that the coefficient $k(f)$ is dimensionless, whereas the coefficient K_0 has the dimension of resistance, since the input current is converted into the output voltage. In some cases, however, it is more convenient to register other electrical quantities at the output of the linear circuit, like the electric charge in the integrator. Then the dimensionality of K_0 must be redefined (Sect. 7.1).

We are interested in the value of the signal at the time it reaches its maximum; adding up all harmonics, we get

$$U_{sA} = U_s(t_m) = K_0 SE_{sA}\,2T_s \int_0^{\infty} k(f) e_s(f) \cos(\omega t_m + \psi_s + \psi_a)\, df\ , \tag{3.32}$$

where ψ_s, ψ_a are the phase-frequency characteristics of, respectively, the signal and the amplifier. From this expression we see that the signal reaches its maximum when all harmonics are in phase, and their amplitudes add up. Hence, at the time t_m all cosine-waves in the integrand should reach their maximum value (unity), whence follows the condition

$$\omega t_m + \psi_s + \psi_a(f) = 0\ , \quad \psi_a(f) = -\omega t_m - \psi_s\ . \tag{3.33}$$

The first term $-\omega t_m$ appears in the spectrum when the pulse is shifted by the amount t_m along the time axis (Sect. 2.1). In radio engineering this is a standard manipulation: the amplifier may be equipped with a delay line to produce any desired time lag t_m; this does not change either the shape or the amplitude of the signal, and cannot affect the signal-to-noise ratio. Of crucial importance is the second term $-\psi_s$: when the pulse is asymmetrical, $\psi_s \neq 0$ (see Sect. 2.1), the amplifier must make up for the phase-frequency distortion of the signal; then all the signal harmonics will join in unison, and it is then that the signal will reach its maximum possible value of

$$U_{sA} = K_0 S E_{sA} 2T_s \int_0^\infty k(f) \mathrm{e}_s(f) df \ . \tag{3.34}$$

The output noise samples are the same at all times, and are equal to the rms value:

$$\sqrt{\overline{U_n^2}} = \sqrt{\int_0^\infty K_{21}^2 \overline{J^2} df} = \sqrt{K_0^2 2qI_n} \sqrt{\int_0^\infty k^2(f)\Theta(f) df} \ , \tag{3.35}$$

and so for the signal-to-noise ratio at $t = t_m$ we get

$$\mathrm{SNR} = \frac{U_{sA}}{U_n} = \frac{SE_{sA}}{\sqrt{2qI_n}} \frac{2T_s \int_0^\infty k(f) \mathrm{e}_s(f) df}{\sqrt{\int_0^\infty k^2(f)\Theta(f) df}} \ . \tag{3.36}$$

Statement of the problem of optimal filtering. There is no need to comment on the obvious dependence of SNR on such parameters of photodetectors and PDAs as sensitivity S and noise density $2qI_n$. Naturally, one goes in for higher S and lower I_n; the achievable benchmarks were discussed in detail in Sect. 1.2, 2.5. We showed that in modern PDAs these values are close to theoretical limit. There is no need to analyze the dependence of SNR on the amplitude E_{sA}: as in any linear system, the output signal and the SNR increase in sympathy with the input signal. The task of the engineer is to obtain the required SNR with the lowest possible amplitude of the optical signal E_{sA}. Other parameters of the signal, which occur in (3.35) (duration T_s and spectrum $\mathrm{e}_s(f)$), are preassigned. So at our disposal we have only one degree of freedom, the choice of the relative frequency response of the amplifier $k(f)$. Let us rewrite our formula for SNR (3.36) as

$$\mathrm{SNR} = \frac{SE_{sA}}{\sqrt{2qI_n}} \frac{u_{sA}}{\sqrt{f_n}} \ , \tag{3.37}$$

$$u_{sA} = 2T_s \int_0^\infty k(f) \mathrm{e}_s(f) df \ , \tag{3.38}$$

$$f_n = \int_0^\infty k^2(f)\Theta(f)df \; . \tag{3.39}$$

Here u_{sA} is the relative amplitude of the output signal; f_n is the effective frequency band of the input noise. The physical meaning of these parameters is clear from their definitions (3.38), (3.39). For example, the relative amplitude u_{sA} depends on the inertiality of the amplifier: if it is a broad-band amplifier, $k(f)=1$, the signal spectrum is not distorted, and

$$u_{sA} = 2T_s\int_0^\infty e_s(f)df = 2T_s\frac{1}{2T_s} = 1 \; . \tag{3.40}$$

Recall that the integral here defines the effective spectral bandwidth of the signal (2.15). If the passband of the amplifier is comparable with the spectral band of the signal, one has to take into account the frequency dependence of the gain $k(f)<1$; as follows from (3.38), the amplitude of the output signal rolls off because of inertiality of the amplifier ($u_{sA}<1$).

We have already used the band f_n in the calculation of the rate of false alarms (Sect. 3.1). Recall that in case of white noise ($\Theta=1$) the expression (3.39) is the commonly recognized definition of the effective noise band of the amplifier. Now the physical meaning of the term $u_{sA}/\sqrt{f_n}$ in (3.37) becomes clear: it is the ratio of the relative amplitude to square root of the effective noise band. When the passband of the amplifier is very broad, $k(f)=1$, the signal increases ($u_{sA}\to 1$), but the effective frequency band f_n also increases, and so does the noise. When the amplifier band is narrow, the noise decreases, but the output signal also falls off ($u_{sA} << 1$). So, one might expect that (and this will be proved shortly) the SNR reaches its maximum for a certain optimal band of the amplifier (or, more precisely, for the optimal frequency response $k(f)$).

In this way, the maximization of SNR reduces to choosing the best frequency response $k(f)$, which brings us to the main problem in the detection theory: the problem of optimal filtering. The term "filtering" pinpoints the essence of the problem: it is the filter and its relative frequency response $k(f)$ that we have to choose. The magnitude of the transfer coefficient K_0 itself is irrelevant to the detection, and does not occur in the expression for SNR. As K_0 increases, the signal and the noise increase in equal proportion, and their ratio remains the same. Of course, K_0 must be large enough to suppress the interference at the output, and to ensure the signal level required by the discriminator.

Observe that the input resistance and capacitance of PDA affect both the noise I_n and the frequency response $k(f)$. The parameters of the input circuit must be chosen

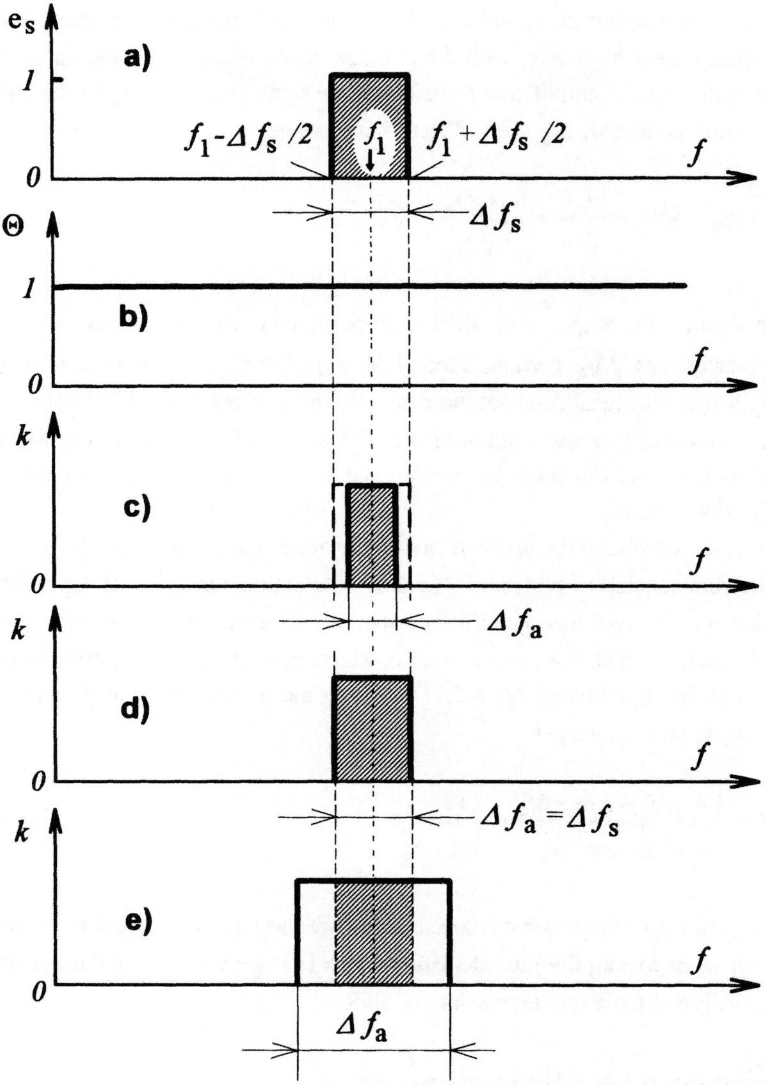

Figure 3.5: In case of the narrow-band signal the passband of optimal filter is equal to the bandwidth of the signal: (a) spectrum of near-harmonic signal; (b) white noise spectrum; (c) narrow passband of amplifier; (d) optimal passband of amplifier; (e) broad passband of amplifier.

primarily from considerations of minimum noise (Sect. 2.5), and the desired frequency response should be shaped by the subsequent stages, using equalizers and feedbacks if necessary (see Sect. 1.3).

Optimal filter in case of white noise. Let us find the frequency response of the optimal filter, moving "from the simple to the complex". We start with the most simple spectra of signal and noise, and gradually advancing towards the more complicated spectra, all the way to those of preassigned arbitrary shape. The simplest case is illustrated in Fig. 3.5: the noise spectrum is white, and the signal is harmonic of frequency f_1 (or, more precisely, near-harmonic, with the constant spectral density in the narrow band, $\Delta f_s \le f_1$). Naturally, the amplification must be at the signal frequency f_1. If the passband of the amplifier is narrow, $\Delta f_a \le \Delta f_s$ (Fig. 3.5a,c), we get

$$\Delta f_a \le \Delta f_s \ , \ \mathrm{SNR} \propto \frac{u_{sA}}{\sqrt{f_n}} \propto \frac{e_s k_1 \Delta f_a}{\sqrt{k_1^2 \Delta f_a}} \propto \sqrt{\Delta f_a} \ . \tag{3.41}$$

Since the signal at the output of the filter is proportional to Δf_a, and the noise to $\sqrt{\Delta f_a}$, the SNR increases as $\sqrt{\Delta f_a}$ until the band of the amplifier exceeds the band of the signal. When $\Delta f_a > \Delta f_s$, the signal does not increase with the increasing amplifier bandwidth; it is only the noise that increases, and therefore SNR starts to fall. Hence it follows that the passband of the amplifier must be made equal to the frequency band of the signal, $\Delta f_a \equiv \Delta f_s$ (Fig. 3.5d,e).

Now let us complicate the problem, and add another signal frequency f_2, so that the signal spectrum consists of two narrow (and by assumption equal) bands Δf_s at different frequencies f_1, f_2, and having different spectral densities e_{s1}, e_{s2} (Fig. 3.6a,b). Arguing by analogy with the precedent case, we conclude that the amplifier band must correspond to the signal band, $\Delta f_a = \Delta f_s$. Denoting the relative gains at f_1 and f_2 by, respectively, k_1 and k_2, we get

$$\mathrm{SNR} \propto \frac{k_1 e_{s1} \Delta f_a + k_2 e_{s2} \Delta f_a}{\sqrt{k_1^2 \Delta f_a + k_2^2 \Delta f_a}} \propto \frac{1 + k_2 e_{s2}}{\sqrt{1 + k_2^2}} \ . \tag{3.42}$$

Since $k(f)$ and $e_s(f)$ are relative characteristics, we have set them equal to unity at the frequency f_1 so as to simplify the calculations, $k_1 = 1$, $e_{s1} = 1$. Let us find the numerator of the derivative of the above expression for SNR:

$$\frac{d\,\mathrm{SNR}}{dk_2} \propto e_{s2}\sqrt{1 + k_2^2} - (1 + k_2 e_{s2}) \frac{k_2}{\sqrt{1 + k_2^2}}$$

$$\propto e_{s2}\left(1 + k_2^2\right) - (1 + k_2 e_{s2}) k_2 = e_{s2} - k_2 \ . \tag{3.43}$$

We see that the derivative $d\,\mathrm{SNR}/dk_2$ goes to zero when $k_2 = e_{s2}$. Thus,

$$k_1 = e_{s1} = 1 \ , \ k_2 = e_{s2} \ . \tag{3.44}$$

This straightforward reasoning brings us to the important conclusion: the optimal filter (such as maximizes the SNR) exists. The gain at the frequency f_2 of such optimal filter is proportional to the spectral density of signal e_{s2} at this frequency.

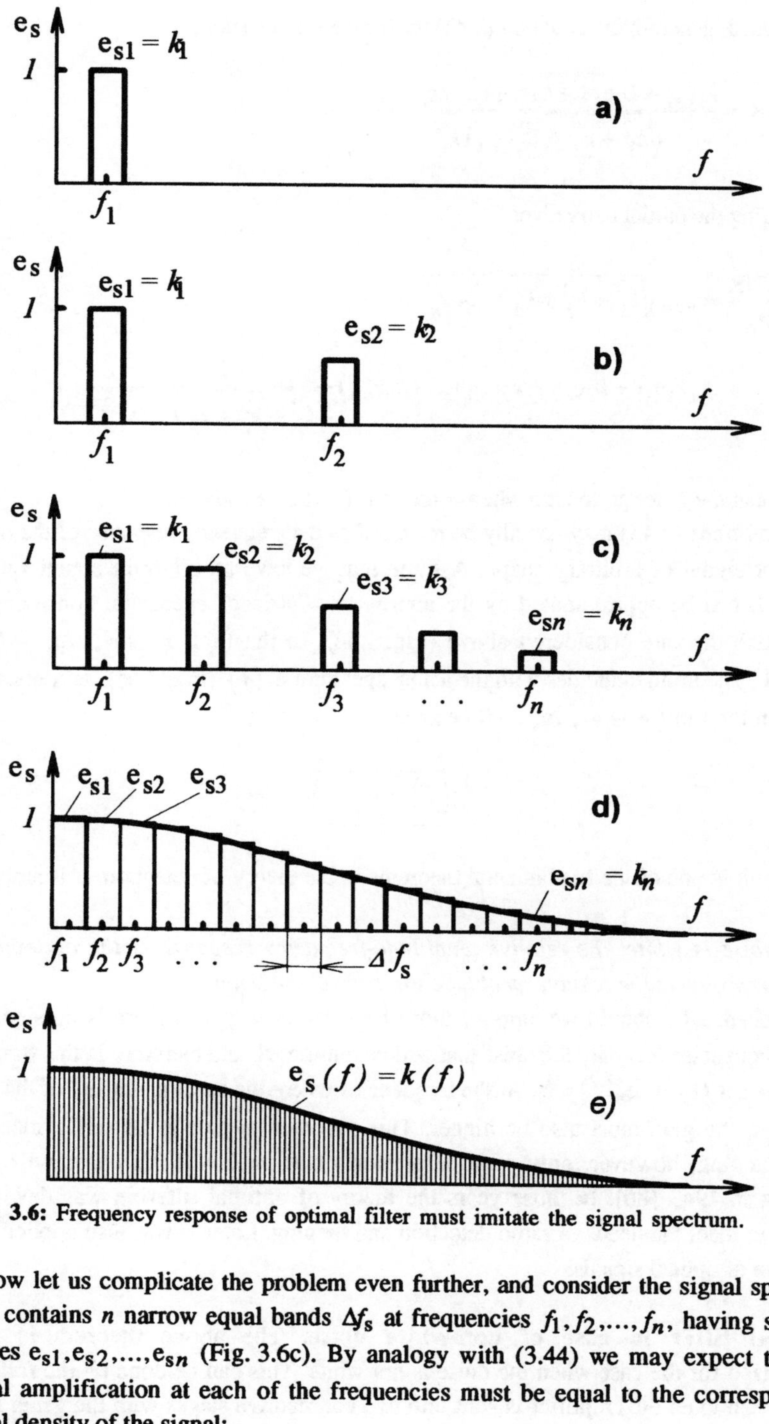

Figure 3.6: Frequency response of optimal filter must imitate the signal spectrum.

Now let us complicate the problem even further, and consider the signal spectrum which contains n narrow equal bands Δf_s at frequencies $f_1, f_2, ..., f_n$, having spectral densities $e_{s1}, e_{s2}, ..., e_{sn}$ (Fig. 3.6c). By analogy with (3.44) we may expect that the optimal amplification at each of the frequencies must be equal to the corresponding spectral density of the signal:

$$k_1 = e_{s1}\ ,\ \ k_2 = e_{s2}\ ,\ \ k_3 = e_{s3}\ ,\ ...,\ k_n = e_{sn}\ . \tag{3.45}$$

And indeed, generalizing equation (3.42) for the case in question,

$$\mathrm{SNR} \propto \frac{k_1 e_{s1} + k_2 e_{s2} + k_3 e_{s3} + \ldots + k_n e_{sn}}{\sqrt{k_1^2 + k_2^2 + k_3^2 + \ldots + k_n^2}}, \tag{3.46}$$

and finding the partial derivatives

$$\frac{\partial\,\mathrm{SNR}}{\partial k_n} \propto e_{sn}\sqrt{k_1^2 + k_2^2 + k_3^2 + \ldots + k_n^2} - \left(k_1 e_{s1} + k_2 e_{s2} + k_3 e_{s3} + \ldots + k_n e_{sn}\right)\frac{k_n}{\sqrt{k_1^2 + k_2^2 + k_3^2 + \ldots + k_n^2}}, \tag{3.47}$$

we see that the latter go to zero when conditions (3.45) are satisfied.

Conditions (3.45) may actually be regarded as the frequency response of the optimal filter for signal of arbitrary shape. Assume that we have an arbitrary signal spectrum $e_s(f)$. It can be approximated by the appropriate "striped" spectrum, consisting of n bands like the one considered above (Fig. 3.6d). In the limit $n \to \infty$, $\Delta f_s \to 0$, the "striped " spectrum coincides with the actual spectrum $e_s(f)$ (Fig. 3.6e), and conditions (3.45) in the limit $n \to \infty$, $\Delta f_s \to 0$ become

$$\boxed{k(f) = e_s(f)} \tag{3.48}$$

This result is one of the fundamental theorems in the theory of detection. It is amazingly simple:

If the noise is white, the relative amplitude-frequency response of the optimal filter exactly reproduces the relative amplitude spectrum of the signal.

And indeed, why should we amplify those frequencies at which there is no signal? At these frequencies it is only the noise that will be reinforced, and therefore in this frequency range we set $k(f) = e_s(f) = 0$. At the frequencies where the spectral density of the signal is higher, the gain must also be higher. This statement seems to be self-evident; this understanding, however, only came after North had published his paper on optimal filtering in 1943 [80]. In those years the theory of optimal filtering was developing rapidly to meet the needs of radio detection and ranging. Later is was also applied to the detection of optical signals.

Optimal filter in case of non-white noise. The above theorem is easily generalized for the case when the noise is not white. This can be done by the following clever technique: the amplifier is split into two consecutive stages with the gains k_{I} and k_{II}, respectively (Fig. 3.7). Then the gain of the first stage k_{I} is adjusted in such a way as to "whiten" the noise at the output. This requires setting $k_{\mathrm{I}} = 1/\sqrt{\Theta(f)}$, so that

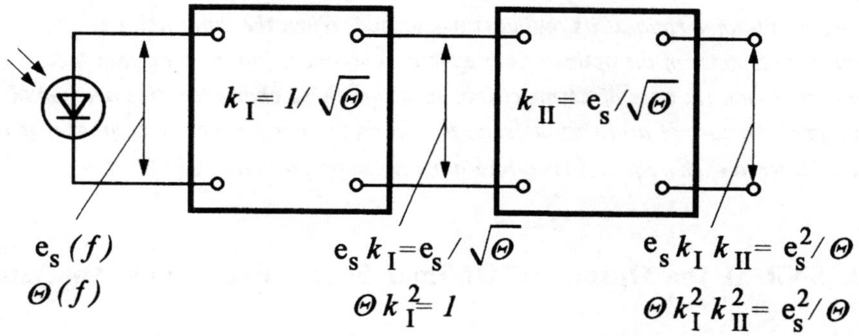

Figure 3.7: To the derivation of frequency response of optimal filter for arbitrary noise spectrum.

$\Theta(f)k_I^2 = 1$. Now at the output of the first stage we have white noise, and the signal spectrum is $e_s(f)k_I(f) = e_s(f)/\sqrt{\Theta(f)}$. For the case of white noise, however, we already know how to choose the optimal filter. Assume that such optimal filter is the second stage. Its frequency response must exactly reproduce the signal spectrum at its input, $k_{II} = e_s(f)/\sqrt{\Theta(f)}$. The total gain of the two stages is

$$k(f) = k_I k_{II} = \frac{1}{\sqrt{\Theta(f)}}\frac{e_s(f)}{\sqrt{\Theta(f)}} = \frac{e_s(f)}{\Theta(f)} . \tag{3.49}$$

Thus, we have obtained the transfer coefficient of optimal filter for arbitrary spectra of signal and noise. The physical meaning of the modified formula for optimal filter is simple: at those frequencies where the noise is stronger, the noise must be suppressed through reducing the gain, even though the signal at these frequencies will also be reduced.

The transfer coefficient is often represented in the complex form. Using the expressions for the magnitude (3.49) and the phase (3.33), we may write [81]

$$\mathbf{K}_{21} = K_{21} \exp j\psi_a = K_0 \frac{e_s(f)}{\Theta(f)} \exp(-j\omega t_m - j\tilde{\psi}_s)$$

$$= K_0 \exp(-j\omega t_m) \frac{e_s(f)}{\Theta(f)} \exp(-j\tilde{\psi}_s) , \tag{3.50a}$$

$$\mathbf{K}_{21} \propto \frac{e_s^*(f)}{\Theta(f)} . \tag{3.50b}$$

The last expression indicates that the complex frequency response of the optimal filter is equal up to a constant-magnitude factor to the complex-conjugate signal spectrum divided by the noise spectrum.

Summary. *There exists such frequency response of the amplifier (optimal filter) that the signal-to-noise ratio at its output is maximized. When the input noise is white, the frequency response of the optimal filter exactly reproduces (up to a constant factor) the signal spectrum (in case of asymmetrical spectrum these characteristics are complex-conjugate). In case of arbitrary noise spectrum* $\Theta(f)$, *the frequency response of the optimal filter at each frequency is corrected in inverse proportion to* $\Theta(f)$.

3.3. SNR at the Output of Optimal Filter (Frequency Analysis)

General expression for SNR at the output of optimal filter. In the preceding section we found expressions (3.37) – (3.39) for SNR at the output of filter-amplifier having an arbitrary frequency response, and expression (3.49) for the frequency response of the optimal filter. Substituting the latter into the former, we get the desired formula:

$$\mathrm{SNR_{OF}} = \frac{SE_{\mathrm{sA}}}{\sqrt{2qI_{\mathrm{n}}}} \frac{u_{\mathrm{sA}}}{\sqrt{f_{\mathrm{n}}}} = \frac{SE_{\mathrm{sA}}}{\sqrt{2qI_{\mathrm{n}}}} \frac{2T_{\mathrm{s}}\int_0^\infty k e_{\mathrm{s}} df}{\sqrt{\int_0^\infty k^2 \Theta df}} = \frac{SE_{\mathrm{sA}}}{\sqrt{2qI_{\mathrm{n}}}} \frac{2T_{\mathrm{s}}\int_0^\infty \frac{e_{\mathrm{s}}^2}{\Theta} df}{\sqrt{\int_0^\infty \frac{e_{\mathrm{s}}^2}{\Theta} df}}, \tag{3.51}$$

or, after reduction,

$$\mathrm{SNR_{OF}} = \frac{SE_{\mathrm{sA}}(2T_{\mathrm{s}})}{\sqrt{2qI_{\mathrm{n}}}} \sqrt{\int_0^\infty \frac{e_{\mathrm{s}}^2}{\Theta} df} . \tag{3.52}$$

These expressions call for some comments.

1. The noise band of optimal filter coincides with the proper noise band of the amplifier with the gain $k = e_{\mathrm{s}}/\sqrt{\Theta}$ (this is the gain of the second stage of our optimal filter, Fig. 3.7). Recall that the noise band of the amplifier itself, defined in case of white input noise, is equal to the area under the curve $k^2(f)$:

$$f_{\mathrm{n}} = \int_0^\infty k^2 df = \int_0^\infty \left(\frac{e_{\mathrm{s}}}{\sqrt{\Theta}}\right)^2 df = \int_0^\infty \frac{e_{\mathrm{s}}^2}{\Theta} df . \tag{3.53}$$

2. The relative amplitude is determined by the same band; more precisely, by the ratio of this band to the effective band of signal spectrum $f_{\mathrm{s}} = 1/2T_{\mathrm{s}}$:

$$u_{sA} = \frac{f_n}{f_s} = 2T_s \int_0^{\infty} \frac{e_s^2}{\Theta} df \ . \tag{3.54}$$

3. Fourier analysis in linear systems allows us to calculate independently each harmonic of signal and noise, and then add up the harmonics (algebraically or geometrically) to find the signal amplitude or the rms noise value. A similar method turns out to be suitable for calculating the signal-to-noise ratio. First we find the signal-to-noise ratio snr for each harmonic separately at the input of the filter. The most that the optimal filter can do is to do no harm — to transmit carefully all these partial ratios to the output. Then the SNR at the output can be found through geometrical summation of all the partial signal-to-noise ratios, in exactly the same way as it is done for the signal and the noise power.

For calculation of snr^2 we take advantage of the results obtained in Sect. 2.2. We showed that, when the observation time T is finite, the continuous spectrum as though splits up into separate harmonics with the frequency step $\Delta f = 1/T$. The spectral density, initially spread over the frequency interval Δf, pulls up and defines the amplitudes of the harmonics — in our present case, those of signal $j_s(f)$ and noise $j_n(f)$; recall equation (2.32):

$$j_s = (SE_{sA} 2T_s) e_s(f) \Delta f \ , \quad \frac{\overline{j_n^2}}{2} = \overline{J^2} \Delta f \ , \quad \overline{J^2} = 2qI_n \Theta(f) \ . \tag{3.55}$$

Hence, for the partial ratio snr^2 of such harmonic we have

$$\mathrm{snr}^2 = \frac{I_s^2}{\frac{\overline{j_n^2}}{2}} = \frac{(SE_{sA} 2T_s e_s(f) \Delta f)^2}{\overline{J^2} \Delta f} = \frac{(SE_{sA} 2T_s)^2}{2qI_n} \frac{e_s^2(f)}{\Theta(f)} \Delta f \ . \tag{3.56}$$

Now we only have to fulfil our promise: add up all these partial ratios, and, letting $\Delta f \to 0$, go from the sum to the integral:

$$\mathrm{SNR_{OF}} = \sqrt{\sum \mathrm{snr}^2} = \frac{SE_{sA} 2T_s}{\sqrt{2qI_n}} \sqrt{\sum \frac{e_s^2(f)}{\Theta(f)} \Delta f} = \frac{SE_{sA} 2T_s}{\sqrt{2qI_n}} \sqrt{\int_0^{\infty} \frac{e_s^2(f)}{\Theta(f)} df} \ . \tag{3.57}$$

These expressions for $\mathrm{SNR_{OF}}$ belong to the fundamental relations of the detection theory. By definition, this value is the highest achievable signal-to-noise ratio for the given spectra $e_s(f)$ and $\Theta(f)$. Striving for an SNR higher than that of the optimal filter is the same as trying to build a perpetuum mobile. In real PDAs the value of SNR can only approach $\mathrm{SNR_{OF}}$. Pursuing the analogy with the engine, we may introduce a characteristic as important as, for example, the thermodynamic efficiency: the coefficient of loss (or simply loss) of real PDA (real filter), defined as the ratio of SNR of the real filter to that of the optimal filter:

$$\mathcal{L} = \frac{\mathrm{SNR}}{\mathrm{SNR_{OF}}} \ . \tag{3.58}$$

The loss coefficient is a most important indicator of quality of PDA as a detection system of weak optical signals.

Calculation of $\mathrm{SNR_{OF}}$ for particular spectra of signal and noise is one of the main problems in the detection theory. Since the formula for $\mathrm{SNR_{OF}}$ is already obtained, further calculations reduce to straightforward mathematical manipulations. The resulting expressions, however, for the general case of complicated spectra $\mathrm{e_s}(f)$ and $\Theta(f)$ are rather cumbersome, and the transparency of physics of optimal filtering is lost. To avoid this, and to be able to trace how and why the value of $\mathrm{SNR_{OF}}$ depends on the parameters of signal and noise, we shall first use a simplified technique:

1. We shall directly calculate both the signal $SE_{\mathrm{sA}}u_{\mathrm{sA}}$ and the noise $\sqrt{2qI_{\mathrm{n}}f_{\mathrm{n}}}$ geometrically as the area under the curves $\mathrm{e_s}k$, Θk^2, and then find their ratio.

2. Following the recommendations of Sect. 2.1, we shall carry out our calculations for the effective rectangular signal spectrum of constant density $SE_{\mathrm{sA}}2T_{\mathrm{s}}$ in the band $1/2T_{\mathrm{s}}$.

3. We shall consider "pure" cases, when one of the noise components (white, high-frequency, or low-frequency) prevails.

Before proceeding with our analysis, we should make the following comments.

The signal at the output is actually $K_0 SE_{\mathrm{sA}}u_{\mathrm{sA}}$, and the noise $K_0\sqrt{2qI_{\mathrm{n}}f_{\mathrm{n}}}$. Since, however, we are concerned with their ratio, it is assumed that $K_0 \equiv 1$ whenever possible. The quantity $SE_{\mathrm{sA}}u_{\mathrm{sA}}$ will nevertheless be termed the output signal, and $\sqrt{2qI_{\mathrm{n}}f_{\mathrm{n}}}$ the output noise. Certainly, to obtain the absolute values of signal and noise at the output, these expressions must be multiplied by K_0. We may also interpret $SE_{\mathrm{sA}}u_{\mathrm{sA}}$ and $\sqrt{2qI_{\mathrm{n}}f_{\mathrm{n}}}$ as the output voltages referred to the input of the amplifier. When we say that "the signal increases with increasing passband" we mean that filter-amplifiers have the same gain K_0.

In general, the gain coefficient K_0 is dimensional. In the present section, for instance, the inputs are currents (SE_{sA} and $\sqrt{2qI_{\mathrm{n}}}$), and the outputs are voltages, so that K_0 has the dimension of resistance [V/A]. Assuming that $K_0 = 1$, we measure the output in amperes. In spite of this, we still refer to the output voltages of signal and noise. For the sake of consistency it may be assumed, however, that we are dealing with dimensionless quantities, normalized to their respective units of measurement.

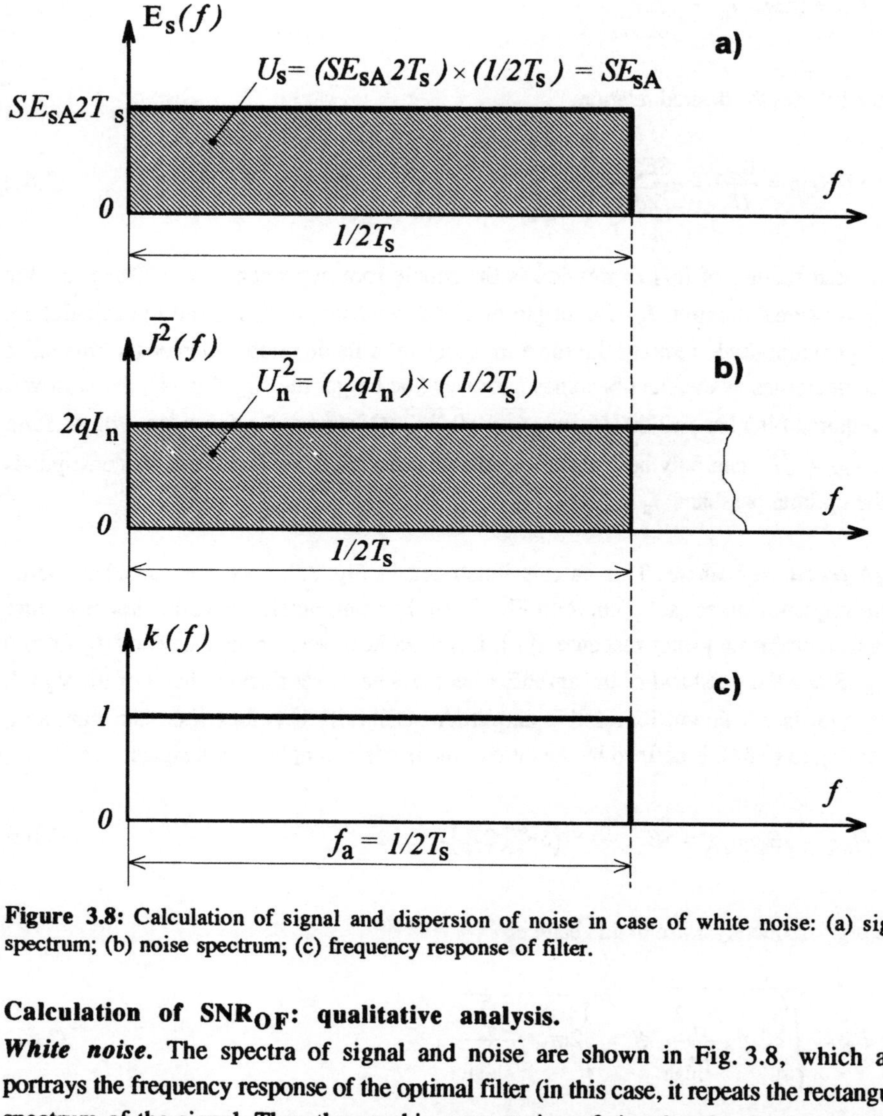

Figure 3.8: Calculation of signal and dispersion of noise in case of white noise: (a) signal spectrum; (b) noise spectrum; (c) frequency response of filter.

Calculation of $\mathbf{SNR_{OF}}$: qualitative analysis.

White noise. The spectra of signal and noise are shown in Fig. 3.8, which also portrays the frequency response of the optimal filter (in this case, it repeats the rectangular spectrum of the signal. Then the graphic computation of signal and noise is especially simple (the relevant areas in the diagram are hatched). As follows from the diagram, such filter transmits all the signal harmonics without loss, and therefore the signal amplitude at the output is not diminished:

$$U_{sA} = \left(SE_{sA}2T_s\right)f_a = \left(SE_{sA}2T_s\right)\frac{1}{2T_s} = SE_{sA} \,. \tag{3.59}$$

The noise in the selected band $f_a = f_n = f_s$ is

$$U_{\mathrm{n}} = \sqrt{2qI_{\mathrm{n}}f_{\mathrm{n}}} = \sqrt{2qI_{\mathrm{n}}\frac{1}{2T_{\mathrm{s}}}}\ ; \tag{3.60}$$

hence follows the desired relation:

$$\mathrm{SNR_{OF}} = \frac{U_{\mathrm{sA}}}{U_{\mathrm{n}}} = \frac{SE_{\mathrm{sA}}}{\sqrt{2qI_{\mathrm{n}}}}\sqrt{2T_{\mathrm{s}}}\ . \tag{3.61}$$

The main feature of this expression is the square-root dependence of $\mathrm{SNR_{OF}}$ on the effective signal duration T_{s}. The origin of this dependence is clear from the calculation: the signal amplitude remains the same irrespective of its duration, whereas the rms noise value decreases as the band becomes narrower (the longer the signal pulse, the narrower the required filter bandwidth), $U_{\mathrm{n}} \propto \sqrt{f_{\mathrm{a}}} \propto \sqrt{1/2T_{\mathrm{s}}}$. Of course, the dependence of the form $\mathrm{SNR_{OF}} \propto \sqrt{T_{\mathrm{s}}}$ can only be realized by adjusting the filter: each value of T_{s} corresponds to the optimal passband $f_{\mathrm{a}} = 1/2T_{\mathrm{s}}$.

High-frequency noise. This case is illustrated in Fig. 3.9. First we consider "pure" high-frequency noise (solid curve in Fig. 3.9b). For simplicity, we again choose a filter with rectangular frequency response $k(f)$; this time, however, we vary its band f_{a} from 0 to f_{s}. Since the passband of the amplifier becomes narrower than the band of the signal, the signal is cut down: its relative amplitude sensitivity becomes less than one, and, according to (3.54), is defined by the ratio of the bands of amplifier and signal:

$$U_{\mathrm{sA}} = SE_{\mathrm{sA}}u_{\mathrm{sA}} = SE_{\mathrm{sA}}\frac{f_{\mathrm{a}}}{f_{\mathrm{s}}} = \left(SE_{\mathrm{sA}}2T_{\mathrm{s}}\right)f_{\mathrm{a}}\ . \tag{3.62}$$

The high-frequency noise in this band is

$$U_{\mathrm{n}} = \sqrt{\int_0^{f_{\mathrm{a}}} 2qI_{\mathrm{n}}\frac{f^2}{f_{\mathrm{high}}^2}df} = \sqrt{2qI_{\mathrm{n}}\frac{f_{\mathrm{a}}^3}{3f_{\mathrm{high}}^2}}\ . \tag{3.63}$$

The areas which define the values of U_{s}, U_{n} are hatched in Fig. 3.9. The signal-to-noise ratio is

$$\mathrm{SNR} = \frac{U_{\mathrm{sA}}}{U_{\mathrm{n}}} \propto \frac{f_{\mathrm{a}}}{\sqrt{f_{\mathrm{a}}^3}} = \frac{1}{\sqrt{f_{\mathrm{a}}}}\ . \tag{3.64}$$

Hence it follows that in theory, by making the band infinitesimally narrow ($f_{\mathrm{a}} \to 0$), one can obtain an infinitely large signal-to-noise ratio. This is easy to explain: the density of high-frequency noise very rapidly tends to zero at low frequencies. The reality, however,

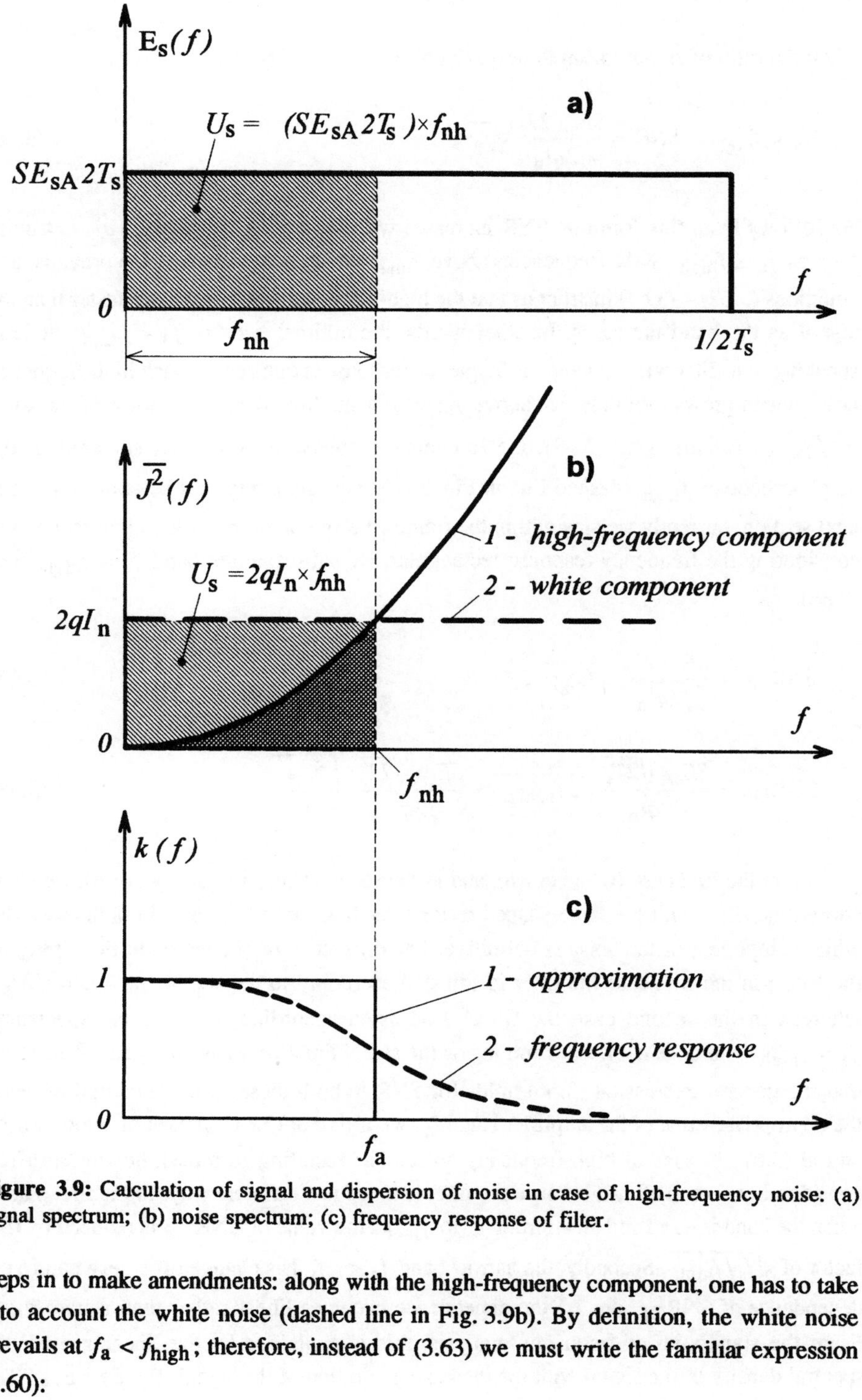

Figure 3.9: Calculation of signal and dispersion of noise in case of high-frequency noise: (a) signal spectrum; (b) noise spectrum; (c) frequency response of filter.

steps in to make amendments: along with the high-frequency component, one has to take into account the white noise (dashed line in Fig. 3.9b). By definition, the white noise prevails at $f_a < f_{high}$; therefore, instead of (3.63) we must write the familiar expression (3.60):

$$U_{\mathrm{n}} = \sqrt{2qI_{\mathrm{n}} f_{\mathrm{a}}} \ . \tag{3.65}$$

Then the ratio of signal (3.62) to noise (3.65) is

$$f_{\mathrm{a}} < f_{\mathrm{high}} \ , \quad \mathrm{SNR} = \frac{SE_{\mathrm{sA}} 2T_{\mathrm{s}}}{\sqrt{2qI_{\mathrm{n}}}} \sqrt{f_{\mathrm{a}}} \ . \tag{3.66}$$

As follows from this formula, SNR increases with increasing passband. This is true as long as $f_{\mathrm{a}} \le f_{\mathrm{high}}$; at the frequencies above f_{high} the high-frequency noise prevails, and equations (3.62) – (3.64) instruct us that the high-frequency noise increases faster than the signal as the band increases. In other words, the optimal band is $f_{\mathrm{a}} = f_{\mathrm{high}}$. Strictly speaking, our filter with rectangular frequency response is optimal for such high-frequency noise which grows infinitely fast above f_{high}. For the high-frequency noise of the form $f^2 / f_{\mathrm{high}}^2$, according to (3.49), the frequency response is somewhat smeared in the neighborhood of f_{high} (dashed line in Fig. 3.9c). This smearing will be analyzed in the next section; currently we stay within the framework of our simple calculation technique: considering the frequency response rectangular, we substitute the band $f_{\mathrm{a}} = f_{\mathrm{high}}$ into (3.66):

$$\mathrm{SNR}_{\mathrm{OF}} = \frac{SE_{\mathrm{sA}} 2T_{\mathrm{s}}}{\sqrt{2qI_{\mathrm{n}}}} \sqrt{f_{\mathrm{high}}} \ , \tag{3.67}$$

$$\mathrm{SNR}_{\mathrm{OF}} = \frac{SE_{\mathrm{sA}} \sqrt{2T_{\mathrm{s}}}}{\sqrt{2qI_{\mathrm{n}}}} \sqrt{2T_{\mathrm{s}} f_{\mathrm{high}}} = \frac{SE_{\mathrm{sA}} \sqrt{2T_{\mathrm{s}}}}{\sqrt{2qI_{\mathrm{n}}}} \sqrt{\frac{f_{\mathrm{high}}}{f_{\mathrm{s}}}} \ . \tag{3.68}$$

So, in the first case (white noise), and in the second (high-frequency noise), we have chosen the filter with the same-shaped rectangular frequency response. In both cases the white component of the noise is definitive. The difference only consists in the tuning of the filter: in the first case the filter is adjusted according to the signal, $f_{\mathrm{a}} = f_{\mathrm{s}} = 1/2T_{\mathrm{s}}$, whereas in the second case the filter is adjusted according to the noise spectrum, $f_{\mathrm{a}} = f_{\mathrm{high}}$ (and does not need trimming as the signal duration T_{s} is changed). It follows that the general expression (3.66) holds for SNR in both these cases: substituting there the appropriate value of the amplifier band f_{a}, we arrive at (3.61) in case of white noise, and at (3.67) in case of high-frequency noise. The resulting formulas, however, differ considerably — and all just because the filter is tuned in a different way. In the first place, when the band is narrowed down from f_{s} to f_{high}, the value of $\mathrm{SNR}_{\mathrm{OF}}$ is reduced by the factor of $\sqrt{f_{\mathrm{s}}/f_{\mathrm{high}}}$. Secondly, the narrow band $f_{\mathrm{a}} \ll f_{\mathrm{s}}$ has changed the very nature of dependence of $\mathrm{SNR}_{\mathrm{OF}}$ on T_{s}: the dependence becomes linear rather than square-root. From the standpoint of frequency analysis, this is explained by the linear growth of spectral density of the signal with the increasing duration of the signal, $\mathrm{E}_{\mathrm{s}}(f) = E_{\mathrm{sA}} 2T_{\mathrm{s}}$;

accordingly, the band $f_a = f_{high}$ being constant, the output amplitude linearly increases (3.62). From the standpoint of time-domain analysis, in case of the narrow-band filter ($f_a << f_s$) the signal may be considered as delta-pulse, and therefore the amplitude at the output is always proportional to the power (area) of the input pulse $E_{sA}T_s$ (Sect. 4.1).

Low-frequency 1/f noise. This case is illustrated in Fig. 3.10; here the frequency response of the filter in principle cannot be rectangular: the noise voltage at the output of such filter will theoretically be infinite because the power of $1/f$ noise at low frequencies tends to infinity as $f \to 0$. Because of this, we have to give up our "rectangular reasoning" in favor of "triangular reasoning": as it turns out, the frequency response of optimal filter is, according to (3.48), triangular:

$$f \le f_s \ , \ \ K(f) = K_0 e_s(f)\frac{1}{\Theta(f)} = K_0\frac{f}{f_{low}} = K_0\frac{f_s}{f_{low}}\frac{f}{f_s} \ ,$$

$$K(f) = K_0' k(f) \ , \ \ K_0' = K_0\frac{f_s}{f_{low}} \ , \ \ k(f) = \frac{f}{f_s} \ . \tag{3.69}$$

The transfer coefficient $K(f)$ is here normalized to its maximum value at $f = f_s$. Since the noise density increases infinitely as the frequency tends to zero, the amplification at these frequencies has also to be reduced to infinitesimal values. The calculation of signal and noise does not become much more complicated: they can once again be calculated as the areas under the curves $k(f)e_s(f)$ and $k^2(f)\Theta(f)$, which are now triangular rather than rectangular. Accordingly, we have

$$U_{sA} = SE_{sA}2T_s\int_0^{f_s} ke_s df = SE_{sA}2T_s\int_0^{f_s}\frac{f}{f_s}df = SE_{sA}2T_s\frac{f_s}{2} = \frac{SE_{sA}}{2} \ , \tag{3.70}$$

$$U_n = 2qI_n\int_0^{f_s} k^2\Theta df = 2qI_n\int_0^{f_s}\left(\frac{f}{f_s}\right)^2\frac{f_{low}}{f}df$$

$$= 2qI_n\frac{f_{low}}{f_s^2}\int_0^{f_s} f df = 2qI_n\frac{f_{low}}{f_s^2}\frac{f^2}{2}\bigg|_0^{f_s} = qI_n f_{low} \ . \tag{3.71}$$

Neither the signal amplitude nor the noise at the output of the filter in question depend on the duration of the input signal. Figure 3.10a shows clearly why this is so for the amplitude of the input signal: the area under $k(f)e_s(f)$, which defines the amplitude of the output signal is always — whatever the signal duration T_s — reduced by half because of the triangular shape of $k(f)$. One may put it in a different way: since in the working frequency band the gain varies gradually from 0 to 1, the average gain is 0.5, and therefore

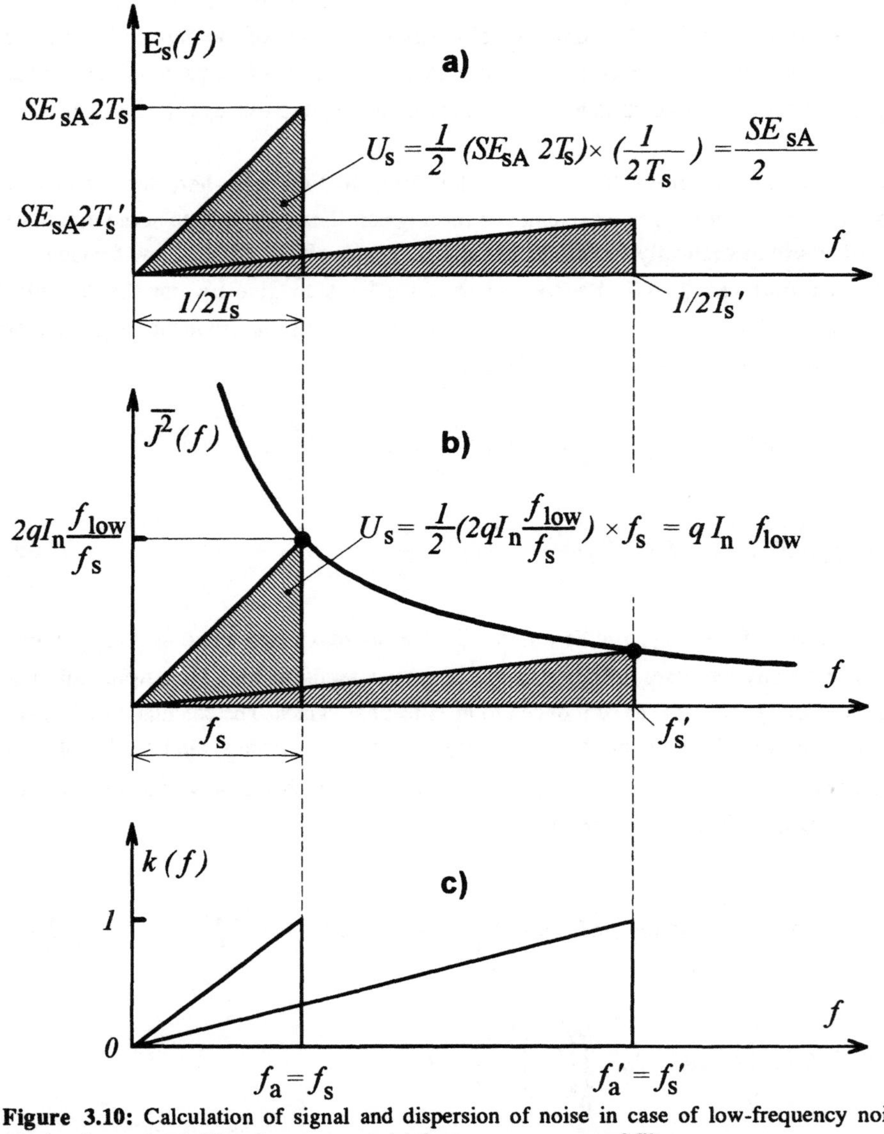

Figure 3.10: Calculation of signal and dispersion of noise in case of low-frequency noise: (a) signal spectrum; (b) noise spectrum; (c) frequency response of filter.

the amplitude is reduced by half. The fact that the output noise voltage does not depend on the input pulse length is illustrated in Fig. 3.10b. When the pulse T_s' is short, the passband of the filter $f_a = f_s'$ broadens; however, the spectral density of the noise at the upper frequency, $\Theta(f_s) = f_{low}/f_s'$, declines proportionally, so that the area under the curve which defines the output noise remains the same. To make this more clear, Fig. 3.10 shows two pulses of different lengths: observe that the areas under $k(f)e_s(f)$ and $k^2(f)\Theta(f)$ for the two are the same. Since the signal and the noise do not depend on T_s, their ratio is also invariant with respect to this parameter:

$$\mathrm{SNR_{OF}} = \frac{U_{\mathrm{sA}}}{U_{\mathrm{n}}} = \frac{SE_{\mathrm{sA}}}{2}\frac{1}{\sqrt{qI_{\mathrm{n}}f_{\mathrm{low}}}} = 0.707\frac{SE_{\mathrm{sA}}}{\sqrt{2qI_{\mathrm{n}}f_{\mathrm{low}}}} \ . \tag{3.72}$$

$\mathrm{SNR_{OF}}$: exact calculation with due account for the shape of signal. So far we have been considering the simplest (rectangular) spectrum of the signal, for the sake of clear and simple analysis. Strictly speaking, however, the rectangular spectrum corresponds to the signal of a quite definite form: bell-shaped, with periodically damped edges of the form $\sin x/x$, $x = \pi t/T_{\mathrm{s}}$ (Fig. 2.1c; see Sect. 2.1). True, in Sect. 2.1 we showed that the spectrum of any bell-shaped signal may be approximated with the rectangular one, and so we may expect that our formulas for $\mathrm{SNR_{OF}}$ are more or less universal. All the same, it will be interesting to see how the value of $\mathrm{SNR_{OF}}$ is changed when the precise shape of the spectrum $\mathrm{e_s}(f)$ is taken into account. How strongly does the shape of the signal (variation of the spectrum $\mathrm{e_s}(f)$) influence the value of $\mathrm{SNR_{OF}}$? To answer these questions, let us carry out exact calculations of $\mathrm{SNR_{OF}}$ for the same three cases of the predominantly white noise, high-frequency noise, and low-frequency noise.

White noise. Let us bring the general expression for SNR (3.52) to the form

$$\mathrm{SNR_{OF}} = \frac{SE_{\mathrm{sA}}}{\sqrt{2qI_{\mathrm{n}}}}2T_{\mathrm{s}}\sqrt{\int_0^\infty \frac{\mathrm{e_s^2}(f)}{\Theta(f)}df} = \frac{SE_{\mathrm{sA}}\sqrt{2T_{\mathrm{s}}}}{\sqrt{2qI_{\mathrm{n}}}}\gamma \ , \tag{3.73}$$

$$\gamma = \sqrt{2T_{\mathrm{s}}\int_0^\infty \frac{\mathrm{e_s^2}(f)}{\Theta(f)}df} = \sqrt{\frac{f_{\mathrm{n}}}{f_{\mathrm{s}}}} = \sqrt{\frac{\int_0^\infty \mathrm{e_s^2}(f)df}{\int_0^\infty \mathrm{e_s}(f)df}} \ . \tag{3.74}$$

In (3.74) we used the fact that the noise is white, $\Theta(f) = 1$, and introduced the effective signal band $f_{\mathrm{s}} = 1/2T_{\mathrm{s}}$; by virtue of the definition (2.15) we replace the latter with the area under the curve $\mathrm{e_s}(f)$. We see that (3.73) differs from the earlier expression (3.61) by a numerical coefficient γ. This coefficient is equal to square root of the ratio of the noise band of optimal filter to the effective signal band f_{s}. Given the definition of the bands f_{n}, f_{s}, the coefficient γ is square root of the ratio between areas under the curves $\mathrm{e_s^2}(f)$, $\mathrm{e_s}(f)$.

These areas, however, differ little. As a matter of fact, in case of the rectangular spectrum we have $\mathrm{e_s^2}(f) = \mathrm{e_s}(f) = 1$, and the two areas are equal, $\gamma = 1$. In the general case, the area under $\mathrm{e_s^2}(f)$ is, of course, smaller, because $\mathrm{e}(f) < 1$ (this is true at least when $\mathrm{e_s}(f) > 0$). For the class of pulses under consideration, the spectral density $\mathrm{e_s}(f)$

falls off quite rapidly in the neighborhood of the frequency $f_s = 1/2T_s$. Because of this, the area under the curves $e_s^2(f)$, $e_s(f)$ is defined primarily by the area in the plateau region, where $e_s^2(f) \approx e_s(f) \approx 1$, the width of this plateau is about $1/2T_s$. In this way, even this qualitative analysis of expressions (3.73), (3.74) allows us to make an important conclusion: for the class of bell-shaped pulses under consideration, the coefficient γ, and hence the value of SNR_{OF}, do not exhibit any strong dependence on the particular shape of the signal (given that the effective signal duration T_s is the same). Let us support this conclusion with calculations for rectangular and square-cosine pulses:

$$\gamma = \sqrt{2T_s \int_0^\infty e_s^2(f)df} = \sqrt{2\int_0^\infty \frac{\sin^2 \pi f T_s}{(\pi f T_s)^2} d(fT_s)} = \sqrt{\frac{2}{\pi}\int_0^\infty \frac{\sin^2 x}{x^2} dx} = 1 \ ,$$

$$SNR_{OF} = \sqrt{2}\frac{SE_{sA}\sqrt{T_s}}{\sqrt{2qI_n}} \tag{3.75}$$

for rectangular pulse, and

$$\gamma = \sqrt{2T_s \int_0^\infty e_s^2(f)df} = \sqrt{\int_0^\infty \frac{\sin^2 2\pi f T_s}{(2\pi f T_s)^2 \left(1 - 4f^2T_s^2\right)^2} d(f2T_s)}$$

$$= \sqrt{\frac{1}{\pi}\int_0^\infty \frac{\sin^2 x}{x^2\left(1 - \frac{x^2}{\pi^2}\right)^2} dx} = \sqrt{\frac{3}{4}} \ ,$$

$$SNR_{OF} = \sqrt{2}\gamma\frac{SE_{sA}\sqrt{T_s}}{\sqrt{2qI_n}} = \sqrt{2}\sqrt{\frac{3}{4}}\frac{SE_{sA}\sqrt{T_s}}{\sqrt{2qI_n}} = 1.22\frac{SE_{sA}\sqrt{T_s}}{\sqrt{2qI_n}} \tag{3.76}$$

for square-cosine pulse (here we have used the tables of integrals [82]).

The result is even better than we could hope for: expression (3.75) for the rectangular pulse exactly coincides with our former expression (3.61), obtained for the effective rectangular spectrum. This expression defines the maximum value of SNR_{OF} in the detection of pulsed signal.

For any other shape of the signal, the energy of the pulse "spreads out" over the time, and some energy goes from the maximum into the tail portions of the pulse, as is the case with the bell-shaped signal, the value of SNR_{OF} can only become smaller. In case of the square-cosine pulse the smearing of the energy is not considerable, and the drop in SNR_{OF} is also small (by 13.5%; cf. (3.75) and (3.76)). As a matter of fact, this reduction of SNR_{OF} could have been predicted using equation (3.61), obtained through

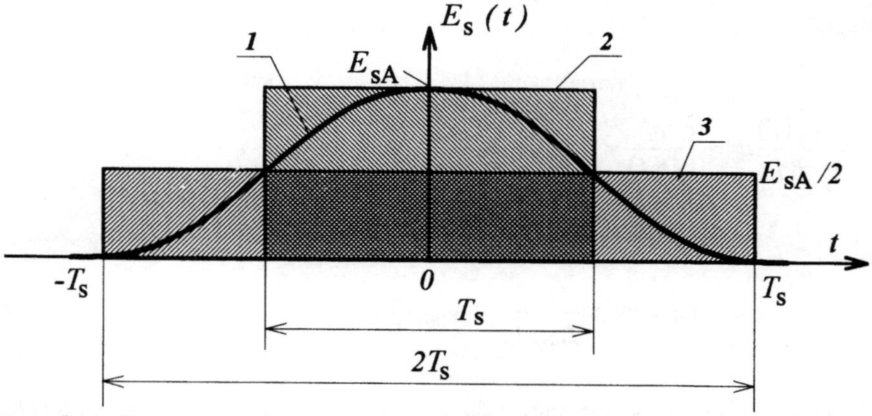

Figure 3.11: Two approximations of bell-shaped pulse.

qualitative analysis in the preceding section. Namely, in accordance with Sect. 1.3, we replace the bell-shaped pulse (curve 1 in Fig. 3.11) with the equivalent rectangular pulse of the same amplitude $E'_{sA} = E_{sA}$ and effective duration $T'_s = T_s$ (curve 2 in Fig. 3.11). However, the equivalent replacement can be accomplished in a different way: if we preserve the pulse length at the foot, $T''_s = 2T_s$, the effective amplitude (from condition of energy conservation) will be $E''_{sA} = E_{sA}/2$ (curve 3 in Fig. 3.11). For these two pulses we get

$$E'_{sA} = E_{sA}\,,\quad T'_s = T_s\,,\quad \mathrm{SNR}'_{\mathrm{OF}} = \frac{SE'_{sA}\sqrt{2T'_s}}{\sqrt{2qI_n}} = \sqrt{2}\,\frac{SE_{sA}\sqrt{T_s}}{\sqrt{2qI_n}}\;; \tag{3.77}$$

$$E''_{sA} = \frac{E_{sA}}{2}\,,\quad T''_s = 2T_s\,,\quad \mathrm{SNR}''_{\mathrm{OF}} = \frac{SE''_{sA}\sqrt{2T''_s}}{\sqrt{2qI_n}} = \frac{SE_{sA}\sqrt{T_s}}{\sqrt{2qI_n}}\,. \tag{3.78}$$

The actual bell-shaped pulse lies somewhere in between these two rectangular extremes, so we may expect that

$$\mathrm{SNR}_{\mathrm{OF}} = \frac{\mathrm{SNR}'_{\mathrm{OF}} + \mathrm{SNR}''_{\mathrm{OF}}}{2} = \frac{1+\sqrt{2}}{2}\,\frac{SE_{sA}\sqrt{T_s}}{\sqrt{2qI_n}} = 1.2\,\frac{SE_{sA}\sqrt{T_s}}{\sqrt{2qI_n}}\,. \tag{3.79}$$

Once again we get very good agreement (up to the third decimal point) with the exact formula (3.76).

High-frequency noise. Assume that the high-frequency noise prevails in the entire spectral region of the signal, $f_{\mathrm{high}} \ll 1/2T_s$. Because of the rapid growth of the noise, $\Theta \propto f^2$, the integrand in (3.52) falls off sharply in the neighborhood of $f \approx f_{\mathrm{high}}$. With respect to the spectrum of the signal, this frequency is quite low, and the spectral density of the signal here is constant, $e_s(f) = 1$. This circumstance facilitates calculation of the integral,

$$\int_0^\infty \frac{e_s^2(f)}{\Theta(f)} df \approx \int_0^\infty \frac{df}{\Theta(f)} = f_{high} \int_0^\infty \frac{d\frac{f}{f_{high}}}{1+\left(\frac{f}{f_{high}}\right)^2}$$

$$= f_{high} \arctan \frac{f}{f_{high}} \Bigg|_0^\infty = f_{high} \frac{\pi}{2} . \tag{3.80}$$

For the signal-to-noise ratio (3.47) we get

$$\mathrm{SNR_{OF}} = \frac{SE_{sA} 2T_s}{\sqrt{2qI_n}} \sqrt{\int_0^\infty \frac{e_s^2(f)}{\Theta(f)} df}$$

$$= \sqrt{2\pi} \frac{SE_{sA} T_s}{\sqrt{2qI_n}} \sqrt{f_{high}} = 2.5 \frac{SE_{sA} T_s}{\sqrt{2qI_n}} \sqrt{f_{high}} . \tag{3.81}$$

We see that in this case the shape of the signal has no effect on the value of $\mathrm{SNR_{OF}}$: the optimal filter transmits only the low frequencies, where the relative spectral density is the same for any shape of the signal, $e_s = 1$. At the same time, those frequencies where the behavior of the spectrum does depend on the shape of the signal ($f \sim f_s = 1/2T_s$ and above) make no contribution to the integral (3.80) because of the large $\Theta(f) \sim f_s^2 / f_{high}^2 >> 1$ in the denominator.

The exact expression (3.81) results in a somewhat larger value of $\mathrm{SNR_{OF}}$ than that given by the approximate formula (3.67). This is explained by the fact that the approximate calculation was based on the rectangular frequency response of the filter (solid curve in Fig. 3.9c). The optimal filter is smoothed out, and exhibits a rapidly declining tail at $f > f_{high}$ (dashed curve in Fig. 3.9c). The inclusion of harmonics at these frequencies leads to a higher value of $\mathrm{SNR_{OF}}$ in the exact expression (3.81).

Low-frequency noise. Substituting the noise spectrum $\Theta = f_{low}/f$ into (3.52), for the rectangular pulse we get

$$\mathrm{SNR_{OF}} = \frac{SE_{sA} 2T_s}{\sqrt{2qI_n}} \sqrt{\int_0^\infty \frac{e_s^2(f)}{\Theta(f)} df} =$$

$$= \frac{SE_{sA} 2T_s}{\sqrt{2qI_n}} \sqrt{\int_0^\infty \frac{\sin^2 \pi f T_s}{\pi^2 f^2 T_s^2} \frac{f}{f_{low}} df} \sim \sqrt{\int_0^\infty \frac{\sin^2 \pi f T_s}{\pi^2 f_{low} T_s^2} \frac{df}{f}} \ . \tag{3.82}$$

At the high frequencies ($f > 1/2T_s$), the integrand falls off very slowly ($\propto 1/f$), and therefore the integral ($\propto \ln f$) diverges, $SNR_{OF} \to \infty$. For a second time we meet with situation when the noise seems not to limit the detection. This, however, is just a speculation: the last expression holds under three conditions (three "infinities"), which cannot be satisfied in practice.

1. Amplification of the optimal filter must carry on to infinity in the frequency range,

$$k(f) = e_s(f)\frac{1}{\Theta} = \frac{\sin \pi f T_s}{\pi f T_s} \frac{1}{\frac{f_{low}}{f}} \propto \sin \pi f T_s \ , \tag{3.83}$$

whereas in reality the band is always limited.

2. As the frequency increases, the noise in our model goes to zero, $\Theta = f_{low}/f \to 0$. In reality, the minimum of the noise will be limited at least by the white-noise component (as was the case with the high-frequency noise component when $f \to 0$).

3. The rectangular pulse that we are considering has infinitely steep edges. Actually, the transition processes always take some time; because of this, the spectrum of the signal falls off faster than $1/f$ as the frequency increases, and the integral in (3.82) converges.

To avoid the trouble associated with the divergent integral, it will suffice to include only the last factor, the finite slope of the edges. For example, for the bell-shaped signal we get

$$SNR_{OF} = \frac{SE_{sA}}{\sqrt{2qI_n}} \sqrt{(2T_s)^2 \int_0^\infty \left[\frac{\sin 2\pi f T_s}{2\pi f T_s \left(1 - \frac{(2\pi f T_s)^2}{\pi^2}\right)} \right]^2 \frac{f df}{f_{low}}}$$

$$= \frac{SE_{sA}}{\sqrt{2qI_n f_{low}}} \sqrt{\int_0^\infty \left[\frac{\sin x}{x\left(1 - \frac{x^2}{\pi^2}\right)} \right]^2 \frac{x dx}{\pi^2}} = 0.6 \frac{SE_{sA}}{\sqrt{2qI_n f_{low}}} \ . \tag{3.84}$$

Here we symbolize $x = 2\pi f T_s$; the value of the integral is taken from tables [82].

Exact calculations of $\mathrm{SNR_{OF}}$ for particular shapes of the signal (rectangular, bell-shaped) entirely confirm our earlier approximate findings, only refining to some extent (within 0...20%) the values of numerical coefficients (the relevant comments on the cases of low-frequency noise and the rectangular pulse have just been made).

Discussion of formulas for SNR. So, we have derived formulas for signal-to-noise ratio in all typical situations. For comparison and quick reference they are displayed in Table 3.2.

Table 3.2. **SNR at the output of optimal filter**

Type of noise	$\mathrm{SNR_{OF}}$	Coefficient a: Approximate (rectangular signal spectrum)	Coefficient a: Exact, Rectangular pulse	Coefficient a: Exact, Bell-shaped pulse
Low-frequency	$a\dfrac{SE_{sA}}{\sqrt{2qI_n f_{low}}}$	0.707	—	0.6
White	$a\dfrac{SE_{sA}\sqrt{T_s}}{\sqrt{2qI_n}}$	1.41	1.41	1.22
High-frequency	$a\dfrac{SE_{sA}T_s}{\sqrt{2qI_n/f_{high}}}$	2	2.5	2.5

These formulas define the guiding principles for the designers of threshold-level optoelectronic systems and PDAs, since they relate the signal-to-noise ratio to the parameters of the signal E_{sA}, T_s (which are determined by the optoelectronic system), and to the parameters of photodetectors and PDAs S, I_n, f_{low}, f_{high}.

The fact that $\mathrm{SNR_{OF}}$ is proportional to the input electrical signal SE_{sA} and inversely proportional to the intensity of input noise $\sqrt{2qI_n}$ is obvious, and is a direct consequence of the linear nature of the system. The dependence of $\mathrm{SNR_{OF}}$ on the duration of input signal T_s, however, is not trivial, and only comes up in the formulas. We have paid due attention to this most important and interesting feature.

Let us consider two simple examples which illustrate how the expressions for $\mathrm{SNR_{OF}}$, and especially the dependence of $\mathrm{SNR_{OF}}$ on T_s, influence the ideology of the optoelectronic system.

Detection of radiation coming from remote stars. In the classical case of white noise, the natural desire is to increase the observation time (the pulse length T_s), because this increases the signal-to-noise ratio, $\mathrm{SNR_{OF}} \propto \sqrt{T_s}$ (3.61). When the signal is too lengthy ($T_s \to \infty$), however, the optimal filter becomes a low-frequency one ($f_a = f_s = 1/2T_s \to 0$). At some point the low-frequency noise may prevail, so that further increase of T_s is useless, because it does nothing to improve the signal-to-noise ratio (3.72). We see that the struggle for a higher $\mathrm{SNR_{OF}}$ in case of large integration time also involves efforts to combat the low-frequency noise.

Laser (LED) systems. The energy of the pulse emitted by the transmitter is limited (because limited is the energy available for laser pumping). This imposes a limit on the energy of the optical pulse coming to the receiver (PDA). Let us write the signal-to-noise ratio as function of the energy of the incoming pulse ($E_{sA}T_s$) in the classical case of white noise (3.37):

$$\mathrm{SNR_{OF}} = a\frac{SE_{sA}\sqrt{T_s}}{\sqrt{2qI_n}} = a\frac{S(E_{sA}T_s)}{\sqrt{2qI_n}\sqrt{T_s}} \,. \tag{3.85}$$

This formula explains the desire to make the pulse in laser systems as short as possible: this is supposed not only to improve the time resolution, but also to increase the signal-to-noise ratio without increasing the pumping energy, $\mathrm{SNR_{OF}} = 1/\sqrt{T_s}$. Very short pulses, however, require a high-frequency broad-band filter-amplifier, and so the high-frequency noise prevails. Equation (3.85) no longer holds; according to (3.67), the signal-to-noise ratio is defined by the energy of the incoming signal ($E_{sA}T_s$) and does not improve as the signal becomes shorter (given that its energy remains the same), to the disappointment of the engineers. The Nature intervenes with its own restrictions: in these examples, the value of $\mathrm{SNR_{OF}}$ is limited by the excessive (with respect to the white noise) components. These are the low-frequency noise in case of low-frequency signal, and the high-frequency noise in case of high-frequency signal.

So we see that the formulas of Table 3.2 are quite useful for the design of optoelectronic systems. The nature of PDA noise may influence the choice of the signal duration (frequency of modulation or scanning). This applies not only to the examples just considered. In Sect. 7.2, devoted to multielement systems, we shall discuss the effects of the noise nature on the choice of number of elements and geometry of the device.

To round off our analysis, we have to make one final step towards generality, and consider the arbitrary noise spectrum that may occur in a typical PDA.

$\mathrm{SNR_{OF}}$ in case of arbitrary noise spectrum. In general, the noise spectrum of PDA may include all three noise components: $\Theta \propto 1/f$, $\Theta = \mathrm{const}$, $\Theta \propto f^2$ (Sect. 2.4). So that the white noise region should not be completely overlapped by low-frequency and

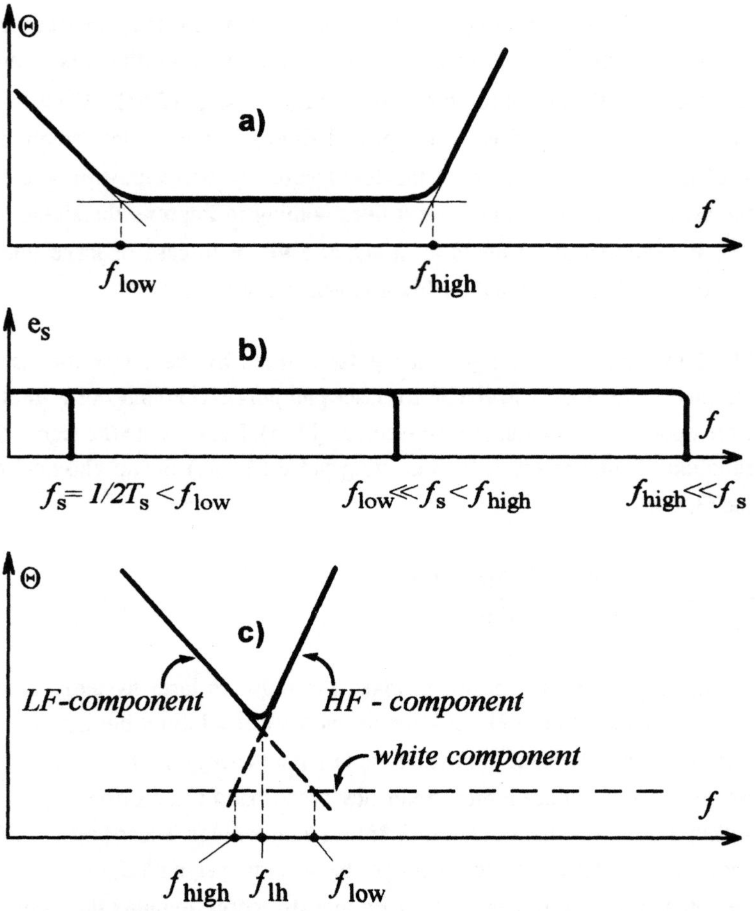

Figure 3.12: Domination of different noise components in the signal frequency band.

high-frequency noises, condition $f_{low} < f_{high}$ must be satisfied (Fig. 3.12a). Exact formulas for $\mathrm{SNR_{OF}}$ in case of such spectrum are quite cumbersome, and will not be quoted here. Instead, we shall try to find convenient practical approximations. We take advantage of the fact that, as the signal frequency band varies according to the signal length, one of the three noise components may become prevalent. It is the low-frequency component in case of narrow-band signal (signal of large duration), when $f_s = 1/2T_s < f_{low}$; the white-noise component in case of medium-band signal ($f_{low} < f_s < f_{high}$), and the high-frequency component in case of broad-band signal ($f_s >> f_{high}$); Fig. 3.12a,b. Accordingly, the value of $\mathrm{SNR_{OF}}$ for these particular ranges of T_s is defined by the appropriate expressions obtained above. Let us first turn to the approximate formulas (3.61), (3.67), (3.72), and rewrite them in a unified form:

$$f_s < f_{low}\,,\quad \mathrm{SNR_{OF}} = \frac{SE_{sA}}{\sqrt{2qI_n}\,\sqrt{2f_{low}}} = \frac{SE_{sA}\sqrt{2T_s}}{\sqrt{2qI_n}}\,\frac{1}{\sqrt{2f_{low}(2T_s)}}\,; \tag{3.86}$$

$$f_{\rm low} < f_{\rm s} < f_{\rm high}\ , \quad {\rm SNR_{OF}} = \frac{SE_{\rm sA}\sqrt{2T_{\rm s}}}{\sqrt{2qI_{\rm n}}}\frac{1}{\sqrt{1}}\ ; \tag{3.87}$$

$$f_{\rm high} < f_{\rm s}\ , \quad {\rm SNR_{OF}} = \frac{SE_{\rm sA}T_{\rm s}}{\sqrt{2qI_{\rm n}}}\sqrt{f_{\rm high}} = \frac{SE_{\rm sA}\sqrt{2T_{\rm s}}}{\sqrt{2qI_{\rm n}}}\frac{1}{\sqrt{1/(f_{\rm high}\,2T_{\rm s})}}\ . \tag{3.88}$$

Now we construct the desired approximation for $\mathrm{SNR_{OF}}$ in case of arbitrary signal duration T_s by generalizing these three formulas. They differ in the last factor, so we combine all three terms under the square root in the denominator:

$$f_{\rm low} < f_{\rm high}\ , \quad {\rm SNR_{OF}} = \frac{SE_{\rm sA}\sqrt{2T_{\rm s}}}{\sqrt{2qI_{\rm n}}}\frac{1}{\sqrt{2f_{\rm low}(2T_{\rm s}) + 1 + 1/(f_{\rm high}\,2T_{\rm s})}}\ . \tag{3.89}$$

With the replacement $1/2T_s \Rightarrow f_s$, this expression can be represented in the form which is very easy to memorize:

$$f_{\rm low} < f_{\rm high}\ , \quad {\rm SNR_{OF}} = \frac{SE_{\rm sA}}{\sqrt{2qI_{\rm n}\left[f_{\rm s} + 2f_{\rm low} + (f_{\rm s}/f_{\rm high})f_{\rm s}\right]}}\ . \tag{3.90}$$

This representation is so transparent that it hardly calls for any comments: the signal-to-noise ratio is nothing else but the ratio of the signal amplitude SE_{sA} to the noise calculated in a certain equivalent band. In case of white noise, this band is equal to the signal band f_s. The contribution of the excess noise is formally included by extending the equivalent band: by the amount of $2f_{low}$ in case of low-frequency noise, and f_s^2/f_{high} in case of high-frequency noise.

As indicated above, the exact calculation of $\mathrm{SNR_{OF}}$ for bell-shaped signal leads to redefinition of numerical coefficients; accordingly, the numerical coefficients in (3.90) should also be corrected. Using the same technique as before, we find:

$$f_{\rm s} < f_{\rm low}\ , \quad {\rm SNR_{OF}} = 0.6\frac{SE_{\rm sA}}{\sqrt{2qI_{\rm n}f_{\rm low}}} = 0.865\frac{SE_{\rm sA}\sqrt{2T_{\rm s}}}{\sqrt{2qI_{\rm n}}}\sqrt{\frac{1}{2.1f_{\rm low}(2T_{\rm s})}}\ ; \tag{3.91}$$

$$f_{\rm low} < f_{\rm s} < f_{\rm high}\ , \quad {\rm SNR_{OF}} = 0.865\frac{SE_{\rm sA}\sqrt{2T_{\rm s}}}{\sqrt{2qI_{\rm n}}}\sqrt{\frac{1}{1}}\ ; \tag{3.92}$$

$f_{high} < f_s$,

$${\rm SNR_{OF}} = 2.5\frac{SE_{\rm sA}T_{\rm s}}{\sqrt{2qI_{\rm n}}}\sqrt{f_{\rm high}} = 0.865\frac{SE_{\rm sA}\sqrt{2T_{\rm s}}}{\sqrt{2qI_{\rm n}}}\sqrt{\frac{1}{0.48/(f_{\rm high}\,2T_{\rm s})}}\ . \tag{3.93}$$

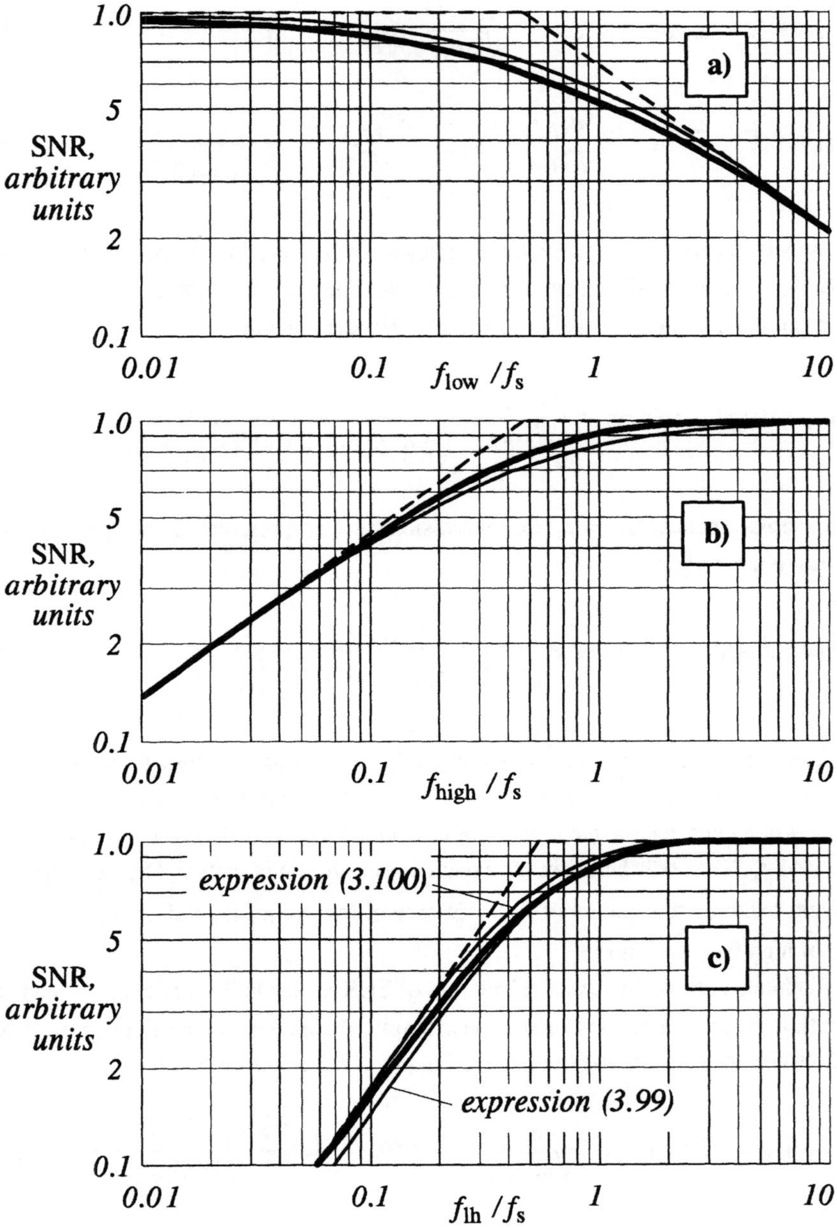

Figure 3.13: SNR_{OF} as function of signal duration (signal frequency band $f_s = 1/2T_s$) for different noise spectra of PDA: (a) low-frequency and white noise; (b) white and high-frequency noise; (c) low-frequency and high-frequency noise. Normalization to the value of SNR_{OF} for the predominant white noise (a,b) or low-frequency noise (c) components. Exact curve (thick line), approximation (thin line), asymptotics (dashed line).

Given this refinement, the general approximation takes on the form

$$f_{\text{low}} < f_{\text{high}} \,,$$

$$\text{SNR}_{\text{OF}} = \frac{0.865 S E_{\text{sA}} \sqrt{2T_{\text{s}}}}{\sqrt{2qI_{\text{n}}}} \frac{1}{\sqrt{2.1 f_{\text{low}}(2T_{\text{s}}) + 1 + 0.48/(f_{\text{high}} 2T_{\text{s}})}} \,. \tag{3.94}$$

As the signal duration T_s is varied, one of the three terms under the root may become prevalent, and then the general expression (3.94) will coincide with the appropriate asymptotic expression (3.91), (3.92), or (3.93). There is only one unsolved question: will this expression hold also in the transition regions, when $f_s \approx f_{\text{low}}$, or $f_s \approx f_{\text{high}}$? This question can be answered by comparing the approximation (3.94) with the exact value of SNR_{OF} as given by (3.52). Such comparison is made in Fig. 3.13. We see that we are lucky again: the simple approximation (3.94) turns out to be accurate enough not only in those regions where we had expected (in the asymptotic regions), but also in the transition regions, the neighborhoods of the characteristic values $f_s = f_{\text{low}}$ and $f_s = f_{\text{high}}$. The error is as small as 7...9%. We say "small" with good reason: the accuracy of measurements of signal and noise in PDAs is usually not better than that, and so this precision of calculations is, as a rule, sufficient. If a higher accuracy is required, one has to use the solid curves in Fig. 3.13 for the transition regions.

To end this section, let us consider the case when the low-frequency and the high-frequency noise components overlap. Although this situation is not likely to occur in real PDAs, it will add to the generality of our analysis (Fig. 3.12c). The region of white noise is gone, and a new characteristic frequency f_{lh} appears, at which the power of the low-frequency noise ($\Theta = f_{\text{low}}/f$) is equal to that of the high-frequency noise ($\Theta = f^2/f_{\text{high}}^2$):

$$\frac{f_{\text{low}}}{f} = \frac{f^2}{f_{\text{high}}^2} \quad \text{at } f = f_{\text{lh}} \;; \quad f_{\text{lh}}^3 = f_{\text{low}} f_{\text{high}}^2 \,. \tag{3.95}$$

When the pulse is long, the low-frequency component prevails, and our earlier findings (3.91) remain in force. However, when the pulse is short ($f_{\text{lh}} < f_s = 1/2T_s$), the calculation of signal and noise must be revised. As we know, the passband of the optimal filter must be delimited by the characteristic frequency at which the high-frequency noise starts to grow sharply (in our present case, f_{lh}). In the selected filter band $[0, f_{\text{lh}}]$ the low-frequency noise component dominates; the frequency response of the filter is triangular, the output noise voltage does not depend on the band, and is given by (3.71), whereas the signal is calculated by analogy with (3.70):

$$U_{\text{sA}} = (S E_{\text{sA}} 2T_{\text{s}}) \frac{f_{\text{lh}}}{2} \,. \tag{3.96}$$

Then for the signal-to-noise ratio we have

$$\mathrm{SNR_{OF}} = \frac{U_{sA}}{U_n} = \frac{SE_{sA}T_s f_{lh}}{\sqrt{qI_n f_{low}}} = 2\frac{SE_{sA}T_s}{\sqrt{2qI_n}}\sqrt{\frac{f_{lh}^2}{2f_{low}}} . \tag{3.97}$$

Comparing this formula with the above expression (3.88) for the high-frequency component, we see that the two expressions coincide if we make the following replacement:

$$f_{high} \Rightarrow \frac{f_{lh}^2}{2f_{low}} . \tag{3.98}$$

With this replacement, the generalized expression (3.94) becomes

$$f_{low} >> f_{high} , \tag{3.99}$$

$$\mathrm{SNR_{OF}} = \frac{0.865 SE_{sA}\sqrt{2T_s}}{\sqrt{2qI_n}\sqrt{2.1 f_{low}(2T_s) + \frac{0.48}{(2T_s)}\frac{2f_{low}}{f_{lh}^2}}} = \frac{0.6 SE_{sA}}{\sqrt{2qI_n f_{low}}\sqrt{1+\frac{0.46}{(f_{lh}2T_s)^2}}} .$$

By trial and error, it is possible to refine the numerical coefficient in this approximation:

$$f_{low} >> f_{high} , \quad \mathrm{SNR_{OF}} = \frac{0.6 SE_{sA}}{\sqrt{2qI_n f_{low}}}\frac{1}{\sqrt{1+\frac{0.307}{(f_{lh}2T_s)^2}}} . \tag{3.100}$$

Like the previous ones, this approximation fits in with the numerical results (Fig. 3.13) to a good accuracy (within 12%).

Summary. *In fact, the results of this section are summarized in Table 3.2. The formulas of Table 3.2 ought to be supplemented by (3.94), (3.100). In all cases of practical importance, the signal-to-noise ratio at the output of optimal filter* $\mathrm{SNR_{OF}}$ *can be calculated by these relatively simple formulas, as long as we know the parameters of the signal* E_{sA}, T_s, *the sensitivity of the photodetector S, and the characteristics of the noise* I_n, f_{low}, f_{high}. *The minor details of the signal shape have little effect on the SNR. Depending on the nature of the noise,* $\mathrm{SNR_{OF}}$ *exhibits different functional dependence on the signal duration: it is linear in case of high-frequency noise* ($\mathrm{SNR_{OF}} \propto T_s$), *and weaker (square-root) in case of white noise* ($\mathrm{SNR_{OF}} \propto \sqrt{T_s}$). *In case of low-frequency noise the dependence of* $\mathrm{SNR_{OF}}$ *on the signal duration entirely disappears.*

3.4. Noise Equivalent Power, Detectivity, and Other Detection Invariants

Detection invariants. Expressions for SNR, obtained in the preceding section, allow us to define the main characteristic of PDA as a detection system of low-level optical signals. Of course, the value of SNR_{OF} itself is not yet such a characteristic, because it is defined in terms of not only the parameters of PDA, but also those of the optical signal; accordingly, one and the same PDA will exhibit different values of SNR_{OF} in optoelectronic systems with different parameters of the signal E_{sA}, T_s. Let us rewrite once again our expressions for SNR_{OF} in such a way that the parameters of the signal are in the numerator, and the parameters of PDA in the denominator:

$$SNR_{OF} = 0.6\frac{SE_{sA}}{\sqrt{2qI_n f_{low}}} = 0.6\frac{E_{sA}}{\frac{\sqrt{2qI_n}}{S}\sqrt{f_{low}}} = 0.6\frac{E_{sA}}{NEP\sqrt{f_{low}}} \tag{3.101}$$

for low-frequency noise,

$$SNR_{OF} = (1.22...1.41)\frac{SE_{sA}\sqrt{T_s}}{\sqrt{2qI_n}} = (1.22...1.41)\frac{E_{sA}\sqrt{T_s}}{\frac{\sqrt{2qI_n}}{S}} = (1.22...1.41)\frac{E_{sA}\sqrt{T_s}}{NEP} \tag{3.102}$$

for white noise, and

$$SNR_{OF} = 2.5\frac{SE_{sA}T_s\sqrt{f_{high}}}{\sqrt{2qI_n}} = 2.5\frac{(E_{sA}T_s)}{\frac{\sqrt{2qI_n}}{S\sqrt{f_{high}}}} = 2.5\frac{(E_{sA}T_s)}{\frac{NEP}{\sqrt{f_{high}}}} \tag{3.103}$$

for high-frequency noise, where

$$NEP = \frac{\sqrt{2qI_n}}{S} \tag{3.104}$$

is the so-called noise equivalent power in unit frequency band, as will be explained shortly. Recall that the coefficients 0.6; 1.22; 2.5 correspond to the bell-shaped pulse, and 1.41; 2.5 to the rectangular pulse (Table 3.2). All these expressions can be summarized as

$$\mathrm{SNR_{OF}} = \frac{E_{\mathrm{sA}} T_{\mathrm{s}}^{\alpha}}{\text{Ѵ}_{\mathrm{DI}}} \,, \tag{3.105}$$

where

$$\alpha = 0 \,, \quad \text{Ѵ}_{\mathrm{DI}} = \frac{\mathrm{NEP}\sqrt{f_{\mathrm{low}}}}{a} \,, \quad a = 0.6 \tag{3.106}$$

for low-frequency noise,

$$\alpha = \frac{1}{2} \,, \quad \text{Ѵ}_{\mathrm{DI}} = \frac{\mathrm{NEP}}{a} \,, \quad a = 1.22 \ldots 1.41 \tag{3.107}$$

for white noise, and

$$\alpha = 1 \,, \quad \text{Ѵ}_{\mathrm{DI}} = \frac{\mathrm{NEP}}{a\sqrt{f_{\mathrm{high}}}} \,, \quad a = 2.5 \tag{3.108}$$

for high-frequency noise.

We see that the desired characteristic of PDA as the detection system of weak optical signals, which is invariant with respect to the parameters of the optical pulse, is the parameter Ѵ_{DI}, which will be referred to as the detection invariant. The physical meaning and the dimensionality of the detection invariant are clear from the formulas. The value of the detection invariant is numerically equal to the intensity of the pulse $\left(E_{\mathrm{sA}} T_{\mathrm{s}}^{\alpha}\right)_{\mathrm{t}}$ for which $\mathrm{SNR_{OF}} = 1$ (see (3.105)). The intensity as low as that is called the "threshold intensity", which is what the subscript "t" stands for. This definition is clear: below the threshold value, the signal is drowned in the noise; when the intensity exceeds the threshold, it is possible to detect the signal.

What is the unit of the threshold intensity? The function $\mathrm{SNR_{OF}} = f(T_{\mathrm{s}})$ is different for different kinds of noise; as a consequence, the dimensionality of the detection invariant is also not the same.

1. In case of low-frequency noise, Ѵ_{DI} is measured in watts: we have $\alpha = 0$, $\text{Ѵ}_{\mathrm{DI}} = \left(E_{\mathrm{sA}} T_{\mathrm{s}}^{0}\right)_{\mathrm{t}} = \left(E_{\mathrm{sA}}\right)_{\mathrm{t}}$, and so the detection invariant may be called the threshold power (see (3.101), (3.106)).

2. In case of high-frequency noise, Ѵ_{DI} is measured in joules: we have $\alpha = 1$, $\text{Ѵ}_{\mathrm{DI}} = \left(E_{\mathrm{sA}} T_{\mathrm{s}}\right)_{\mathrm{t}}$, and so the detection invariant may be called the threshold energy (see (3.103), (3.108)).

3. In case of white noise, the threshold intensity is neither power, nor energy: it has the dimension of $\left[W\cdot s^{1/2}\right]=\left[W\cdot Hz^{-1/2}\right]$ — since, according to (3.102), $\mathbf{V}_{DI}=\left(E_{sA}T_s^{1/2}\right)_t$. In such case, however, for the detection invariant one takes not the quantity $\mathbf{V}_{DI}=\left(E_{sA}T_s^{1/2}\right)_t$ itself, but the quantity NEP (3.104), (3.107), which differs from the former by the coefficient *a*. This "noise equivalent power" is defined as the threshold power in unit frequency band. The forthcoming analysis of this and related parameters will prove the expedience of such definition, and explain why it is necessary to relate the threshold power to unit frequency band ($Hz^{-1/2}$).

Noise equivalent power and its family. We have defined NEP with reference to the pulsed modulation of optical signal. In IR technology, however, the threshold parameters (including NEP) are defined and measured under conditions of sine-wave modulation. Let us also consider these ways of defining the threshold parameters,

So, assume that the signal is harmonic. To select this signal, the output of PDA is connected to a narrow-band measuring amplifier with the known noise band f_n. This amplifier is used for measuring the signal and the noise at the output of the PDA. The noise is measured with a square-law detector, which reads the rms value $\propto\sqrt{\overline{i_n^2}}$. The signal may be measured either with the square-law detector, which takes the effective value $\propto\sqrt{\overline{I_s^2}}$, or with the amplitude detector, which determines the signal amplitude $\propto I_{sA}$. Accordingly, the incoming optical signal must be described in terms of its effective magnitude $\sqrt{\overline{E_s^2}}$, or in terms of its amplitude E_{sA}. By definition, the threshold power Φ_t is calculated as the magnitude of optical signal for which the signal is equal to the noise:

$$I_s=SE_{st}=i_n\ ,\quad \Phi_t=E_{st}=\frac{i_n}{S}=\frac{\sqrt{2qI_nf_n}}{S}\ . \tag{3.109}$$

If the PDA noise is white, then (3.109) may be regarded as the definition of PDA threshold power in its own noise band f_n. Then the measuring amplifier must be a broad-band one (meaning that its passband must be broader than that of the PDA).

The threshold power Φ_t is a very convenient parameter for the designer of optoelectronic systems. By definition, its value is numerically equal to the threshold power of a particular PDA in case of the sine-wave signal, $\Phi_t=E_{st}$. Knowing the value of Φ_t, one can calculate the signal-to-noise ratio given the power of the signal, $SNR_{OF}=E_{sA}/\Phi_t$. This parameter, however, cannot be used for comparing the quality of different PDAs. This statement can be illustrated with the example of IR PDAs of spectral ranges III – V. As demonstrated in Sect. 2.5, in such PDAs it is possible to realize the shot noise of the photodetector current which is ultimately due to the thermal background, $I_n = I_d = j_d A_e$, where j_d is the density of the current, and A_e is the area of photosensitive element. Substituting this value of the noise current into the threshold power definition (3.109), we get

$$\Phi_t = \frac{\sqrt{2qI_n f_n}}{S} = \frac{\sqrt{2qj_d A_e f_n}}{S} . \qquad (3.110)$$

Now which PDA is better: one with the PSE area of 0.1 mm^2, or 10 mm^2? This is a rhetorical question: better is the photodetector which has the lower current density j_d (in case of IR PD, it should be close to the density of the background current). The PSE area A_e — as well as, for that matter, the band f_n, and the angle of vision, which determines the amount of background radiation falling on the photodetector — is not by any means an indicator of quality of PDA. After all, these parameters are chosen so as to meet the technical requirements to the optoelectronic system. According to this criterion, however, the threshold power of PDA with the PSE area of 10 mm^2 is ten times worse than that for the one with the PSE area of 0.1 mm^2, while the "quality" of photodetectors is actually the same. This circumstance prompts us to look for the more objective criteria of quality, eliminating the dependence on A_e, f_n, and the angle of view. Equation (3.110) shows the way of doing this: the value of Φ_t must be referred to the square root of the frequency band and the area. In this manner we obtain the following two parameters: the noise equivalent power in unit frequency band,

$$\mathrm{NEP} = \frac{\Phi_t}{\sqrt{f_n}} = \frac{\sqrt{2qI_n}}{S} \left[\mathrm{W \cdot Hz^{-1/2}}\right] , \qquad (3.111)$$

and the specific noise equivalent power (reduced to unit frequency band and unit area of photosensitive element) [83],

$$\mathrm{NEP}^* = \frac{\mathrm{NEP}}{\sqrt{A_e}} = \frac{\Phi_t}{\sqrt{A_e f_n}} = \frac{\sqrt{2qj_d}}{S} \left[\mathrm{W \cdot Hz^{-1/2} \cdot cm^{-1}}\right] . \qquad (3.112)$$

To eliminate the dependence on the angle of vision of IR PDA, we must substitute in (3.112) the density of the background thermal radiation current $j_d = j_{db}$ calculated for the solid angle of 2π steradian (hemisphere).

Let us use (3.111) for calculating the noise equivalent power of the optimal filter in unit frequency band. For the rectangular optical signal, the noise band f_n of such filter is $f_n = 1/2T_s$, and hence

$$\mathrm{NEP} = \frac{\Phi_t}{\sqrt{f_n}} = \frac{E_{st}}{\sqrt{f_n}} = \left(E_{sA}\sqrt{2T_s}\right)_t = \frac{\sqrt{2qI_n}}{S} . \qquad (3.113)$$

We see that the value of NEP of the optimal filter defines the intensity of the threshold-level signal as given by the expression $\left(E_{sA}\sqrt{2T_s}\right)_t$. Here E_{st} is the threshold power of the signal, and the factor of $\sqrt{2T_s}$ occurs because the threshold is referred to the unit band. So, once again we make sure that the invariant $\left(E_{sA}\sqrt{T_s}\right)_t$, which has been introduced for the case of pulsed modulation, coincides (up to a constant a) with the noise equivalent power in unit band NEP, which now is defined for the case of sine-wave modulation. For the rectangular pulse the constant a is the largest, and is equal to $\sqrt{2}$; in case of the bell-shaped pulse $a = 1.22$.

Our introduction to the family of threshold parameters is not over yet. Along with those described above, their reciprocals are also widely used in IR technology [83]:

detectivity as defined for a filter with the given noise band f_n,

$$D = \frac{1}{\Phi_t} = \frac{S}{\sqrt{2qI_n f_n}} \left[\mathrm{W}^{-1}\right]; \tag{3.114}$$

detectivity in unit frequency band,

$$D_1 = D\sqrt{f_n} = \frac{1}{\mathrm{NEP}} = \frac{S}{\sqrt{2qI_n}} \left[\mathrm{W}^{-1} \cdot \mathrm{Hz}^{1/2}\right]; \tag{3.115}$$

specific detectivity (in unit frequency band per unit area of photosensitive element),

$$D^* = D_1\sqrt{A_e} = D\sqrt{A_e f_n} = \frac{1}{\mathrm{NEP}^*} = \frac{S}{\sqrt{2qj_d}} \left[\mathrm{W}^{-1} \cdot \mathrm{Hz}^{1/2} \cdot \mathrm{cm}\right]. \tag{3.116}$$

Is "detectivity" a good name for this parameter? Let us use (3.102) for calculating the SNR at the output of optimal filter for the "unit" optical pulse — that is, the pulse of unit amplitude ($E_{sA} = 1\,\mathrm{W}$) and unit length ($T_s = 1\,\mathrm{s}$):

$$\mathrm{SNR_{OF}} = a\frac{S}{\sqrt{2qI_n}}\left(E_{sA}\sqrt{T_s}\right) = aD_1 \cdot 1\,\mathrm{W} \cdot \sqrt{1\,\mathrm{s}} = aD_1 \cdot \mathrm{W} \cdot \mathrm{Hz}^{-1/2}. \tag{3.117}$$

The ability to detect the signal is characterized by the signal-to-noise ratio. The numerical value of D_1 characterizes this ability: it is equal to SNR in case of detection of the signal of unit amplitude and unit length up to the constant factor a (where $a = 1.22 \ldots 1.41$ in case of pulsed modulation, and $a = 1$ in case of sine-wave modulation). The more "talented" the PDA, the higher D_1, and the higher the signal-to-noise ratio at the output of PDA (with one and the same input signal).

The specific detectivity, along with the specific NEP, may be called "the invariant of invariants": these two parameters do not depend on the signal duration T_s (the band of

PDA), on the PSE area A_e, on the angle of vision (if referred to the hemisphere). The specific detectivity is not only the most popular detection invariant, but also the most popular parameter of photodetectors and PDAs, especially in the infrared. The inventor of the parameter D^* R. Clark Jones, who introduced this quantity in 1953 [83], has joined Coulomb and Ampère, Ohm and Watt in the physical pantheon: his name is given to the unit of specific detectivity, [Jones] = $\left[\mathrm{W}^{-1}\cdot\mathrm{cm}\cdot\mathrm{Hz}^{1/2}\right]$. The history of development of photodetectors and PDAs is the perpetual striving for higher detectivity D^*, higher sensitivity S (the numerator in (3.116) for D^*), and lower noise (the denominator).

All the same, the specific parameters should not be idolized. By tradition, they are still applied to the silicon and germanium photodetectors, and silicon and germanium-based PDAs. However, the dark current of photodiodes based on these materials is very low, and is often caused not by the bulk generation, but by the leakage current across the perimeter; the total noise in such PDAs may be determined by the noise of the amplifier. Then condition $I_n = I_d = j_d A_e$ is violated, and there is no reason to refer to unit PSE area.

What are the ultimate achievable values of detectivity in unit frequency band? Naturally, they are determined by the limits of sensitivity (given that the quantum yield is 100%), and the lowest achievable dark current (see Table 1.1). In spectral ranges I and II the dark current density is limited by thermal generation, and varies strongly from one specimen to another. Using the values from Table 1.1, we may get the following estimates:

Spectral range I, silicon PD, $\lambda = 1\,\mu\mathrm{m}$, $d = 0.5\,\mathrm{mm}$:

$$D_1 = \frac{S}{\sqrt{2qj_dA_e}} = \frac{0.5}{\sqrt{2\cdot 1.6\cdot 10^{-19}\cdot 3\cdot\left(10^{-10}...10^{-7}\right)\cdot 2\cdot 10^{-3}}}$$

$$\approx 3\cdot 10^{13}...10^{15}\ \mathrm{W}^{-1}\cdot\mathrm{Hz}^{1/2}\,; \qquad (3.118)$$

Spectral range II, germanium PD, $\lambda = 1.5\,\mu\mathrm{m}$, $d = 0.5\,\mathrm{mm}$:

$$D_1 = \frac{1.2}{\sqrt{2\cdot 1.6\cdot 10^{-19}\cdot (1...3)\cdot 10^{-4}\cdot 2\cdot 10^{-3}}}$$

$$\approx (3...5)\cdot 10^{12}\ \mathrm{W}^{-1}\cdot\mathrm{Hz}^{1/2}\,. \qquad (3.119)$$

It is very difficult to obtain a higher value of D_1 than that indicated above for the silicon PD, even with cooling and/or a smaller PSE area: the amplifier noise will interfere. The limit of detectivity in unit frequency band for spectral ranges III – V can be calculated with good enough precision, and not just evaluated.

Specific detectivity of background-limited infrared photodetector. In the IR spectral ranges III – V one can achieve the minimum current density as permitted by the background thermal radiation ($j_d = j_{db}$), and the background noise becomes prevalent (Sect. 1.2, 2.5). The regime dominated by the fluctuations of thermal background radiation is called the BLIP (background-limited infrared photodetector) mode. Accordingly, the detectivity (specific detectivity) in this mode of operation is referred to as BLIP detectivity, and symbolized D_{BLIP} (or D^*_{BLIP}).

BLIP detectivity is calculated by (3.115), (3.116), with the replacement $I_n \Rightarrow I_{db}\mathfrak{F}$, $j_d \Rightarrow j_{db}\mathfrak{F}$. The noise-factor has to be taken into account when λ'' is large, and the grouping of photons gives rise to over-Poisson noise (Sect. 2.3)

The BLIP specific detectivity D^*_{BLIP} is calculated primarily for the idealized model, when the thermal background radiation falls on the photodetector from the hemisphere (2π steradian), the density of this radiation obeys Planck's law of blackbody radiation, and the photodetector registers all the incident photons without loss. We have already used this model in Sect. 1.2 for calculating the background current caused by thermal background radiation, and in Sect. 2.3 for calculating the noise of such current (see (2.111)). In Sect. 1.2 we quoted the formula for sensitivity at given wavelength (1.18). Now we only have to substitute the results (2.111) and (1.18) into (3.116) to obtain the desired expression for specific BLIP detectivity at the maximum of sensitivity $S(\lambda'') = S''$:

$$D^*_{BLIP} = \frac{S''}{\sqrt{2qj_{db}\mathfrak{F}}} = \frac{S''}{\sqrt{\overline{J^2}}}$$

$$\approx \frac{q\lambda''}{\mathrm{h}c\sqrt{2q^2\sigma'\mathrm{T}^3\exp(-x'')\left(0.416x''^2+0.83x''+1.37\right)}}$$

$$= \frac{\lambda''\exp\frac{x''}{2}}{\mathrm{h}c\sqrt{2\sigma'\mathrm{T}^3\left(0.416x''^2+0.83x''+1.37\right)}}$$

$$= 1.82\cdot 10^9\cdot\left(\frac{293}{\mathrm{T}}\right)^{3/2}\frac{\lambda''\exp\frac{x''}{2}}{\sqrt{0.416x''^2+0.83x''+1.37}}\,. \tag{3.120}$$

Here λ'' is expressed in μm, and D^*_{BLIP} in $\mathrm{W}^{-1}\cdot\mathrm{cm}\cdot\mathrm{Hz}^{1/2}$. In transition to numerical values we used (1.18), (1.55).

Equation (2.111) can also be used for obtaining the exact formula for D^*_{BLIP} in terms of the integral $\mathrm{In}_2(2)$, which may be calculated from (1.35). It turns out, however, that the analytical approximation (3.120) differs from the exact formula by no more than 2%.

Observe that our expression (3.120) has been written for the ideal PD for the spectral

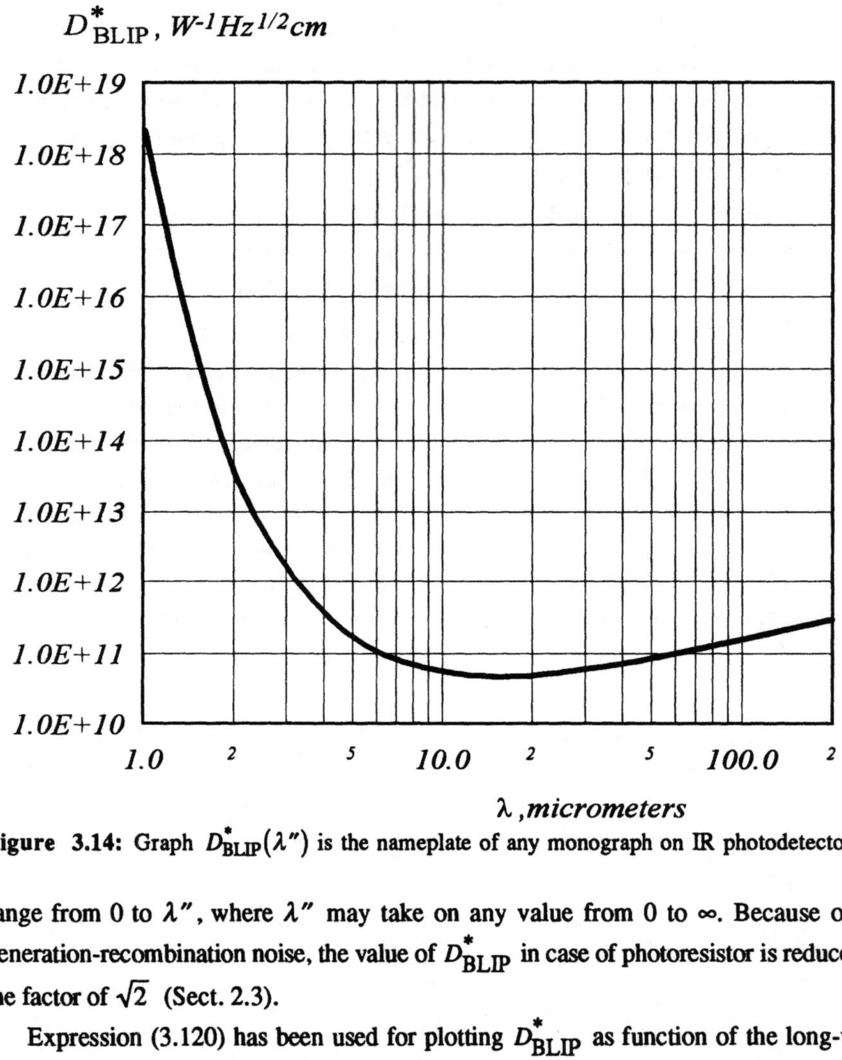

Figure 3.14: Graph $D^*_{BLIP}(\lambda'')$ is the nameplate of any monograph on IR photodetectors.

range from 0 to λ'', where λ'' may take on any value from 0 to ∞. Because of the generation-recombination noise, the value of D^*_{BLIP} in case of photoresistor is reduced by the factor of $\sqrt{2}$ (Sect. 2.3).

Expression (3.120) has been used for plotting D^*_{BLIP} as function of the long-wave limit λ'' in Fig. 3.14. Such graph is a trademark of every monograph on IR photodetectors. BLIP detectivity exhibits a very strong dependence on λ'', especially in the range from 3 to 10 μm. In this range it falls by one-and-a-half orders of magnitude, which is explained by the sharp increase in the background current. The variation of the background current, and, as a consequence, the change in the BLIP detectivity, become less considerable in the neighborhood of the thermal radiation maximum ($\lambda_0 = 12.7 \mu m$ at 293 K). Obviously, in the limit, when the spectral range is sufficiently broad ($\lambda \geq 3\lambda_0$), the photodetector will register almost all thermal photons, and its background current density j_{db} tends to a constant (maximum possible) value. When the background current is constant, the value of D^*_{BLIP} is proportional to the sensitivity, which, in turn, is proportional to λ'' (see (3.120)). So, while at $\lambda'' < \lambda_0$ the detectivity was falling as λ'' increased, at $\lambda'' >> \lambda_0$ it grows in sympathy with λ''. Accordingly, in the neighborhood of λ_0 there is a minimum of D^*_{BLIP}. Calculations indicate that this minimum occurs at the following wavelength and has the following value:

$$\lambda'' = 1.11\lambda_0 = 14.5\mu m\,, \quad D^*_{BLIP} = 4.9\cdot 10^{10}\, W^{-1}\cdot cm\cdot Hz^{1/2}\,. \tag{3.121}$$

Now let us consider an ideal PD whose spectral range is limited not only from above (λ''), but also from below (λ'). The problem here consists not so much in finding the analytical dependence of D^*_{BLIP} on λ'', as in the complexity of this dependence. The background current, and hence its noise, is determined by the difference of values calculated for $\lambda = \lambda''$ and $\lambda = \lambda'$ (Sect. 1.2):

$$D^*_{BLIP}(\lambda', \lambda'') = \frac{S''}{\sqrt{\overline{J^2(\lambda',\lambda'')}}} = \frac{S''}{\sqrt{\overline{J^2(\lambda'')} - \overline{J^2(\lambda')}}}\,. \tag{3.122}$$

So, into the expression for D^*_{BLIP} we must substitute two approximations for the background current noise (2.113), for $\lambda = \lambda''$ and $\lambda = \lambda'$. Since, however, in Fig. 3.14 we have a diagram which portrays the function $D^*_{BLIP}(\lambda'')$, we can use it for finding the value of D^*_{BLIP} for any λ' and λ''. This follows from (3.122):

$$\frac{1}{\left[D^*_{BLIP}(\lambda',\lambda'')\right]^2} = \frac{\overline{J^2(\lambda'')} - \overline{J^2(\lambda')}}{S''^2} = \frac{\overline{J^2(\lambda'')}}{S''^2} - \frac{S'^2}{S''^2}\frac{\overline{J^2(\lambda')}}{S'^2}$$

$$= \frac{1}{\left[D^*_{BLIP}(\lambda'')\right]^2} - \frac{\lambda'^2}{\lambda''^2\left[D^*_{BLIP}(\lambda')\right]^2}\,. \tag{3.123}$$

Here we have used the fact that $S \propto \lambda$. The values of BLIP specific detectivity, calculated from (3.123) for the IR spectral ranges under consideration, are presented in Table 3.3. In these examples the dependence on the short-wave limit is very weak; it only is manifest in the spectral range V: imposing a limit on the sensitivity of photodetector in the region from 0 to 8 μm raises the value of D^*_{BLIP} by 9...24%, because the number of thermal photons in this range ($1.8\cdot 10^{17}$) is comparable with their total number in the ranges 0...10 μm and 0...14 μm ($4.2\cdot 10^{17}$ and 10^{18} per square centimeter, respectively; see (1.47), (1.54)).

Table 3.3. **Background-limited specific detectivity**

Spectral Range	III	IV		V		
Wavelength, μm	2...3	3...5		8...10	8...12	8...14
Photodetector type	PD	PD	PR	PD	PD	PR
	Photodetector with ideal spectral curve					
λ_m, μm	3	5		10	12	14
$S_{\lambda m}$, A/W	2.42	4.03		8.06	9.68	11.3
D^*_{BLIP}, $W^{-1}cmHz^{1/2}$	$1.75\cdot10^{12}$	$1.77\cdot10^{11}$	$1.25\cdot10^{11}$	$7.2\cdot10^{10}$	$5.81\cdot10^{10}$	$3.83\cdot10^{10}$
	Photodetector with typical spectral curve					
λ_m, μm	2.71	4.4		9.45	10.8	12.2
$S_{\lambda m}$, A/W	2.14	3.5		7.46	8.57	9.68
D^*_{BLIP}, $W^{-1}cmHz^{1/2}$:						
exact value	$1.38\cdot10^{12}$	$1.49\cdot10^{11}$	$1.05\cdot10^{11}$	$6.63\cdot10^{10}$	$5.06\cdot10^{10}$	$3.19\cdot10^{10}$
approximation	$1.55\cdot10^{12}$	$1.54\cdot10^{11}$	$1.1\cdot10^{11}$	$6.67\cdot10^{10}$	$5.14\cdot10^{10}$	$3.28\cdot10^{10}$

BLIP specific detectivity as function of background temperature. As the background temperature increases, the background current increases sharply, and the BLIP detectivity falls. Its relative temperature coefficient ξ_D is uniquely determined by the relative temperature coefficient of the background current ξ, which follows directly from the definition of D^*_{BLIP} (3.120). Let us take the logarithm of this expression,

$$\ln D^*_{BLIP} = \ln\frac{S''}{\sqrt{2qj_{db}\mathcal{F}}} = \ln\frac{S''}{\sqrt{2q\mathcal{F}}} - \frac{1}{2}\ln j_{db}\ , \tag{3.124}$$

and then find its derivative with respect to the temperature,

$$\frac{1}{D^*_{BLIP}}\frac{dD^*_{BLIP}}{dT} \approx -\frac{1}{2j_{db}}\frac{dj_{db}}{dT}\ . \tag{3.125}$$

Here we have assumed that the temperature derivative of the noise-factor is small, because small is the range of variation of this parameter, $\mathcal{F} = 1...1.37$ (Sect. 2.3).

The left-hand side of (3.125) is nothing else but the temperature coefficient ξ_D, while the right-hand side is $-\xi/2$, which means that

$$\xi_D = -\frac{\xi}{2}\ . \tag{3.126}$$

Hence it follows that all approximations and numerical estimates for ξ (1.62), (1.67), obtained in Sect. 1.2 in the analysis of the temperature dependence of background current, apply also to ξ_D — with the opposite sign, and divided by two.

This discussion proves once again that we are right to choose the specific detectivity as the main criterion of quality of IR PDAs, as the most important parameter in the theory of detection of low-level optical signals. The BLIP specific detectivity of the ideal photodetector is a fundamental quantity, and does not depend on the particular properties of either the photodetector or the PDA: a perfect IR PDA carefully carries on to the output that value of SNR which was at its optical input.

Given that the thermal background is the same, the value of D^* of a real PDA is less than that of a perfect one. In optoelectronic systems, however, the thermal background may be limited artificially (as compared to the idealized situation considered above), and then the value of D^*_{BLIP} may increase. Let us consider the main factors which influence the values of j_{db}, D^*_{BLIP} under real conditions (Sect. 1.2).

Calculation of BLIP specific detectivity with due account for the actual spectral characteristic of photodetector. To carry out this calculation we must, as before, determine the effective number of photons n_{eff} registered by the photodetector. The task is complicated by the fact that now one must take into account the dependence of the quantum yield on the wavelength $\eta(\lambda)$, and perform numerical integration of the density of photoactive pairs $n(\lambda)\eta(\lambda)$. In Sect. 1.2 we showed, however, that n_{eff} is practically equal to the number of thermal photons in the range from λ' to λ'', the short-wave (λ') and the long-wave (λ'') limits of photodetector's spectral range being determined at half-maximum. This simplifies the calculation, allowing us to avoid the integration, and define the dark current $j_{db} = qn_{eff}$, as before, with the aid of the familiar analytical approximations (1.53), (1.47).

Now let us find the value of D^*_{BLIP} by each of the two techniques, using the exact and the approximate values of n_{eff} for the typical spectral ranges III – V. In these ranges the noise is Poissonic, and so we may assume that $\mathfrak{F} = 1$. The value of D^*_{BLIP} will be determined at the sensitivity maximum $\lambda = \lambda_m$. In real photodetectors we have $\lambda_m < \lambda''$, as different from the ideal photodetector, for which $\lambda_m = \lambda''$. We choose a typical spectral characteristic $S(\lambda)$ (curve 3 in Fig. 1.7), and define the limits of the spectral range $\lambda'...\lambda''$ (from 2...3 to 8...14 μm). The parameters λ_m and $\eta(\lambda)$, required for our calculation, will be taken from that curve.

The results of the calculations are presented in Table 3.3; we see that

1. The inaccuracy of the approximate calculation of D^*_{BLIP} for the range 3...14 μm is within 3...5% (at least for spectral curves $S(\lambda)$ which exhibit a sufficiently steep decline at the long-wave end, like curve 3 in Fig. 1.7);

2. BLIP specific detectivity of real photodetectors is 8...22% lower than that of ideal photodetectors, which is explained by the displacement of the sensitivity maximum towards shorter wavelengths.

In these calculations we assumed that the quantum yield at the sensitivity maximum is equal to one, whereas in real PDAs we usually have $\eta = 0.5...0.8$ (see Sect. 1.2). Since $j_d = j_{db} \propto \eta$, we also have $S \propto \eta$, and, according to (3.112),

$$D^*_{BLIP} \propto \frac{S}{\sqrt{j_{db}}} \propto \sqrt{\eta} = \sqrt{0.5...0.8} \approx 0.7...0.9 \; . \tag{3.127}$$

In other words, the actual value of D^*_{BLIP} is usually 10...30% less because of the reduced sensitivity.

Effects of the field of view. The effects of the field of view can be assessed in exactly the same way, in terms of j_{db}. According to (3.116),

$$D^*_{BLIP} \propto \frac{1}{\sqrt{j_{db}}} = \frac{1}{\sqrt{qn_{eff}\big|_{\alpha=\pi} \sin^2 \alpha}} \propto \frac{1}{\sin \alpha} \; . \tag{3.128}$$

When the field of view is made smaller, the detectivity D^*_{BLIP} increases (as compared to the system which looks into the hemisphere) owing to the reduced background radiation. In the wide-angle systems ($2\alpha \sim 54°$) the increase is by a factor of 2.2; a considerable increase (from 6 to 11-fold) can only be achieved in very narrow-angle systems ($2\alpha = 10...20°$).

From arguments developed above it follows that the struggle for the high detectivity D^*_{BLIP} is the struggle for the low current density j_{db}. As indicated in Sect. 1.2, the technical requirements to the optoelectronic system restrict our freedom to reduce j_{db}: neither the field of view, nor the spectral range can be made smaller than the specified limits; also, it is far not always possible to cool the optical parts so as to reduce their thermal emission, since this complicates the system and greatly increases power consumption.

Threshold power and energy. What is the lowest optical pulse amplitude that can be detected by PDA? One of the main tasks of the present book consists in calculating this value, and trying to make it as small as possible. To avoid possible confusion, we should review some of the definitions.

For the designer of optoelectronic system, the lowest detectable power is often understood as such power $E_{sA\,min}$ that ensures compliance with the preassigned values of probabilities P_{fa}, P_{tm} [84].

For the designer of PDA, the lowest detectable power is the threshold optical signal E_{st}, which we have defined earlier as the optical power at which the signal is equal to the noise. According to Sect. 3.1, these two powers are related as

$$E_{sA\,min} = \mathrm{SNR}\, E_{st} \approx (6...12) \cdot E_{st} \ . \tag{3.129}$$

Hereinafter we are going to deal with E_{st}, advising the optoelectronics engineers to take due account of the relation (3.129).

Threshold power in case of white noise. When the noise is white, and the passband of the amplifier is arbitrary, it is easy to calculate E_{st} knowing the specific parameters $\mathrm{NEP}^* = 1/D^*$. The first step obviously consists in going over from the specific values to the values referred to unit bandwidth, $\mathrm{NEP} = \mathrm{NEP}^* \sqrt{A_e}$ (see (3.112). Problems sometimes arise in the next step, in the calculation of E_{st}. This task is made easy, however, by the detection theory: from the general expression (3.37), which defines the SNR at $E_{sA} = E_{st}$,

$$\mathrm{SNR} = \frac{S E_{st} u_{sA}}{\sqrt{2qI_n f_n}} = \frac{E_{st} u_{sA}}{\mathrm{NEP}\sqrt{f_n}} = 1 \ , \tag{3.130}$$

it follows that

$$E_{st} = \frac{\mathrm{NEP}\sqrt{f_n}}{u_{sA}} = \frac{\Phi_t}{u_{sA}} \ . \tag{3.131}$$

The noise band f_n and the relative pulse sensitivity u_{sA} are calculated from (3.38), (3.39), (3.32). We see that, in general, the threshold optical power E_{st} is somewhat greater than the threshold power Φ_t because of the frequency roll-off of the pulsed signal, $u_{sA} < 1$. In case of rectangular pulse, however, the optimal filter does not cut down the amplitude, $u_{sA} = 1$ (Sect. 3.3), and the threshold optical power E_{st} is equal to the power Φ_t, calculated in the passband of the filter — which is already known to be equal to the signal bandwidth $1/2T_s$ (3.60):

$$E_{st} = \frac{\mathrm{NEP}\sqrt{f_n}}{u_{sA}} = \frac{\mathrm{NEP}}{\sqrt{2T_s}} \ . \tag{3.132}$$

The same relation follows from (3.102) when the numerical constant is selected equal to $1.41 = \sqrt{2}$.

Optimal filtering of the bell-shaped pulse leads to a similar relation: one only has to replace the constant $a = \sqrt{2}$ in (3.132) with $a = 1.22$. Real PDAs may lose further

3.5...11% to the optimal filter (Sect. 4.1), so the threshold optical power may be evaluated as (see (3.102)):

$$E_{st} \approx (1.03...1.1)\frac{\text{NEP}}{a\sqrt{T_s}} = \frac{1.03...1.1}{1.22...\sqrt{2}}\frac{\text{NEP}}{\sqrt{T_s}} \approx (0.7...0.9)\frac{\text{NEP}}{\sqrt{T_s}} \; . \tag{3.133}$$

Multielement PDAs are characterized both by the noise equivalent power in case of sine-wave modulation, and by the threshold optical power in case of pulsed modulation. Charge integration devices, and especially CCDs, are often characterized by "watts per element" corresponding to SNR = 1. This number is nothing else but E_{st}. In this case our last equation comes in especially handy, since it relates E_{st} with NEP in case of white noise (for charge integration devices, the signal length T_s in this formula is replaced by the integration time; see Sect. 7.1).

Arbitrary noise spectrum. When the noise spectrum is not white, the detectivity is defined as function of the frequency, $D_1(f)$. In principle, this characteristic is informative: as follows from formula (3.52) for SNR, the function $D_1(f)$ can be used for calculating E_{st} if the filter is optimal:

$$\text{SNR}_{\text{OF}} = \frac{SE_{st}}{\sqrt{2qI_n}}\sqrt{\int_0^\infty \frac{e_s^2}{\Theta(f)}df} = E_{st}\sqrt{\int_0^\infty \left[\frac{S}{2qI_n\Theta(f)}\right]^2 e_s^2 df}$$

$$= E_{st}\sqrt{\int_0^\infty [D_1(f)]^2 e_s^2 df} = 1 \; , \tag{3.134}$$

whence

$$E_{st} = \frac{1}{\sqrt{\int_0^\infty [D_1(f)]^2 e_s^2 df}} = \frac{1}{\sqrt{\int_0^{1/2T_s} [D_1(f)]^2 e_s^2 df}} \; . \tag{3.135}$$

The second equation in (3.135) corresponds to signal spectrum of rectangular shape; then E_{st} is determined by the area under the curve D_1^2 in the frequency range from 0 to $1/2T_s$. Equation (3.135) is universal enough; it holds also for an inertial photodetector. In case of non-optimal filter and slow photodetector, the signal and the noise of the PDA (3.31), (3.32) must be calculated separately in order to calculate E_{st}, and the knowledge of the detectivity $D_1(f)$ does not help. Then, however, our knowledge of the spectrum will come to the rescue: after all, the spectrum is not entirely arbitrary, since we know the low-frequency and the high-frequency noise components.

High-frequency noise. This noise prevails in the high-frequency PDAs (first of all, in laser systems). The detective properties in case of high-frequency noise and optimal filter are described in terms of the threshold energy (3.103), (3.108); accordingly, the threshold power is inversely proportional to the duration of the signal. Recall that the high-frequency noise component is characterized by the frequency f_{high} at which its power becomes equal to the power of white noise:

$$2qI_{\text{n}} = \overline{e_{\text{n}}^2} C_{\text{in}}^2 \omega_{\text{high}}^2 \,, \quad f_{\text{high}} = \frac{\omega_{\text{high}}}{2\pi} = \frac{\sqrt{2qI_{\text{n}}}}{2\pi e_{\text{n}} C_{\text{in}}} \,. \tag{3.136}$$

Substituting this frequency in the expression for the threshold energy (3.108), we reduce it to the form

$$\left(E_{\text{sA}} T_{\text{s}}\right)_{\text{t}} = \frac{\text{NEP}}{a\sqrt{f_{\text{high}}}} = \frac{\sqrt{2qI_{\text{n}}}}{aS}\sqrt{\frac{2\pi e_{\text{n}} C_{\text{in}}}{\sqrt{2qI_{\text{n}}}}} = \frac{\sqrt{\sqrt{2qI_{\text{n}}}\, 2\pi e_{\text{n}} C_{\text{in}}}}{aS} \,. \tag{3.137}$$

This formula includes both noises which interfere with the detection of the signal: the white noise $2qI_{\text{n}}$, and the high-frequency noise, whose intensity is defined by the product $e_{\text{n}} C_{\text{in}}$. This expression is marked by the weak dependence of the threshold energy on the power of the white noise (as the root of the fourth power). For the typical parameters of PDA based on silicon PD ($I_{\text{n}} = 10^{-8} \ldots 10^{-6}\,\text{A}$, $e_{\text{n}} = 1 \ldots 3\,\text{nV} \cdot \text{Hz}^{-1/2}$, $C_{\text{in}} = 3 \ldots 10\,\text{pF}$, $S = 0.5\,\text{A/W}$) we get the following numerical estimate of the threshold energy:

$$\left(E_{\text{sA}} T_{\text{s}}\right)_{\text{t}} = \frac{\sqrt{\sqrt{2 \cdot 1.6 \cdot 10^{-19} \cdot \left(10^{-8} \ldots 10^{-6}\right)} \cdot 2\pi \cdot (1 \ldots 3) \cdot 10^{-9} \cdot (0.3 \ldots 1) \cdot 10^{-11}}}{2.5 \cdot 0.5}$$
$$\approx 2.6 \cdot \left(10^{-17} \ldots 10^{-16}\right) \text{J} \,;$$

$$Q_{\text{t}} = S\left(E_{\text{sA}} T_{\text{s}}\right)_{\text{t}} \approx 1.3 \cdot \left(10^{-17} \ldots 10^{-16}\right) \text{C} \,. \tag{3.138}$$

This estimate shows that a photodetector with optimal filtering is capable of detecting (on the noise level) from 100 to 1000 photoelectrons. The avalanche PD improves the threshold energy, while a broader-band filter makes it worse (a broader-band filter may be required to ensure better reproduction of the shape of the signal; Sect. 5.3).

Low-frequency noise. When the low-frequency noise prevails, the threshold power E_{st} is a detection invariant (3.101), (3.106). The uncontrollable nature of the flicker noise defies reliable numerical evaluations. For the same reason, it is difficult to name the type of PDA for which the value of E_{st} could be a priory chosen as the detection invariant.

Invariants, invariants, invariants... Detection invariants show the ability to multiply and reproduce like a fruit fly. Let us give an example. So far we have been interested in the *total flux of radiation* detectable by a PDA. Accordingly, we introduced the parameter Φ_t and related parameters NEP, NEP*. In some systems, however, it is the sensitivity to the *flux density* (illumination, irradiance) that is important. In literature one may encounter a pertinent parameter called noise equivalent flux density (NEFD), or noise equivalent irradiance (NEI), NEFD = NEI = Φ_t/A_e $\left[\mathrm{W/cm^2}\right]$.

Notation and terms are not firmly established. In the above example one and the same quantity, measured in $\left[\mathrm{W/cm^2}\right]$ has two names, NEFD and NEI. The specific noise equivalent power NEP* is often pronounced as NEP-star. There also are cases when one and the same notation is used for different specific quantities. We use notation NEP for noise equivalent power in unit frequency band, NEP $= \Phi_t/\sqrt{f_n}$ $\left[\mathrm{W}/\sqrt{\mathrm{Hz}}\right]$. Some authors, however, used this abbreviation for noise equivalent power in fixed frequency band (in place of our Φ_t measured in watts). Of course, it is not the number of terms but their expedience that is important: it must be a real invariant, independent of such parameters as A_e, f_n, α, and capable of serving as a reliable criterion of quality in the comparison between different PD and PDAs.

Summary. *The detective properties of PDA are characterized by the threshold intensity of the optical pulse, for which the signal-to-noise ratio at the output of filter-amplifier is equal to one. Depending on the nature of the noise, the threshold intensity of the pulse is measured in different physical units: in the classical case of white noise, it is the quantity* $\left(E_{sA}\sqrt{T_s}\right)_t$; *when the high-frequency noise prevails, it is the threshold energy* $\left(E_{sA}T_s\right)_t$; *in case of low-frequency noise, it is the threshold power* E_{st}. *The threshold intensity of the pulse, measured in the units of* $\left(E_{sA}T_s^{\alpha}\right)_t$, *is uniquely defined by the parameters of the PDA alone (sensitivity of the photodetector S, white noise power* $2qI_n$, *frequency limits of the low-frequency (*f_{low}*) and the high-frequency (*f_{high}*) noise components), and is equal to (up to the constant factor of* $a = 0.6...2.5$*):*

— *the noise equivalent power* $\Phi_t = \sqrt{2qI_n}/S = a\left(E_{sA}\sqrt{T_s}\right)_t$ *in case of while noise;*

— *the threshold energy* $\Phi_t/\sqrt{f_{high}} = a\left(E_{sA}T_s\right)_t$ *in case of high-frequency noise;*

— *the threshold power* $\Phi_t\sqrt{f_{low}} = aE_{st}$ *in case of low-frequency noise.*

Along with the noise equivalent power Φ_t, *widely used in the IR technology is the inverse parameter called the detectivity. It is convenient to characterize the quality of PDA in terms of specific noise equivalent power, reduced to the unit frequency band (in case of white noise), and to unit area of photosensitive element (when prevalent is the shot noise of the current which is proportional to the area of PSE). In the limit, with IR PDAs it is possible to obtain such values of the threshold signal which cease to be dependent on the properties of the photodetector and the PDA, and are limited only by the fluctuations of thermal background radiation.*

Chapter 4

OPTIMAL FILTERING OF PHOTO SIGNALS (TIME-DOMAIN ANALYSIS)

In the previous chapter we carried out the analysis of optimal filtering in the frequency domain. This chapter is devoted to the time-domain analysis. In this case the input signal is defined as a function of the time; for PDA, $I_s(t) = SE_{sA}e_s(t)$ (there is no need to expand the signal into its harmonics). The noise is also defined by a function of the time, the correlation function $\mathcal{K}(\Delta t)$ (which in Sect. 2.2 was interpreted as the statistical-average noise pulse). In place of the frequency response, the amplifier is characterized by the pulse response $\mathrm{H}(\Delta t)$, or by the transfer characteristic. Functions $E_{sA}e_s(t)$, $\mathcal{K}(\Delta t)$, and $\mathrm{H}(\Delta t)$ are sufficient for calculating the shape of the output signal and the correlation function of the output noise, which can then be used for calculating the ratio of the signal amplitude to the rms noise at the output, and for maximizing this ratio. In general, this calculation is cumbersome, and so far we preferred to rely on the frequency analysis. The time-domain analysis can be much simplified, however, by using the approximate technique, based on considering the sequence of samples of signal and noise, and representing the noise as the sequence of pulses of random amplitude with the average duration τ_n. This is sufficient for deriving all the main laws that have already been found through the frequency analysis in Sect. 3.2 and 3.3, formulating the rules for the choice of optimal filter, defining the achievable values of $\mathrm{SNR_{OF}}$ and the dependence of the signal-to-noise ratio on the signal duration for different types of noise. A pragmatic reader may skip this section altogether: he is not going to find here any basically new formulas. The time-domain analysis, however, will add to our understanding of the physical principles of optimal filtering, and will clarify the methods of signal processing in matrix PDAs with integration from the standpoint of optimal filtering (see Sect. 6.1 and 6.2). Some calculations — for example, calculations of the shape of the output signal — also become more economical (Sect. 4.2).

4.1. Pulse Characteristic of Optimal Filter

First of all, let us recall the definition of the pulse response of linear network, and its linkage with the frequency-dependent transfer coefficient, emphasizing the analogies between the frequency and the time-domain techniques.

Passage of signal through filter-amplifier. Three simple test signals are used for describing the inertial properties of the amplifier: harmonic signal, delta pulse, and stepwise pulse. The response to each of these test signals defines, respectively, the frequency response, the pulse response, and the transient response of the network. Knowing any of these characteristics, it is easy to calculate the response of the linear network to the input signal of arbitrary form. The calculation technique is universal: the input signal is represented as a sum of simple test signals, and the response of the network to each of these is found. This response is defined by the corresponding characteristic (frequency, pulse, or transient response). Adding up the partial responses, we get the resulting response at the output of the amplifier. This was our procedure in the frequency analysis (Fourier analysis), where the harmonics served as test signals. Let us now by analogy consider the method of pulse response (we are not going to deal with the transition response here, since the pulse response gives a much better insight into the optimal filtering in PDAs).

Decomposition of input signal into delta pulses. It takes some imagination to realize how the signal of arbitrary shape is reconstructed from the sum of its harmonics. Decomposition into delta pulses is much easier to grasp: the signal is built up of a sequence of short rectangular pulses of length $T_\delta << T_s$ each (Fig. 4.1a). Let us select one such rectangular elementary pulse I_δ (hatched):

$$I_\delta(t-\tau) = SE_s(\tau)e_\delta(t-\tau) = I_s(\tau)e_\delta(t-\tau) = I_s(\tau)T_\delta \frac{e_\delta(t-\tau)}{T_\delta}, \tag{4.1}$$

$$\delta(t-\tau) = \lim_{T_\delta \to 0} \frac{e_\delta(t-\tau)}{T_\delta}, \tag{4.2}$$

where $e_\delta(t-\tau)$ is the relative form of the elementary rectangular pulse, and τ is the location of the middle of the pulse on the time axis. Letting $T_\delta \to 0$, we get from our elementary rectangular pulse e_δ/T_δ a delta pulse (see definition in Sect. 2.3), whose amplitude is infinite at $t=\tau$, and the area is equal to one, $(e_\delta/T_\delta)T_\delta = 1$. Graphic decomposition of the signal into elementary rectangles (Fig. 4.1a) corresponds to the following analytical representation:

$$I_s(t) = \sum I_\delta(t-\tau) = \sum I_s(\tau)e_\delta(t-\tau) = \sum I_s(\tau)\frac{e_\delta(t-\tau)}{T_\delta}T_\delta$$

$$= \int_{-\infty}^{\infty} I_s(\tau)\delta(t-\tau)d\tau\ . \tag{4.3}$$

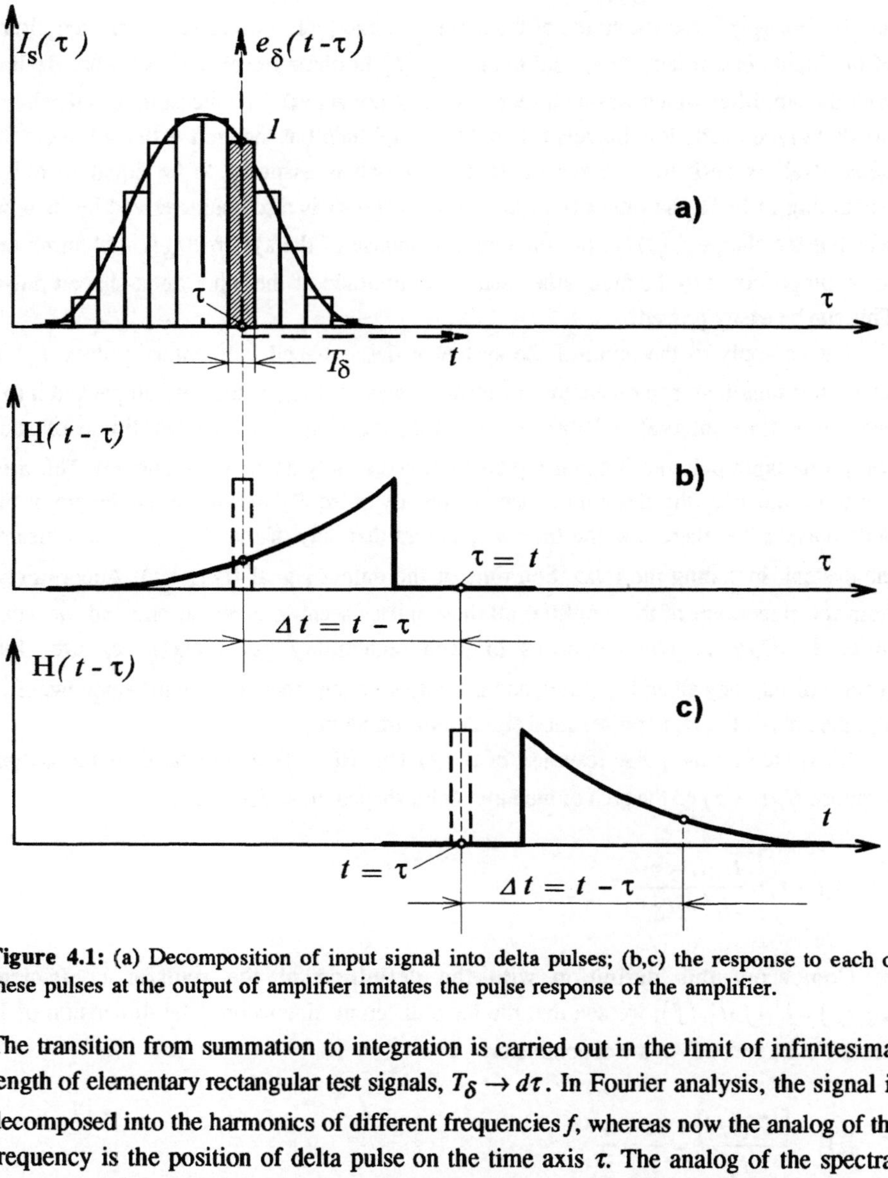

Figure 4.1: (a) Decomposition of input signal into delta pulses; (b,c) the response to each of these pulses at the output of amplifier imitates the pulse response of the amplifier.

The transition from summation to integration is carried out in the limit of infinitesimal length of elementary rectangular test signals, $T_\delta \to d\tau$. In Fourier analysis, the signal is decomposed into the harmonics of different frequencies f, whereas now the analog of the frequency is the position of delta pulse on the time axis τ. The analog of the spectral density $I_s(f) = SE_s(f)$ is the value of the input signal itself $I_s(\tau) = SE_s(\tau)$ at the time τ, which characterizes the intensity of the delta pulse.

Pulse response. By definition, the transfer characteristic is the response of the system to the input test signal of unit intensity. Applying to the amplifier input the delta-shaped test signal (the current pulse $I_\delta(t-\tau)$), at the output we get the response ($U_\delta(t-\tau)$). Then we must relate it to the input signal intensity. Now the question is what must be the unit of measurement of the intensity of the input signal. The answer might seem obvious: since the input signal is current, its intensity should be measured in amperes. This is what we did in the Fourier analysis when calculating the transfer coefficient of the amplifier $K_{21}(f)$: the amplitude of the output voltage (volts) we related to the amplitude of the input current (amperes), and found $K_{21}(f)$ in ohms (recall that we were dealing with the amplifier which was a current to voltage converter). The measure of intensity of the delta-type excitation, however, is not the amplitude but the area. After all, the delta pulse itself is defined in terms of its area, which is assumed to be equal to unity. According to (4.1), the intensity of the pulse $I_\delta(t-\tau)$ is also characterized by its area, which is the charge $I_s(\tau)T_\delta$. So, the output response of the system $U_\delta(t-\tau)$ turns out to be proportional to the area rather than the amplitude of the input delta-shaped pulse. This can be easily proved.

Let us apply to the input of the system n delta-shaped elementary pulses $I_\delta(t)$, superimposing them one on another in arbitrary way, as long as they are all packed into a very short time interval $\Delta T = nT_\delta << T_s$. Both the shape and the amplitude of such composite input pulse are arbitrary; predetermined is only its area (the charge). This area is n times the area (the charge) of each elementary pulse. By definition, the length of the delta pulse is T_δ; therefore, the time nT_δ is less than any of the characteristic times of the system, including the relaxation time of the output signal $U_\delta(t-\tau)$. Accordingly, from the standpoint of the amplifier all these pulses seem to arrive at one and the same time, $\tau + nT_\delta \approx \tau$. The responses to each elementary pulse $U_\delta(t-\tau)$ are thus synchronized; they all add together, and the output voltage increases by the same factor of n as the area of the input pulse, quod erat demonstrandum.

So we define the pulse response of the system $\mathrm{H}(t-\tau)$ as the ratio of the output response $U_\delta(t-\tau)$ to the area of the input delta-shaped pulse $I_s(\tau)T_\delta$:

$$\mathrm{H}(t-\tau) = \frac{U_\delta(t-\tau)}{I_s(\tau)T_\delta} . \tag{4.4}$$

Comparing this definition with the definition of the transfer coefficient $K_{21}(f) = U_s(f)/\mathrm{I}_s(f)$, we see that the have different dimensions: the dimension of H has an "extra second" in the denominator:

$$[\mathrm{H}] = \frac{[K_{21}(f)]}{[\mathrm{s}]} = \frac{\Omega}{\mathrm{s}} = \frac{1}{\mathrm{F}} .$$

If the delta pulse is applied to the input of the system, then the output signal is nothing else but the pulse response. Setting the area $I_s(\tau)T_\delta = 1$ in (4.1) – (4.4), we get

$$I_\delta(t) = \delta(t-\tau) \ ; \ \ \mathrm{H}(t-\tau) \equiv U_\delta(t-\tau) . \tag{4.5}$$

Now, can the area of the current pulse be considered equal to dimensionless unity if this area actually represents the charge (coulombs)? The answer is yes, if only we go over to dimensionless units, which means that I_δ, T_δ, U_δ denote not the corresponding physical quantities themselves, but their numerical values expressed in the selected units of measurement (amperes, seconds, volts, etc.). Further on we shall meet with expressions similar to (4.5), without specifying each time that dimensionless units are used.

Let us end this subsection with the example of pulse response (Fig. 4.1b,c). This response is a function of the time interval $\Delta t = t - \tau$ between the sampling time t and the time of arrival of the input delta pulse τ. The independent variable in Fig. 4.1b is τ, and in Fig. 4.1c the independent variable is t. Naturally, the two graphs are mirror-symmetrical, because τ and t in $\mathrm{H}(t-\tau)$ have opposite signs. In future we shall need the function $\mathrm{H}(\tau)$ for calculating the output signal at the specified time instant t, when delta pulses arrive at the input at different instants τ. The function $\mathrm{H}(t) = U_\delta(t)$ of the current time t is usually found experimentally by applying the delta pulse to the input at a certain time τ.

In general, the amplifier distorts the signal. In the example shown in Fig. 4.1c we observe smearing of the pulse edge, and a time lag (the maximum of the response is displaced with respect to the time of arrival of the input signal τ). In the Fourier analysis the counterparts of these two kinds of distortion are the amplitude and the phase-frequency distortions.

The linkage between pulse response and frequency response is clear from the following experiment. Sending the delta pulse to the input of filter-amplifier, we get the pulse response $\mathrm{H}(t-\tau) = U_\delta(t-\tau)$ at the output (Fig. 4.2). The spectrum of the delta pulse is white, and, from the standpoint of Fourier analysis, the harmonics of all frequencies f of the same amplitude are supplied to the input. Each of these harmonics is amplified $K_{21}(f)$-fold, and at the output we get the spectrum of harmonics $K_{21}(f)$. In other words, we get the signal $U_\delta(t-\tau) = \mathrm{H}(t-\tau)$ at the output, which in the parlance of Fourier analysis is the spectrum $K_{21}(f)$. This proves the known statement of the theory of linear systems: the frequency response $K_{21}(f)$ of such system is the spectrum of its pulse response $\mathrm{H}(t-\tau)$. Hence it follows that, if the frequency response is known, the pulse response can be calculated with the aid of inverse Fourier transform:

$$\mathrm{H}(\Delta t) = \int_0^\infty e_\delta(f) K_{21}(f) \cos(\omega \Delta t + \psi_\mathrm{a}) df = 2\int_0^\infty K_{21}(f) \cos(\omega \Delta t + \psi_\mathrm{a}) df \ . \qquad (4.6)$$

The filtering properties of the amplifier are characterized not by the absolute but rather by the relative transfer characteristics. It would be logical to define the relative characteristics (pulse response and frequency response) with reference to one and the same amplifier. Its gain (transfer coefficient) is by the factor of K_0 less than that of the actual one — in other words, its relative gain is equal to one at the plateau of the frequency curve. For such amplifier, we have

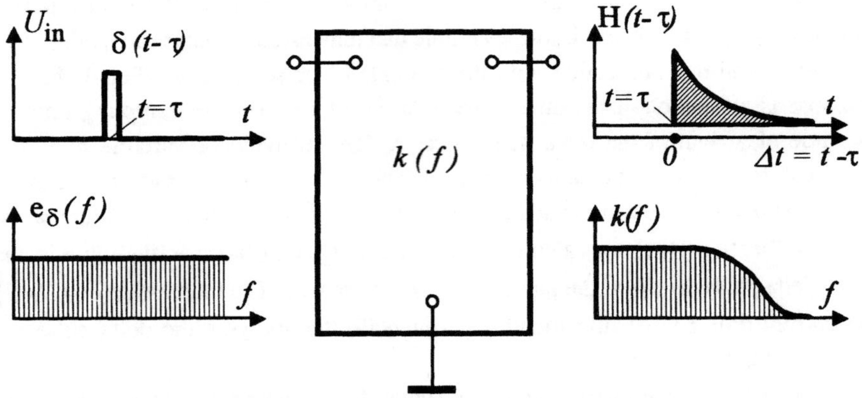

Figure 4.2: Frequency response of amplifier is the spectrum of its pulse response.

$$k(f)=\frac{K_{21}(f)}{K_0}\,, \quad \mathrm{h}(\Delta t)=\frac{\mathrm{H}(\Delta t)}{K_0}\,. \tag{4.7}$$

From this definition and (4.4) it follows that the characteristic $\mathrm{h}(\Delta t)$ is dimensional: its dimension is s^{-1}. The linkage between the relative characteristics follows from (4.6), (4.7):

$$\mathrm{h}(\Delta t)=2\int_0^{\infty} k(f)\cos(\omega\Delta t+\psi_a)df\,. \tag{4.8}$$

Shape of the output signal. So, we have decomposed the input signal $I_s(\tau)$ into the delta pulses $I_\delta(\tau)=I_{sA}e_s(\tau)\delta(t-\tau)$, and found the response $U_\delta(t-\tau)\sim \mathrm{H}(t-\tau)$ to each such pulse. The responses have the same shape, and differ only in their amplitude and position on the time axis (since the areas and the times of arrival of the input delta pulses are not the same), Now we only have to sum up all these responses and find the signal at the output of the amplifier:

$$U_s(t)=\sum U_\delta(t-\tau)=\sum I_s(\tau)T_\delta\,\mathrm{H}(t-\tau)=\int_{-\infty}^{\infty} I_s(\tau)\mathrm{H}(t-\tau)d\tau\,, \tag{4.9}$$

$$U_s(t)=\int_{-\infty}^{t} I_s(\tau)\mathrm{H}(t-\tau)d\tau\,,\quad u_s(t)=\int_{-\infty}^{t} e_s(\tau)\mathrm{h}(t-\tau)d\tau\,. \tag{4.10}$$

Here we have used the definition of the pulse response (4.4); the transition from summation to integration is carried out in the limit $T_\delta \to d\tau \to 0$.

Convolution (4.9), (4.10) in the time-domain analysis replaces the inverse Fourier transform, and allows one to find the output signal from the known input signal $I_S(\tau) = SE_{SA}e_S(\tau)$ and the pulse response $\mathrm{H}(\Delta t)$. Both representations of the convolution are useful. A physically realizable system cannot react to the input excitation before it has occurred; therefore, for such system we have $\mathrm{H}(\Delta t) = 0$ when $\tau > t$, $\Delta t < 0$. Then the upper limit of integration in (4.9) is changed from infinity to t, and this brings us straight to (4.10), where we have written the convolution for both the absolute (U_S, I_S, H) and the relative (u_S, e_S, h) quantities.

Now we have learned enough about the pulse response to proceed with our analysis. First of all we shall try to find the pulse response of the optimal filter, taking advantage of what we know about its frequency response.

The pulse response of optimal filter in case of white noise repeats the form of the signal. We start with white noise and symmetrical signal. The frequency response of the optimal filter $k(f)$ is not only the spectrum of the pulse response $\mathrm{h}(\Delta t)$, but at the same time the spectrum of the signal $e_S(t)$ (see Sect. 3.2). Since, however, the spectra of functions $\mathrm{h}(\Delta t)$ and $e_S(t)$ coincide, the functions themselves also coincide (in form):

$$\mathrm{h}(\Delta t) = Be_S\left(t - t_{lag}\right) = \frac{e_S\left(\Delta t - t_{lag}\right)}{T_S}, \tag{4.11}$$

where $B = \text{const}$.

Observe that the constant of proportionality here is chosen equal to the effective signal duration. It cannot be otherwise: the spectral density of $\mathrm{h}(\Delta t)$ at the zero frequency is $k(f) = k(0) = 1$, and in Sect. 2.1 we demonstrated that it must be equal to the area of the pulse (in this case, the pulse $\mathrm{h}(\Delta t)$). This requirement is only satisfied when the value of the constant is as indicated above:

$$\int_{-\infty}^{\infty} \mathrm{h}(\Delta t)dt = \frac{1}{T_S}\int_{-\infty}^{\infty} e_S\left(\Delta t - t_{lag}\right)dt = \frac{1}{T_S}T_S \equiv 1\,. \tag{4.12}$$

Equations (4.11), (4.12) involve the time lag t_{lag}, which is arbitrary (see Sect. 3.2) and does not affect the signal-to-noise ratio: it only defines the moment when this ration attains its maximum.

So, we have formulated another amazing property of the optimal filter:

the shape of the pulse response of the optimal filter precisely imitates the shape of the input signal to which this filter is tuned,

in exactly the same way as the frequency response of the filter imitates the signal spectrum.

This conclusion holds also when the input signal is asymmetrical, with one correction. Since the frequency response of the filter is complex conjugated with the signal spectrum, $k(f) = e_s^*(f)$ (see Sect. 3.2), the pulse response is a mirror reflection of the input signal:

$$h(\Delta t) = \frac{e_s(-\Delta t - t_{lag})}{T_s} \quad . \tag{4.13}$$

Unfortunately, this elegant result only holds in case of white noise: when the noise spectrum is arbitrary, the frequency response of the optimal filter differs from the signal spectrum, and $k(f) = e_s^*(f)/\Theta(f)$. Substituting this response into (4.8), we get

$$h(\Delta t) = 2\int_0^\infty k(f)\cos(\omega\Delta t + \psi_a)df = 2\int_0^\infty \frac{e_s(f)}{\Theta(f)}\cos(\omega\Delta t + \psi_a)df \quad . \tag{4.14}$$

In case of the high-frequency noise the roll-off of the integrand starts at the lower frequencies $f_{high} < f_s$, where the spectral density of the signal is still constant, $e_s(f) = 1$, $\psi_a(f) \approx 0$. Because of this, the pulse response does not depend on the signal spectrum,

$$h(\Delta t) \approx 2\int_0^\infty \frac{1}{1+\left(\frac{f}{f_{high}}\right)^2}\cos\omega\Delta t df = \frac{1}{2\tau_{high}}\exp\left(-\frac{|\Delta t|}{\tau_{high}}\right) , \tag{4.15}$$

$$\tau_{high} = \frac{1}{2\pi f_{high}} \quad .$$

The integral in (4.15) was taken from tables in (82). The low-frequency roll-off at $f \approx f_{high}$ is also responsible for the larger effective time constant of the filter τ_{high}.

Now we only have to find the pulse characteristic of the optimal filter in case of the low-frequency noise. The input signal is assumed to be bell-shaped, since the optimization problem for the rectangular signal and low-frequency noise is ill-posed (see Sect. 3.3). Let us try to guess the shape of the pulse response $h(\Delta t)$. After all, what is the difference between this filter and the optimal filter in case of white noise? The frequency response of both filters imitates the spectrum of the signal (which currently is square-cosine); in case of the low frequency noise, however, the gain of the filter falls in inverse proportion to $\Theta(f)$. Exaggerating a little, we may say that this filter differs only in that it does not transmit the permanent component (and the lowest harmonics, when $f_{low} << 1/2T_s$). If, however, the frequency response of the filters $k(f)$ is the same, the

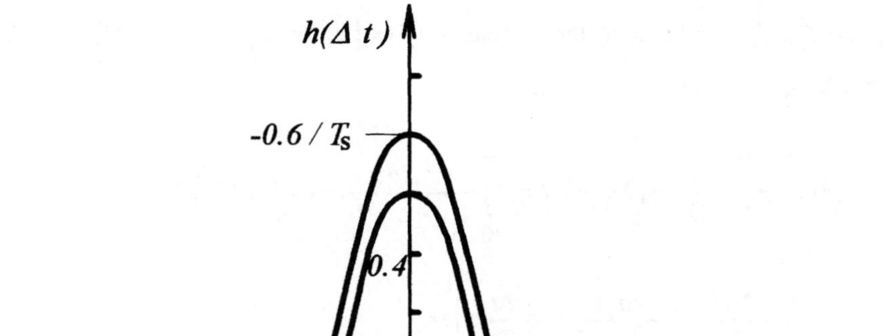

Figure 4.3: Pulse response in case of low-frequency noise is shifted downwards (with respect to the same in case of white noise), and exhibits negative tails.

pulse response $h(\Delta t)$ must also be the same. The response $h(\Delta t)$ in case of white noise is known: it imitates the bell-shaped signal. Accordingly, the pulse response $h(\Delta t)$ of the optimal filter in case of the low-frequency noise is also the same bell-shaped signal. How then does the pulse response $h(\Delta t) \sim e_S(\Delta t)$ reflect the fact that it does not transmit the permanent component? As regards equation (4.10), this means that the constant input signal $e_S(t) = 1$ at the input must produce $u_S = 0$ at the output:

$$u_S(t) = \int_{-\infty}^{\infty} e_S(\tau) h(t-\tau) d\tau = \int_{-\infty}^{\infty} h(t-\tau) d\tau \equiv 0 \ . \tag{4.16}$$

It follows that the curve $h(t-\tau) \sim e_S(t-\tau)$ must be moved downward in such a way as to make the area under this curve equal to zero. This reasoning brings us to the pulse response of the optimal filter (curve 1 in Fig. 4.3). Here the area under the curve $h(\Delta t)$ is by convention made equal to zero on the interval $[-T_S, T_S]$.

Is our guess correct? Substituting into (4.14) the spectra of the noise $\Theta(f) = f_{low}/f = x'/x$ and the signal $e_s(f) = \left(\pi^2 \sin x\right) / \left[x\left(\pi^2 - x^2\right)\right]$; $x = 2\pi f T_s$, $x' = 2\pi f_{low} T_s$, we get:

$$h(\Delta t) = 2\int_0^\infty e_s \frac{1}{\Theta} \cos(2\pi f \Delta t) df = 2\int_0^\infty \frac{\pi^2 \sin x}{x\left(\pi^2 - x^2\right)} \cos\left(\frac{\Delta t}{T_s} x\right) \frac{x}{x'} df$$

$$= \frac{2\pi^2}{x' 2\pi T_s} \int_0^\infty \frac{\sin x}{\pi^2 - x^2} \cos\left(\frac{\Delta t}{T_s} x\right) dx$$

$$= \frac{1}{2 f_{low} T_s} \frac{1}{T_s} \int_0^\infty \frac{\sin x}{\pi^2 - x^2} \cos\left(\frac{\Delta t}{T_s} x\right) dx \ . \qquad (4.17)$$

Recall that the gain (and hence $h(\Delta t)$) in case of the low-frequency noise is additionally normalized to the value of $f_s/f_{low} = 1/2 f_{low} T_s$ (3.69). With such normalization, (4.17) becomes

$$h(\Delta t) = \frac{1}{T_s} \int_0^\infty \frac{\sin x}{\pi^2 - x^2} \cos\left(\frac{\Delta t}{T_s} x\right) dx \ . \qquad (4.18)$$

The pulse response according to (4.18) is also plotted in Fig.4.3 (curve 2). Numerical calculation confirms the surmised behavior of $h(\Delta t)$: it coincides in form with the square-cosine signal pulse $e_s(t)$. True, the downward displacement is somewhat smaller (cf. curves 1 – 3), and the coincidence is observed not for all times (although it holds for a quite large interval $|\Delta t| < 0.75 T_s$, which is of the order of the signal length). At large times ($|\Delta t| > 0.75 T_s$), the response slowly relaxes towards zero. This feature was not foreseen, and the reason is clear: the large times correspond to low frequencies, and it is at the low frequencies that our digression from the actual dependence $k(f) = e_s(f)/\Theta(f)$ was the largest when we assumed that $k(f) \approx e_s(f)$.

The basic difference of this pulse characteristic from those considered above consists in the presence of negative backswings. It follows that the input disturbances at the corresponding intervals are added with a negative weight. In other words, the filter performs not only arithmetical addition, but also arithmetical subtraction.

More on the decision-making rule concerning the presence or absence of the signal. Discussing this rule in Sect. 3.1, we passed over in silence the question why the decision is taken by the logical device on the basis of just one sample, even though this sample is taken when the signal reaches its maximum. It might seem that such algorithm is not optimal, because it fails to use the entire information carried by the input signal which is nonzero over a sufficiently long time (about $(1...2)T_s$). From general considerations, a better algorithm for the logical device might consist in taking samples over the entire time interval of the signal $U_s(t)$, superimpose them according to certain rules, and use this composite sample as the basis for making the decision. This would ensure that all information about the signal (the power of the signal) is used.

The time-domain representation (convolution (4.10)) indicates that there is no need to realize this algorithm in the decision-taking device, since the desired superposition of the signal samples is prepared by the optimal filter itself. It adds up all the samples with the appropriate weighted coefficient $h(\Delta t)$. This is why the logical device only has to compare the maximum value of the output signal with the preassigned threshold value (see Sect. 3.1).

In case of the time-domain analysis the problem of optimal filtering is reduced to the optimization of the weighted coefficient — the pulse response $h(\Delta t)$ — so as to "scoop up" from the mixture of signal and noise as much signal and as little noise as possible. Let us perform this optimization for our familiar menagerie of noises: the white noise, the high-frequency noise, and the low-frequency noise.

White noise. Assume that a rectangular signal arrives at the input (Fig. 4.4a). The white noise is approximated (see Sect. 2.2) as a sequence of pulses of arbitrary amplitude and average duration τ_n (Fig. 4.4b). The spectral density of the power of such noise may be considered constant ($\overline{J^2(f)} = 2qI_n$) in the effective noise band $f_n = 1/2\tau_n$, so that the mean square amplitude of each kth noise pulse, according to (2.44), is

$$\overline{i_{nk}^2} = \overline{J^2} f_n = 2qI_n \frac{1}{2\tau_n} . \tag{4.19}$$

Altogether, $N = T_s/\tau_n$ noise pulses arrive within the time T_s. To obtain complete information about the signal and the noise over the time T_s, we must take $n = N$ samples, evenly spaced with the interval τ_n between the samples (Fig. 4.4c). Assume that the filter adds up the readings from all n samples; then the values of the signal and the noise at the output are, respectively,

$$\sum_{k=1}^{n} I_{sk} = nSE_{sA} ; \tag{4.20}$$

$$\sum_{k=1}^{n} i_{nk} = i_{n1} + i_{n2} + ... + i_{nn} . \tag{4.21}$$

Since the noise is the sum of n random independent terms having the same mean square value (4.19), we have

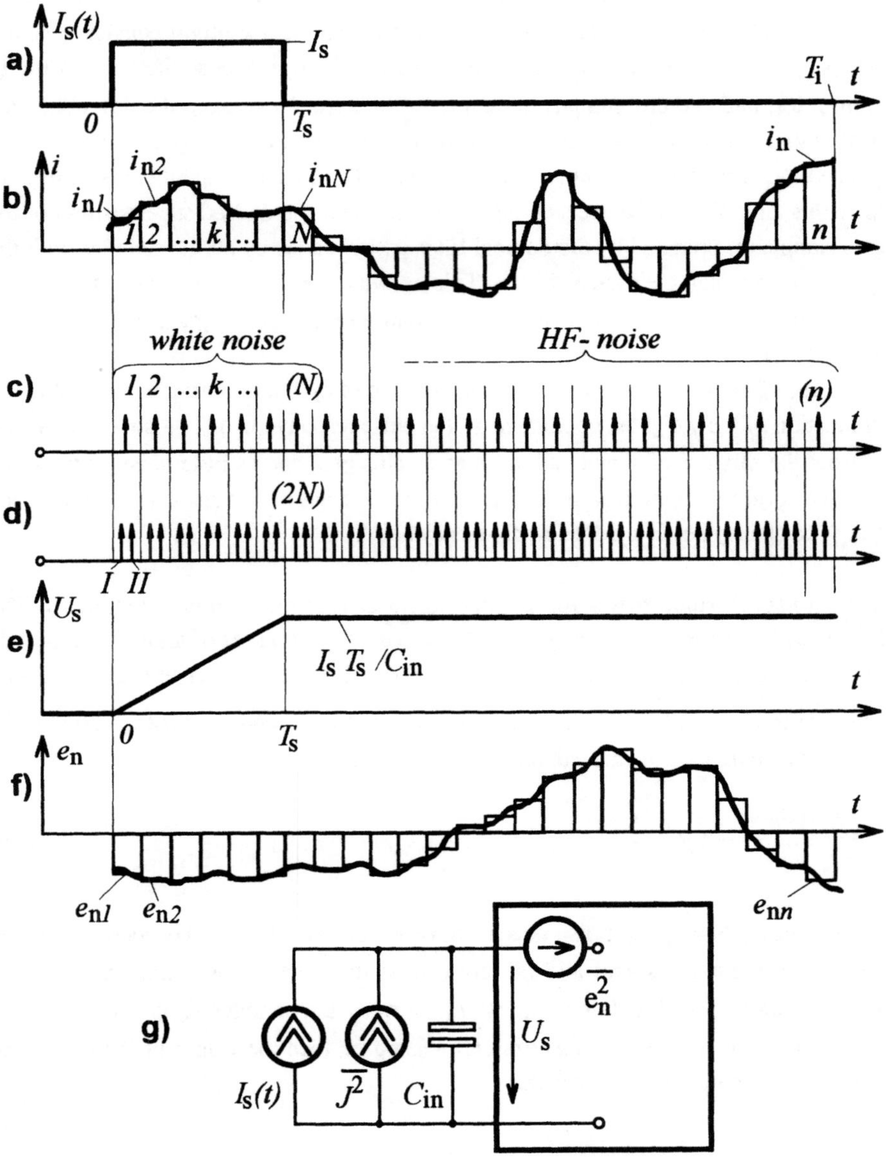

Figure 4.4: Samples of signal and noise: (a) signal; (b) noise current; (c) samples (one for each noise pulse); (d) doubling the number of samples does not improve the SNR; (e) signal accumulated by the input capacitance; (f) noise emf; (g) input equivalent diagram of PDA.

$$\sum_{k=1}^{n} \overline{i_{nk}^2} = n\overline{i_n^2} = n\frac{2qI_n}{2\tau_n} \; . \tag{4.22}$$

The signal-to-noise ratio is

$$\mathrm{SNR}=\frac{\sum_{k=1}^{n} I_{\mathrm{s}k}}{\sqrt{\sum_{k=1}^{n}\overline{i_{\mathrm{n}k}^{2}}}}=\frac{nSE_{\mathrm{sA}}}{\sqrt{n\frac{2qI_{\mathrm{n}}}{2\tau_{\mathrm{n}}}}}=\frac{SE_{\mathrm{sA}}\sqrt{2\tau_{\mathrm{n}}n}}{\sqrt{2qI_{\mathrm{n}}}}\,. \tag{4.23}$$

Substituting here the number of samples $n = N = T_{\mathrm{s}}/\tau_{\mathrm{n}}$, we get

$$\mathrm{SNR}=\frac{SE_{\mathrm{sA}}\sqrt{2T_{\mathrm{s}}}}{\sqrt{2qI_{\mathrm{n}}}}\,, \tag{4.24}$$

which is the familiar expression for $\mathrm{SNR_{OF}}$ in case of rectangular signal and white noise. Equations (4.20) – (4.24) and Fig. 4.4a-d supply answers to a number of questions.

What is the optimal observation time? We have chosen it equal to the signal pulse length, $T_{\mathrm{i}} = T_{\mathrm{s}}$. When $T_{\mathrm{i}} < T_{\mathrm{s}}$, the number of samples n is smaller, and the value of $\mathrm{SNR_{OF}}$ decreases, since $\mathrm{SNR_{OF}} \sim \sqrt{n}$ (4.23). When $T_{\mathrm{i}} > T_{\mathrm{s}}$, we integrate only the noise, and $\mathrm{SNR_{OF}}$ also decreases. Since, however, $\mathrm{SNR_{OF}} \sim \sqrt{n}$, can it be worthwhile to increase the number of samples within the limits of the signal duration, so that we had two, three, or more samples per each noise pulse (Fig. 4.4): $n = 2N$, $n = 3N$, $n = 4N$, etc. From Fig. 4.4d we see that the samples i_{nI} and i_{nII} within one noise pulse are completely correlated (similarly to the signal samples, measured is one and the same amplitude of one and the same noise pulse); accordingly, when these samples are added together, this will multiply by the same factor both the signal $I_{\mathrm{sI}} + I_{\mathrm{sII}} = 2SE_{\mathrm{sA}}$, and the noise $i_{\mathrm{nI}} + i_{\mathrm{nII}} = 2i_{\mathrm{n}}$; $\sqrt{\overline{(i_{\mathrm{nI}}+i_{\mathrm{nII}})^2}} = \sqrt{\overline{(2i_{\mathrm{n}})^2}} = 2\sqrt{\overline{i_{\mathrm{n}}^2}}$, so that the SNR is, unfortunately, not improved. In practice, however, one has to make the interval between the samples very small, much less than τ_{n}. Although the signal-to-noise ratio will not improve, we shall be liberated from the fanciful model. When the samples are taken often enough, they need not be synchronized with the noise pulses (this can be pictured in Fig. 4.4c,d, but cannot be accomplished in practice). Instead, the samples will be automatically averaged over the noise pulses of varying length, imposed one on another.

What is the rule of superposition of samples for the signal of arbitrary shape $e_{\mathrm{s}}(t)$? In case of such signal, we add the samples with the weight h_k. Then (4.20)–(4.23) will be modified as follows:

$$\sum_{k=1}^{n}\mathrm{h}_k I_{\mathrm{s}k} = SE_{\mathrm{sA}}\left(\mathrm{h}_1 e_{\mathrm{s}1} + \mathrm{h}_2 e_{\mathrm{s}2} + \ldots + \mathrm{h}_n e_{\mathrm{s}n}\right)\,; \tag{4.25}$$

$$\sum_{k=1}^{n} i_{\mathrm{n}k} = \mathrm{h}_1 i_{\mathrm{n}1} + \mathrm{h}_2 i_{\mathrm{n}2} + \ldots + \mathrm{h}_n i_{\mathrm{n}n}\,; \tag{4.26}$$

$$\sum_{k=1}^{n} \overline{i_{nk}^2} = \overline{i_n^2}\left(h_1^2 + h_2^2 + \ldots + h_n^2\right) = \frac{2qI_n}{2\tau_n}\left(h_1^2 + h_2^2 + \ldots + h_n^2\right) ; \tag{4.27}$$

$$\mathrm{SNR} = \frac{SE_{sA}}{\sqrt{2qI_n}} \sqrt{2\tau_n n} \; \frac{h_1 e_{s1} + h_2 e_{s2} + \ldots + h_n e_{sn}}{\sqrt{n\left(h_1^2 + h_2^2 + \ldots + h_n^2\right)}} . \tag{4.28}$$

Exactly the same fraction (the last term on the right-hand side) was obtained in the frequency analysis of $\mathrm{SNR_{OF}}$ (3.46), where in place of h_n, e_{sn} we had k_n, $e_{sn}(f_n)$. By analogy with (3.46) we may conclude that the maximum of the fraction in question is attained when $h_k = e_{sk}$ — that is, when $h(\Delta t) \sim e_s(\Delta t)$. Thus, we have once again proved that the pulse response of the optimal filter in case of white noise repeats the shape of the useful signal; this time the proof is totally "honest", since it does not involve the results of the frequency analysis, or the frequency response of the optimal filter.

High-frequency noise. The high-frequency noise component occurs in case of the capacitive input impedance (Fig. 4.4g), when the noise emf generator is replaced by noise current generator connected in parallel with the photodetector (see Sect. 2.4). Now we shall directly refer to the diagram in Fig. 4.4g. The realization of the emf noise $U_n(t)$, like that of the current noise, will be approximated by a sequence of rectangular pulses (Fig. 4.4f). By analogy with (4.19), the mean square value $\overline{U_{n1}^2}$ (of the sample at any time) is related to the spectral density $e_n^2(f)$ as

$$\overline{U_{n1}^2} = \frac{\overline{e_n^2}}{2\tau_n} . \tag{4.29}$$

As follows from Fig. 4.4g, the signal current I_s charges the input capacitance, and creates the constant voltage $U_s = I_s T_s / C_{in}$ (Fig. 4.4e). It is because of the integration by the capacitance that the output signal and the ratio of the output signal to the noise depends on the area (the energy) of the input signal $E_{sA} T_s$. The particular shape of the signal in no way affects the result of the integration. These features have also been discussed in connection with the frequency analysis; we pointed out that for an infinitely long pulse the signal-to-noise ratio $\mathrm{SNR_{OF}}$ tends to infinity (see Sect. 3.3). These arguments translate into the language of formulas as follows:

$$U_s = \frac{nSE_{sA}T_s}{C_{in}} , \tag{4.30}$$

$$U_n = U_{n1} + U_{n2} + \ldots + U_{nn} , \tag{4.31}$$

$$\overline{U_{\mathrm{n}}^2} = n\overline{U_{\mathrm{n1}}^2} = \frac{n\overline{e_{\mathrm{n}}^2}}{2\tau_{\mathrm{n}}} , \tag{4.32}$$

$$\mathrm{SNR} = \frac{U_{\mathrm{s}}}{\sqrt{\overline{U_{\mathrm{n}}^2}}} = \frac{nSE_{\mathrm{sA}}T_{\mathrm{s}}}{C_{\mathrm{in}}}\sqrt{\frac{2\tau_{\mathrm{n}}}{n\overline{e_{\mathrm{n}}^2}}} = \frac{SE_{\mathrm{sA}}T_{\mathrm{s}}\sqrt{2\tau_{\mathrm{n}}n}}{C_{\mathrm{in}}e_{\mathrm{n}}} , \tag{4.33}$$

$$\lim_{n\to\infty} \mathrm{SNR} = \infty . \tag{4.34}$$

Once again we repeat the result of Sect. 3.3: when only the high-frequency noise is present, the band of the amplifier must be made infinitely narrow (which means that the observation time is infinitely large, $T_{\mathrm{i}} \sim 1/2f_{\mathrm{a}}$) — then the value of $\mathrm{SNR_{OF}}$ becomes infinitely large. In Sect. 3.3 we also showed that the value of $\mathrm{SNR_{OF}}$ is actually delimited by the white noise component — let us also take this into account in the diagram shown in Fig. 4.4g. Consider again the realization $i_{\mathrm{n}}(t)$ (Fig. 4.4b). The noise pulses $i_{\mathrm{n}k}$ are stored by the input capacitance C_{in} in exactly the same way as the signal pulse, and create at the input of the amplifier a random voltage jump $i_{\mathrm{n}k}\tau_{\mathrm{n}}/C_{\mathrm{in}}$. This voltage is sampled at each round of sampling. Since the optimal filter tuned to the rectangular signal pulse adds up all samples over the time T_{i}, the first noise pulse (according to the numbering in Fig. 4.4b) will be polled and added up n times, the second $(n-1)$ times, and the last one will enter only once. Accordingly, the sum of the voltage samples by the end of the integration period T_{i} will be

$$U_{\mathrm{n}} = \frac{i_{\mathrm{n1}}\tau_{\mathrm{n}}n}{C_{\mathrm{in}}} + \frac{i_{\mathrm{n2}}\tau_{\mathrm{n}}(n-1)}{C_{\mathrm{in}}} + \frac{i_{\mathrm{n3}}\tau_{\mathrm{n}}(n-2)}{C_{\mathrm{in}}} + \ldots + \frac{i_{\mathrm{n}n}\tau_{\mathrm{n}}}{C_{\mathrm{in}}} , \tag{4.35}$$

whereas the dispersion of the sum of these n independent terms is given by

$$\overline{U_{\mathrm{n}}^2} = \frac{\overline{i_{\mathrm{n}}^2}\tau_{\mathrm{n}}^2}{C_{\mathrm{in}}^2}\left[n^2 + (n-1)^2 + (n-2)^2 + \ldots + 2^2 + 1^2\right] . \tag{4.36}$$

Let us evaluate the sum of the series on the right-hand side of this equation. If all n members had been the same (and equal to n^2), the sum of the series would have been n^3. The first terms differ little from n^2; they are the largest, and are definitive for the sum — one might expect therefore that the sum of this series is of the order of n^3, although less than that. From tables in [82] we find that the sum of the series in question is $n^3/3$. Then (4.36), with due account for (4.19), is reduced to

$$\overline{U_{\mathrm{n}}^2} = \frac{\overline{i_{\mathrm{n}}^2}\tau_{\mathrm{n}}^2}{C_{\mathrm{in}}^2} \cdot \frac{n^3}{3} = \frac{2qI_{\mathrm{n}}}{2\tau_{\mathrm{n}}} \cdot \frac{\tau_{\mathrm{n}}^2}{C_{\mathrm{in}}^2} \cdot \frac{n^3}{3} = \frac{2qI_{\mathrm{n}}\tau_{\mathrm{n}}n^3}{2 \cdot 3 \cdot C_{\mathrm{in}}^2} \ . \tag{4.37}$$

Because of the double integration (by the input capacitance, and by the optimal filter itself), the power component of the current noise grows with the increasing observation time much faster ($n^3 \sim T_{\mathrm{i}}^3$) than does the power component of the emf noise ($n \sim T_{\mathrm{i}}$; see (4.32)). At a certain $n = m$ the two components become equal:

$$\frac{2qI_{\mathrm{n}}\tau_{\mathrm{n}}m^3}{2 \cdot 3 \cdot C_{\mathrm{in}}^2} = \frac{\overline{e_{\mathrm{n}}^2}m}{2\tau_{\mathrm{n}}} \ , \quad \tau_{\mathrm{n}}^2 m^2 = \frac{3\overline{e_{\mathrm{n}}^2}C_{\mathrm{in}}^2}{2qI_{\mathrm{n}}} \ ,$$

$$T_{\mathrm{i}} = \tau_{\mathrm{n}}m = \frac{\sqrt{3}e_{\mathrm{n}}C_{\mathrm{in}}}{\sqrt{2qI_{\mathrm{n}}}} = \frac{\sqrt{3}}{2\pi f_{\mathrm{high}}} \ , \tag{4.38}$$

where we have used the definition of f_{high} (2.169). Inclusion of the shot noise has made the observation time finite, but it is still much larger than the signal duration T_{s}. This follows from (4.38), if we recall that we are considering the case when the high-frequency noise dominates in the signal frequency range:

$$f_{\mathrm{high}} << f_{\mathrm{s}} = \frac{1}{2T_{\mathrm{s}}} \ ; \quad T_{\mathrm{i}} = \frac{\sqrt{3}}{2\pi f_{\mathrm{high}}} >> T_{\mathrm{s}} \ . \tag{4.39}$$

The obtained values of m and T_{i} are optimal. When the number of samples is smaller ($n < m$), the emf noise prevails, which means the the number of samples (4.33) must be increased in order to improve the signal-to-noise ratio. However, as soon as the number of samples exceeds m (4.38), the signal-to-noise ratio will start to decrease, since this region is dominated by the white noise, and at $T_{\mathrm{i}} > T_{\mathrm{s}}$ we shall only accumulate additional noise.

Having defined the optimal observation time T_{i}, we can easily calculate also the signal-to-noise ratio. For calculating the signal we employ equation (4.30). To find the noise, we use the formula for the shot noise (4.37), doubling the noise power, since with the chosen integration time T_{i} the contributions of the shot noise and the emf noise to the total noise power are equal. Substituting $T_{\mathrm{i}} = m\tau_{\mathrm{n}}$ from (4.38) into these equations, we get

$$\mathrm{SNR_{OF}} = \frac{U_{\mathrm{s}}}{\sqrt{2\overline{U_{\mathrm{n}}^2}}} = \frac{\dfrac{mSE_{\mathrm{sA}}T_{\mathrm{s}}}{C_{\mathrm{in}}}}{\sqrt{\dfrac{2(2qI_{\mathrm{n}})\tau_{\mathrm{n}}m^3}{6C_{\mathrm{in}}^2}}} = \frac{SE_{\mathrm{sA}}T_{\mathrm{s}}}{\sqrt{2qI_{\mathrm{n}}}}\sqrt{\frac{3}{\tau_{\mathrm{n}}m}} =$$

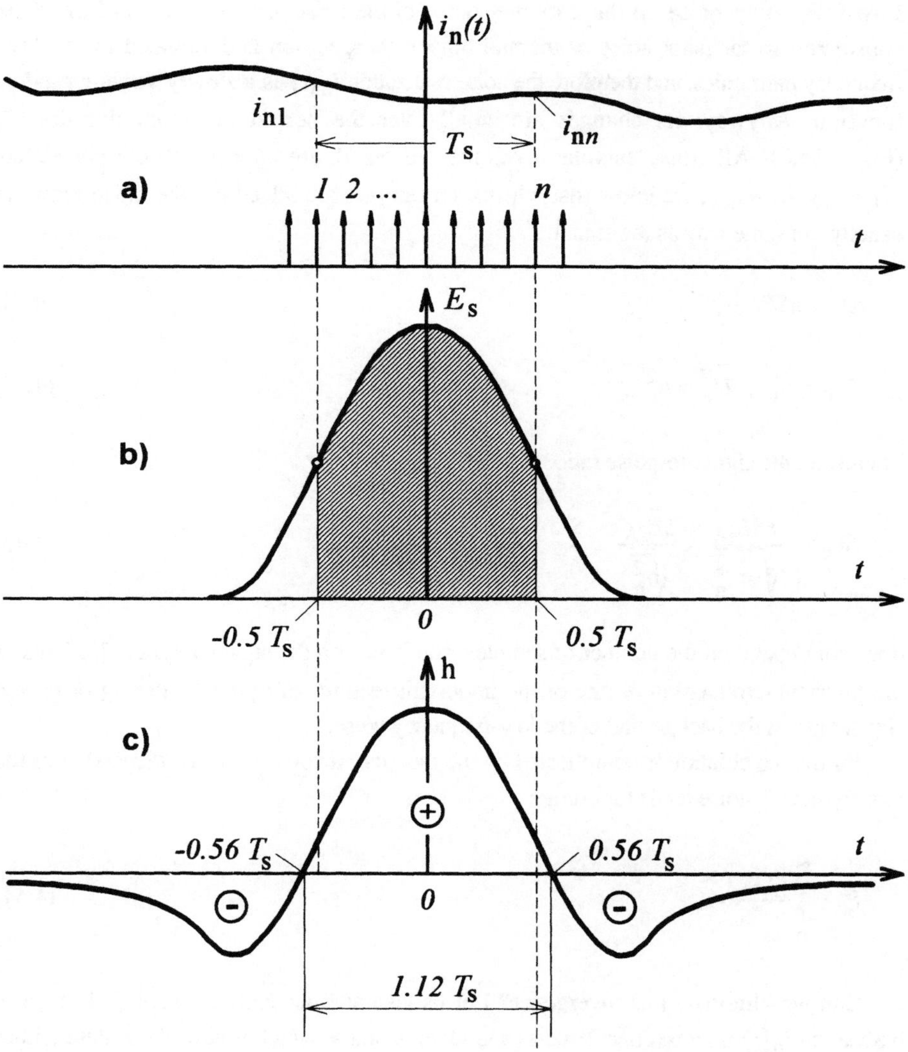

Figure 4.5: (a) Low-frequency noise fluctuates slowly, and (b) serves as pedestal for the signal; (c) optimal filter eliminates the pedestal by virtue of its pulse response.

$$= \frac{SE_{sA}T_s}{\sqrt{2qI_n}}\sqrt{\frac{3}{\sqrt{3}}2\pi f_{high}} = \sqrt[4]{3}\sqrt{2\pi}\frac{SE_{sA}T_s}{\sqrt{2qI_n}}\sqrt{f_{high}}\ . \tag{4.40}$$

Now compare this result with the exact expression (3.81), obtained in the frequency analysis. In (4.40) we have an extra factor of $\sqrt[4]{3} = 1.31$, and so the error is as small as 31%, despite quite considerable simplifying assumptions concerning both the noise (which was represented by a sequence of rectangular pulses of length τ_n) and the signal processing (the samples over the interval T_i were added up without the corresponding weight coefficients $h(\Delta t)$).

Low-frequency noise. In this case the source of the noise itself is different from those considered so far (shot noise or thermal noise): its spectrum is dominated by the low-frequency harmonics, and therefore the noise realization $i_n(t)$ is a slowly varying random function. Anyway, its changes are small over the period of signal duration T_s (Fig. 4.5a,b). All noise samples over the period T_s are almost entirely correlated, $i_{n1} \approx i_{n2} \approx \ldots \approx i_{nn}$; we know that when such samples are added up, the noise grows in exactly the same way as the signal:

$$U_s = nSE_{sA} , \tag{4.41}$$

$$U_n = ni_n , \quad \overline{U_n^2} = n^2 \overline{i_n^2} . \tag{4.42}$$

Therefore, the signal-to-noise ratio

$$\mathrm{SNR} = \frac{nSE_{sA}}{\sqrt{n^2 \overline{i_n^2}}} = \frac{SE_{sA}}{\sqrt{\overline{i_n^2}}} \tag{4.43}$$

does not depend on the number of samples, and hence on the signal duration T_s. This is the physical explanation of one of the important features of optimal filtering of pulsed signal against the background of the low-frequency noise.

Further calculation is complicated by the fact that, theoretically, the dispersion of the low-frequency noise tends to infinity:

$$\overline{i_n^2} \sim \int_0^\infty \frac{df}{f} \to \infty . \tag{4.44}$$

Can we eliminate this divergence? Let us take another look at Fig. 4.5. The noise realization $i_n(t)$ may be considered as a random quantity, which is almost constant within the period of action of the signal (the effective duration of the signal T_s). Obviously, from the mixture of the signal with the noise "pedestal", taken over the interval $[-0.5T_s, \ 0.5T_s]$, we must subtract this "pedestal". The pedestal must be measured when the signal is absent (or almost absent) — that is, when $t < -0.5T_s$, $t > 0.5T_s$. The optimal filter performs exactly these necessary operations, as follows from its pulse response, which is faithfully reproduced in Fig. 4.5c from Fig. 4.3: this filter adds up the samples with positive weight on the interval $[-0.56T_s, \ 0.56T_s]$, and with the negative weight outside of this interval.

To proceed with our calculation and find the dispersion $\overline{i_n^2}$, we need to know the correlation function. Unfortunately, the relevant mathematics is cumbersome. When we come to deal with matrix PDAs with integration and subtraction of frames (see Sect. 6.1)

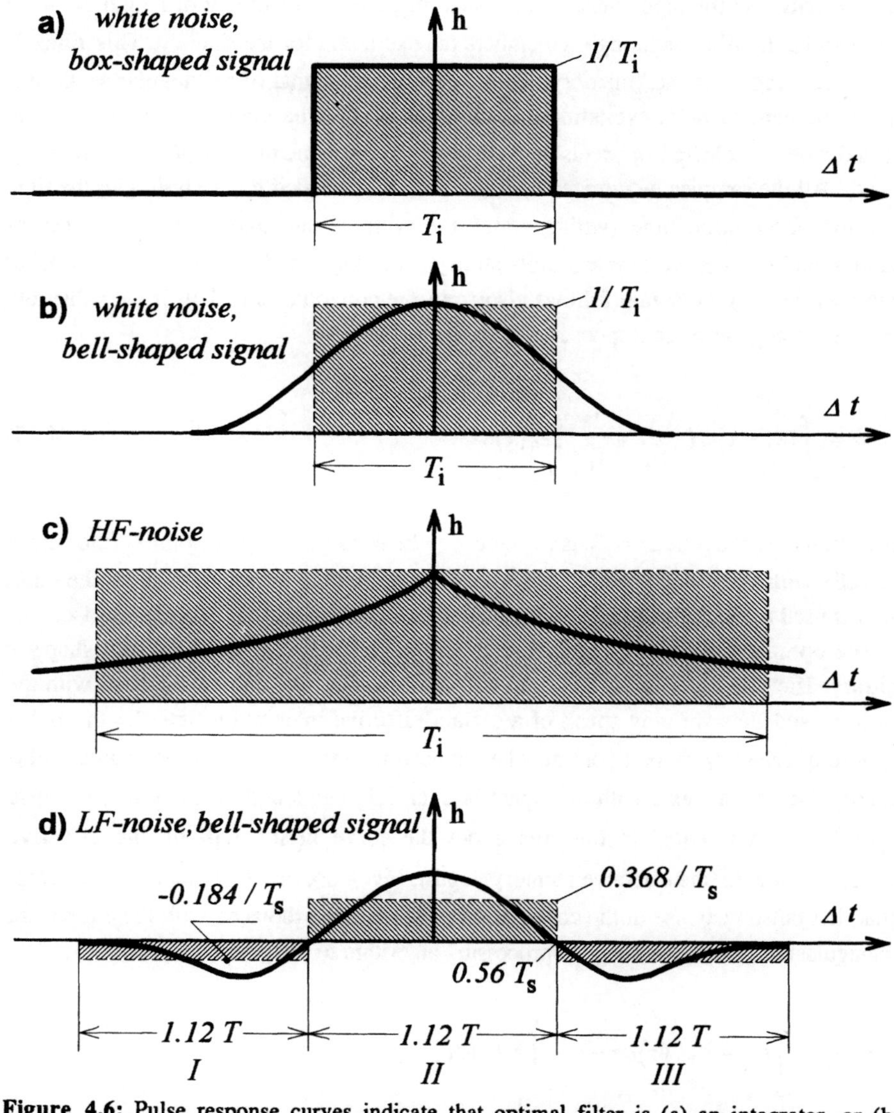

Figure 4.6: Pulse response curves indicate that optimal filter is (a) an integrator, or (b,c) equivalent to integrator; (d) three times so in case of low-frequency noise.

— operations similar to those just considered — we shall return to the definition of $\overline{i_n^2}$ and the signal-to-noise ration. The formulas for SNR we are going to obtain in this connection are similar to those for the optimal filter (Sect. 3.3).

Optimal filter-integrator as memory element. The time-domain analysis gives a better insight into the process of optimal filtering. The statement that the pulse response of the optimal filter imitates the shape of the signal implies that in the classical case of white noise and rectangular signal the pulse response is also rectangular (Fig. 4.6a). The filter with such response may be regarded as a memory element: a brief delta-excitation at the input causes a persistent response at the output of length T_i — in other words, the filter as though remembers the excitation for the time T_i. This time T_i may be referred to as the "memory depth". The optical signal of rectangular shape is a sum of elementary delta-excitations on the interval T_s. The depth of memory of the optimal filter is selected in precisely such a way as to remember all these excitations, $T_i = T_s$. All the samples memorized over the time T_i are added up; accordingly, the time T_i is the observation time (within which the samples are taken), and the depth of memory, and the interval of integration, storage, addition. The fact that the optimal filter integrates (stores) the signal follows also from the convolution (4.10), if we substitute there the rectangular pulse response:

$$u_s = \int_{-\infty}^{t} \mathrm{h}(t-\tau)e_s(\tau)d\tau = \frac{1}{T_i}\int_{t-T_i}^{t} e_s(\tau)d\tau \ . \tag{4.45}$$

Integration over the time $T_i = T_s$ is the essence, the dead certainty of optimal filtering: it is exactly with this integration time (memory depth) that the information about the input signal is used to the full extent, without accumulating (memorizing) excessive noise.

The optimal filter may be regarded as an integrator also when the signal shape is arbitrary. The difference is that the filter will store (add up) the input excitations with the weight h, and thus we may speak of a certain effective integration time T_i: the pulse response of arbitrary shape $\mathrm{h}(\Delta t)$ may be replaced with the equivalent rectangular pulse $\mathrm{h}_{eq}(\Delta t)$, like it was done with the signal in Sect. 2.1. The length of the effective pulse $\mathrm{h}_{eq}(\Delta t)$, which is equal to the area under the curve $\mathrm{h}(\Delta t)$, will be the effective integration time T_i (the effective memory depth); Fig. 4.6b. For evaluation of the output signal, the pulse response in the convolution (4.10) may be replaced with the equivalent (rectangular) response, reducing the processing algorithm to integration:

$$u_s(t) = \int_{-\infty}^{t} \mathrm{h}(t-\tau)e_s(\tau)d\tau \approx \frac{1}{T_i}\int_{t-T_i}^{t} e_s(\tau)d\tau \ . \tag{4.46}$$

For white noise we have $\mathrm{h}(\Delta t) \sim e_s(\Delta t)$, and so the effective integration time is equal to the effective signal duration, $T_i = T_s$, like in case of the rectangular signal.

The optimal filter actually remains an integrator also in case of the high-frequency noise, only the effective integration time becomes much greater than the signal duration. It was precisely the operation of integration (summation of samples) that was performed by the filter in case of the high-frequency noise. The optimal integration time was found to be $T_i = \sqrt{3}/2\pi f_{high} = 1.7\tau_a >> T_s$ (4.38). Strictly speaking, the optimal filter should add up the samples with the weight h; when, however, the actual function $\mathrm{h}(\Delta t)$ is replaced with the equivalent rectangle $\mathrm{h}_{eq}(\Delta t)$ (Fig. 4.6c), the processing is reduced to integration. Since the front and the rear edges of the pulse $h(\Delta t)$ fall off exponentially with the time constant τ_{eq} (4.15) (Fig. 4.6c), the length of this pulse, which defines the effective integration time, is

$$T_i = 2\tau_{eq} = \frac{2}{2\pi f_{high}} = \frac{1}{\pi f_{high}} \,. \qquad (4.47)$$

This value differs from that given by (4.38) by as little as 15%.

In case of the low-frequency noise the optimal filter also performs the operation of integration, but not only that. The pulse response (Fig. 4.6d) cannot be replaced by one equivalent rectangle: since the permanent component does not pass through, the area under the curve $h(\Delta t)$ is zero, and so is the interval T_i. Each of the three segments of $h(\Delta t)$ (one positive segment II, and two negative segments I, III) must be replaced with appropriate rectangles of the same polarity. This equivalent replacement can be accomplished in different ways; the reader will certainly appreciate the approximation $h_{eq}(\Delta t)$ (dashed line in Fig. 4.6d) which demonstrates with special clarity the optimal signal processing in case of the low-frequency noise. This approximation is modeled after the standard signal processing technique in case of white noise: the main interval II is used for integrating the signal for the the time almost equal to the signal duration. The novelty is represented by the side intervals I and III, which are placed side-by-side with the main interval in such a way as to preserve the magnitude of the noise "pedestal" accumulated over these intervals. The effective time of integration is the same as that for the main interval, and the value of $h_{eq}(\Delta t)$ is half that large. Accordingly, the contributions from the permanent component from interval II and from intervals I and III cancel out after summation. This is the main distinguishing feature of the filter in case of the low-frequency noise. This approximation of $h_{eq}(\Delta t)$ corresponds to the following representation of the convolution:

$$u_s(t) = \int_{-\infty}^{\infty} h(t-\tau)e_s(\tau)d\tau$$

$$= -\frac{1}{2}\int_{t-1.68T_s}^{t-0.56T_s} e_s(\tau)d\tau + \int_{t-0.56T_s}^{t+0.56T_s} e_s(\tau)d\tau - \frac{1}{2}\int_{t+0.56T_s}^{t+1.68T_s} e_s(\tau)d\tau \,. \qquad (4.48)$$

We see that, whatever the nature of the noise, the device performs integration (effective integration), and has memory. This is not surprising: memory is inseparable from intelligence, and the detection system is an intelligent device (however primitive): after all, it takes its own decision about the presence or absence of the signal. In general, the detection system may memorize the following information:

1. Current samples of the signal mixed with the noise;

2. The shape of the detected signal in case of white noise, which is described by the frequency response (transition response):

$$k(f) = e_s(f)\ , \ \ h(\Delta t) = \frac{e_s(\Delta t)}{T_s}\ ; \tag{4.49}$$

3. Information about the noise spectrum $k(f) = e_s(f)/\Theta(f)$, which is contained in the frequency response in case of non-white noise;
4. The value of the threshold U_t, which is used for comparison with the output response in order to take the decision whether the signal is there or not.

The concept of optimal filtering as the process of integration (in combination with subtraction in case of the low-frequency noise) is especially useful for the analysis of signal processing in multielement PDAs, CCDs, CIDs. Integration in these devices is the optimal (or near-optimal) way of handling the signal. This issue is so important that we shall discus it in detail in Sect. 6.1 and 6.2.

Summary. *The pulse response of the optimal filter in case of white noise imitates the shape of the signal to be detected (being its mirror image when the signal is asymmetrical). Accordingly, the pulse response is rectangular when rectangular is the signal. This implies that the optimal filter is an integrator with the integration time* $T_i = T_s$. *As a matter of fact, the optimal signal processing, as a first approximation, may be represented by the operation of integration in the most general case. When the noise is white, the effective integration time* T_i *is equal to the effective signal duration* T_s *irrespective of the shape of the signal. In case of the high-frequency noise the time* T_i *is much greater than the signal duration, and is determined by the relevant frequency limit of the high-frequency noise* $T_i \approx 1/\pi f_{high}$. *In case of the low-frequency noise the results of integration over the side intervals are subtracted from the signal integrated over the main interval, which eliminates the slow low-frequency noise fluctuation.*

The detection system as a logical device (which takes the decision concerning the presence or absence of the signal) must have memory. Integration by the optimal filter, which stores all the necessary information about the mixture of signal with the noise, is one of the constituents of such memory.

The time-domain analysis has allowed us to formulate all the main laws of optimal filtering previously derived through the frequency analysis: the rules for choosing the filter depending on the nature of the noise, the shape of the signal, and the formulas for the achievable values of signal-to-noise ratio.

4.2. Shape of Signal at the Output of Optimal Filter

Early experiments with the detection of low-level signals are now part of the history. Trying to make the signal as large as possible, the engineers used a broad-band filter [4], and so there was no question about the shape of the output signal: it faithfully reproduced that of the incoming signal. With a broad-band filter it was possible, of course, to maximize the output signal, but the noise was unjustifiably high. As demonstrated in Sect. 3.2, the band of the optimal filter must be limited. In case of white noise it is the same as the band of the useful signal T_s; in case of the high-frequency noise the passband of the filter is much narrower, and is determined by the frequency limit of the noise, $f_{high} << f_s$ (Sect. 3.3). As a consequence, one may expect that the output signal will be protracted. In case of white noise the protraction is about 1.4-fold (this rule-of-thumb estimate follows from the semiempirical expression $T_{0.5} = \sqrt{T_s^2 + T_a^2}$ with $T_a \equiv T_s$). In case of the high-frequency noise the protraction is especially large, because the relaxation time, corresponding to the optimal filter band, is of the order of $1/2\pi f_{high}$, and is much greater than the signal length T_s.

The same conclusion was also made in Sect. 4.1. A broad-band filter at any time t reproduces only the instantaneous value of the input signal, and does not memorize its previous values. The optimal filter must have memory to remember (that is, to drag on) each instantaneous excitation for the time T_i. Accordingly, we again anticipate the increase in the effective duration of the output signal for the time T_i, which, depending on the nature of the noise, is either T_s, or $1/2\pi f_{high}$ (Sect. 4.1). But then, why should we "anticipate" if we know both the frequency response and the transition response of the optimal filter, which are quite sufficient for calculating the shape of the output signal given the shape of the input signal.

The output signal of the optimal filter is always a symmetrical signal. This interesting feature of the optimal filter is the unambiguous consequence of its frequency response, which is a complex conjugate of the signal spectrum (3.50):

$$k(f) = \frac{e_s^*(f)}{\Theta} \; ; \; |k(f)| = \frac{|e_s(f)|}{\Theta} \; ; \; \psi_a(f) = -\psi_s(f) - \omega t_m \; . \tag{4.50}$$

Substituting this response into the inverse Fourier transform (3.32), we get

$$u_s(t) = \int_0^\infty e_s(f) k(f) \cos(\omega t + \psi_s + \psi_a) df = \int_0^\infty \frac{e_s^2}{\Theta} \cos \omega \Delta t \, df \; , \tag{4.51}$$

where $\Delta t = t - t_m$.

This formula may be explained as follows. In order to get the maximum signal (4.50), we adjust the frequency response of the optimal filter in such a way that at a certain time $t = t_m$ all the harmonics at the output should reach their maximum (amplitude) values (see Sect. 3.2). Then the phase shifts of the signal and the amplifier cancel out, and do not enter the expression for $u_s(t)$. Cosine is an even function, so the

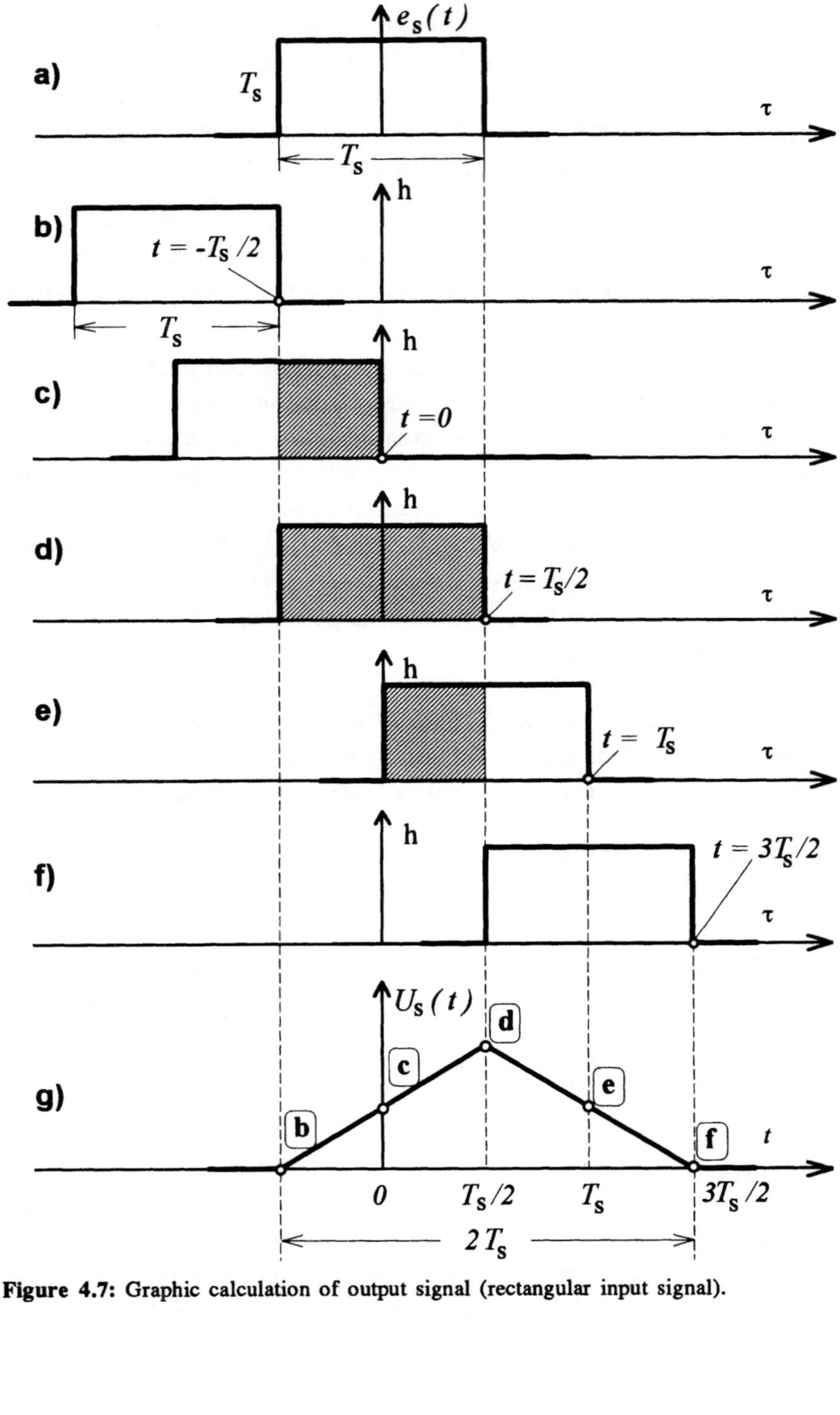

Figure 4.7: Graphic calculation of output signal (rectangular input signal).

displacement from the maximum t_m to the right or to the left by the same amount $|\Delta t| = |t - t_m|$ reduces the signal by one and the same factor (see (4.51)).

The output signal can also be calculated according to (4.10) on the basis of the pulse response. This method is especially simple and transparent in case of white noise: then the pulse response imitates the shape of the signal, and (4.10) is reduced to

$$u_s(t) = \int_{-\infty}^{\infty} e_s(\tau) h(t-\tau) d\tau = \frac{1}{T_s} \int_{-\infty}^{\infty} e_s(\tau) e_s(t-\tau) d\tau \ . \tag{4.52}$$

A similar integral (see Sect. 2.2) has been used for defining the correlation function for the noise (2.49), so we see that the output signal of the optimal filter adjusted for the white noise spectrum is actually the correlation function of the input signal. As usual, we begin with the calculation for the case of white noise, and use equation (4.52).

White noise. As follows from the convolution (4.52), the shape of the output signal may be computed graphically. When the input signal is rectangular, the procedure is as simple as the ABC (Fig. 4.7).

1. Set the form of the signal $e_s(\tau)$, which currently is rectangular (Fig. 4.7a)
2. Select the pulse response $h(t-\tau)$. For the optimal filter it coincides with the shape of the input signal, and is thus also a rectangle (Fig. 4.7b). The outline circle in this diagram marks the reading time of the output signal $\tau = t$ (the current time t).
3. To find the response at different times τ, we must move the pulse response from left to right along the τ-axis, Fig. 4.71b-d.
4. To find the magnitude of the output signal, we only have to calculate the area of overlap of the two rectangles: the signal and the pulse response (this area is hatched in Fig. 4.7). This follows from the convolution (4.52), as well as from physical considerations: the filter-integrator stores up the input excitation over the time interval T_s (in the preceding section we called this the memory depth).

If the reading is taken before the signal has arrived at the input, $t \le -T_s/2$ (Fig. 4.7a,b), the output response is, of course, zero. The next pulse response (Fig. 4.7c) corresponds to the sampling time $t = 0$. By this time the PDA receives half of the signal energy. The output response of our filter-integrator is equal to this energy (hatched area). It follows that the output signal grows linearly with the time on the interval $[-T_s/2,\ T_s/2]$, and reaches its maximum at $t = T_s/2$. This reading is taken directly at the time when the action of the input pulse is over, and the pulse itself is completely stored in the "memory" of the filter (Fig. 4.7d). If the reading is taken at a later time, $t > T_s/2$, the output response starts to decline linearly (Fig. 4.7e). At $t \ge 3T_s/2$ the output signal is zero again, since the reading is taken too late, and the memory depth of the filter is not sufficient to remember the past (long past) input excitation (Fig. 4.7f). In this way, the shape of the output signal is triangular (Fig. 4.7g).

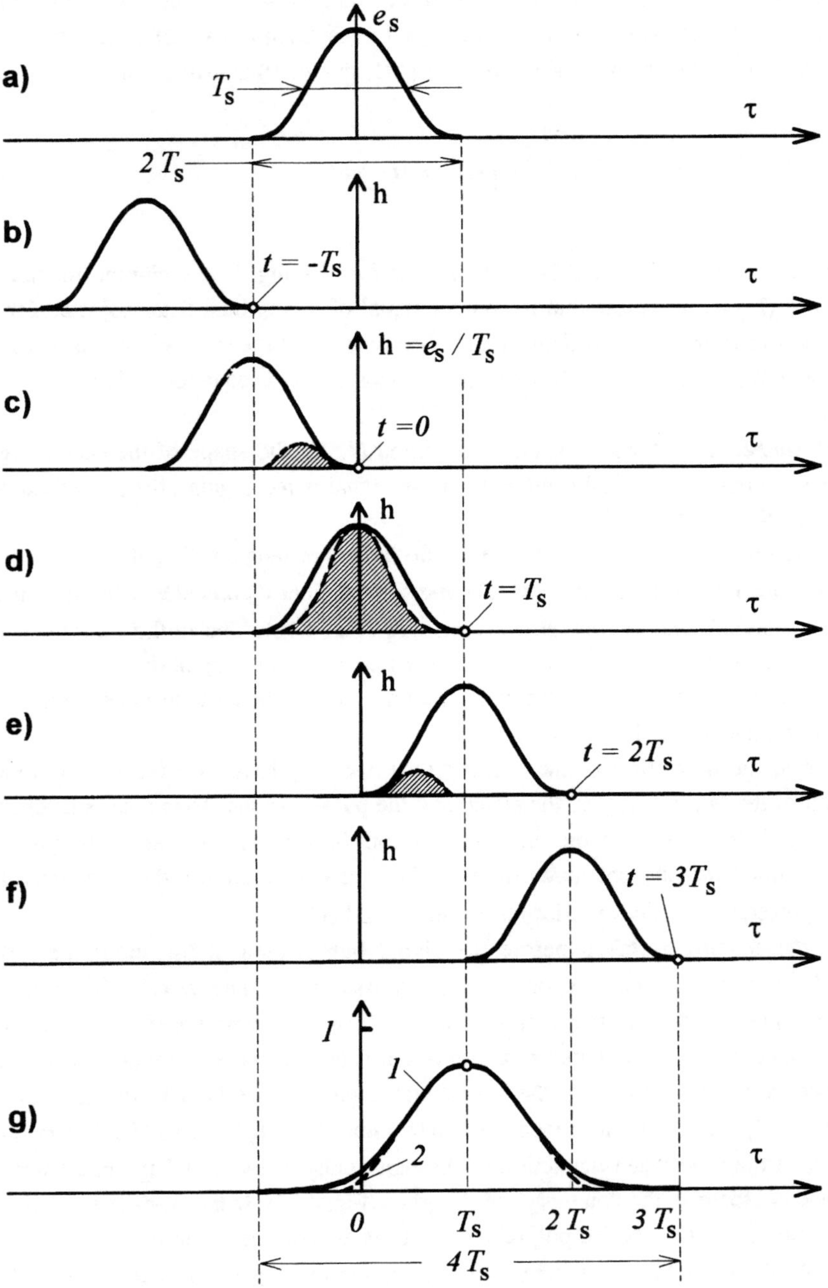

Figure 4.8: Graphic calculation of output signal (square-cosine input signal).

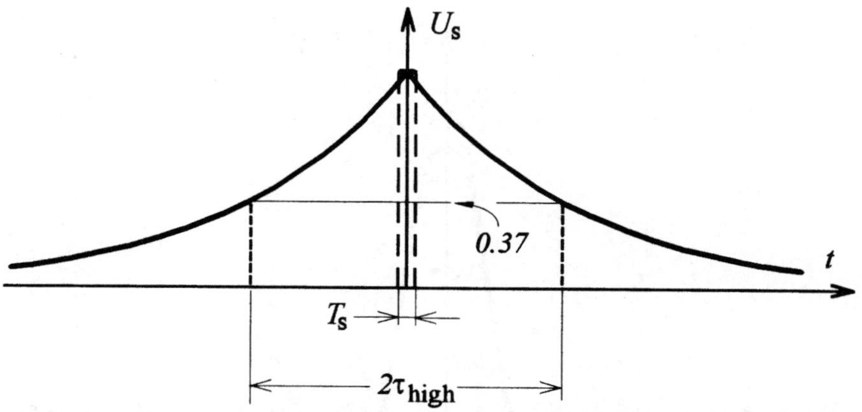

Figure 4.9: In case of high-frequency noise the signal at the output of optimal filter relaxes with a large time constant.

Graphical method is suitable for calculating the output signal irrespective of the shape of the input pulse (for example, for the square-cosine pulse, Fig. 4.8). In general, the output signal, according to (4.52), is defined by the area under the curve $e_s(\tau)e_s(t-\tau)$ (hatched area in Fig. 4.8). Because of this, the signal becomes nonzero as soon as the curves $e_s(\tau)$ and $e_s(t-\tau)$ start to overlap, and reaches its maximum when the two curves coincide. The output signal calculated in this way is shown in Fig. 4.8g, curve 1. The optimal filter keeps bringing pleasant surprises to us: the shape of the output signal is almost square-cosine (for comparison, the square-cosine is plotted in Fig. 4.8g, dashed curve 2). Some difference is only observed at the tails of the two curves. Of course, the optimal filter cannot defy the laws of nature: because of the limited bandwidth, the input bell-shaped pulse is slightly extended: its length (at half-amplitude) is increased by the factor of 1.28, and the amplitude decreased by almost the same factor (more precisely, by the factor of 1.33). This confirms our expectations concerning the prolongation of the response.

High-frequency noise. When the noise is not white, we have to use the inverse Fourier transform (4.51). Into this formula we substitute the noise spectrum $\Theta = 1 + f^2/f_{high}^2$. The fall-off of the integrand e_s^2/Θ is caused by the term Θ at the low frequencies ~ f_{high} where the signal spectrum is still constant, $e_s(f) \approx 1$. Hence,

$$u_s(\Delta t) = \int_0^\infty \frac{e_s^2}{\Theta} \cos \omega \Delta t df \approx \int_0^\infty \frac{\cos \omega \Delta t}{1 + \frac{f^2}{f_{high}^2}} df = \frac{1}{2\tau_{high}} \exp\left(-\frac{|\Delta t|}{\tau_{high}}\right), \quad (4.53)$$

where $\tau_{high} = 1/2\pi f_{high}$.

Formula (4.53) was used for plotting the signal at the output of the filter optimized for the high-frequency noise in question (Fig. 4.9). Again, the expected prolongation of the response is confirmed.

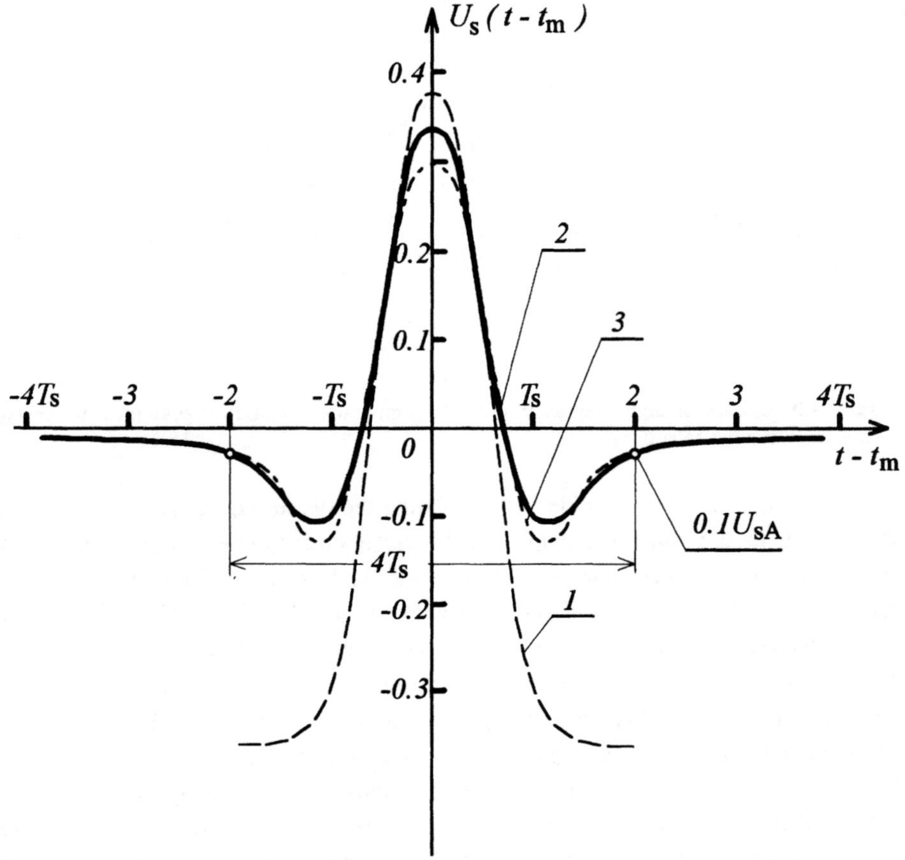

Figure 4.10: The signal profile at the output of optimal filter in case of low-frequency noise resembles the pulse response curve.

We meet with function (4.53) for the second time: in Sect. 4.1 we obtained it for the pulse response (4.15) of the filter in question. The similarity is understandable: the integrands in (4.15) and (4.53) only differ in the terms $e_s(f)$, $e_s^2(f)$; when, however, $f_{high} << f_s$, these terms are about equal: $e_s \approx e_s^2 \approx 1$. This is the mathematical reason. The physical reason is also clear: the input pulse is much shorter than the time constant of the filter τ_{eff}, and may therefore be considered a delta-pulse. Because of this, the signal at the output of the filter in question reproduces the pulse response of the filter, and does not at all depend on the shape of the short input signal.

Low-frequency noise. As usual, the filter in case of such noise is tuned to the bell-shaped signal (the problem of optimization for rectangular signal in case of the low-frequency noise is ill-posed; see Sect. 3.3). In Sect. 4.1 we have made a good guess at the pulse response of such filter. Now let us try to guess the shape of the output signal. Again, we start from the fact that the main feature of the optimal filter in case of the low-frequency noise is that it does not transmit the permanent component. Accordingly, from the square-cosine signal with the effective duration $1.28T_s$, which is observed at the output of the filter tuned to the white noise, we must subtract the permanent component. This brings us to curve 1 in Fig. 4.10. The rigorous calculation is carried out with the aid of the inverse Fourier transform (4.51). Substituting here the noise spectrum $\Theta = f_{low}/f$ and the spectrum of the bell-shaped signal $e_s = \left(\pi^2 \sin x\right)/\left[x\left(\pi^2 - x^2\right)\right]$, $x = 2\pi f T_s$, we get:

$$U_s = 2SE_{sA}T_sK_0 \int_0^\infty \frac{e_s^2}{\Theta} \cos \omega\Delta t \, df = \frac{SE_{sA}K_0}{\pi} \int_0^\infty \frac{e_s^2}{\Theta} \cos\left(x \frac{\Delta t}{T_s}\right) dx \ ,$$

$$U_s = SE_{sA} \frac{K_0}{\pi} \pi^4 \int_0^\infty \frac{\sin^2 x}{x^2\left(\pi^2 - x^2\right)^2} \frac{x}{x'} \cos\left(x \frac{\Delta t}{T_i}\right) dx$$

$$= SE_{sA}K_0' \left\{ \pi^2 \int_0^\infty \frac{\sin^2 x}{x\left(\pi^2 - x^2\right)^2} \cos\left(x \frac{\Delta t}{T_i}\right) dx \right\} . \tag{4.54}$$

In accordance with Sect. 3.3, we have introduced here the gain K_0' (3.69):

$$\frac{\pi K_0}{x'} = \frac{\pi K_0}{2\pi f_{low} T_s} = K_0 \frac{f_s}{f_{low}} = K_0' \ , \quad x' = 2\pi f_{low} T_s \ . \tag{4.55}$$

The gain coefficient in case of the low-frequency noise (see Sect. 3.3) must be normalized to the value of K_0'. Because of this, the expression in braces in (4.54) describes the relative shape of the signal at the output of the filter with unit gain $K_0' = 1$. Curve 2 in Fig. 4.10 is plotted from these considerations. As follows from the diagram, we have correctly guessed the shape of the positive peak of the output signal. This is the main thing: it is the positive peak that activates the discriminator. The negative outliers are due to the differentiating property of the filter. We were not able to determine their shape; it is defined by the decline of $k(f)$ in the region of low frequencies, and not only by the filtering of the permanent component, as was assumed in the construction of curve 1. The negative outbursts slowly relax to zero, and the length of the signal at half-amplitude is about $4T_s$.

We discern a remarkable regularity: the bell-shaped signal is a test for determining the relative shape of the pulse response of the optimal filter tuned to such signal. As soon as we supply this bell to the input of the filter, its output signal exactly (almost exactly) reproduces the relative shape of the pulse response. This statement applies not only to the high-frequency noise (when the rule holds for any shape of the input signal), not only for the case of white noise (when $e_s(t)$, $h(\Delta t)$, and $u_s(t)$ are all cosine waves like triplet brothers), but also for the case of the low-frequency noise. To prove this, we have reproduced in Fig. 4.10 (curve 3) the pulse response of filter for low-frequency noise from Fig. 4.4, reducing its amplitude by half, and stretching it by one-third along the time axis. The behavior of curves 2 and 3 is quite unanimous.

Summary. *The shape of the signal at the output of optimal filter is always symmetrical. In case of white noise the output signal is the correlation function of the input signal, and is triangular in the classical case of the rectangular input pulse. In case of the low-frequency noise, the front and rear edges of the output signal exhibit negative outbursts, since the optimal filter in this case has differentiating properties. In case of the high-frequency noise, the output signal does not depend on the shape of the input signal, and reproduces the relative profile of the pulse response of the filter. Incidentally, when the square-cosine signal is applied to the input, the relative shape of the pulse response will be reproduced at the output irrespective of the nature of the noise, whether it is low-frequency or white noise (in the last case the input and the output signals, as well as the pulse response, all have practically the same square-cosine profile).*

The limited passband of the optimal filter leads to protraction of the output signal. In case of white and low-frequency noise its length at 0.1-amplitude usually increases 1.5...2-fold, or by about 30% at half-amplitude. When the high-frequency noise prevails (which is characterized by its frequency limit f_{high}*), the passband of the optimal filter is much narrower than the bandwidth of the input signal, and the output signal is protracted especially strongly: it relaxes exponentially with the time constant* $1/2\pi f_{high}$.

Chapter 5

SIGNAL PROCESSING IN REAL PHOTODETECTING ASSEMBLIES

Now we know how the signal must be processed in ideal photodetecting assemblies: in the preceding two chapters we have discussed in detail the methods of optimal processing (filtering) of the signal, the ways used to separate the signal from the noise. This, however, is theory. How is the signal processed in practice, in real PDAs? To what extent is the real filtering inferior to the idealized model? These questions constitute the topic of the present chapter.

5.1. Quasi-Optimal Filtering

The theory of detection has clearly formulated the task of the designer of the photodetecting device: it consists in maximizing the signal-to-noise ratio. This requires not only choosing a highly sensitive photodetector, not only reducing the noise, but also constructing an amplifier with the frequency response of the optimal filter. This difficult problem gives rise to a number of questions.

Question one. How does one implement the desired frequency response of the amplifier? The calculation of frequency response from the circuit diagram is trivial. However, there is no general solution to the inverse problem: the synthesis of a circuit having a specified (arbitrary) frequency response. Of course, in particular cases it is possible to design (by trial and error) a circuit with the desired characteristics. The classical example is the circuit of the optimal filter for the detection of rectangular signal mixed with white noise (Fig. 5.1) [4, 85, 86]. As demonstrated in Sect. 4.1, the optimal circuit in this case is the integrator with the integration time $T_i = T_s$. To make sure that the integration takes place only over the specified interval $[t,\ t - T_i]$, the signal from the output of the integrator is split into two channels; one part remains unchanged, and the other is passed through the delay line. Then the two signals are supplied to the inputs of the differential amplifier. The order of manipulations may be changed (the integrator may stand after the differential amplifier). The realization of even this seemingly simple circuit meets with problems: it is not so easy to construct the delay line; the capacity of the integrator is limited.

Approximate techniques developed for the synthesis of optimal filters for bell-shaped signals lead to sophisticated and cumbersome circuitry [85].

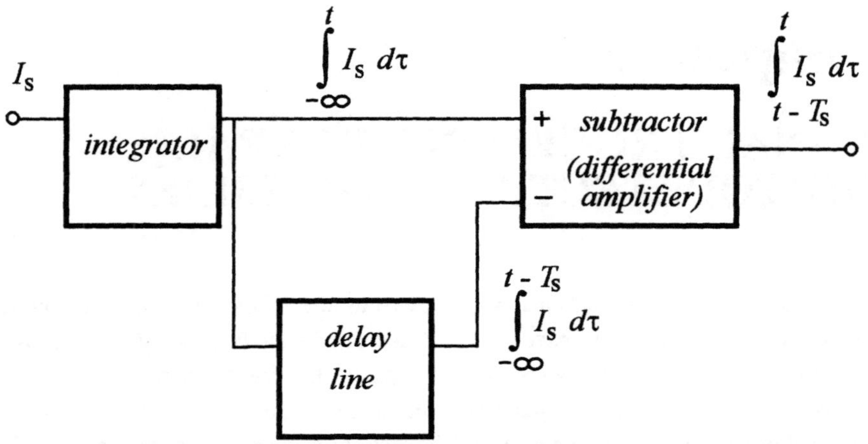

Figure 5.1: Circuit diagram of optimal filter for detection of rectangular signal against the background of white noise.

Question two. Is it always possible to realize the desired optimal filter? In case of white noise the pulse response of the optimal filter must imitate the shape of the signal, $h(\Delta t) = e_s(\Delta t)/T_s$ (see Sect. 4.1). If, for instance, the signal is symmetrical and rectangular, the pulse response $h(\Delta t)$ is symmetrical and rectangular (Fig. 5.2a). Recall that $h(\Delta t)$ is the response to the delta-pulse which arrives at the time $\tau = t = 0$ (Fig. 5.2b). The situation is unrealistic: the response at the output $h(\Delta t)$ appears before the signal has arrived at the input! Obviously, a filter with such pulse response cannot really exist. There is a way out, however: since the argument of the pulse response of the optimal filter Δt is defined up to an arbitrary constant added time lag t_{lag} (4.13), we must move the pulse $h(\Delta t)$ to the right along the time axis until its origin lines up with the zero time $\Delta t = 0$. For the case in question (see Fig. 5.2c) we have $t_{lag} = T_s/2$, $h(\Delta t) = e_s(\Delta t - 0.5T_s)/T_s$. The optimal filter shown in Fig. 5.1 has exactly this kind of frequency response. If the delta-pulse arrives at the input at $t = 0$, the signal proportional to the charge of the pulse appears immediately at the output of the integrator. This signal is transmitted further through the non-inverting channel of the differential amplifier. A similar signal is passed through the delay line, and with the time lag of T_s is supplied to the second (inverting) input of the differential amplifier, which subtracts one signal from the other of the same magnitude, and produces zero at its output.

In case of symmetrical pulse of arbitrary shape $e_s(t)$, the time lag t_{lag} must be not less than half the signal length measured at the pulse base. For the bell-shaped pulse this means that $t_{lag} \geq T_s$ (in Fig. 5.2d we have chosen the minimum allowable lag $t_{lag} = T_s$). It was for the filters with such "displaced" characteristics that the shape of the output signal was calculated in Sect. 4.1 (Fig. 4.7, 4.8).

Now, what if the function $e_s(t)$ has no "beginning", and only tends to zero asymptotically at $t \to -\infty$? This situation is encountered in the theory when the diffraction in the optical system is taken into account, which diffuses the boundary of the light spot. Then the signal is approximated by the symmetrical Gaussian curve, which

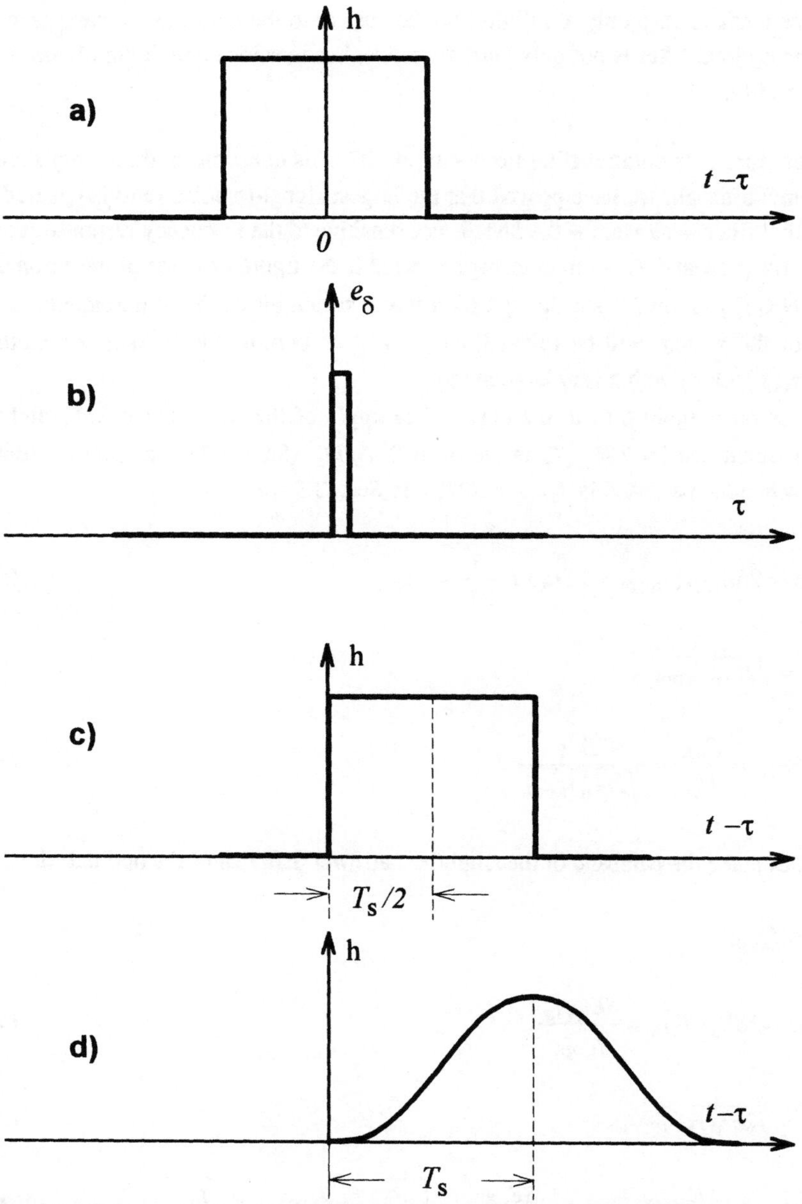

Figure 5.2: Pulse response curves of optimal filter: (a) physically unfeasible; (b) input delta pulse; (c,d) physically feasible.

gradually approaches zero at $|t| \to \infty$. It is obviously not possible to line up the nonexistent origin of the curve $h(\Delta t) = e_s(t - t_0)/T_s$ with the point $\Delta t = 0$; the pulse response $h(\Delta t)$ is always greater than zero, even at $\Delta t < 0$. This is a physical nonsense: the signal has to appear at the output even before it arrives at the input of the filter. With a similar situation we meet when the noise spectrum is dominated by the low-frequency or the high-frequency noise component. As shown in Sect. 4.1, the pulse response in such case tends to zero only at infinity. So the answer to the question we have asked is clear: the optimal filter is not only hard to realize, but in some cases it simply does not exist in reality.

Question three. Is optimal filter necessary at all? This question might at first seem a blasphemy: after all, we have proved that the largest signal-to-noise ratio is ensured by the optimal filter. And what if the SNR is not sensitive to the frequency response profile $k(f)$, to the passband f_a — in other words, what if the figure of merit of the resonance curve $\mathrm{SNR}(f_a)$ is low? Then this question is not so pointless: the requirements to the tuning of the system will be relaxed, and it will be permissible to satisfy condition $k(f) = e_s(f)/\Theta(f)$ with a very large margin.

Let us once again turn to the classical example of the white noise $2qI_n$ and the constant signal density $2SE_{sA}T_s$ in the band $0...1/2T_s$ (Sect. 3.3). For precise tuning (that is, when the passband is $f_{a\,opt} = 1/2T_s$), in Sect. 3.3 we found that

$$U_{sA} = 2SE_{sA}T_s f_{a\,opt} = 2SE_{sA}T_s \frac{1}{2T_s} = SE_{sA} \ , \tag{5.1}$$

$$U_n = \sqrt{2qI_n f_{a\,opt}} \ , \tag{5.2}$$

$$\mathrm{SNR_{OF}} = \frac{U_{sA}}{U_n} = \frac{SE_{sA}}{\sqrt{2qI_n f_{a\,opt}}} \ . \tag{5.3}$$

Now let us make the passband of the amplifier narrower than that of the optimal filter:

$$f_a < f_{a\,opt}:$$

$$U_{sA} = 2SE_{sA}T_s f_a = \frac{SE_{sA} f_a}{f_{a\,opt}} \ , \tag{5.4}$$

$$U_n = \sqrt{2qI_n f_a} \ , \tag{5.5}$$

$$\mathrm{SNR} = \frac{SE_{sA} f_a}{f_{a\,opt}\sqrt{2qI_n f_a}} = \frac{SE_{sA}}{\sqrt{2qI_n f_{a\,opt}}}\sqrt{\frac{f_a}{f_{a\,opt}}} = \mathrm{SNR_{OF}}\sqrt{\frac{f_a}{f_{a\,opt}}} \ . \tag{5.6}$$

Making the passband of the amplifier broader than that of the optimal filter, we get:

$$f_a > f_{a\,opt}:$$

$$U_{sA} = SE_{sA} \ , \tag{5.7}$$

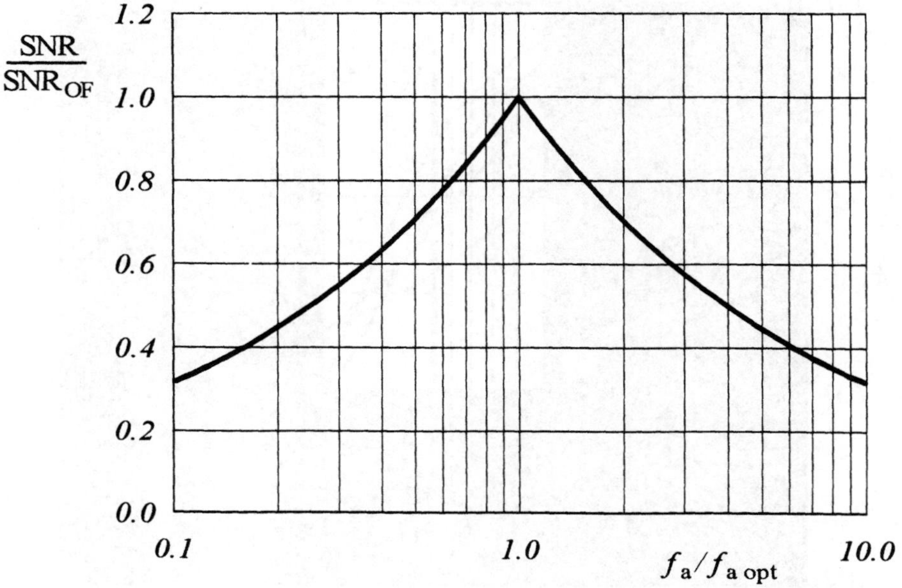

Figure 5.3: Signal-to-noise ratio exhibits weak dependence on the passband of filter-amplifier.

$$U_n = \sqrt{2qI_n f_a}\ , \tag{5.8}$$

$$\mathrm{SNR} = \frac{SE_{sA}}{\sqrt{2qI_n f_a}} = \frac{SE_{sA}}{\sqrt{2qI_n f_{a\,opt}}}\sqrt{\frac{f_{a\,opt}}{f_a}} = \mathrm{SNR_{OF}}\sqrt{\frac{f_{a\,opt}}{f_a}}\ . \tag{5.9}$$

Equations (5.6), (5.9) were used for plotting SNR in Fig. 5.3 as a function of the amplifier passband f_a. The value of SNR shows a quite weak dependence on the passband: it varies as the square root of f_a. The physics of this is clear: As the passband narrows down, the signal decreases in sympathy with f_a, and the noise also decreases as $\sqrt{f_a}$, so that $\mathrm{SNR} \propto \sqrt{f_a}$. When $f_a > f_{a\,opt}$, the noise increases in proportion with $\sqrt{f_a}$, whereas the signal keeps its maximum possible value.

The fact that $\mathrm{SNR}(f_a)$ is a slow function allows one to vary the amplifier band f_a without sacrificing the SNR too much. For example, when the band is varied by a factor of four (from $0.5 f_{a\,opt}$ to $2 f_{a\,opt}$), the change in the SNR is less by an order of magnitude (about 30%).

Since the value of SNR is not sensitive to the variation of the passband f_a within such broad limits, it is certainly not sensitive to the shape of the frequency response. This is explained by the diagram in Fig. 5.4, which shows the frequency response curves of the three stepwise filters that we have just considered, with the passbands $f_{a\,opt}$, $f_{a\,opt}/2$, $2f_{a\,opt}$ (curves 1 – 3). The same diagram shows also the typical frequency

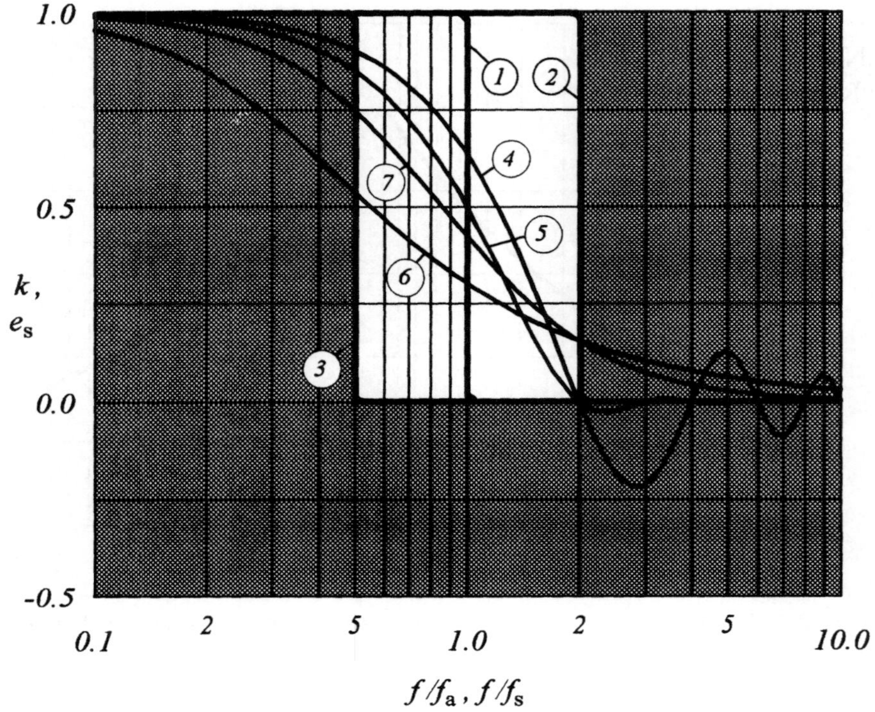

Figure 5.4: Spectra of optical signal and frequency response of amplifiers may vary, but the signal-to-noise ratio changes little: (1,2,3) rectangular frequency response curves; (4) spectrum of rectangular-shaped signal; (5) bell-shaped signal spectrum; (6) frequency response of amplifier with one RC– section; (7) same for amplifier with two *RC*–sections.

responses and spectra considered previously (or to be considered shortly); they are the spectra of rectangular (4) and bell-shaped (5) pulses, and the frequency profiles of the simplest filters with one (6) and two (7) *RC*-sections. Compared with the four-fold swing of the passband of the filters just considered (curves 2, 3), the difference between the passband profiles and spectral curves (4 – 7) does not seem too dramatic. So, if even the four-fold variation of the passband (curves 2, 3) caused a moderate (within 30%) deterioration of the SNR, we should expect a much smaller effect with the filters having a closer-lying frequency curves (4 – 7). On the basis of these arguments we can make not just one but two conclusions at once:

1. If in place of the optimal filter (having, for example, the frequency response 4 or 5) we use a filter with a different response which is close to the optimal one (for example, 6 or 7), the loss in the SNR will be not considerable;
2. The signal-to-noise ratio itself shows a sufficiently weak dependence on the minute details of the signal profile, since the spectra of such signals are close enough to one another (curves 1, 4, 5).

As a matter of fact, the last statement is already familiar to us (see Sect. 3.3): the SNR for the signals of different shape (rectangular, square-cosine, $(\sin x)/x$) varies within no more than 15% (the spectra of such signals are represented in Fig. 5.4 by curves 1, 4, 5). It was precisely because the SNR is not sensitive to the minute details of the signal shape that we could use the approximate technique in Sect. 3.3, replacing the actual spectrum of the rectangular pulse with the stepwise spectrum, and still obtaining the same values of SNR both through the rigorous calculations based on the exact spectrum of the rectangular pulse, and by the approximate technique based on the stepwise spectrum (Table 3.2). The signal shape in Sect. 3.3 was varied by varying the spectrum.

So, we have answered our last question. The frequency response of the real detection system need not exactly reproduce that of the optimal filter; it can be chosen within a good margin, as long as the characteristic frequencies (effective passbands) of the real and the optimal filters are close enough [87, 88]. More precisely, the characteristic frequencies of the real filter must be adjusted so as to maximize the signal-to-noise ratio. Since such real filter is not much inferior to the optimal filter in terms of its SNR, it is called quasi-optimal (near-optimal).

It would be hard to overestimate the importance of this conclusion for the designers of detection systems, including the engineers of PDAs. Should this be otherwise (that is, should the value of SNR exhibit a strong dependence on the tuning of the system), one can hardly imagine all the problems that would have been encountered. Not only because it is difficult to implement the desired optimal frequency response: the fact is that, in principle, all the parameters of the problem may be loose, such as the parameters of the optical signal itself (for example, the diameter of light spot, and hence the signal length, may vary because of the varying aberration across the field of the optical system), the parameters of the noise spectrum (because of variations of background radiation and temperature), the parameters of the amplifier (because of the temperature and the time drift of capacitors and resistors of the filter). If the resonance curve $\mathrm{SNR}(f_a)$ had had a sharp peak, all these factors would have been upsetting the system, causing a dramatic drop in the SNR. It would have been necessary to design sophisticated adaptive filters to cope with even small variations of the parameters. Fortunately, all these problems are happily avoided just because the SNR is not too sensitive to the tuning of the filter.

The history of IR technology and radars knows examples when the engineers concerned with the detection of low-level signals attempted to realize the frequency response of the optimal filter at all costs. As a rule, this only made the system more complicated and expensive, and less reliable. The deterioration of performance was a too high price for adding a few per cent to the SNR (as compared to the simple quasi-optimal system). As a matter of fact, the sophisticated circuitry often introduced additional noise, so that even this small gain was lost.

Of special importance is the quasi-optimal filtering for the multichannel systems. Technically, it would be simply not feasible to equip each of the 10^2 or 10^3 channels (to say nothing of systems having $10^4 \ldots 10^6$ channels) with the cumbersome optimal filters.

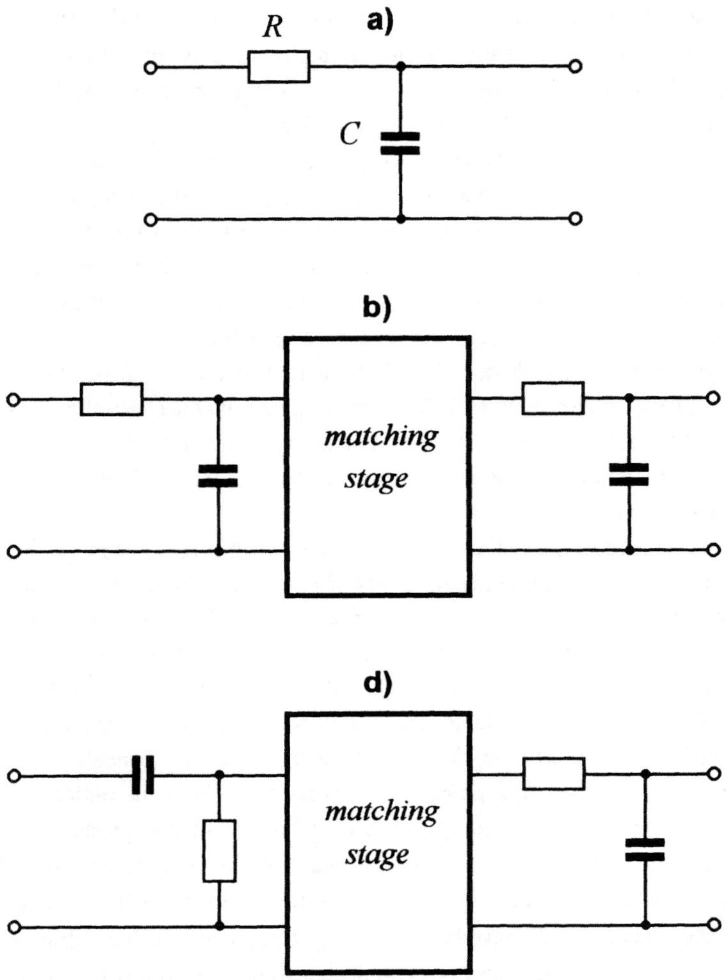

Figure 5.5: Schematics of quasi-optimal filters: (a) with one integrating section; (b) with two integrating sections; (c) with differentiating and integrating sections.

So, instead of trying to construct the optimal filter, we have to answer the question how to design a quasi-optimal filter. For traditional PDAs, the approach to the synthesis of quasi-optimal filter is obvious: it must be as simple as possible! In radio engineering there are no simpler filters than *RC*-circuits. As a rule, one or two sections are enough for the PDA [89, 90]. The time constant must be adjusted in such a way as to make the passband of the *RC*-filter equal to that of the optimal filter. In case of white and high-frequency noise the passband of the filter must be limited from above; this is done with the integrating *RC*-circuits. The diagram in Fig. 5.5a shows one integrating *RC*-circuit. Two sections ensure a steeper roll-off of the frequency response (Fig. 5.5b). In case of the low-frequency noise, the band must be limited from below as well as from above, and so at least two sections are required (one integrating, and one differentiating;

Fig. 5.5c). The time constants of the two sections are made equal (if the $1/f$-noise is observed within the entire characteristic band of the signal, $f_s = 1/2T_s$). For the time being, we shall restrict ourselves to these examples of quasi-optimal filters.

Summary. *The optimal filter is not only hard to synthesize; it often is physically not feasible. Fortunately, there is no need to reproduce exactly the frequency response of the optimal filter, since the signal-to-noise ratio shows a relatively weak dependence on the shape of the frequency response of the filter. It is sufficient to construct a filter whose passband is matched to that of the optimal filter, and has a sufficiently steep roll-off of the frequency response. Such filter is called quasi-optimal. In PDAs it is realized with the simplest RC-circuits, and one or two sections are enough: integrating sections in case of white noise and high-frequency noise, and integrating plus differentiating sections in case of low-frequency noise.*

5.2. Loss in SNR in Quasi-Optimal Filter

The order of further analysis is obvious:
— first we consider the passage of the signal and the noise through quasi-optimal filters proposed above and based on RC-circuits;
— then we select the time constant of RC-circuits (the effective passband) so as to maximize the SNR at the output (it is under this condition that the filters become quasi-optimal)
— finally, we compare the resulting SNR with the best possible SNR achieved with optimal filtering (in other words, we determine the loss in SNR in quasi-optimal filter).

Let us carry out this analysis first for integrating RC-circuits (Fig. 5.5a,b), and then for the circuit with additional differentiating RC-section (Fig. 5.5c).

Passage of signal through the one-section integrating RC-filter. According to the diagram in Fig. 5.5a, the frequency response of such filter is

$$k(f) = \frac{\dfrac{1}{j\omega C}}{\dfrac{1}{j\omega C} + R} = \frac{1}{1 + j\omega CR} = \frac{1}{1 + j\omega\tau_f}, \quad \tau_f = RC . \tag{5.10}$$

Since $k(f)$ is the spectrum of the pulse response, the latter can be found with the aid of the inverse Fourier transform (4.14):

$$h(\Delta t) = 2\int_0^\infty |k(f)| \cos(\omega\Delta t + \psi_a) df = \frac{1}{\tau_f} \exp\left(-\frac{\Delta t}{\tau_f}\right), \quad \psi_a = -\arctan \omega\tau_f . \tag{5.11}$$

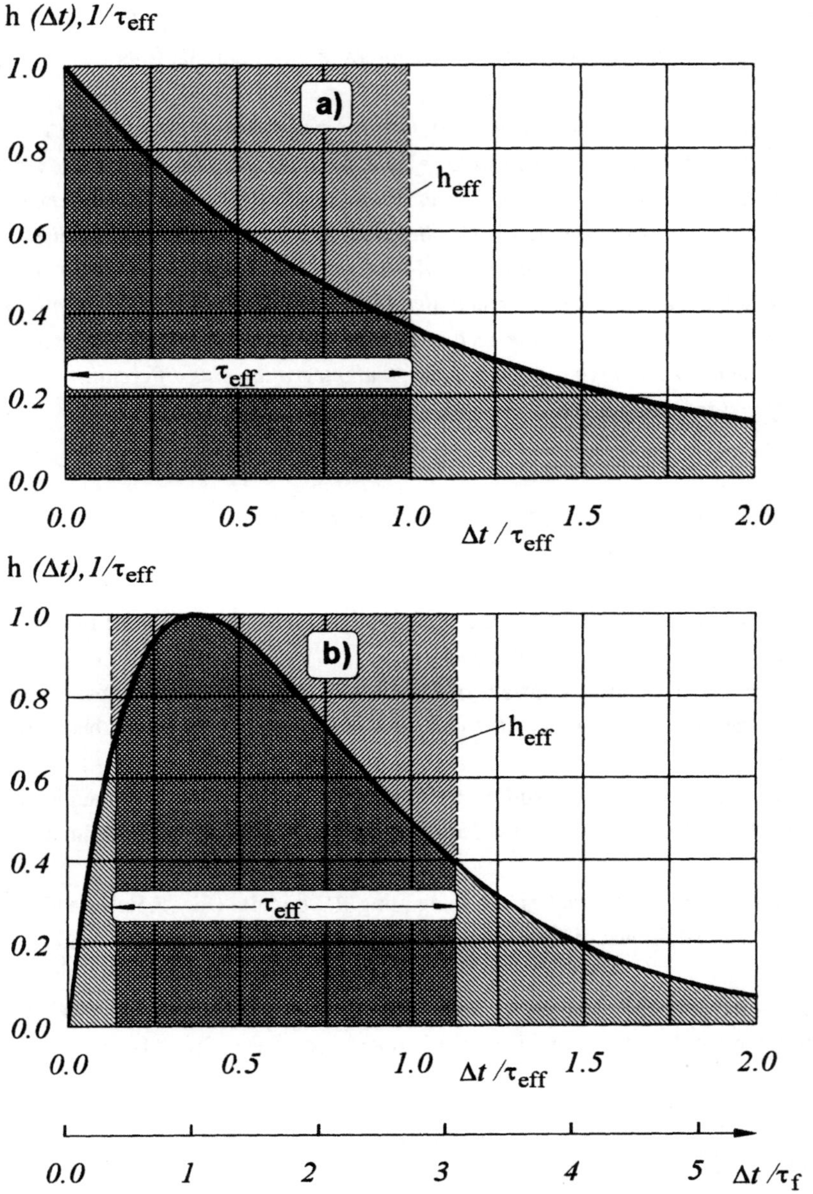

Figure 5.6: Pulse response curves of filters: (a) with one, and (b) with two integrating sections.

In this example we could as well do without the calculation of the pulse response: the exponential nature of relaxation in a system with one time constant is universally known, and is a textbook example of a transition process.

One of the main parameters of the filter-amplifier is its effective time constant (or the effective time of observation or integration). By analogy with the signal, we defined this

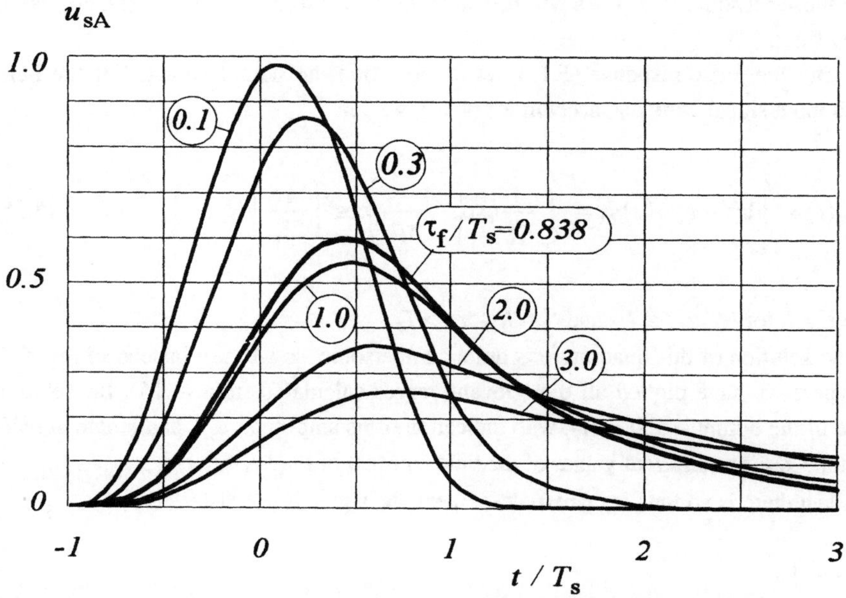

Figure 5.7: Signal profile at the output of one-section filter for different time constants.

characteristic parameter in Sect. 4.1 as the length of the equivalent rectangular pulse response h_{eq}. Let us replace the real pulse response of the one-section filter with the equivalent one (Fig. 5.6a). Since the areas and the amplitudes of the two characteristics are equal, and

$$h_{eq}\,\tau_{eff} = \int_0^{\infty} h(\Delta t)d\Delta t \;, \quad h_{eq} = h(t_m) \;, \quad t_m = 0 \;, \tag{5.12}$$

for the effective time constant we get

$$\tau_{eff} = \frac{1}{h_{eq}}\int_0^{\infty} h(\Delta t)d\Delta t = \frac{1}{h(t_m)}\int_0^{\infty}\frac{1}{\tau_f}\exp\left(-\frac{\Delta t}{\tau_f}\right)d\Delta t = \frac{1}{h(t_m)} \;, \tag{5.13}$$

$$h(t_m) = \frac{1}{\tau_f} \;, \quad \tau_{eff} = \frac{1}{h(t_m)} = \tau_f \;. \tag{5.14}$$

The effective time constant of the one-section filter turns out to be exactly equal to its RC-value. Accordingly, the effective frequency band of the amplifier in question is $f_a = 1/2\tau_{eff} = 1/2\tau_f$ (in the same way as the effective bandwidth of the signal was found to be $f_s = 1/2T_s$).

It is interesting that the value of τ_{eff} is directly expressed in terms of the maximum of the pulse response (5.14). As will be shown later, this formula is universal, and holds for any filter.

Using the pulse response (5.11), let us now find the output signal. For the bell-shaped input signal, with due account for (4.10), we get

$$u_s(t) = \int_{-\infty}^{t} \mathrm{h}(t-\tau)e_s(\tau)d\tau = \int_{-T_s}^{t'} \frac{1}{\tau_f}\exp\left(-\frac{t-\tau}{\tau_f}\right)\cos^2\left(\frac{\pi\tau}{2T_s}\right)d\tau \,, \tag{5.15}$$

where $t' = t$ for $-T_s \le t \le T_s$, and $t' = T_s$ for $t > T_s$.

The solution of this quadrature is quite cumbersome, so we turn instead to Fig. 5.7, 5.8, where we have plotted all the relevant curves calculated from (5.15): the relative profile of the output signal $u_s(t)$, with indication of its amplitude u_{sA} and length at half-amplitude T_{05}, for different values of the time constant of the RC-filter τ_f.

When there is no time lag (inertialess filter), the signal is not distorted:

$\tau_f = 0$:

$$u_{sA} = e_{sA} = 1 \,, \quad T_{05} = T_s \,. \tag{5.16}$$

In the other extreme, when the time constant is very large, the output signal $u_s(t)$ reproduces the shape of the pulse response. Then the value of $u_s(t)$ is proportional to the area of the input signal, which is $e_{sA}T_s = T_s$ if $e_{sA} = 1$. Hence,

$$u_s(t) = (e_{sA}T_s)\mathrm{h}(t) = T_s\frac{1}{\tau_f}\exp\left(-\frac{t}{\tau_f}\right). \tag{5.17}$$

The amplitude and the duration of such signal are, respectively,

$\tau_f \gg T_s$:

$$u_{sA} = u_s(0) = \frac{T_s}{\tau_f} \,, \quad \frac{u_s(T_{05})}{u_{sA}} = \exp\left(-\frac{T_{05}}{\tau_f}\right) = \frac{1}{2} \,, \quad T_{05} = \tau_f \ln 2 \,. \tag{5.18}$$

The limiting expressions (5.16), (5.18) help us to find approximations for $u_{sA}(\tau_f)$, $T_{05}(\tau_f)$ in the entire range of variation of the filter time constant τ_f. Let us add the signal lengths and the inverse amplitudes from (5.16), (5.18):

$$\frac{1}{u_{sA}(\tau_f)} = 1 + \frac{\tau_f}{T_s} = \frac{T_s + \tau_f}{T_s} \,, \quad u_{sA}(\tau_f) = \frac{T_s}{T_s + \tau_f} \,; \tag{5.19}$$

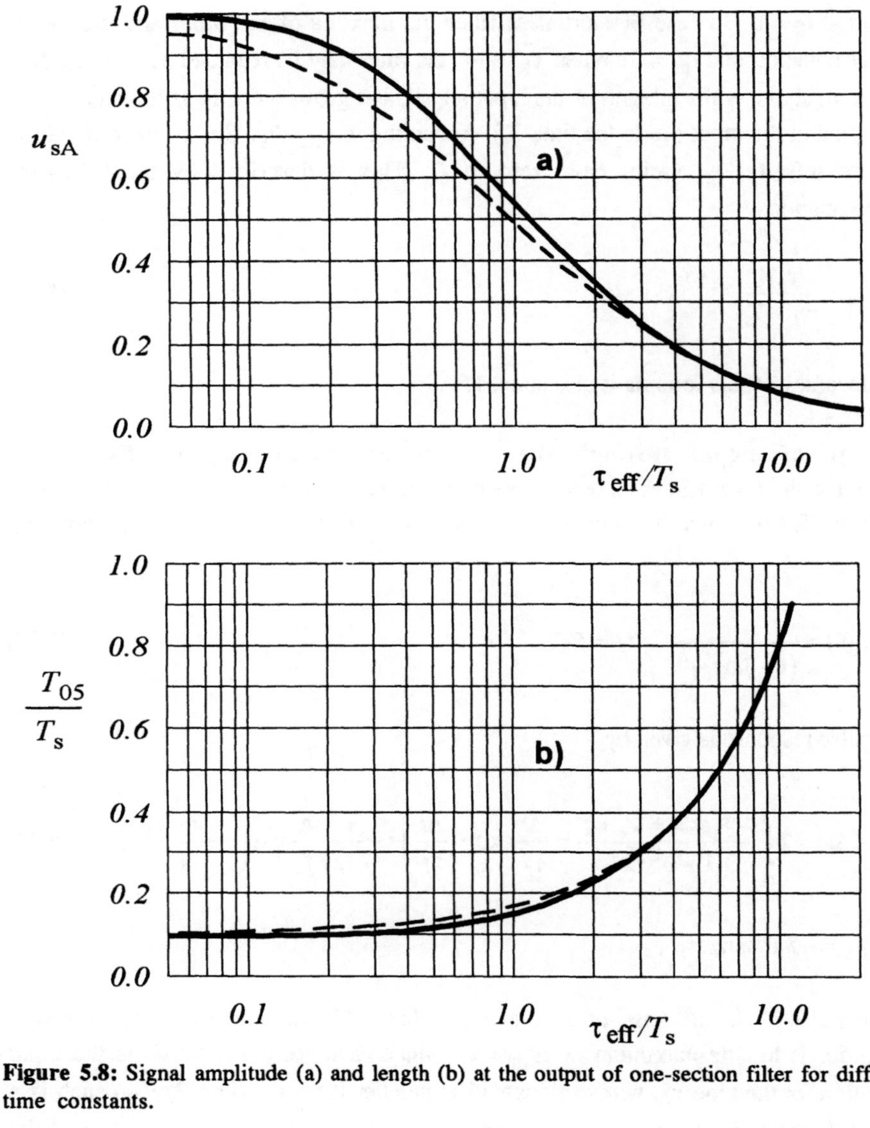

Figure 5.8: Signal amplitude (a) and length (b) at the output of one-section filter for different time constants.

$$T_{05} = T_s + \tau_f \ln 2 \ . \tag{5.20}$$

These formulas are constructed in such a way that in the limits $\tau_f \to 0$ and $\tau_f \to \infty$ they coincide, respectively, with (5.16) and (5.18). Whether they coincide, however, with the actual functions $u_{sA}(\tau_f)$, $T_{05}(\tau_f)$ in the neighborhood of $\tau_f \approx T_s$, is a matter of luck. Figure 5.8 testifies that luck is on our side: the accuracy of approximations is quite acceptable (within 10...15%), and so expressions (5.19), (5.20) may be used in our further calculations.

In a similar way we find the approximation for t_m, the location of the maximum of the output signal on the time axis. We position the input signal so that its maximum occurs at $t_0 = 0$. In case of inertialess filter the maxima of the input and the output signals coincide, and $t_m = 0$. When $\tau_f \to \infty$, the filter may be regarded as an integrator, its output signal is the integral of the input signal, and grows steadily as long as there is the signal at the input (up to the time T_s). Accordingly, $t_m = T_s$. So, as τ_f varies from zero to infinity, t_m varies from zero to T_s. This is described by the following approximation:

$$t_m = \frac{\tau_f T_s}{\tau_f + T_s} . \tag{5.21}$$

The accuracy of this formula is also about 15%.

Passage of signal through the two-section integrating *RC*-filter. We follow the same guidelines as before; since the frequency response of each *RC*-circuit is given by (5.10), to get the same for two consecutive sections we have to multiply it by itself,

$$k(f) = \frac{1}{(1 + j\omega\tau_f)^2} , \quad \tau_f = RC . \tag{5.22}$$

The pulse response is given by

$$\mathrm{h}(\Delta t) = 2\int_0^\infty \frac{\cos(\omega\Delta t + \psi_a)df}{1 + \omega^2\tau_f^2} = \frac{\Delta t}{\tau_f^2}\exp\left(-\frac{\Delta t}{\tau_f}\right) \equiv \frac{1}{(e\tau_f)}\frac{\Delta t}{\tau_f}\exp\left(-\frac{\Delta t - \tau_f}{\tau_f}\right), \tag{5.23}$$

$$\psi_a = -2\arctan\omega\tau_f ,$$

where $e \cong 2.72$ is the base of natural logarithms. This characteristic is shown in Fig. 5.6b. It has its maximum at $\Delta t_m = \tau_f$. Since each section remembers the input excitation for the time τ_f, two sections will remember it for the time $2\tau_f$ — such is a rule-of-thumb estimate for the time constant (the effective integration time) τ_{eff} of the two-section filter. Its exact value is found by the familiar method: either by replacing the actual pulse response with the equivalent rectangle (Fig. 5.6b), or with the aid of (5.13):

$$\mathrm{h}(\Delta t_m) = \mathrm{h}(\tau_f) = \frac{1}{\tau_f}e^{-1} , \quad \tau_{eff} = \frac{1}{\mathrm{h}(\Delta t_m)} = e\tau_f . \tag{5.24}$$

We see that the exact value $e\tau_f$ is somewhat higher than the approximate estimate $2\tau_f$.

The output signal for the bell-shaped input pulse is calculated as

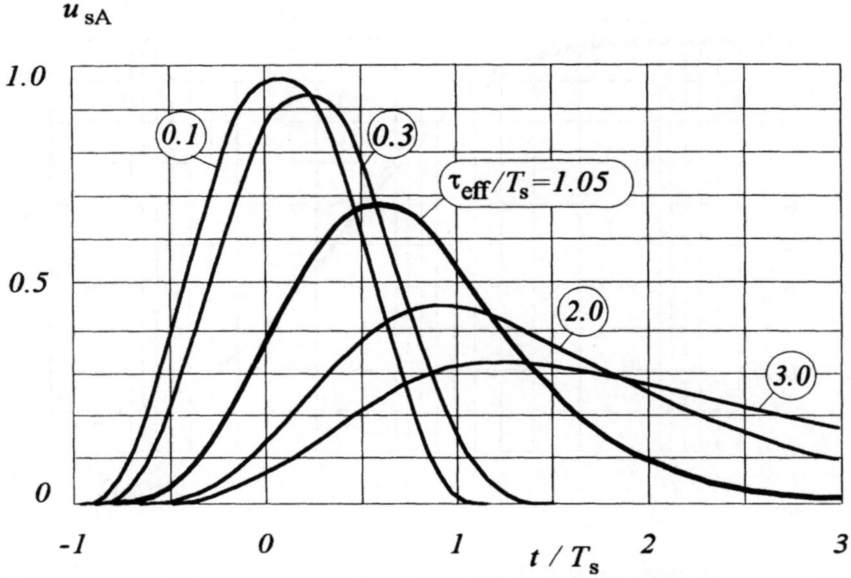

Figure 5.9: Signal profile at the output of two-section filter for different time constants.

$$u_s(t) = \int_{-T_s}^{t'} \frac{t-\tau}{\tau_f^2} \exp\left[-\frac{(t-\tau)}{\tau_f}\right] \cos^2\left(\frac{\pi\tau}{2T_s}\right) d\tau \ , \tag{5.25}$$

where $t' = t$ for $-T_s \le t \le T_s$, and $t' = T_s$ for $t > T_s$.

This quadrature is even less manageable than that for the one-section RC-filter. The results of numerical calculation of the output signal profile, its amplitude and duration, are presented in Fig. 5.9, 5.10. Expression (5.16) still holds for the amplitude and the duration of the signal when the time constant of the RC-circuit is small, $\tau_f << T_s$, when the signal passes through the filter without distortion. At the output of the inertial filter we now have:

$$\tau_f >> T_s \ ,$$

$$u_s(t) = (e_{sA} T_s)\, \mathrm{h}(t) = T_s \frac{t}{\tau_f^2} \exp\left(-\frac{t}{\tau_f}\right) = \frac{T_s}{(e\tau_f)}\left[\frac{t}{\tau_f} \exp\left(-\frac{t-\tau_f}{\tau_f}\right)\right] . \tag{5.26}$$

The amplitude value is achieved at $t_m = \tau_f$:

$$\tau_f >> T_s \ , \quad u_{sA} = u_s(\tau_f) = \frac{T_s}{(e\tau_f)} \ . \tag{5.27}$$

The length of signal at half-amplitude is calculated with the aid of (5.26):

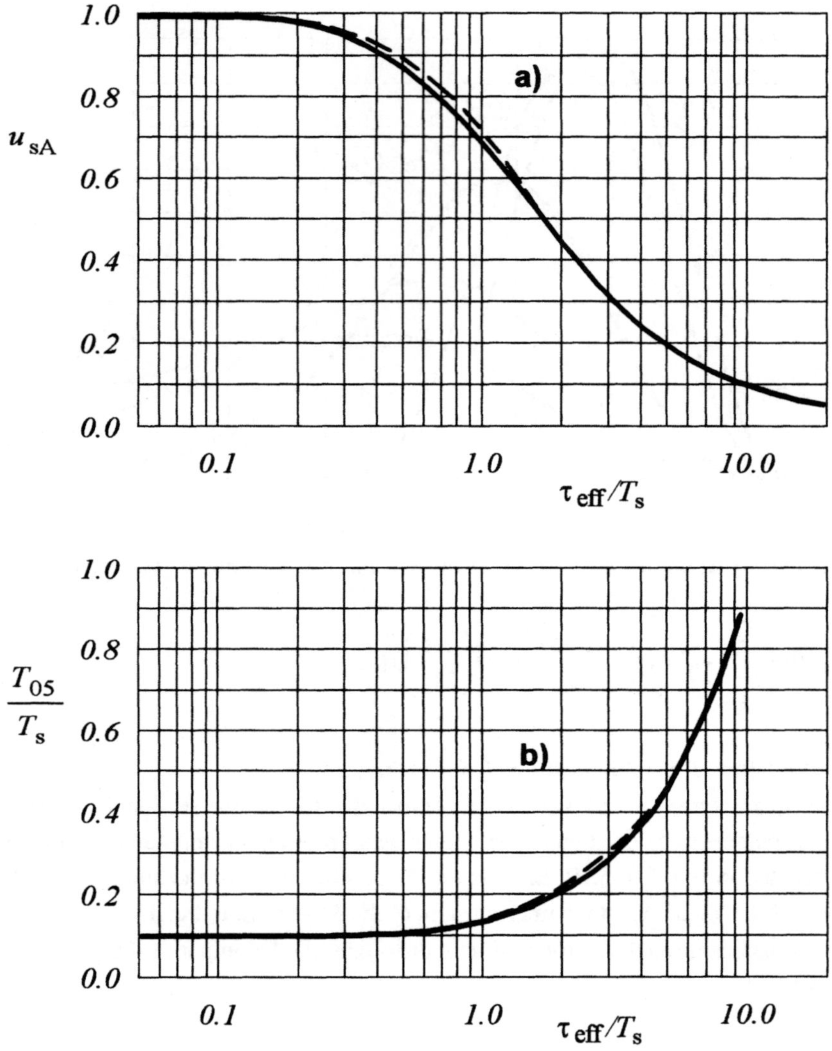

Figure 5.10: Signal amplitude (a) and length (b) at the output of two-section filter for different time constants.

$$\tau_f >> T_s \ , \quad T_{05} = \sqrt{6}\tau_f = 2.45\tau_f \ . \tag{5.28}$$

We see that T_{05} is somewhat less than the effective time constant of the filter $\tau_{eff} = e\tau_f = 2.72\tau_f$.

The approximations for $u_{sA}(\tau_f)$, $T_{05}(\tau_f)$, $t_m(\tau_f)$ can now be written as follows:

$$u_{sA} = \frac{T_s}{\sqrt{T_s^2 + \tau_{eff}^2}} = \frac{T_s}{\sqrt{T_s^2 + (e\tau_f)^2}} \ , \tag{5.29}$$

$$T_{05} = \sqrt{T_s^2 + 6\tau_f^2} \ , \tag{5.30}$$

$$t_m = \tau_f \left(1 + \frac{T_s}{\sqrt{T_s^2 + (4\tau_f)^2}} \right) , \tag{5.31}$$

At $\tau_f \to 0$, $\tau_f \to \infty$ these approximations coincide with the corresponding limits (5.16), (5.27), (5.28). Equation (5.31) hardly calls for any comment: when the time constant is small, each RC-section delays the signal (the maximum of the signal) by the time τ_f (5.21); accordingly, two sections delay the signal by $2\tau_f$. When the time constant is large, the maximum is delayed by τ_f. All this follows from (5.31).

The approximations (5.29) and (5.30) are plotted in Fig. 5.10; they practically coincide with the exact curves (the mismatching does not exceed 1...3%).

Noise at the output of the filter. The expression for the rms noise value at the output of arbitrary filter-amplifier was quoted in Sect. 3.2 (see (3.35)):

$$u_n = \sqrt{\int_0^\infty \overline{J^2(f)}\, k^2(f) df} \ . \tag{5.32}$$

Now we are considering the white and the high-frequency noise components:

$$\overline{J^2(f)} = 2qI_n + \omega^2 C_{in}^2 e_n^2 = 2qI_n + (2\pi C_{in} e_n)^2 f^2 \ . \tag{5.33}$$

Substituting this spectral density of the input noise power into (5.32), for the output noise we get

$$u_n = \sqrt{2qI_n \int_0^\infty k^2(f) df + (2\pi C_{in} e_n)^2 \int_0^\infty f^2 k^2(f) df} \ . \tag{5.34}$$

We see that the calculation of the white and the high-frequency noise components in the entire frequency band of the amplifier reduces to computing the integrals in (5.34). These integrals define the noise bands of the filter. It would be interesting to compare them with other characteristic frequencies of the filter; let us therefore digress for a while from the calculation of the output noise, and compile an inventory of the characteristic frequencies which define the passage of signal and noise through the amplifier.

Table 5.1. Characteristic frequencies

Parameters	Filter with rectangular frequency response	Filter integrator	One-section filter	Two-section filter
$k(f)$	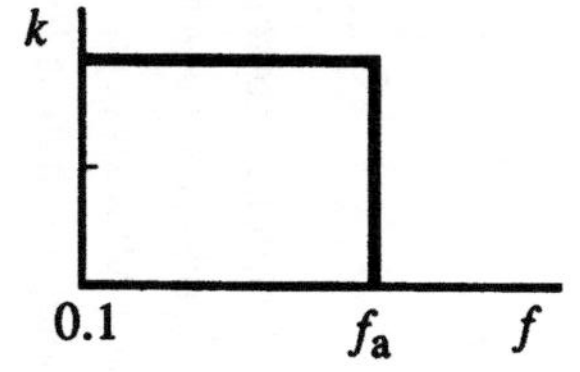	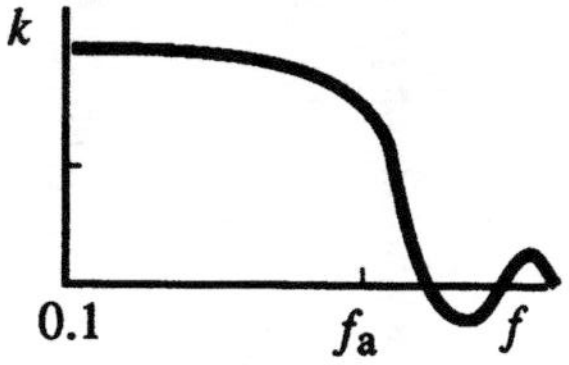	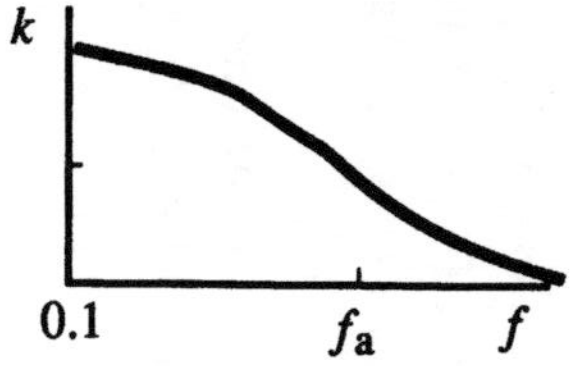	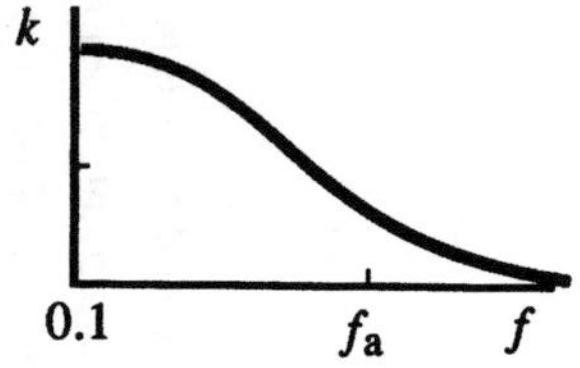
	$\begin{cases} 1 \text{ at } f < f_a \\ 0 \text{ at } f > f_a \end{cases}$	$\frac{\sin \pi f T_i}{\pi f T_i}$	$\frac{1}{\sqrt{1+\omega^2\tau_f^2}}$	$\frac{1}{1+\omega^2\tau_f^2}$
$h(\Delta t)$	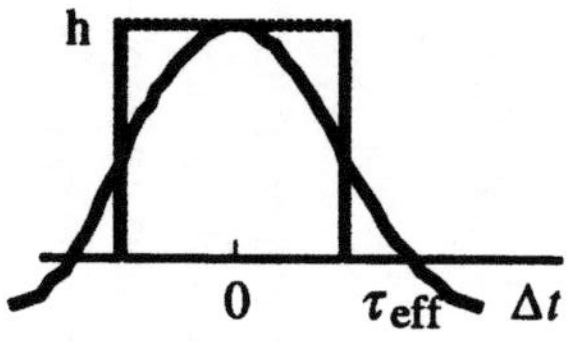	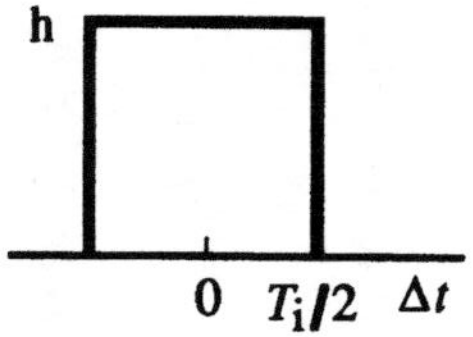	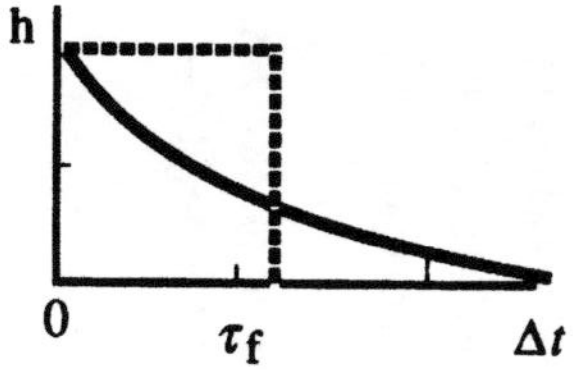	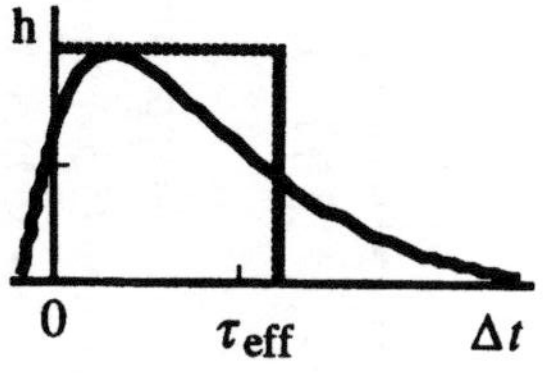
	$\frac{\sin(\pi\Delta t/\tau_{eff})}{\tau_{eq}(\pi\Delta t/\tau_{eff})}$	$\begin{cases} 1/T_i \text{ at } \lvert\Delta t\rvert < T_i/2 \\ 0 \text{ at } \lvert\Delta t\rvert > T_i/2 \end{cases}$	$\frac{1}{\tau_f}\exp\left(-\frac{\Delta t}{\tau_f}\right)$	$\frac{\Delta t}{\tau_f^2}\exp\left(-\frac{\Delta t}{\tau_f}\right)$

Table 5.1. (continued)

Parameters	Filter with rectangular frequency response	Filter integrator	One-section filter	Two-section filter
τ_{eff}	τ_{eff}	T_i	τ_f	$e\tau_f$
f_a	$1/2\tau_{eff}$	$1/2\tau_{eff}$	$1/2\tau_{eff}$	$1/2\tau_{eff}$
f_{nw}	$1/2\tau_{eff}$	$1/2\tau_{eff}$	$1/4\tau_{eff}$	$\frac{1}{8\tau_f} = \frac{e}{4}\left(\frac{1}{2\tau_{eff}}\right)$
$\Gamma_1 = \frac{f_{nw}}{f_a}$	1	1	0.5	$\frac{e}{4} = 0.68$
f_{nh}	$f_a/3$	∞	∞	$\frac{e^3}{4\pi^2} f_a$
$\Gamma_2 = \frac{f_{nh}}{f_a}$	1/3	∞	∞	$\frac{e^3}{4\pi^2} = 0.51$
f_{07}	$1/2\tau_{eff} = f_a$	$\frac{0.443}{\tau_{eff}} = 0.886 f_a$	$\frac{0.159}{\tau_{eff}} = 0.318 f_a$	$\frac{0.28}{\tau_{eff}} = 0.55 f_a$

Characteristic frequencies (bands). Table 5.1 shows the characteristics of a number of filter-amplifiers. Apart from the one-section and two-section filters, it also includes filters with rectangular pulse response and rectangular frequency response. These are well familiar to us from the preceding discussion: the rectangular frequency response was used as an approximation for calculating the signal-to-noise ratio (Sect. 3.3), and the rectangular pulse response is the characteristic of filter-integrator; recall that integration is the main method of signal processing in matrices. So, altogether we have selected four examples. These examples are sufficient for demonstrating how the relative profile of the frequency response affects the relationship between the characteristic frequencies which are responsible for the passage of signal and noise through the amplifier.

Effective time constant and effective frequency band of filter-amplifier (with respect to the signal). These parameters characterize the passage of the signal. In Sect. 4.1 we defined the time constant τ_{eff} as the length of the rectangular pulse response $\text{h}_{\text{eq}}(\Delta t)$ which is equivalent to the actual characteristic $\text{h}(\Delta t)$ (see Table 5.1). The constant τ_{eff} is the sluggishness of the amplifier, and the effective observation time, as well as the effective integration time. In Sect. 4.1 we also showed that the relative frequency response of the amplifier $k(f)$ is the spectrum of its relative pulse response $\text{h}(\Delta t)$. But then, exactly the same relationship exists between the characteristics of the signal pulse: function $\text{e}_{\text{s}}(f)$ is the spectrum of function $e_{\text{s}}(t)$. It follows that the characteristics $\text{h}(\Delta t)$, $k(f)$ should exhibit the same features as the signal pulse $e_{\text{s}}(t)$ and its spectrum $\text{e}_{\text{s}}(f)$, as derived in Sect. 2.1.

1. The area under the curves $\text{h}(\Delta t)$, $\text{h}_{\text{eq}}(\Delta t)$ is equal to the maximum value of the frequency response (the value at zero frequency):

$$\int_{-\infty}^{\infty} \text{h}(\Delta t)d\Delta t = \int_{-\infty}^{\infty} \text{h}_{\text{eq}}(\Delta t)d\Delta t = \text{h}(\Delta t_{\text{m}})\tau_{\text{eff}} \equiv k(0) \equiv 1 \ . \tag{5.35}$$

Equation (5.35) implies that τ_{eff} can be defined not only as the duration of $\text{h}_{\text{eq}}(\Delta t)$, but also directly in terms of the maximum value $\text{h}(\Delta t_{\text{m}})$:

$$\tau_{\text{eff}} = \frac{1}{\text{h}(\Delta t_{\text{m}})} \ . \tag{5.36}$$

Previously we obtained equation (5.36) for the special case of the one-section RC-filter (5.13); this time it holds for an arbitrary filter. The values calculated according to (5.36) are presented in Table 5.1.

2. The second feature is reciprocal to the first: the area under the curve $k(f)$ is equal to the maximum value of the pulse response:

$$\int_{-\infty}^{\infty} k(f)df = 2\int_{0}^{\infty} k(f)df = 2k(0)f_a = 2f_a = \mathrm{h}(\Delta t_m) = \frac{1}{\tau_{eff}} . \tag{5.37}$$

Hence, by analogy with equation (2.15) for the signal, we get the definition of the effective band of filter-amplifier:

$$f_a = \frac{1}{2\tau_{eff}} . \tag{5.38}$$

From (5.37) it follows that the band f_a is defined by the area under the curve $k(f)$, and defines the "transmission window" which characterizes the ability of the filter to transmit the spectrum of the signal of the specified width. Recall that this vivid definition holds when the characteristic $\mathrm{h}(\Delta t)$ is symmetrical — that is, for the filters of types I and II (see Sect. 2.1). When the pulse response is not symmetrical (RC-filters in the case in question), the integrals in (5.37) must be redefined to include (by analogy with (2.18)) the phase-frequency characteristic into the integrand. Equation (5.38), however, which defines the effective passband f_a in terms of the effective time constant τ_{eff}, still holds.

Noise band of white noise. Contribution of the white noise component into the output noise is given by the first term under the root in (5.34):

$$u_n = \sqrt{2qI_n \int_{0}^{\infty} k^2(f)df} = \sqrt{2qI_n f_{nw}} , \tag{5.39}$$

$$f_{nw} = \int_{0}^{\infty} k^2(f)df .$$

From (5.39) it follows that the noise band for white noise f_{nw} — that is, the effective band in which the white noise reaches the output — is defined by the area under the curve $k^2(f)$ rather than the curve $k(f)$ as it was in case of the signal. The band f_{nw} is widely used in radio engineering (see Sect. 3.1 and 3.2). Let us calculate its value for each of the four filters under consideration:

Filter with rectangular frequency response:

$$f_{nw} = \int_{0}^{\infty} k^2(f)df = \int_{0}^{f_a} 1 \cdot df \equiv f_a ; \tag{5.40}$$

Filter-integrator with the integration time T_i:

$$f_{nw} = \int_0^\infty k^2(f)df = \int_0^\infty \left[\frac{\sin \pi f T_i}{\pi f T_i}\right]^2 df = \frac{1}{\pi T_i}\int_0^\infty \frac{\sin^2 x}{x^2}dx = \frac{1}{\pi T_i}\frac{\pi}{2} = \frac{1}{2T_i}; \tag{5.41}$$

One-section RC-filter:

$$f_{nw} = \int_0^\infty k^2(f)df = \int_0^\infty \frac{df}{1+\omega^2\tau_f^2} = \frac{1}{2\pi\tau_f}\int_0^\infty \frac{dx}{1+x^2}$$

$$= \frac{1}{2\pi\tau_f}\arctan x\Big|_0^\infty = \frac{1}{2\pi\tau_f}\frac{\pi}{2} = \frac{1}{4\tau_f}; \tag{5.42}$$

Two-section RC-filter:

$$f_{nw} = \int_0^\infty k^2(f)df = \int_0^\infty \frac{df}{\left(1+\omega^2\tau_f^2\right)^2} = \frac{1}{2\pi\tau_f}\int_0^\infty \frac{dx}{\left(1+x^2\right)^2} = \frac{1}{2\pi\tau_f}\frac{\pi}{4} = \frac{1}{8\tau_f}. \tag{5.43}$$

These values of f_{nw} are also shown in Table 5.1; we notice that the noise band f_{nw} is equal to or less than the effective band of the amplifier for the signal:

$$\Gamma_1 = \frac{f_{nw}}{f_a}, \quad \Gamma_1 \le 1. \tag{5.44}$$

This fact is easy to explain for those filters for which $k(f) \ge 0$, $\psi_a(f) = 0$. For such filters the effective band for the signal is defined by the area under the curve $k(f)$, which cannot be less than that under $k^2(f)$ (the noise band), since $k(f) \le 1$. The steeper the decline of the frequency response of the filter, the less the noise band differs from the effective band. For the filters under consideration the ratio between these bands is $\Gamma_1 = 0.5...1$.

Noise band for the high-frequency noise. The high-frequency noise component is defined by the second term under the root in (5.34):

$$u_n = \sqrt{(2\pi C_{in} e_n)^2 \int_0^\infty f^2 k^2(f)df}. \tag{5.45}$$

The situation is now somewhat more complicated than in case of white noise, when the spectral density is constant — it can be factored out from under the integral, and we only need to deal with the integral $\int_0^\infty k^2(f)df$, which has the dimension of frequency and defines the noise band f_{nw} (5.39). The spectral density of the high-frequency noise increases as f^2, and so its integral has the dimension of $[f^3]$ (5.45). Because of this, it cannot be directly taken for the definition of the noise band.

Let us therefore choose the value of the spectral density of high-frequency noise at a certain fixed frequency. Let this be the frequency f_a; by convention, we assume that the spectrum is homogeneous, with the fixed spectral density $\overline{J^2} = (2\pi C_{in} e_n f_a)^2$. This spectral density $\overline{J^2}$ in a certain band f_{nh} must give the actual dispersion of the output noise (5.45):

$$u_n = \sqrt{(2\pi C_{in} e_n f_a)^2 f_{nh}} = \sqrt{(2\pi C_{in} e_n)^2 \int_0^\infty f^2 k^2(f) df} , \tag{5.46}$$

whence follows the definition of the noise band for the high-frequency noise:

$$f_{nh} = \frac{1}{f_a^2} \int_0^\infty f^2 k^2(f) df . \tag{5.47}$$

Now everything is fine with the dimensions, and we may calculate the values of f_{nh} for the same four filters:

Filter with rectangular frequency response:

$$f_{nh} = \frac{1}{f_a^2} \int_0^\infty f^2 k^2(f) df = \frac{1}{f_a^2} \int_0^{f_a} f^2 df = \frac{1}{f_a^2} \frac{f_a^3}{3} = \frac{f_a}{3} ; \tag{5.48}$$

Filter-integrator with the integration time T_i:

$$f_{nh} \propto \int_0^\infty f^2 k^2(f) df = \int_0^\infty f^2 \left[\frac{\sin \pi f T_i}{\pi f T_i} \right]^2 df = \frac{1}{(\pi T_i)^3} \int_0^\infty \sin^2 x dx \to \infty ; \tag{5.49}$$

One-section RC-filter:

$$f_{nh} \propto \int_0^\infty f^2 k^2(f) df = \int_0^\infty \frac{f^2 df}{1 + \omega^2 \tau_f^2} = \frac{1}{(2\pi\tau_f)^3} \int_0^\infty \frac{x^2 dx}{1 + x^2} \to \infty ; \tag{5.50}$$

Two-section RC-filter:

$$f_{\text{nh}} = \frac{1}{f_{\text{a}}^2}\int_0^\infty \frac{f^2 df}{\left(1+\omega^2\tau_{\text{f}}^2\right)^2} = \frac{1}{f_{\text{a}}^2(2\pi\tau_{\text{f}})^3}\int_0^\infty \frac{x^2 dx}{\left(1+x^2\right)^2}$$

$$= \frac{1}{f_{\text{a}}^2(2\pi\tau_{\text{f}})^3}\frac{\pi}{4} = \frac{1}{(2\pi f_{\text{a}}\tau_{\text{f}})^2}\frac{1}{8\tau_{\text{f}}} \ . \tag{5.51}$$

By analogy with (5.44), we may introduce coefficient Γ_2 as the ratio of two characteristic frequencies:

$$\Gamma_2 = \frac{f_{\text{nh}}}{f_{\text{a}}} = \frac{\int_0^\infty f^2 k^2(f) df}{f_{\text{a}}^3} \ . \tag{5.52}$$

The values of f_{nh} (5.48) – (5.51) and Γ_2 (5.52) are also presented in Table 5.1.

The frequency f_{nh} is a specific parameter of PDA. The noise power at the output of the PDA will be finite only if the frequency response rolls off quickly enough ($k \sim 1/f^n$):

$$\overline{u_{\text{n}}^2} \propto \int_0^\infty k^2 f^2 df = \int_0^\infty \frac{1}{f^{2n}} f^2 df = \int_0^\infty \frac{df}{f^{2(n-1)}} \ . \tag{5.53}$$

The integral in (5.53) converges when $2(n-1) > 1$ — that is, when $n > 1.5$. Filters 2 and 3 do not satisfy this condition, and are therefore not suitable for PDAs with the high-frequency noise. The frequency response of filters 1, 4 falls off steeply enough ($n > 1.5$), and the frequency f_{nh} is finite. The values of Γ_2 for filters 1, 4 are close enough: $\Gamma_2 = 0.33...0.51$.

The integrals above are similar to those introduced by Personick [91] in the analysis of noise in PDAs for lightwave communication systems.

Limiting frequency. This concept is widely used in radio engineering: the amplification at this frequency falls to a specified limit (for example, 0.7 of the maximum). The values of limiting frequency f_{07} are also shown in Table 5.1.

General expression for SNR in quasi-optimal filter. Let us return to (5.34) for the dispersion of noise, and replace the integrals with the noise bands as defined above:

$$U_{\text{n}} = \sqrt{2qI_{\text{n}}f_{\text{nw}} + (2\pi C_{\text{in}}e_{\text{n}}f_{\text{a}})^2 f_{\text{nh}}} = \sqrt{2qI_{\text{n}}\Gamma_1 f_{\text{a}} + (2\pi C_{\text{in}}e_{\text{n}})^2\Gamma_2 f_{\text{a}}^3} \ . \tag{5.54}$$

This is a working expression which permits calculating the dispersion of noise at the output of the amplifier knowing the effective amplifier band f_a (or the time constant $\tau_{eff} = 1/2f_a$). The precise shape of the frequency curve of the filter has a relatively small effect on the noise, provided that the passband f_a remains the same. The coefficients Γ_1, Γ_2 in the above examples vary within relatively narrow limits ($\Gamma_1 = 0.5...1$; $\Gamma_2 = 0.33...0.51$), and so the relative variations of the noise are as small as $u_n \propto \sqrt{\Gamma_1} = 0.7...1$, and $u_n \propto \sqrt{\Gamma_2} = 0.58...71$, respectively (with due account for our remark concerning the sufficiently steep roll-off of the frequency response).

The signal amplitude as function of the effective time constant of RC-filters is given by (5.19) and (5.29). So, now we may write the desired expression for the signal-to-noise ratio:

$$\text{SNR} = \frac{SE_{sA}u_{sA}}{U_n} = \frac{SE_{sA}u_{sA}}{\sqrt{2qI_n\Gamma_1 f_a + (2\pi C_{in}e_n f_a)^2\Gamma_2 f_a}}, \quad f_a = \frac{1}{2\tau_{eff}}. \tag{5.55}$$

When the band f_a is too narrow, the signal-to-noise ratio is low because the numerator (the signal u_{sA}) is small; when the band is too broad, the SNR falls because of the large denominator (the noise). Accordingly, there must be a certain optimized bandwidth f_a which corresponds to the highest possible SNR. From the qualitative considerations (see the beginning of the present section, Fig. 5.4), the guidelines for such optimization are clear: the bandwidth must be equal to the effective passband of the optimal filter $f_{a\,opt}$. In case of white noise this band coincides with the frequency band of the signal f_s, and so

$$f_a = f_s, \quad \tau_a = T_s; \tag{5.56}$$

when the high-frequency noise prevails, we get

$$f_a = \int_0^\infty k_{opt}(f)df = \int_0^\infty \frac{e_s(f)}{\Theta(f)}df \approx \int_0^\infty \frac{df}{1+\frac{f^2}{f_{high}^2}} = f_{high}\arctan\frac{f}{f_{high}}\Bigg|_0^\infty = \frac{\pi}{2}f_{high},$$

$$\tau_{eff} = \frac{1}{2f_a} = \frac{1}{\pi f_{high}}. \tag{5.57}$$

Now we only have to reconfirm our more or less obvious choice of f_a with calculations (with perhaps better precision), and define the values of SNR achievable in quasi-optimal filters with such passbands.

White noise. Consider a one-section RC-filter. First we use the approximation (5.19) for the amplitude of the output signal u_{sA}. Substituting it into the SNR (5.55), and retaining only the white-noise component, we get:

$$\mathrm{SNR} = \frac{SE_{sA}}{\sqrt{2qI_n \Gamma_1 f_a}} \frac{T_s}{T_s + \tau_f} = \frac{SE_{sA}}{\sqrt{2qI_n \frac{1}{2} \cdot \frac{1}{2\tau_f}}} \frac{T_s}{T_s + \tau_f}$$

$$= \frac{SE_{sA} T_s}{\sqrt{2qI_n}} \frac{2}{\left(\frac{T_s}{\sqrt{\tau_f}} + \sqrt{\tau_f} \right)} . \tag{5.58}$$

Here we have noted that $\Gamma_1 = 1/2$, $f_a = 1/2\tau_{\mathrm{eff}} = 1/2\tau_f$ for the one-section filter. The binomial in parentheses has been considered in detail in Sect. 2.5; it has its extremum when the terms are equal:

$$\frac{d}{d\tau_f}\left(\frac{T_s}{\sqrt{\tau_f}} + \sqrt{\tau_f} \right) = -\frac{T_s}{2\sqrt{\tau_f^3}} + \frac{1}{2\sqrt{\tau_f}} = 0 ;$$

$$\frac{T_s}{\sqrt{\tau_f}} \equiv \sqrt{\tau_f} , \quad \tau_a = \tau_f = T_s . \tag{5.59}$$

From (5.58) we get the following SNR for the filter with such optimal time constant:

$$\max\{\mathrm{SNR}\} = \frac{SE_{sA}\sqrt{T_s}}{\sqrt{2qI_n}} , \tag{5.60}$$

and, dividing (5.58) by (5.60), we get the tuning curve for the one-section filter:

$$\frac{\mathrm{SNR}}{\max\{\mathrm{SNR}\}} = \frac{2}{\sqrt{\frac{T_s}{\tau_f}} + \sqrt{\frac{\tau_f}{T_s}}} . \tag{5.61}$$

In this way, the prefabricated approximation for the amplitude of the output signal allowed us to simplify all calculations and do them in the analytical form. It should be remembered, however, that at $\tau_f = T_s$ this approximation underestimates the signal amplitude by 9% (or, more precisely, 8.7% — see Fig. 5.8). Because of this, expression (5.60) needs to be corrected:

$$\max\{\mathrm{SNR}\} = 1.09 \frac{SE_{sA}\sqrt{T_s}}{\sqrt{2qI_n}} . \tag{5.62}$$

Strictly speaking, the tuning of the filter must also be corrected. Numerical calculations based on the exact value of the amplitude of the output signal (5.15) defines the optimal time constant of the filter as $\tau_f = 0.84T_s$. This correction is, however, of purely theoretical interest: the tuning curve is so smooth that the value of SNR increases by a mere 0.45% as τ_f is changed from T_s to $0.84T_s$ (the more precise value of numerical coefficient in (5.62) is 1.092).

The calculation for the two-section filter is similar: once again we use the approximation for the amplitude of the output signal, which this time is (5.29). Substituting this into (5.55), we get the value of SNR in case of the prevalent white noise:

$$\mathrm{SNR} = \frac{SE_{\mathrm{sA}}}{\sqrt{2qI_n\Gamma_1 f_a}} \frac{T_s}{\sqrt{T_s^2 + \tau_{\mathrm{eff}}^2}} = \frac{SE_{\mathrm{sA}}}{\sqrt{2qI_n \frac{e}{4} \cdot \frac{1}{2\tau_{\mathrm{eff}}}}} \frac{T_s}{\sqrt{T_s^2 + \tau_{\mathrm{eff}}^2}}$$

$$= \sqrt{\frac{4}{e}} \frac{SE_{\mathrm{sA}}\sqrt{T_s}}{\sqrt{2qI_n}} \sqrt{\frac{2}{\left(\frac{T_s}{\tau_{\mathrm{eff}}} + \frac{\tau_{\mathrm{eff}}}{T_s}\right)}} . \tag{5.63}$$

Again, the calculation of the optimal time constant of the filter reduces to finding the extremum of the expression in parentheses of the form $(x + 1/x)$. The minimum of this binomial occurs when the two terms are equal (see Sect. 2.5):

$$\frac{T_s}{\tau_{\mathrm{eff}}} = \frac{\tau_{\mathrm{eff}}}{T_s}, \quad \tau_{\mathrm{eff}} = e\tau_f = T_s . \tag{5.64}$$

Accordingly, the maximum value of SNR in (5.63) is achieved when the time constant of each section of the filter is $\tau_f = T_s/e$:

$$\max\{\mathrm{SNR}\} = \sqrt{\frac{4}{e}} \frac{SE_{\mathrm{sA}}\sqrt{T_s}}{\sqrt{2qI_n}} = 1.21 \frac{SE_{\mathrm{sA}}\sqrt{T_s}}{\sqrt{2qI_n}} . \tag{5.65}$$

According to (5.63), (5.65), the tuning curve of the two-section filter is given by

$$\frac{\mathrm{SNR}}{\max\{\mathrm{SNR}\}} = \sqrt{\frac{2}{\frac{T_s}{\tau_{\mathrm{eff}}} + \frac{\tau_{\mathrm{eff}}}{T_s}}} . \tag{5.66}$$

The approximation used for the amplitude of the output signal of the two-section filter is even more precise than that for the one-section filter: at $\tau_{\mathrm{eff}} = T_s$ it overestimates the signal by as little as 2.7% (Fig. 5.10). A stern reader may reduce appropriately the coefficient in (5.65):

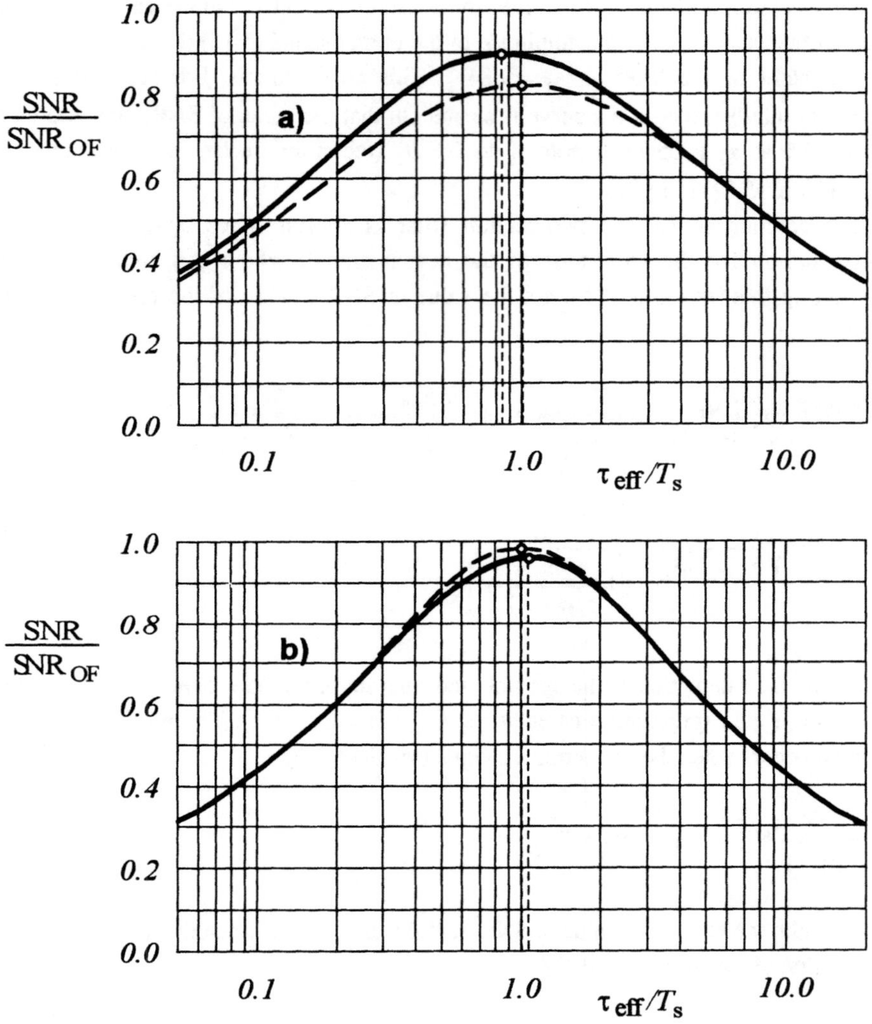

Figure 5.11: Smooth tuning curves of quasi-optimal filters for white noise: (a) one-section filter; (b) two-section filter.

$$\max\{\text{SNR}\} = 1.18\frac{SE_{\text{sA}}\sqrt{T_{\text{s}}}}{\sqrt{2qI_{\text{n}}}}\ ; \tag{5.67}$$

this value is achieved at a somewhat larger value of the effective time constant than the value given by the approximate calculation, $\tau_{\text{eff}} = 1.05T_{\text{s}}$.

What can be our comment on these results? Of course, we can hardly wait to compare them with the values for the optimal filter. Observe that the expression for the signal-to-noise ratio of the optimal filter (3.76) differs from (5.62), (5.67) only in the numerical coefficient, $\sqrt{3/2} = 1.22$. Accordingly, the loss coefficient is

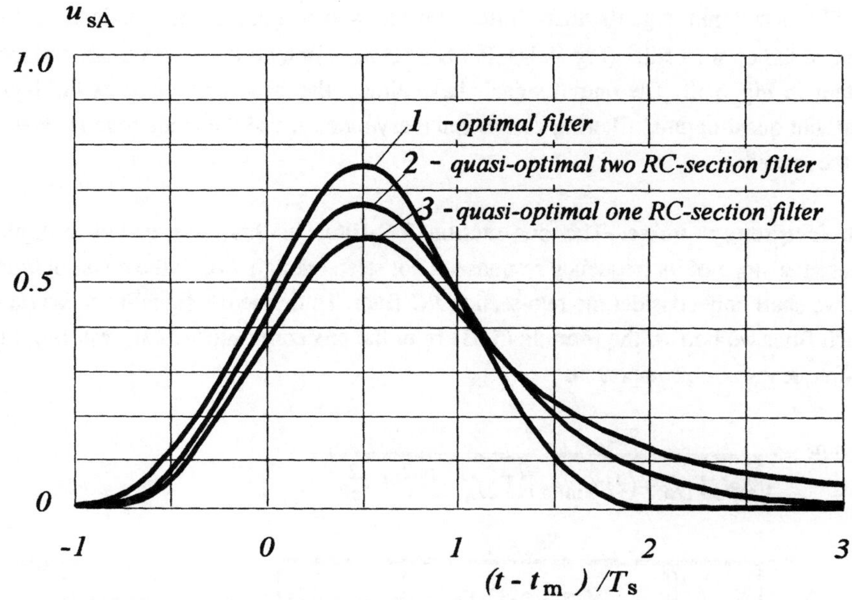

Figure 5.12: The reader will hardly distinguish the signals at the output of filter viewing them separately on the screen of oscilloscope.

$$\mathcal{L} = \frac{\max\{\text{SNR}\}}{\text{SNR}_{OF}} = \frac{1.09}{1.22} = 0.89 \tag{5.68}$$

for the one-section RC-filter, and

$$\mathcal{L} = \frac{1.18}{1.22} = 0.965 \tag{5.69}$$

for the two-section RC-filter.

Of course, after our qualitative analysis it would be hard to surprise us with such low values of the loss. But still, the result exceeds our expectations: the SNR of the two-section RC-filter is just 3.5% short of the SNR_{OF}! "Quasi-optimal" seems to be an inappropriate name for such a filter.

As expected, the maximum of signal-to-noise ratio is achieved when the effective time constant of the filter is equal (almost equal) to the length of the signal (5.59), (5.64), or, which is the same, when the effective band of the filter is equal to the bandwidth of the signal. The tuning curves of the RC-filters are very flat (see Fig. 5.11 and (5.61), (5.66)). So, depending on the particular task (see Sect. 5.3), the PDA engineer may vary the time constant within sufficiently broad limits without compromising the SNR too much.

The profiles of the signal at the output of one-section and two-section filters were shown in Fig. 5.7, 5.9 (thick curve corresponds to the optimal value of τ_f). The two output signals differ little from one another, and from the signal at the output of the optimal filter (Fig. 4.8g). For easy comparison all these curves are brought together in Fig. 5.12. The output signals have almost the same bell shape as the input signal, but quasi-optimal filtering causes some asymmetry, and the front edge is steeper than the rear edge.

High-frequency noise. The one-section RC-filter in this case is not suitable, because the slope of its frequency response is not steep enough. So, as the quasi-optimal filter we shall only consider the two-section RC-filter. To define the signal-to-noise ratio of such filter we borrow the formula (5.63) from the preceding subsection, restoring the high-frequency noise component:

$$\mathrm{SNR} = \frac{SE_{sA}}{\sqrt{2qI_n\Gamma_1 f_a + (2\pi C_{in}e_n)^2\Gamma_2 f_a^3}}\,\frac{T_s}{\sqrt{T_s^2+\tau_{eff}^2}}$$

$$= \frac{SE_{sA}T_s}{\sqrt{2qI_n\Gamma_1\left\{\left[f_a + \frac{(2\pi C_{in}e_n)^2}{2qI_n}\frac{\Gamma_2}{\Gamma_1}f_a^3\right]\left(T_s^2+\tau_{eff}^2\right)\right\}}}\,. \tag{5.70}$$

Using (2.169) and Table 5.1, and taking into account that $f_a = 1/2\tau_{eff}$, $\tau_{eff} >> T_s$, we replace $2qI_n/(2\pi C_{in}e_n)^2 \Rightarrow f_{high}^2$, $\Gamma_2/\Gamma_1 \Rightarrow e^2/\pi^2$ and transform the expression in braces to

$$\left\{\left[\frac{1}{2\tau_{eff}} + \frac{1}{f_{high}^2}\frac{e^2}{\pi^2}\frac{1}{(2\tau_{eff})^3}\right]\left(\tau_{eff}^2\right)\right\} = \left\{\frac{\tau_{eff}}{2} + \frac{1}{f_{high}^2}\frac{e^2}{\pi^2}\frac{1}{8\tau_{eff}}\right\}$$

$$= \left\{\frac{1}{2}\left(\tau_f e + \frac{e^2}{(2\pi f_{high})^2}\frac{1}{e\tau_f}\right)\right\} = \left\{\frac{e}{2\pi f_{high}}\left[\frac{1}{2}\left(\tau_f\omega_{high} + \frac{1}{\tau_f\omega_{high}}\right)\right]\right\}. \tag{5.71}$$

Once again we get a familiar binomial of the form $x + 1/x$, whose extremum occurs when the two terms are equal — that is, when equal are the high-frequency and the white noise components. Accordingly, the time constant of each section, and the effective time constant of the filter as a whole must satisfy the following conditions:

$$\tau_f\omega_{high} = \frac{1}{\tau_f\omega_{high}}\,, \quad \tau_f = \frac{1}{\omega_{high}}\,, \quad \tau_{eff} = e\tau_f = \frac{e}{2\pi f_{high}}\,. \tag{5.72}$$

Deviation from the predicted value of the effective time constant of quasi-optimal filter (5.57) is as small as $e/2 = 1.36$ times.

Substituting the result of (5.71) into (5.70), and the value of $\Gamma_1 = e/4$, for the value of SNR we get:

$$\mathrm{SNR} = \frac{S E_{\mathrm{sA}} T_{\mathrm{s}}}{\sqrt{2 q I_{\mathrm{n}} \frac{e}{4} \frac{e}{2\pi f_{\mathrm{high}}}}} \sqrt{\frac{2}{\tau_{\mathrm{f}} \omega_{\mathrm{high}} + \frac{1}{\tau_{\mathrm{f}} \omega_{\mathrm{high}}}}}$$

$$= \frac{2}{e} \sqrt{2\pi} \frac{S E_{\mathrm{sA}} T_{\mathrm{s}} \sqrt{f_{\mathrm{high}}}}{\sqrt{2 q I_{\mathrm{n}}}} \sqrt{\frac{2}{\tau_{\mathrm{f}} \omega_{\mathrm{high}} + \frac{1}{\tau_{\mathrm{f}} \omega_{\mathrm{high}}}}} . \tag{5.73}$$

When the filter is tuned up ($\tau_{\mathrm{f}} \omega_{\mathrm{high}} = 1$), the last square root is equal to one, and the maximum value of the signal-to-noise ratio is, therefore,

$$\max\{\mathrm{SNR}\} = \frac{2}{e} \sqrt{2\pi} \frac{S E_{\mathrm{sA}} T_{\mathrm{s}}}{\sqrt{2 q I_{\mathrm{n}}}} \sqrt{f_{\mathrm{high}}} , \tag{5.74}$$

whereas the tuning curve has the same form as that for the two-section filter in case of white noise:

$$\frac{\mathrm{SNR}}{\max\{\mathrm{SNR}\}} = \sqrt{\frac{2}{\tau_{\mathrm{f}} \omega_{\mathrm{high}} + \frac{1}{\tau_{\mathrm{f}} \omega_{\mathrm{high}}}}} . \tag{5.75}$$

Expression (5.74) differs from the corresponding formula for the optimal filter (3.77) by the numerical coefficient $2/e = 0.735$. We see that it is more difficult for the two-section RC-filter to cope with the high-frequency noise than with the white noise; the loss of 26.5% in the SNR is, however, quite acceptable. The tuning curve is shown in Fig. 5.13.

A few words on the shape of the output signal. Since the passband of the quasi-optimal filter is much narrower than the frequency band of the signal, the latter may be considered a delta-pulse, and then the output signal imitates the pulse response (this situation is similar to the case of optimal filtering, Sect. 4.2). The pulse response of the two-section RC-filter is shown in Fig. 5.6b; its profile differs considerably from the bell shape: the rear edge is stretched, and the length at half-amplitude is much greater than the length of the input signal and can be calculated by (5.28), (5.72):

$$T_{\mathrm{s}} f_{\mathrm{high}} << 1 , \quad \frac{T_{05}}{T_{\mathrm{s}}} = \frac{2.45 \tau_{\mathrm{f}}}{T_{\mathrm{s}}} = \frac{2.45}{T_{\mathrm{s}} \left(2\pi f_{\mathrm{high}}\right)} = \frac{0.39}{T_{\mathrm{s}} f_{\mathrm{high}}} . \tag{5.76}$$

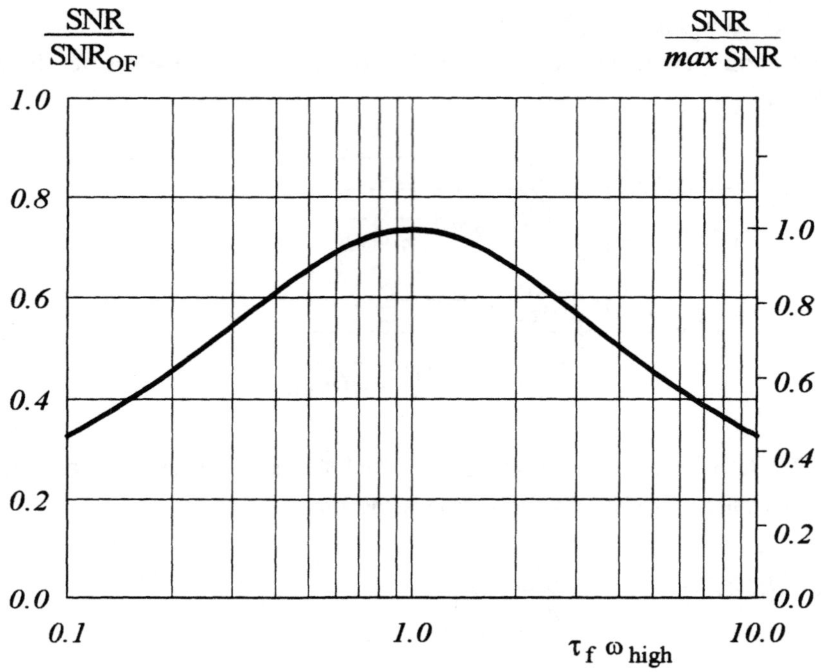

Figure 5.13: Smooth tuning curve of quasi-optimal filters for high-frequency noise.

The two-section RC-filter can be used as quasi-optimal filter also in the general case, when the white and the high-frequency components are comparable. Equation (5.70), which includes both these components, allows us to find the condition of tuning, and the coefficient of loss in the signal-to-noise ratio. The approximate expressions are found from the limiting formulas for the prevalent white noise (5.64), (5.67), or the prevalent high-frequency noise (5.72), (5.74), using the same technique as used for deriving (3.102) in case of optimal filtering:

$$\tau_{\text{eff}} = T_s + \frac{e}{2\pi f_{\text{high}}} , \tag{5.77}$$

$$\text{SNR} = \frac{0.835 S E_{sA} \sqrt{2T_s}}{\sqrt{2qI_n \left(1 + \dfrac{0.82}{f_{\text{high}} 2T_s}\right)}} . \tag{5.78}$$

The inaccuracy of these formulas is about 30%. Since the method of construction of the approximate formulas for SNR is the same for both optimal (3.102) and quasi-optimal filtering (5.78), it is not surprising that the expressions have the same structure, and differ only in the numerical coefficients (0.865 and 0.48 for optimal filter, and 0.835 and 0.82 for quasi-optimal filter).

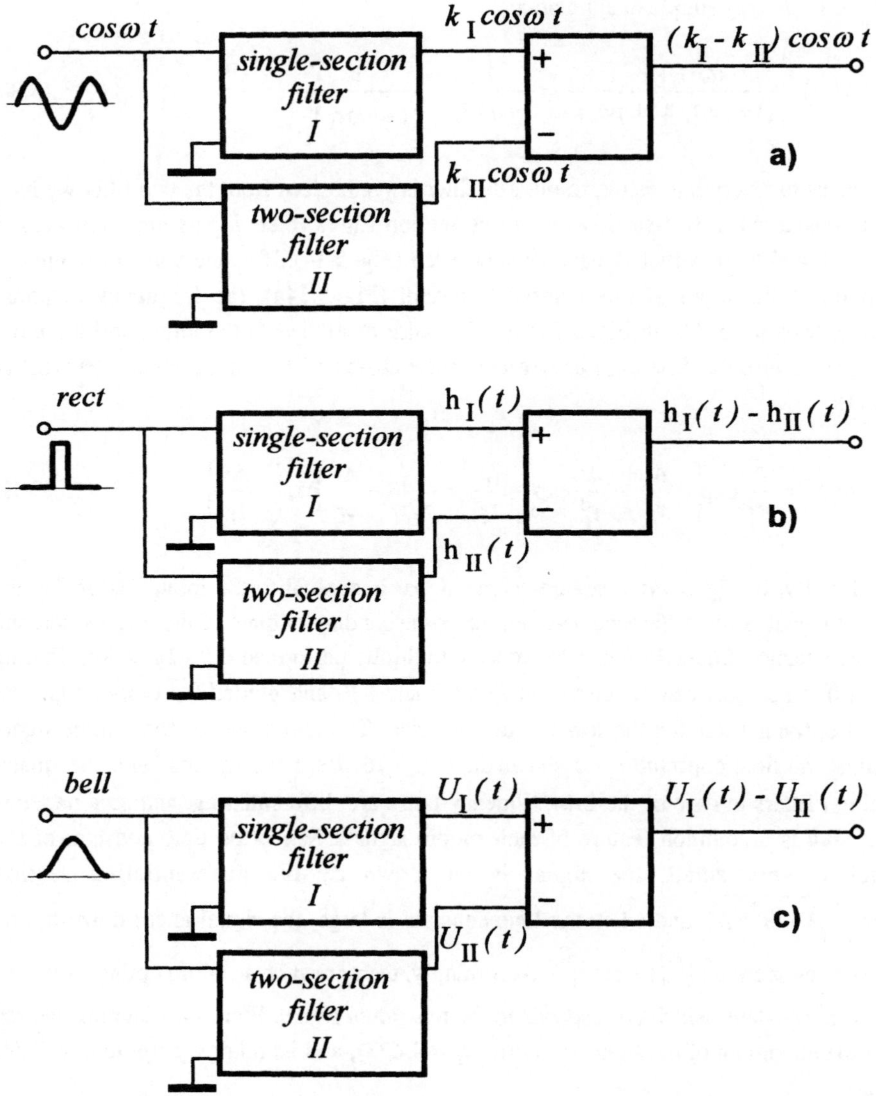

Figure 5.14: Equivalent diagram of a filter with consecutive differentiating and integrating sections.

Low-frequency noise. In this case the method of filtering is different from those just considered, because one has to take care of not only the high-frequency end of the frequency curve, but also of the low-frequency end. This brings us to the circuit which includes both integrating and differentiating *RC*-sections (Fig. 5.5c). The frequency response of the filter is composed of the frequency characteristics of the two sections:

$$k(\omega) = \frac{j\omega\tau_f}{1 + j\omega\tau_f} \cdot \frac{1}{1 + j\omega\tau_f} \ . \tag{5.79}$$

To simplify the analysis, we assume that the time constants τ_f of the two sections are the same. A very simple manipulation,

$$k(\omega)=\frac{j\omega\tau_f+1-1}{(1+j\omega\tau_f)(1+j\omega\tau_f)}=\frac{1}{1+j\omega\tau_f}-\frac{1}{(1+j\omega\tau_f)^2}\ , \tag{5.80}$$

brings us to a very interesting result. The filter is synthesized from the two filter we have just considered, a two-section and a one-section filters: their inputs are connected in parallel, and their output signals are subtracted (Fig. 5.14). If a sine-wave test signal is applied to the input of this composite circuit (Fig. 5.14a), the frequency response corresponds to (5.80). In Fig. 5.14b a delta-pulse is applied to the input, and the pulse response is also the difference between the pulse characteristics of the same filters (5.11), (5.23):

$$\mathrm{h}(\Delta t)=\frac{1}{\tau_f}\exp\left(-\frac{\Delta t}{\tau_f}\right)-\frac{\Delta t}{\tau_f^2}\exp\left(-\frac{\Delta t}{\tau_f}\right)=\frac{1}{\tau_f}\left(1-\frac{\Delta t}{\tau_f}\right)\exp\left(-\frac{\Delta t}{\tau_f}\right). \tag{5.81}$$

Finally, in Fig. 5.14c a square-cosine signal is applied to the input. Naturally, the output signal is the difference between the corresponding outputs of the one-section and the two-section filters. In Fig. 5.15 we have faithfully performed the subtraction (finding the difference between the curves in Fig. 5.7 and 5.9) and plotted the output signal of quasi-optimal filter for the low-frequency noise. The amplitude of the output signal against the time constant τ_f is shown in Fig. 5.16. Both the optimal and the quasi-optimal filters in case of the low-frequency noise are differentiating, and so a negative outswing is a common feature of their output signals. When the time constant of the filter is very small, the signal is cut down by the differentiating section ($\omega\tau_f/\sqrt{1+\omega^2\tau_f^2}$), and when the time constant is large, the signal is cut down by the integrating section ($1/\sqrt{1+\omega^2\tau_f^2}$). Accordingly, there must be a certain optimal value of the time constant, which corresponds to the maximum signal. From the diagrams we see that the maximum of the signal occurs at $\tau_f=0.67T_s$, and its relative amplitude is 0.34; hence

$$U_{sA}(t_m)=0.34SE_{sA}\ . \tag{5.82}$$

Here we have violated our tradition to obtain all the principal results analytically. It is easy, however, to get a rule-of-thumb estimate of the maximum signal value without resorting to numerical calculations. This estimate will take just five steps and Fig. 5.17, which represents consecutively the signals at the input (Fig. 5.17a) and at the outputs of the one-section and the two-section filters (Fig. 5.17b,c).

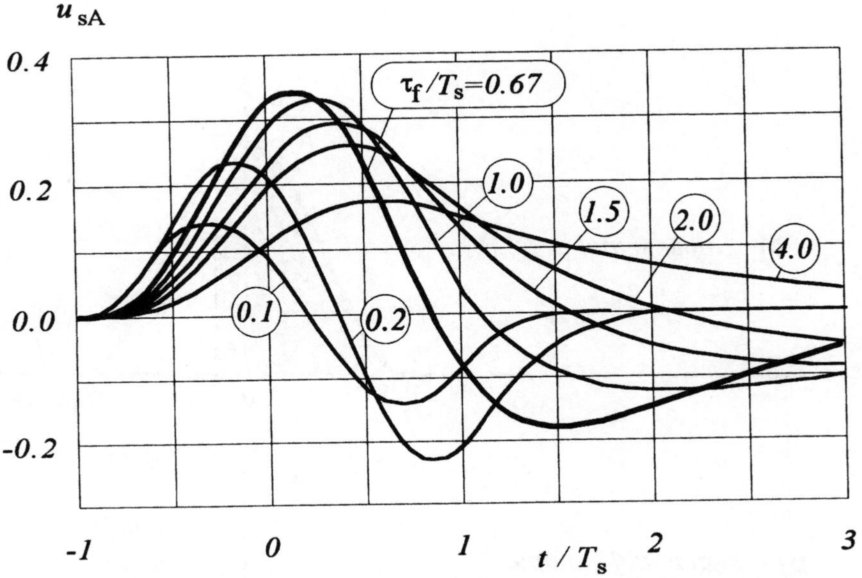

Figure 5.15: Signal profile at the output of filter with differentiating and integrating sections for different τ_f (bell-shaped input signal).

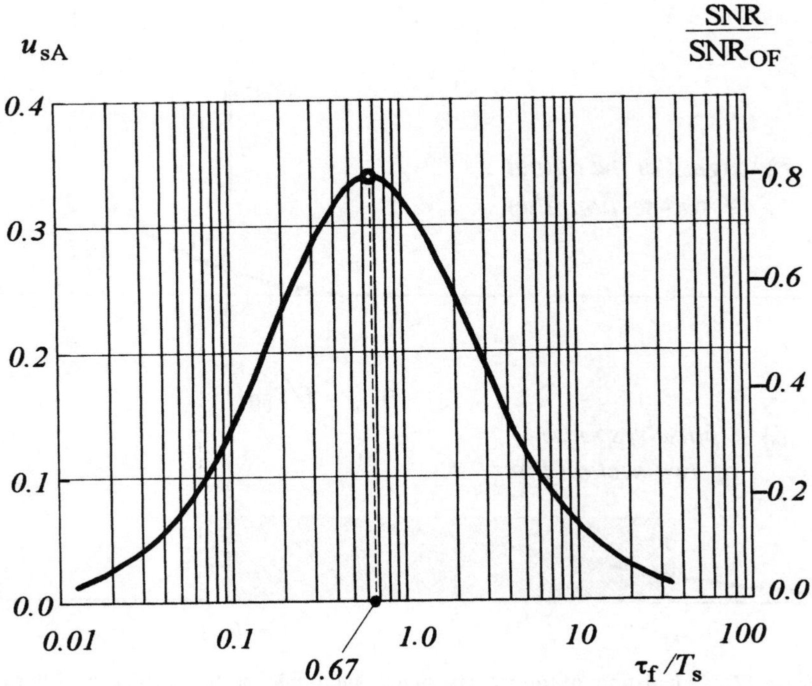

Figure 5.16: Signal amplitude at the output of filter with differentiating and integrating sections as function of τ_f and the tuning curve SNR(τ_f) coincide.

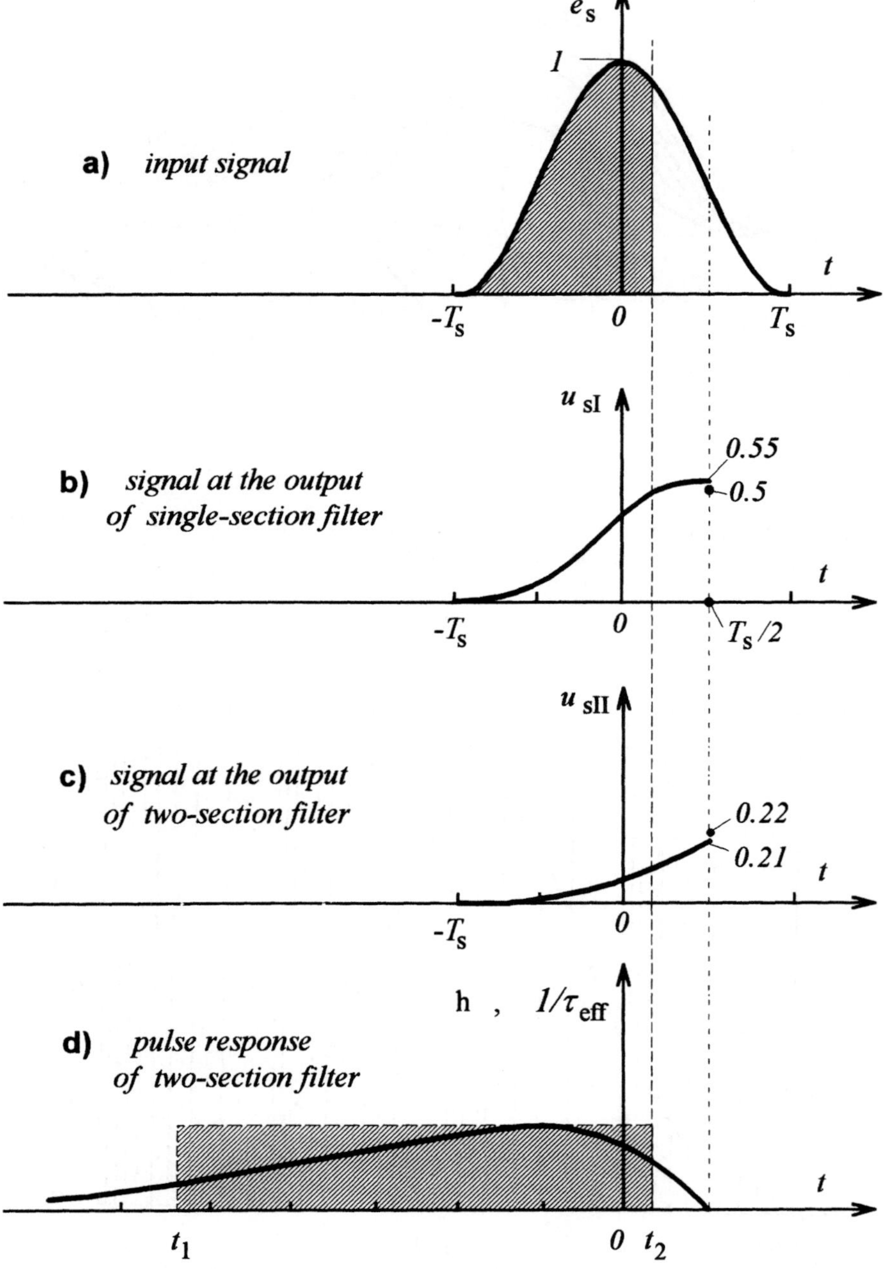

Figure 5.17: Calculation of maximum signal amplitude at the output of filter with differentiating and integrating sections.

1. Select the time constant. This can be done without referring to the calculated curves $u_s(t, \tau_f)$: the general rule of matching the passband of the optimal filter with the bandwidth of the signal dictates that $\tau_f = T_s$.

2. According to (5.19), the signal at the output of the one-section filter at $\tau_f = T_s$ is

$$u_{sA\,I} = \frac{T_s}{\tau_f + T_s} = \frac{T_s}{T_s + T_s} = \frac{1}{2} \ . \tag{5.83}$$

3. The maximum value is achieved at the time (5.21)

$$t = t_m = \frac{\tau_f}{\tau_f + T_s} T_s = \frac{T_s}{T_s + T_s} T_s = \frac{1}{2} T_s \ . \tag{5.84}$$

4. It might seem that the most difficult thing is to find the response of the two-section filter at the same moment $t = 0.5T_s$. This estimate is made easy, however, by the technique described in Sect. 4.2. For this purpose we use the pulse response — or, more precisely, its rectangular approximation shown in Fig. 5.17d. As a matter of fact, in the first approximation the filter may be considered as the integrator over the time τ_{eff}. In particular, the output signal for the moment $t = 0.5T_s$ is the integral of the input signal over the interval $[t_1, t_2]$ (Fig. 5.17d) — that is, the sought-for amplitude is proportional to the hatched area in Fig. 5.17a. This area is about 60% of the entire area of the input pulse $e_{sA}T_s = T_s$; therefore,

$$u_{s\,II}(t_m) \approx \mathrm{h}(t_m)0.6T_s = \frac{1}{\tau_{eff}} 0.6T_s = \frac{0.6T_s}{e\tau_f} = 0.22 \ . \tag{5.85}$$

5. The maximum of the output signal is approximately equal to

$$u_{sA} \approx u_{sA\,I} - u_{s\,II}(t_m) \approx 0.5 - 0.22 = 0.28 \ . \tag{5.86}$$

Now our tradition is maintained: we managed to get an analytical order-of-magnitude estimate for the amplitude of the output signal (cf. (5.82), (5.86)).

Next we must calculate the noise at the output of the filter in question:

$$\overline{U_n^2} = \int_0^\infty \overline{J^2} k^2 df = \int_0^\infty \left(2qI_n \frac{f_{low}}{f}\right)\left(\frac{\omega\tau_f}{1+\omega^2\tau_f^2}\right)^2 df =$$

$$= 2qI_n f_{low} \int_0^\infty \left(\frac{x}{1+x^2}\right)^2 \frac{dx}{x} = 2qI_n f_{low} \frac{1}{2} \int_0^\infty \frac{dx^2}{\left(1+x^2\right)^2} =$$

$$= qI_{\text{n}}f_{\text{low}} \int_0^{\infty} \frac{dy}{(1+y)^2} = qI_{\text{n}}f_{\text{low}} \frac{1}{1+y}\bigg|_{\infty}^{0} = qI_{\text{n}}f_{\text{low}} \ . \tag{5.87}$$

In case of the excess input $1/f$-noise the output noise does not at all depend on the passband of the filter (the time constant τ_{f}; see Sect. 3.3).

And finally, the ratio of the signal (5.82) to the noise (5.87) is

$$\text{SNR} = \frac{U_{\text{sA}}}{U_{\text{n}}} = \frac{0.34SE_{\text{sA}}}{\sqrt{qI_{\text{n}}f_{\text{low}}}} = \frac{0.48SE_{\text{sA}}}{\sqrt{2qI_{\text{n}}f_{\text{low}}}} \ . \tag{5.88}$$

From (5.80), (5.88) it follows that the loss coefficient of the quasi-optimal filter in case of the low-frequency noise is

$$\mathcal{L} = \frac{\text{SNR}}{\text{SNR}_{\text{OF}}} = \frac{0.48}{0.6} = 0.8 \ , \tag{5.89}$$

and is less than in case of the high-frequency noise.

The tuning curve $\text{SNR}(\tau_{\text{f}})$ in Fig.5.16 retraces the tuning curve for the signal, since the noise does not depend on the time constant τ_{f}. This curve is no exception: it is flat like all other tuning curves.

This analysis has been carried out under the assumption that the time constants of the differentiating section τ_{fd} and the integrating section τ_{f} are the same. Of course, we could carry out cumbersome calculations, making the two time constants independent from each other and trying to reduce the loss coefficient through separate optimization. But then, the top prize — the improvement on the SNR — will not exceed 20%. And secondly... secondly, this calculation reveals that making the two time constants equal to each other is the optimal choice, and further reduction of the loss coefficient with such arrangement is not possible. Of course, this applies to the case of "pure" excess noise, which prevails in the entire signal spectrum ($f_{\text{low}} > 1/2T_{\text{s}}$). If the excess noise is noticeable only at the relatively low frequencies $f_{\text{low}} << 1/2T_{\text{s}}$, then the differentiating section must only cut off these frequencies ($\tau_{\text{fd}} \approx 1/2\pi f_{\text{low}}$). Since the suppressed band (in the neighborhood of the zero frequency) is narrow compared to the signal bandwidth, such filtering has little effect on the output signal and the white noise component: they remain practically the same as in case of white noise and one-section integrating filter.

How many sections are required for the quasi-optimal filter? We proposed quasi-optimal filters for all noise spectra likely to occur in PDAs, and defined the coefficient of loss for such filters. We have also answered the question concerning the number of sections: one or two, and we know what kind of sections these should be in each case. What then makes us ask the same question again? It is the frequency of false alarms, which has not yet been estimated. Let us recall the definition of this frequency

(3.25). For the arbitrary input spectrum Θ, the spectrum at the output of PDA is $\Theta_{out} = k^2\Theta$, and so for the characteristic frequency which defines the rate of false alarms we get:

$$f_{char} = \sqrt{\frac{\int_0^\infty f^2 k^2 \Theta df}{f_n}}, \quad f_n = \int_0^\infty k^2 \Theta df \,. \tag{5.90}$$

The frequency f_n in the denominator is the noise band f_{nw} in case of the predominantly white noise, and is proportional to the noise band f_{nh} in case of the predominantly high-frequency noise. These noise bands have already been calculated before (see (5.42), (5.43), (5.51)). Now let us find the integral in the numerator of (5.90):

$$\int_0^\infty f^2 k^2 \Theta df = \int_0^\infty f^2 \frac{df}{1+\omega^2\tau_f^2} \int_0^\infty df \to \infty \tag{5.91}$$

in case of white noise and one-section filter;

$$\int_0^\infty f^2 k^2 \Theta df = \int_0^\infty \frac{f^2 df}{\left(1+\omega^2\tau_f^2\right)^2} = \frac{1}{(2\pi\tau_f)^3}\int_0^\infty \frac{x^2 dx}{\left(1+x^2\right)^2} = \frac{1}{(2\pi\tau_f)^3}\frac{\pi}{4} \tag{5.92}$$

in case of white noise and two-section filter; and

$$\int_0^\infty f^2 k^2 \Theta df = \int_0^\infty f^2 \frac{1}{\left(1+\omega^2\tau_f^2\right)^2}\frac{f^2}{f_{high}^2} df \sim \int_{f_1}^\infty df \to \infty \tag{5.93}$$

in case of high-frequency noise and two-section filter.

So, have these three integrals been worth computing to yield discouraging results? Two integrals diverge, and it is only for the white noise and two-section filter that the frequency of false alarms is finite. The divergence is due the term $f^2\Theta$ under the integral; accordingly, the filter (which is represented by k^2) must balance out the fast increase caused by this term. The integral will converge when $f^2\Theta^2 k^2 < B/f$. In case of the white noise this condition is satisfied by a filter which contains at least two sections,

$$f \to \infty \,, \quad f^2\Theta^2 k^2 = f^2 \cdot 1 \cdot \frac{1}{\left(1+\omega^2\tau_f^2\right)^2} \sim \frac{1}{f^2} \,, \tag{5.94}$$

and in case of the high-frequency noise the filter must have three or more sections:

$$f \to \infty \, , \quad f^2 \Theta^2 k^2 = f^2 \cdot \frac{f^2}{f_{\mathrm{high}}^2} \cdot \frac{1}{\left(1+\omega^2 \tau_{\mathrm{f}}^2\right)^3} \sim \frac{1}{f^2} \, . \tag{5.95}$$

Does this imply that the arguments developed so far are bankrupt, and it is not possible after all to use the one-section filer in case of the white noise, and two-section filter in case of the high-frequency noise? Are our calculations for these filters futile? The situation is improved by the fact that the additional section, which ensures the steeper high-frequency decline of the transfer coefficient of the filter, may have a small time constant τ_{f1} as compared with the time constant of the main RC-sections τ_{f}. The calculation according to (5.80) indicates that, given such a small value of the time constant τ_{f1}, the characteristic frequency f_{char} increases by the factor of $(0.4...0.7)\sqrt{\tau_{\mathrm{f}}/\tau_{\mathrm{f1}}}$ (as compared with its value corresponding to the symmetrical sections, $\tau_{\mathrm{f}} = \tau_{\mathrm{f1}}$). So, even with $\tau_{\mathrm{f}}/\tau_{\mathrm{f1}} = 10...100$, the frequency of false alarms increases 4...7-fold. As shown in Sect. 3.1, such an increase is not too dramatic, and can be compensated by raising the discrimination threshold by as little as 4 to 20%. At the same time, the value of the characteristic frequency for false alarms f_{char} of the two-section and three-section symmetrical filter (5.89) coincides with both f_{a} and f_{n} to within 30%.

Hence it follows that the time constant of the additional section can be selected with considerable amount of freedom, and can be much less than the time constant of the main sections. It is harder for the designer to make the system broad-band enough, whereas the fall-off of the transfer coefficient at high frequencies ($f \approx 2\pi\tau_{\mathrm{f1}}$) may be due to the sluggishness of the integrated circuits, to stray capacitances. Because of this, it is often not necessary to include an additional RC-section in the PDA: as a rule, this issue is resolved simultaneously with the inhibition of parasitic oscillations. Since the time constant of the main sections is much greater than that of the additional section, the characteristic frequencies f_{a}, f_{nw}, f_{nh} are defined by the main sections, and all our calculations of the signal, the noise, and the signal-to-noise ratio remain in force.

Summary. *The loss in SNR in quasi-optimal filter is very small. Such filter works best in case of white noise. The most-simple one-section* RC*-filter loses 11% to the optimal filter in terms of SNR, and the loss is just 3.5% for a two-section filter, so that "quasi" seems to be an insult to its virtues: it practically is no worse than the optimal filter. In case of the excess (low-frequency) and high-frequency noise the loss in SNR is somewhat higher: 20 and 26.5%, respectively.*

5.3. Broad-Band Amplification

Optimal and quasi-optimal filtering makes it necessary to delimit the frequency band: in case of white noise the passband must correspond to the bandwidth of signal spectrum, and in case of high-frequency noise the passband has to be much narrower, so as to block the frequencies at which this noise prevails. The signal at the output of filters with limited passband is always protracted as compared with the input signal. Some optoelectronic systems, however, impose restrictions on the allowable distortion of the signal. First of all, this concerns systems dealing with pulse sequences. One example is the detection of closely located point sources. The protracted relaxation of the signal from the first target may conceal the signal from the nearby second target (especially if the first target is bright, and the second is not). Another example is the lightwave communication system: to cope with the required transmission speed, the relaxation time of the signal pulse must be limited, so that the signal should not be confused with the next-going one. A third example is the output device of the CCD, which must be able to register within the preassigned sampling time each separate pulse from the sequence of charge packets (see Sect. 7.2).

Limitations on the allowable protraction of the output signal may arise also when a single optical pulse is to be detected, if the detection also involves measuring the position of the pulse on the time axis [92]. This problem is considered in greater detail in Sect. 7.2; at this point we just note that the PDAs for such systems are usually designed on the principles of the detection system, and the position of the pulse on the time axis is determined by the instant it sets off the discriminator, Fig. 5.18. The front edge — and hence the duration of the output signal, which is usually related to the slope — have to be delimited, because this affects the accuracy of measurement: when the signal is weak, the discriminator is likely to come into action at the time the signal reaches its maximum, $t' \approx t_m$; when the signal is strong, the discriminator operates at the time the signal arrives, $t'' \approx t''' \approx 0$.

In such cases, as a rule, it becomes necessary to use a broad-band amplifier whose effective passband is wider than that of the optimal (quasi-optimal) filter. This brings us to the theme of the present section: to find the required time constant of RC-section τ_f given the allowable distortion of the output signal (for example, the allowable signal length at half-amplitude T_{05}). The passband of the broad-band amplifier, like that of the quasi-optimal filter, will be formed with the aid of RC-sections. We shall also have to calculate the coefficient of loss in the SNR of such broad-band amplifier as compared to the SNR of the optimal filter. All this can be done on the basis of formulas and diagrams of Sect. 5.2. In the broad-band systems, APDs come to the fore, and once again we have to deal with the problem of optimizing their operating regime, analyzing their advantages over the ordinary PDs. Such analysis has been carried out in Sect. 2.5 with reference to the unit band. Now we have to repeat it for the broad band of the amplifier.

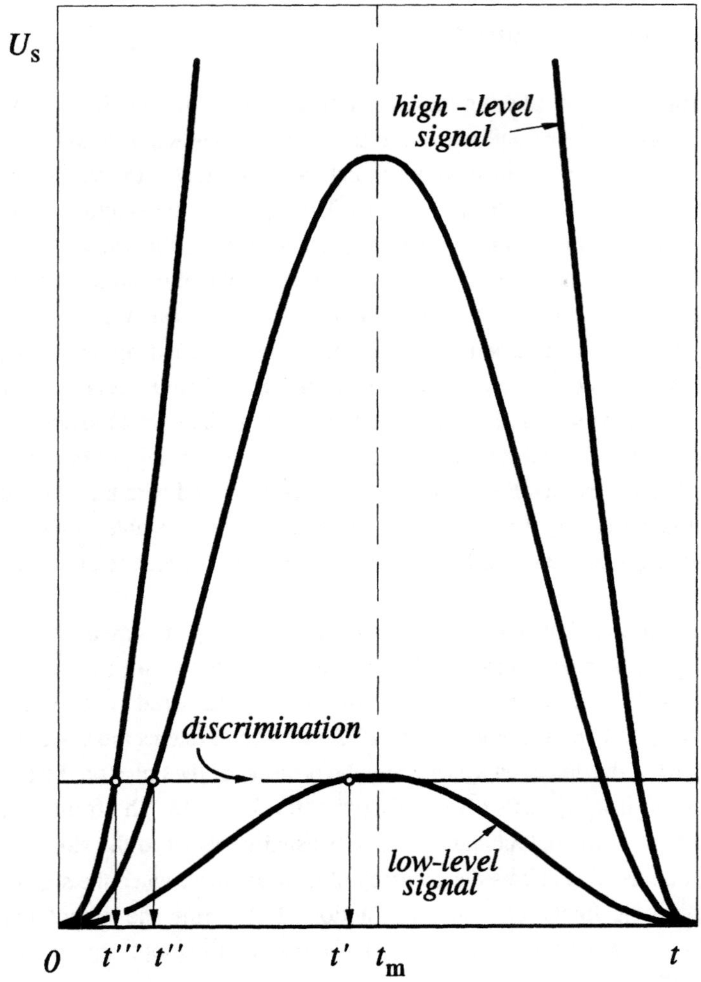

Figure 5.18: Steepness of the front edge affects the precision of measurement of the time of signal arrival.

White noise. The effective time constant of quasi-optimal filter in case of white noise is not too large: it is equal to (or slightly less than) the length of the input signal, and therefore the protraction of the output signal should be moderate. Approximations (5.20), (5.30) allow us to obtain an analytical estimate for the duration of the output signal at half-amplitude:

one-section RC-*filter*:

$$\tau_f = 0.84T_s\ ;\ \ T_{05} = T_s + \tau_f \ln 2 = T_s + 0.84T_s \ln 2 = 1.58T_s\ ; \tag{5.96}$$

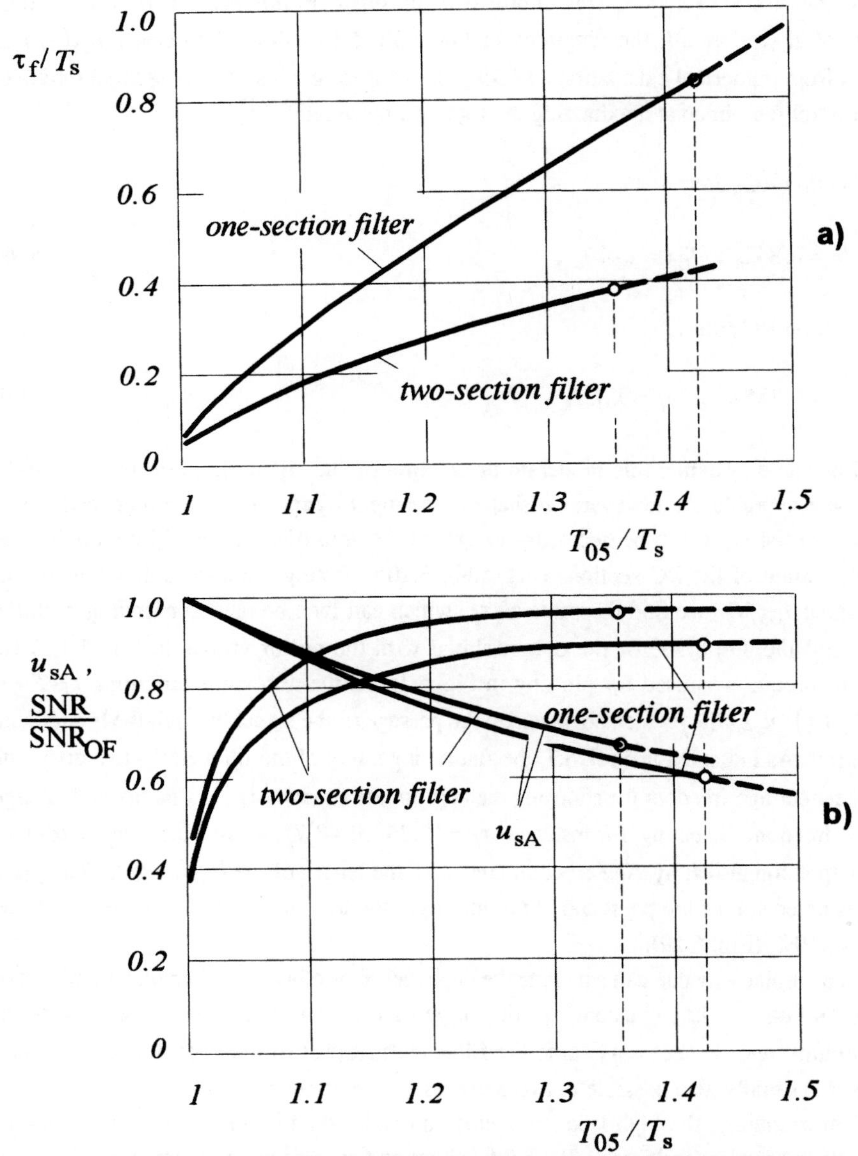

Figure 5.19: (a) Time constant of RC–section of the filter τ_f/T_s is calculated on the basis of allowable signal protraction T_{05}/T_s; (b) the price of the broad-band amplification is the loss in the SNR.

two-section RC-*filter*:

$$\tau_f = \frac{T_s}{e}\ ;\ T_{05} = \sqrt{T_s^2 + 6\tau_f^2} = \sqrt{T_s^2 + \frac{6T_s^2}{e^2}} = 1.34T_s\ . \tag{5.97}$$

Unfortunately, the approximation for the one-section filter has a rather poor accuracy in the neighborhood of $\tau_f = T_s$ (Fig. 5.8b). It is not difficult, however, to find the precise value of T_{05}: after all, the diagrams in Fig. 5.8b, 5.10b show the curves $T_{05}(\tau_f)$ as found from numerical calculations. Making use of these curves, for the optimal tuning of filter-amplifier aimed at maximizing the signal-to-noise ratio we get:

one-section RC-*filter*:

$$\tau_f = 0.84T_s \ ; \ \ T_{05} = 1.42T_s \ ; \tag{5.98}$$

two-section RC-*filter*:

$$\tau_f = 0.385T_s \ ; \ \ T_{05} = 1.35T_s \ . \tag{5.99}$$

For some systems, this moderate protraction of the signal duration (by 30...40%) may be acceptable. If, however, the shape of the input signal has to be reproduced with a higher precision, it will be necessary to extend the amplifier passband by reducing the time constant of the RC-sections (Fig. 5.8b, 5.10b). Fixing the allowable extension of the signal T_{05}/T_s, we find the value of τ_f, which can then be used for finding both the SNR and the amplitude of the output signal with the aid of graphs in Fig. 5.8, 5.10. This technique was used for plotting in Fig. 5.19 all the necessary functions $\tau_f(T_{05})$, $\mathrm{SNR}(T_{05})$, $u_{sA}(T_{05})$ which characterize the passage of the signal through the broad-band amplifier. As indicated in Sect. 5.1, the flat tuning curve of the filter $\mathrm{SNR}(\tau_f)$ gives the engineer enough freedom for choosing the appropriate value of τ_f. Let us take advantage of this freedom. Selecting, for instance, $\tau_f = (0.35...0.48)T_s$, we increase the duration of the output signal just by 10% as compared with the length of the input signal. The price of this extension of the passband is not too high: the loss in the SNR will only increase by 14...10% (Fig. 5.19).

In a similar way one can calculate the dependence of other time-domain parameters of the signal on the time constant τ_f (the slope of front and rear edges, the location of maximum), and, if necessary, tune the filter with respect to these parameters. In such cases it is usually also possible to avoid excessive loss in the value of SNR.

Unfortunately, the high-frequency noise prevails when it comes to the detection of short (laser) pulses (see Sect. 2.3), and the situation becomes not so favorable.

High-frequency noise. Consider the case when the high-frequency noise component dominates in the signal spectrum $0...f_s$: it prevails over the white noise component starting at a low frequency f_{high}. From the beginning we introduce the ratio f_s/f_{high}: it will figure in all the main formulas. For the sake of conspicuity we shall do our numerical estimates with reference to a typical problem: the detection of short laser pulse, $T_s = 10^{-8}...10^{-7}$ s. In accordance with Table 2.2, we have $f_{high} = 2 \cdot 10^2 ... 4 \cdot 10^4$ Hz, and

$$f_s = \frac{1}{2T_s} = \frac{1}{2\cdot\left(10^{-8}...10^{-7}\right)} = 5\cdot\left(10^{6}...10^{7}\right)\text{Hz} \ ,$$

$$\frac{f_s}{f_{high}} = 120...2.5\cdot 10^5 \ . \tag{5.100}$$

The high-frequency noise dominates in the entire signal band, leaving less than one-hundredth of the bandwidth to the white noise!

The desired frequency response of the amplifier will be formed with a two-section RC-filter (one RC section cannot secure filtering of the signal from its mixture with the high-frequency noise, Sect. 5.2). In case of quasi-optimal filtering the passband of such filter had to be restricted to the low-frequency range $0...f_{high}$ where the white noise prevails, which resulted in the considerable protraction of the output signal (5.76). According to (5.76),

$$\frac{T_{05}}{T_s} = \frac{0.39}{T_s f_{high}} = \frac{0.78}{(2T_s)f_{high}} = 0.78\frac{f_s}{f_{high}}$$

$$= 0.78\left(120...2.5\cdot 10^5\right) \approx 100...2\cdot 10^5 \ . \tag{5.101}$$

In (5.101) we already encounter the ratio f_s/f_{high}, which currently indicates that the input signal is protracted practically by the factor of f_s/f_{high}. This means that the signal becomes longer not by 30...40%, as was the case when the signal was separated from the white noise, but by two orders of magnitude, and more! Because of this, in the fast PDAs one has to give up the idea of tuning the filter to the frequency f_{high}, and tune it to the signal band f_s instead. We choose the time constant of the filter $\tau_f = \gamma T_s$ in such a way as to ensure the allowable protraction of the signal $\left[T_{05}/T_s\right]$. Accordingly, the value of γ does not depend on the nature of the noise, and is found from Fig. 5.19a, like in case of the white noise. We can also use the good enough approximation (5.30), by which

$$\left[\frac{T_{05}}{T_s}\right] = \sqrt{\frac{T_s^2}{T_s^2} + \frac{6\tau_f^2}{T_s^2}} = \sqrt{1+6\gamma^2} \ ,$$

$$\gamma = \frac{\tau_f}{T_s} = \sqrt{\frac{\left[\frac{T_{05}}{T_s}\right]^2 - 1}{6}} = \sqrt{\frac{(1.1...1.4)^2 - 1}{6}} \approx 0.2...0.4 \ . \tag{5.102}$$

The numerical value of γ is obtained here for systems which admit the protraction of the signal by 10...40%. The effective passband of the filter with such time constant is

$$f_a = \frac{1}{2\tau_{eff}} = \frac{1}{2e\tau_f} = \frac{1}{2e\gamma T_s} \approx \frac{f_s}{e\gamma} = \frac{f_s}{e(0.2...0.4)} \approx (0.9...1.8) f_s \; . \qquad (5.103)$$

Owing to such a broad band, the relative amplitude of the output signal u_{sA} is sufficiently large. Its value can also be found from Fig. 5.19 or approximation (5.29):

$$U_{sA} = SE_{sA} u_{sA} \; ,$$

$$u_{sA} = \frac{T_s}{\sqrt{T_s^2 + (e\tau_f)^2}} = \frac{1}{\sqrt{1 + (e\gamma)^2}} = \frac{1}{\sqrt{1 + [e(0.2...0.4)]^2}} \approx 0.68...0.9 \; . \qquad (5.104)$$

When the broad-band filter-amplifier is tuned to the signal band, the particulars of the noise spectrum only come up in the calculation of the rms noise value. The broadening of the band results in the dramatic increase of the noise (as compared with the quasi-optimal filtering). From (5.54) and (5.103) we get:

$$U_n = \sqrt{(2\pi C_{in} e_n)^2 f_a^3 \Gamma_2} = \sqrt{\frac{2qI_n}{f_{high}^2} \left(\frac{1}{2e\gamma T_s} \right)^3 \frac{e^3}{4\pi^2}}$$

$$= \frac{\sqrt{2qI_n}}{4\gamma \sqrt{\pi\gamma} \sqrt{2\pi} f_{high} T_s^{3/2}} \; . \qquad (5.105)$$

Here we have used the definition of the characteristic frequency of the high-frequency noise $\left(2\pi f_{high} C_{in} e_n\right)^2 = 2qI_n$, and the value of Γ_2 is taken from Table 5.1. For the ratio of the signal (5.104) to the noise (5.105) we get:

$$\mathrm{SNR} = \frac{SE_{sA} u_{sA}}{U_n} = \left(4\gamma \sqrt{\pi\gamma} f_{high} T_s^{3/2} \right) \frac{SE_{sA} u_{sA}}{\sqrt{2qI_n}}$$

$$= \left\{ 4\gamma \sqrt{\pi\gamma} u_{sA} \sqrt{T_s f_{high}} \right\} \left[\frac{\sqrt{2\pi} SE_{sA} T_s \sqrt{f_{high}}}{\sqrt{2qI_n}} \right] . \qquad (5.106)$$

This formula deserves a most thorough analysis. We notice a new (the fourth in a row) power-law dependence of SNR on the duration of the input signal. In case of optimal (quasi-optimal) filtering and the prevalent low-frequency noise, the value of SNR does not at all depend on the length of the signal, SNR ~ $E_{sA} T_s^0$; SNR ~ $E_{sA}\sqrt{T_s}$ when the white noise prevails, and SNR ~ $E_{sA} T_s$ when the high-frequency noise component is predominant (see Sect. 3.4). Now, when the high-frequency noise prevails, and the passband of the amplifier has to be made broad, we come to the "three-halves power law",

$\mathrm{SNR} \sim E_{\mathrm{sA}} T_{\mathrm{s}}^{3/2}$. This strong dependence is explained by the fact that the spectral density of the noise power itself grows with the frequency as f_{a}^2, the total power in the entire passband of the amplifier is $\overline{U_{\mathrm{n}}^2} \sim f_{\mathrm{a}}^3$, and so the rms noise value is $U_{\mathrm{n}} = \sqrt{\overline{U_{\mathrm{n}}^2}} \sim f_{\mathrm{a}}^{3/2}$. The passband of the broad-band amplifier is tuned to the signal spectrum, $f_{\mathrm{a}} \sim 1/T_{\mathrm{s}}$ (5.103), and therefore the three-halves power law holds also with respect to the signal duration T_{s}.

The nature of dependence $\mathrm{SNR} \sim E_{\mathrm{sA}} T_{\mathrm{s}}^n$ governs the choice of the detection invariant for the PDA. Recall that in Sect. 3.4 we introduced the threshold power $[E_{\mathrm{sA}}]_{\mathrm{t}}$ for the low-frequency noise, the noise equivalent power $\left[E_{\mathrm{sA}}\sqrt{T_{\mathrm{s}}}\right]_{\mathrm{t}}$ for the white noise, and the threshold energy $[E_{\mathrm{sA}} T_{\mathrm{s}}]_{\mathrm{t}}$ for the high-frequency noise. Now we may also introduce the detection invariant V_{DI} as the intensity of optical pulse for which the signal-to-noise ratio is equal to one. From (5.106) it follows that the threshold intensity is a particular quantity $\left[E_{\mathrm{sA}} T_{\mathrm{s}}^{3/2}\right]_{\mathrm{t}}$, which is related to the noise equivalent power in unit frequency band as

$$V_{\mathrm{DI}} = \left[E_{\mathrm{sA}} T_{\mathrm{s}}^{3/2}\right]_{\mathrm{t}} = \frac{\sqrt{2qI_{\mathrm{n}}}}{aSf_{\mathrm{high}}} = \frac{\mathrm{NEP}}{af_{\mathrm{high}}} , \tag{5.107}$$

where

$$a = 4\pi\gamma\sqrt{2\gamma}u_{\mathrm{sA}} .$$

As indicated in Sect. 3.4, in laser-based systems the pumping energy is usually limited, so that the energy of the signal is $E_{\mathrm{sA}} T_{\mathrm{s}} = \mathrm{const}$. Then, in the classical context of white noise and optimal filtering, it is very profitable to reduce the length of the signal and thus improve the detection properties (and not only the time resolution), since

$$\mathrm{SNR} \sim E_{\mathrm{sA}}\sqrt{T_{\mathrm{s}}} = \frac{E_{\mathrm{sA}} T_{\mathrm{s}}}{\sqrt{T_{\mathrm{s}}}} = \frac{\mathrm{const}}{\sqrt{T_{\mathrm{s}}}} . \tag{5.108}$$

In case of the prevalent high-frequency noise and the constant energy of the signal, the length of the signal does not affect the signal-to-noise ratio, $\mathrm{SNR} \sim E_{\mathrm{sA}} T_{\mathrm{s}} = \mathrm{const}$. The three-halves law completely spoils the game: now, with the high-frequency noise and a broad-band amplifier, the shortening of the pulse (under the condition of constant energy) instead of facilitating the detection makes it difficult:

$$\mathrm{SNR} \sim E_{\mathrm{sA}} T_{\mathrm{s}}^{3/2} = E_{\mathrm{sA}} T_{\mathrm{s}}\sqrt{T_{\mathrm{s}}} = \mathrm{const}\sqrt{T_{\mathrm{s}}} . \tag{5.109}$$

If the designers of optoelectronic system stay on the classical track, and expect to reduce the threshold energy by shortening the optical pulse, they are going to be disappointed: for the fast action of the system they will have to pay with the increasing threshold energy of the pulse. The inexorable three-halves law is also a letdown for the engineers of multielement systems. Increasing the number of elements in the PDA with the aim of improving the detection, one may run into a quandary: when the number of elements is very large, the sampling time T_s has to be made small, and the threshold energy (the threshold charge) starts to increase when the noise of the output broad-band amplifier begins to dominate. We shall meet with this effect in Sect. 7.2.

Now is the time to pay for the fast action of our amplifier. The expression (5.106) for the SNR is represented as the product of two terms with a purpose. The term in brackets is the signal-to-noise ratio of the optimal filter $\mathrm{SNR_{OF}}$ (see (3.81)). Accordingly, the other term in braces directly calls the price we have to pay: it is the coefficient of loss in the signal-to-noise ratio of the broad-band filter-amplifier as compared to that of the optimal filter:

$$\mathcal{L} = \frac{\mathrm{SNR}}{\mathrm{SNR_{OF}}} = 2\gamma\sqrt{2\pi\gamma u_{sA}}\sqrt{2T_s f_{\mathrm{high}}} \approx (0.4\ldots1.2)\sqrt{\frac{f_{\mathrm{high}}}{f_s}} \approx 10^{-3}\ldots10^{-1} \; . \quad (5.110)$$

One can hardly put up with such deterioration if the detective properties of the system are important. Because of this, one of the main tasks of the broad-band PDA engineers is to combat the high-frequency noise (by making f_{high} as high as possible). The ways of doing this have been discussed in Sect. 2.5; the main method of suppression of the high-frequency noise of the amplifier consists, however, in using the avalanche multiplication, which will be considered in the next chapter.

Summary. *In case of optimal (quasi-optimal) filtering the passband is always limited, which results in the protraction of the output signal. In some optoelectronic systems this cannot be tolerated, and the passband has therefore to be made broader than that of the optimal filter. True, in case of white noise and optimal (quasi-optimal) filtering the protraction of the signal is moderate (about* 30...40%*), and by increasing the band 2...4–fold it is possible, as a rule, to secure the desired fast action without compromising the SNR too much (*10...15%*). When the high-frequency noise of the amplifier is strong (such noise begins to dominate over the white noise from the frequency* f_{high}*), however, the passband of the optimal (quasi-optimal) filter* $0\ldots f_{\mathrm{high}}$ *is much narrower than the band of the signal* $0\ldots f_s$*. This results in the protraction of the signal by the factor of* f_s/f_{high}*, which usually amounts to several orders of magnitude. Extension of the passband of the amplifier to match the signal band* f_s *ensures the required fast action, but the added contribution from the high-frequency noise spoils the SNR, which is reduced by the factor of about* f_s/f_{high}*.*

Chapter 6

DETECTION OF WEAK OPTICAL SIGNALS WITH PHOTODETECTING ASSEMBLIES BASED ON AVALANCHE PHOTODIODES

In the very beginning of the book (Sect. 1.2) we observed that the internal gain of photodetector is required to "pull" the signal and noise of photodetector itself out of the noise of amplifier. In Sect. 2.5 we calculated the optimal multiplication of avalanche photodiode for the most simple case, when the spectral density of the noise is fixed. Now we are going to consider the general case of detection of pulsed signal, when the frequency-dependent high-frequency noise component is noticeable. We shall see that there is no alternative for the avalanche photodiode in the detection of short pulsed signals in threshold systems (Sect. 6.1). In some cases, the avalanche photodiode is able to detect an optical signal so weak that one has to take into account the fluctuation of the number of quanta in this signal (Sect. 6.2). In a special mode (photon count mode) the avalanche photodiode is capable of detecting single photons, approaching the fundamental quantum detection threshold (Sect. 6.3)

6.1. Broad-Band Amplification and Quasi-Optimal Filtering in Photodetecting Assemblies Based on Avalanche Photodiodes

Avalanche multiplication in case of broad-band amplification. First of all, let us evaluate the limits of capability of the APD: what is the possible gain in the SNR of a broad-band PDA if we replace an ordinary PD with the avalanche photodiode. For the sake of clarity we return to "rectangular reasoning", assuming that the bands of the signal and the broad-band amplifier are rectangular, and $f_a = f_s = 1/2T_s$ (Sect. 3.3). Then $u_{sA} = 1$, and at the output of the PDA with an ordinary PD (without multiplication) we have

$$U_s = SE_{sA} ,$$

$$U_n = \sqrt{2qI_0 f_a + \int_0^{f_s} 2qI_0 \frac{f^2}{f_{high}^2}\, df} \approx \sqrt{2qI_0 \frac{f_s^3}{3f_{high}^2}} , \tag{6.1}$$

and so the signal-to-noise ratio is

$$\mathrm{SNR}_{\mathrm{PDA-PD}} = \frac{SE_{\mathrm{sA}}}{\sqrt{2qI_0 \dfrac{f_{\mathrm{s}}^3}{3f_{\mathrm{high}}^2}}} \ . \tag{6.2}$$

Now let us calculate the SNR for the APD-based PDA. We idealize the model: the avalanche does not make any noise ($\alpha = 0$, Sect. 2.3), the multiplication is not limited; we managed to eliminate thermal noise and all leakage currents, so that the white-noise component only depends on the multiplied current of the photodetector. In this idealized PDA with a perfect APD the signal and the noise of the photodetector are multiplied by the factor of M, when the multiplication factor is large enough, the noise of the photodetector will become larger than the noise of the amplifier:

$$U_{\mathrm{s}} = MSE_{\mathrm{sA}} \ ,$$
$$U_{\mathrm{n}} = \sqrt{2qI_0M^2f_{\mathrm{s}} + 2qI_0\frac{f_{\mathrm{s}}}{3f_{\mathrm{high}}^2}} \approx M\sqrt{2qI_0f_{\mathrm{s}}} \ , \tag{6.3}$$

and the signal-to-noise ratio at the output of the PDA becomes

$$\mathrm{SNR}_{\mathrm{PDA-APD}} = \frac{MSE_{\mathrm{sA}}}{M\sqrt{2qI_0f_{\mathrm{s}}}} = \frac{SE_{\mathrm{sA}}}{\sqrt{2qI_0f_{\mathrm{s}}}} \ . \tag{6.4}$$

As a matter of fact, equations (6.2) and (6.4) both define the signal-to-noise ratio for PDA based on the ordinary PD; the difference is that the first of these takes care of the high-frequency noise of the amplifier, whereas the second includes only the noise of the PD itself. This clearly demonstrates the advantages of the avalanche multiplication: it allows "dragging out" both the signal and the noise of the photodetector from the noise of the amplifier (we discussed this in Sect. 1.2 and 2.5).

So, the PDA does not stand up to its task: it has failed to realize the intrinsic signal-to-noise ratio of the photodetector, spoiling it with the noise of its amplifier. APD came to the rescue: the ideal avalanche completely suppresses the noise of the amplifier. This gives a large gain in the SNR:

$$\frac{\mathrm{SNR}_{\mathrm{PDA-APD}}}{\mathrm{SNR}_{\mathrm{PDA-PD}}} = \frac{\sqrt{2qI_0\dfrac{f_{\mathrm{s}}^3}{3f_{\mathrm{high}}^2}}}{\sqrt{2qI_0f_{\mathrm{s}}}} = \frac{f_{\mathrm{s}}}{\sqrt{3}f_{\mathrm{high}}} = 0.58\frac{f_{\mathrm{s}}}{f_{\mathrm{high}}} \ . \tag{6.5}$$

Once again, the ratio $f_{\mathrm{s}}/f_{\mathrm{high}}$ comes up in (6.5). Perfect multiplication allows improving the signal-to-noise ratio by the factor equal to the ratio of the frequency f_{s} to the frequency f_{high} (up to a numerical coefficient). There is no need to explain the origin of such dependence: all is clear from (6.2), (6.4). The coefficient in (6.5) depends on the shape of the frequency response of the amplifier. For the two-section RC-filter of interest, from (6.5) and Table 5.1 we find:

$$\frac{\mathrm{SNR}_{\mathrm{PDA-APD}}}{\mathrm{SNR}_{\mathrm{PDA-PD}}} = \frac{\sqrt{2qI_0\Gamma_2 \dfrac{f_a^3}{f_{high}^2}}}{\sqrt{2qI_0\Gamma_1 f_a}} = \sqrt{\frac{\Gamma_2}{\Gamma_1}\frac{f_a}{f_{high}}}$$

$$= \sqrt{\frac{e^2}{\pi^2}\frac{f_s}{e\gamma f_{high}}} = \frac{f_s}{\pi\gamma f_{high}} = \frac{f_s}{\pi\cdot(0.2...0.4)f_{high}}$$

$$= (0.8...1.6)\frac{f_s}{f_{high}} \approx 100...4\cdot 10^5 \ . \qquad (6.6)$$

The avalanche photodiode is so much superior to the ordinary PD that there is really no choice: it is only the APD that must be used in the threshold-level broad-band systems.

Unfortunately, in real APDs the noise grows faster than the signal as the multiplication increases; because of this, further increase of the multiplication becomes unfavorable when the noise of the APD exceeds considerably the noise of the amplifier: the signal-to-noise ratio will only decline. The optimal operating regime (the multiplication coefficient M) is selected with due account for this feature of the APD (see Sect. 2.5). The change from the narrow band to the broad band does not affect the technique used for establishing the operating regime: the multiplication of the APD is increased by adjusting the bias, until the noise power of the APD becomes greater than the initial noise power (the noise power of the amplifier) by a certain factor n, where $n = 1...3$. The multiplication factor in the optimal regime is calculated from this condition. First we rewrite the expression for the noise (6.3), taking into account the noise of the APD: we introduce the noise coefficient $æ$ (Sect. 1.2), and, for the sake of generality, include the thermal and the shot noise components of the noise of the amplifier:

$$U_n = \sqrt{2qI_0 M^{2+æ} f_a\Gamma_1 + \overline{U_{na}^2}} \ ,$$

$$\overline{U_{na}^2} = 2qI_a f_a\Gamma_1 + \frac{4\kappa T}{R} f_a\Gamma_1 + \frac{2qI_n}{f_{high}^2} f_a^3\Gamma_2 \ , \qquad (6.7)$$

whence

$$2qI_0M_{\text{qua}}^{2+\text{æ}}f_a\Gamma_1 = n\overline{U_{\text{na}}^2} \ . \tag{6.8}$$

The subscript at the multiplication factor reminds us that the above-described method leads to the quasi-optimal operating regime of the APD. The technique for establishing the optimal regime is much more complicated, whereas the possible gain in the SNR is inconsiderable (see Sect. 2.5).

With due account for (6.7), (6.8), the expression for the signal-to-noise ratio takes on the form

$$\text{SNR}_{\text{PDA-APD}} = \frac{M_{\text{qua}}SE_{\text{sA}}u_{\text{sA}}}{\sqrt{2qI_0M_{\text{qua}}^{2+\text{æ}}f_a\Gamma_1\left(1+\dfrac{1}{n}\right)}} = \frac{SE_{\text{sA}}u_{\text{sA}}}{\sqrt{2qI_0M_{\text{qua}}^{\text{æ}}f_a\Gamma_1\left(1+\dfrac{1}{n}\right)}} \ . \tag{6.9}$$

As always, the reality does not stand up to the ideal (compare (6.4) and (6.9)): the additional noise of the APD gives rise to the extra factor of $M_{\text{qua}}^{\text{æ}}$, and, as a consequence, to the incomplete suppression of the noise of the amplifier (factor $M_{\text{qua}}^{\text{æ}/2}$). Accordingly, the gain in the SNR due to replacing the PD with the APD will be less by the factor of $\sqrt{M_{\text{qua}}^{\text{æ}}(1+1/n)}$ compared to (6.6) for the idealized APD. To obtain a numerical estimate, we must first evaluate the coefficient M_{qua}. From (6.7), (6.8), Table 5.1 with $\text{æ} = 0.2$, $n = 2$, $I_a \to 0$, $R_L \to \infty$, we find that, theoretically, this coefficient must be equal to

$$M_{\text{qua}} = \left[\frac{n2qI_0f_a^3\Gamma_2}{f_{\text{high}}^2\,2qI_0f_a\Gamma_1}\right]^{\frac{1}{2+\text{æ}}} = \left(\frac{n\Gamma_2}{\Gamma_1}\right)^{\frac{1}{2+\text{æ}}}\left(\frac{f_a}{f_{\text{high}}}\right)^{\frac{2}{2+\text{æ}}} \approx 80\ldots1.5\cdot10^5 \ . \tag{6.10}$$

Here we have to acknowledge yet another deviation of real APDs from the ideal: their multiplication factor is limited (in silicon-based APDs, as a rule, $M_{\max} \approx 100\ldots200$, see Sect. 1.2), which is the most annoying inconvenience. With such values of $M_{\max}$, æ, n, the correction factor is $\sqrt{M_{\max}^{\text{æ}}(1+1/n)} \approx 2$. Accordingly, when a reasonably low multiplication is required ($M_{\text{qua}} \le 200$), the SNR of the real silicon APDs is one-half that of the ideal noiseless APD. If, however, the high-frequency noise is very strong, the real APD will not be able to secure sufficient multiplication to suppress this noise (being unable to reach $M_{\text{qua}} > 200$). In this case the replacement of the PD with the APD simply increases the signal by the factor of $M_{\max}$ against the background of the constant noise of the amplifier.

Avalanche multiplication in case of optimal (quasi-optimal) filtering. While in case of broad-band amplification the passband of the filter is strictly fixed and is selected from considerations of the allowable distortion of the signal profile, in case of optimal and quasi-optimal filtering the passband is governed by the frequency limit f_{high}. As the multiplication factor increases, this frequency goes up, because the multiplied shot noise of the photodetector "jams" the high-frequency noise up to the higher frequencies:

$$I_{\text{eq}} = I_0 M^{2+æ} ,$$

$$f_{\text{high}} = \frac{\sqrt{2qI_0M^{2+æ}}}{2\pi C_{\text{in}} e_{\text{n}}} = f_{\text{h}0} M^{1+æ/2} , \tag{6.11}$$

$$f_{\text{h}0} = \frac{\sqrt{2qI_0}}{2\pi C_{\text{in}} e_{\text{n}}} .$$

This difference in the selection of the passband of the filter is responsible for the fact that the avalanche plays different roles in these two modes of operation. We have just demonstrated that in case of the broad-band amplifier the avalanche multiplication is capable of improving the SNR by one or two orders of magnitude, because the value of SNR without multiplication is dramatically low owing to the large contribution of the high-frequency noise power. In this respect, the merits of avalanche multiplication in case of optimal filtering are quite modest. To some extent the filter itself eliminates the high-frequency noise by suppressing the higher frequencies where this noise prevails, thus preventing a drastic fall in the SNR like that observed with the fixed broad band amplification.

Let us repeat this in the language of formulas. The method of choosing the quasi-optimal values of M_{qua} remains the same. As long as the multiplication factor is less than the quasi-optimal value, and the high-frequency noise prevails, expressions (3.81) and (5.74) hold; so, after some manipulations, and with due account for (6.11), we get:

$$\text{SNR} = \frac{aMSE_{\text{sA}}T_{\text{s}}}{\sqrt{2qI_{\text{eq}}}}\sqrt{f_{\text{high}}} = \frac{aMSE_{\text{sA}}T_{\text{s}}}{\sqrt{2qI_0M^{2+æ}}}\sqrt{\frac{\sqrt{2qI_0M^{2+æ}}}{2\pi C_{\text{in}} e_{\text{n}}}}$$

$$= \frac{aSE_{\text{sA}}T_{\text{s}}M^{0.5-æ/4}}{\sqrt{2\pi C_{\text{in}} e_{\text{n}}\sqrt{2qI_0}}} , \tag{6.12}$$

where $a = \sqrt{2\pi}$ for optimal filtering, and $a = 2\sqrt{2\pi}/e$ for quasi-optimal filtering. While in case of the broad-band amplifier the SNR increases proportionally to the coefficient M, the SNR in case of optimal filtering increases much slower — as $\sqrt[4]{M}...\sqrt{M}$. When the quasi-optimal value of the multiplication factor is reached, the noise "whitens out", and for the SNR we get the same (or almost the same) ultimate values as in case of the broad-band amplification (6.4), (6.9). Now what is the maximum gain in the SNR that can be obtained through the avalanche multiplication? As before, let us evaluate this gain using the rectangular approximations for the signal and the noise. While under the assumption of ideal avalanche the maximum values of SNR for the broad-band and the optimal filters turned out to be almost the same, the initial values of SNR (without multiplication) differ considerably. Recall Sect. 3.3: the spectral density of the signal in the optimal frequency band $0 \ldots f_{\mathrm{high}}$ is $2E_{\mathrm{sA}}T_{\mathrm{s}}$, and that of the noise power is $2qI_0$, so that the signal, the noise, and their ratio in the entire band are, respectively:

$$U_{\mathrm{s}} = 2SE_{\mathrm{sA}}T_{\mathrm{s}}f_{\mathrm{high}}\,, \quad U_{\mathrm{n}} = \sqrt{2qI_0 f_{\mathrm{high}}}\,,$$

$$\mathrm{SNR}_{\mathrm{PDA-PD}} = \frac{2SE_{\mathrm{sA}}T_{\mathrm{s}}f_{\mathrm{high}}}{\sqrt{2qI_0 f_{\mathrm{high}}}} = \frac{2SE_{\mathrm{sA}}T_{\mathrm{s}}\sqrt{f_{\mathrm{high}}}}{\sqrt{2qI_0}}\,. \tag{6.13}$$

This initial value of $\mathrm{SNR}_{\mathrm{PDA-PD}}$ is much better than that in case of the broad-band amplification (6.2). Because of this, the gain achieved through replacing the PD with the APD is not so dramatic: dividing (6.4) into (6.13), and replacing $1/2T_{\mathrm{s}} = f_{\mathrm{s}}$, we get:

$$\frac{\mathrm{SNR}_{\mathrm{PDA-APD}}}{\mathrm{SNR}_{\mathrm{PDA-PD}}} = \frac{SE_{\mathrm{sA}}}{\sqrt{2qI_0 f_{\mathrm{s}}}} \cdot \frac{\sqrt{2qI_0}\,f_{\mathrm{s}}}{SE_{\mathrm{sA}}\sqrt{f_{\mathrm{high}}}} = \sqrt{\frac{f_{\mathrm{s}}}{f_{\mathrm{high}}}}\,. \tag{6.14}$$

As before, the gain is expressed in terms of the ratio $f_{\mathrm{s}}/f_{\mathrm{high}}$, but this time it is proportional to the square root of this ratio rather than to its first power as it was in case of the broad-band amplification (6.5).

The tuning of the filter (extension of its passband $0 \ldots f_{\mathrm{high}}$ as the multiplication factor increases) results also in the reduced duration of the output signal. When the multiplication factor is less than quasi-optimal, and the high-frequency noise prevails, the length of the signal according to (5.101), (6.11) decreases in inverse proportion to the multiplication factor, or even slightly faster:

$$T_{05} = 0.78\frac{f_{\mathrm{s}}}{f_{\mathrm{high}}} = 0.78\frac{f_{\mathrm{s}}}{f_{\mathrm{h0}}M^{1+\varkappa/2}}\,. \tag{6.15}$$

When the noise eventually "whitens out", the length of the output signal is just 28% greater than that of the input signal (Sect. 4.2 and 5.1). It follows that the avalanche straightens out the drawbacks displayed by the PD-based PDA (the extremely poor SNR in case of the broad-band amplification, and the sluggishness and inferior SNR in case of optimal filtering).

Summary. *Ideal avalanche multiplication suppresses the high-frequency noise completely, thus making up for the drawbacks of the PD-based PDA, and allows realizing the intrinsic SNR of the photodetector. In case of broad-band amplification the loss in the SNR is compensated, and this ratio is improved by the same factor of*

f_s/f_{high}. *In case of optimal filtering the protraction of the signal by the factor of* f_s/f_{high} *is eliminated, and the SNR is improved by the factor of approximately* $\sqrt{f_s/f_{high}}$ *(this gain is less than in case of the broad-band amplification, when the value of SNR without multiplication showed an especially dramatic fall).*

In real APDs there is the noise of the avalanche, the multiplication factor is limited, and the high-frequency noise of the amplifier cannot be completely suppressed. Nevertheless, it is possible to improve the signal-to-noise ratio by one or two orders of magnitude by using APDs instead of PDs.

6.2. Quantization and Fluctuation of Signal

Every designer dreams of reducing the noises of PDA to a minimum. Let this dream come true, and there is nothing to limit the opportunities: neither the hardware requirements, nor the technologies. Then it is theoretically possible to make the noise as low as desired by making the photosensitive site arbitrarily small, reducing the background and the temperature, increasing the input impedance and the transconductance of the transistor (see Sect. 2.5). Accordingly, the magnitude of the lowest detectable signal, as calculated by the formulas of Sect. 3.3 – 6.1, will also be arbitrarily small. What will then delimit the lowest detectable optical power? The fundamental restriction is imposed by the quantum nature of the radiation: the optical pulse may not contain less than one photon, and the energy of the pulse $E_{sA}T_s$ cannot be less than the energy of this photon $h\nu$. The ideal detector-counter converts one photon into one electron, so that the minimum charge carried by the photocurrent signal pulse $(SE_{sA}T_s)_{min}$ is equal to the charge of electron q. Using the expression for the sensitivity (1.18), we get:

$$(SE_{sA}T_s)_{min} = q\ ,$$
$$(E_{sA}T_s)_{min} = h\nu = \frac{hc}{\lambda} = \frac{q}{S} = 1.24\frac{q}{\lambda} = \frac{2\cdot 10^{-19}}{\lambda}\ , \tag{6.16}$$

where $E_{sA}T_s$ is measured in joules, and λ in micrometers.

So, the lowest detectable energy of the pulse $(E_{sA}T_s)_{min}$ is theoretically delimited not by the properties of the PDA, but by the quantum nature of the radiation itself; for the spectral range 0.5...14 μm this quantum limit is $(4\ldots0.15)\cdot10^{-19}$ J. Of course, in the majority of real PDAs the lowest detectable energy is far from the quantum limit: the requirements imposed by the optoelectronic system do not allow reducing the area of the photosensitive site, the background radiation, the temperature; limited is the performance of electronic components, the electric power dissipation — and hence the current, the transconductance, and the noise emf e_n (Sect. 2.5). Because of this, it is not possible to reduce without limit the noise level of the PDA. Nevertheless, PDAs based on silicon APDs, which by avalanche multiplication suppress the amplifier noise, while the noise

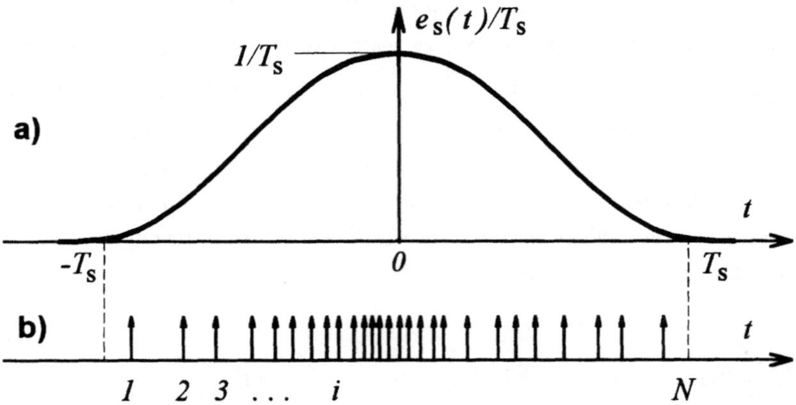

Figure 6.1: The relative profile of optical pulse defines the density of probability $e_s(t)/T_s$ for the time of arrival of the *i*th photon t_i.

of the APD itself is very low, come close to the quantum limit and are capable of detecting if not a single quantum of radiation, then several (or several dozen) quanta. A similar sensitivity can be obtained with specially designed silicon CCDs. Let us consider the peculiarities of registration of such small number of quanta.

Quantization of the signal. The signals as low as that can no longer be described by the deterministic function $E_s(t)$ — the precisely known time-dependent optical power (Fig. 6.1a). Now we must take into account the fact that the optical signal consists of individual photons:

$$E_s(t) = \mathrm{h}\,\nu\left[\delta(t-t_1)+\delta(t-t_2)+...+\delta(t-t_N)\right] . \tag{6.17}$$

Such signal is shown in Fig. 6.1b. Because of the statistical nature of the sources of radiation, and the losses on the way from the source to the photodetector, the optical signal is a random process: random are the times of arrival of each of the photons $t_1, t_2, ..., t_N$, and the number of photons in the pulse N. As with any random process, the parameters of each particular pulse (each realization) cannot be known in advance. The designer of the optoelectronic system only knows the statistical parameters of the signal. These are essentially the same characteristics as those used for describing the deterministic signal: the relative profile $e_s(t)$, the signal duration T_s, the energy (power) of the signal [91]. In place of the energy of particular pulse (particular realization) $E_{sA}T_s$, which is not known a priori, we must introduce the mean energy $\overline{E_{sA}T_s}$. It is convenient to describe the pulses in terms of the number of quanta N; the energy of quantum being $\mathrm{h}\,\nu$, we get:

$$N = \frac{E_{sA}T_s}{\mathrm{h}\,\nu} , \quad \overline{N} = \frac{\overline{E_{sA}T_s}}{\mathrm{h}\,\nu} . \tag{6.18}$$

Since $\overline{E_{sA}T_s}$ is now the mean energy, it may also fall below the quantum limit (when the mean number of quanta in the pulse is less than one, $\overline{N} < 1$). Of course, each particular realization contains a whole number of photons, $N = 0,1,2,3,\ldots$. If, however, the pulses with $N = 0$ prevail, and the pulses with $N = 1$ are not likely to occur (the probability of their occurrence is $p(1) << 1$), whereas the realizations with $N \geq 2$ are even less probable, then we come to the situation when $\overline{N} < 1$:

$$\overline{N} = \sum_{N=0}^{\infty} Np(N) \approx 0 \cdot p(0) + 1 \cdot p(1) \approx p(1) << 1 \; . \tag{6.19}$$

The relative shape of the optical signal $e_s(t)$ also acquires statistical meaning. The value of $e_s(t)dt/T_s$ is nothing else but the probability that any one of the N incident photons arrives within the interval $[t, t + dt]$. The total probability to find this photon at any time t must be equal to one. Using the expression for the effective duration of the signal (2.4) we may write this necessary condition as

$$\frac{1}{T_s}\int_{-\infty}^{\infty} e_s(t)dt = \frac{1}{T_s}T_s = 1 \; . \tag{6.20}$$

With the aid of the thus defined density of probability $e_s(t)/T_s$ for the times of arrival of photons it can be demonstrated that in the limit of large number of photons N we return to the familiar description of the signal in terms of time-dependent optical power, $E_s(t) = E_{sA}e_s(t)$. Indeed, if each of N photons falls into the interval $[t, t + dt]$ with the probability $e_s(t)dt/T_s$, then, on the average, for the pulses consisting of N photons, the number of photons falling into this interval is $dN = Ne_s(t)dt/T_s$, and their combined energy is $h\nu\, dN$. Then the optical power at the time t is

$$E_s(t) = h\nu\frac{dN}{dt} = h\nu\frac{Ne_s(t)}{T_s} = E_{sA}T_s\frac{e_s(t)}{T_s} = E_{sA}e_s(t) \; , \tag{6.21}$$

where the replacement $h\nu N \Rightarrow E_{sA}T_s$ has been made in accordance with (6.18). The random deviations of N from its mean value $\overline{N}$ (and hence the fluctuation of E_{sA} in (6.21)) are characterized by the dispersion $\overline{\Delta N^2} = \overline{\left(N - \overline{N}\right)^2}$. For a Poisson process we have $\overline{\Delta N^2} = \overline{N}$ [55, 56]; accordingly, the relative fluctuations $\sqrt{\overline{\Delta N^2}}/\overline{N} = 1/\sqrt{\overline{N}}$ become smaller and smaller as $\overline{N}$ increases, and the realizations $E_{sA}T_s$, E_{sA} differ less and less from their mean values $\overline{E_{sA}T_s}$, $\overline{E_{sA}}$, and the amplitude E_{sA} in (6.21) becomes a practically deterministic quantity, $E_{sA} \approx \overline{E_{sA}}$. It is this circumstance that allowed us to drop the averaging bar in all the preceding sections, assuming that $\overline{E_{sA}} \approx E_{sA}$.

Detection of random signal. So far we have been considering signals whose distinguishing features were all strictly defined, and all calculations have been carried out with reference to these predetermined features (characteristics). Now we see that the signal is random, the parameters of each pulse falling on the PDA are not known, and known are only the statistical characteristics. Because of this, the error of detection (the target miss probability P_{tm}) can only be calculated in terms of the statistical characteristics, the mean signal amplitude $\overline{E_{sA}}$. Since the relative deviations of the amplitude of each separate pulse from their mean values become negligibly small when N is large, the calculated value of P_{tm} remains valid for practically each pulse realization. When $\overline{N}$ is small, however, the pulses with a small number of quanta N (as compared with the mean number $\overline{N}$) occur more frequently, and of course the probability of missing the pulse with $N < \overline{N}$ increases. In this way, the fluctuation of the signal gives rise to additional error: the target miss probability increases (as compared with that calculated for the mean amplitude).

Let us first consider a "pure" situation: the photons of the signal are counted without loss, and no additional noise is introduced, so that the target miss is only caused by the fluctuation of the signal. Consider a sequence of pulses with the mean number of photons $\overline{N}$. The random number of photons N in each separate pulse obeys the Poisson statistics, and therefore the probability of realization of the pulse with N photons is [56]:

$$p(N) = \frac{e^{-\overline{N}}\,\overline{N}^{N}}{N!} \; . \tag{6.22}$$

The detector may decide that the signal is there if it receives at least one quantum. If the pulse realization contains no photons at all ($N = 0$), the decision is "no signal". Accordingly, the target miss probability is equal to the probability of realization of the pulse with $N = 0$. Substituting this into (6.22), we get

$$P_{tm} = p(0) = \exp\left(-\overline{N}\right) \; . \tag{6.23}$$

From (6.23) it follows that the target miss probability in the one-quantum limit $\overline{N} = 1$ is $P_{tm} = \exp(-1) = 0.37$, and that it tends to unity when $\overline{N} << 1$, as discussed above: $P_{tm} = \exp\left(-\overline{N}\right) \approx 1 - \overline{N} \approx 1$. Such a high probability of error, as a rule, cannot be tolerated in optoelectronic systems.

Expression (6.23) also permits solving the inverse problem: finding the value of $\left[\overline{N}\right]$ which secures the desired probability P_{tm}:

$$\left[\overline{N}\right] = \ln\frac{1}{P_{tm}} = \ln\frac{1}{0.5\cdot\left(10^{-3}\ldots10^{-9}\right)} = 2.3\log\left[2\cdot\left(10^{3}\ldots10^{9}\right)\right] \approx 7\ldots21 \; . \tag{6.24}$$

The numerical estimate here corresponds to the typical values of P_{tm} (see Sect. 3.1). The optical energy of such pulses is, according to (6.28),

$$\left(\overline{E_{sA}T_s}\right)_{min} = h\,\nu\overline{N} = \frac{2\cdot 10^{-19}\left[\overline{N}\right]}{\lambda}$$

$$= \frac{2\cdot 10^{-19}\cdot(7...21)}{0.5...14} \approx (1...80)\cdot 10^{-19}\ \mathrm{J}\,. \qquad (6.25)$$

We see that the lowest detectable optical energy of the signal (the mean value) is by the factor of 7...20 higher than the quantum limit; it is only then that the probability of detection satisfies the requirements to the system.

The ratio of the signal to the noise (of the signal itself) is

$$\mathrm{SNR} = \frac{\left[\overline{N}\right]}{\sqrt{\overline{\Delta N^2}}} = \frac{\left[\overline{N}\right]}{\sqrt{\overline{N}}} = \sqrt{\overline{N}} = \sqrt{7...21} \approx 2.6...4.6\,. \qquad (6.26)$$

From the standpoint of detection the Poisson distribution turns out to be better than Gaussian. In the hypothetical case of fluctuations obeying the Gaussian law, the probability P_{tm} should have been calculated by expressions similar to (3.10), (3.11):

$$P_{tm} = \frac{1}{2}\left\{1 - \Phi\left[\frac{\overline{N} - N_t}{\sqrt{2\overline{\Delta N^2}}}\right]\right\} \approx \frac{\sqrt{\overline{\Delta N^2}}\exp\left[-\frac{\left(\overline{N} - N_t\right)^2}{2\overline{\Delta N^2}}\right]}{\sqrt{2\pi}\left(\overline{N} - N_t\right)}\,. \qquad (6.27)$$

Here the probability P_{tm} is determined primarily by the strong exponential term. Setting $N_t = 1$, $\overline{\Delta N^2} = \overline{N}$, we get the following order-of-magnitude estimate:

$$P_{tm} \sim \exp\left[-\frac{\left(\overline{N} - 1\right)^2}{2\overline{N}}\right] \approx \exp\left(-\frac{\overline{N}}{2}\right)\,. \qquad (6.28)$$

The comparison between (6.23) and (6.28) reveals that in the hypothetical case of the Gaussian fluctuations the desired value of the probability P_{tm} would have been only secured by the signal twice as strong as that required in the real case of the Poisson distribution with respect to the number of quanta N. The calculation according to (6.26) slightly corrects this coefficient, which is found to be equal to 1.7...1.8 for the accepted value of $P_{tm} = 0.5\cdot\left(10^{-3}...10^{-9}\right)$.

So, we have defined the lowest (mean) energy of the signal (6.24), (6.25) as delimited by the signal itself (the fluctuations of the signal). Now is the time for the PDA to come

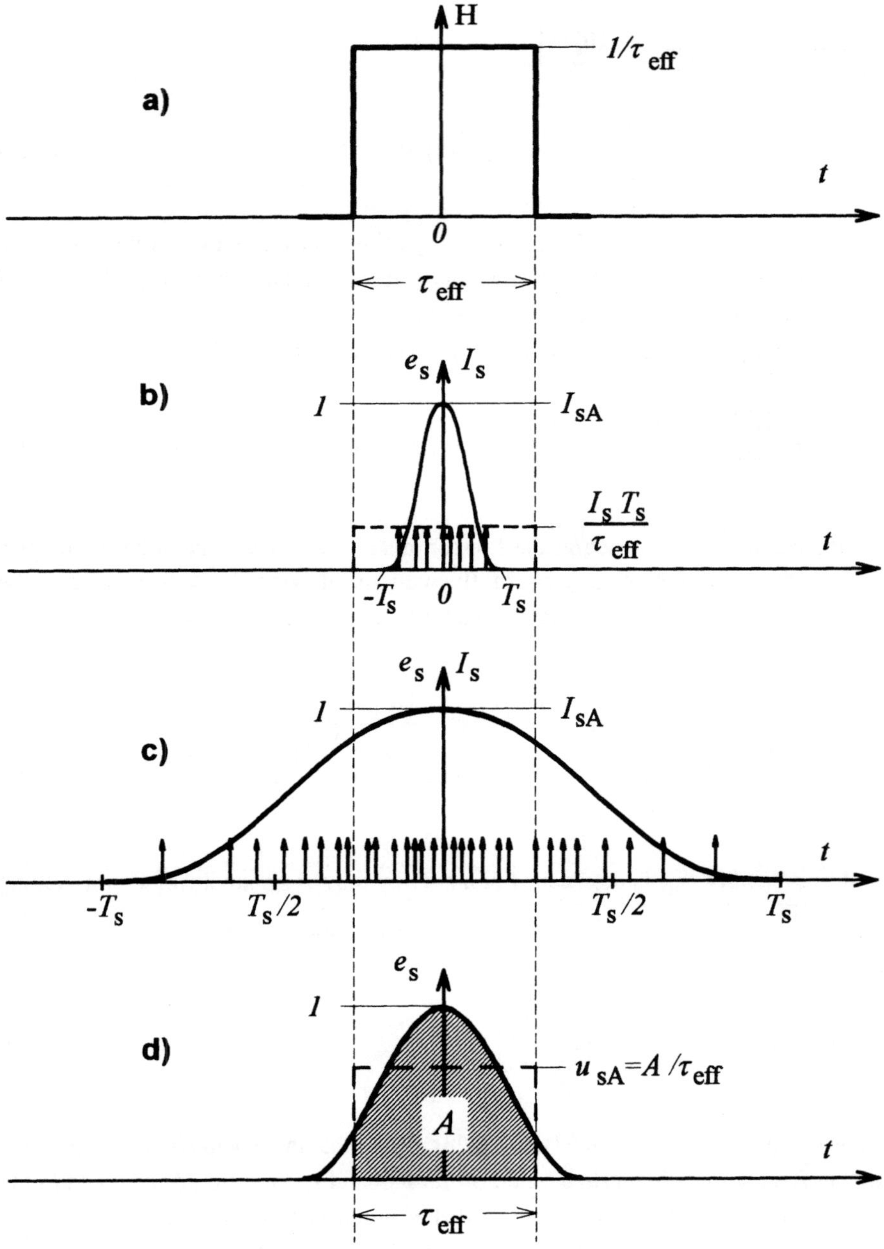

Figure 6.2: To the calculation of signal and noise at the output of PDA: (a) pulse response of PDA; (b) passband of PDA is narrower than the frequency band of the signal; (c) passband of PDA is broader than the frequency band of the signal (bands are comparable); (d) arbitrary relationship between the passband of PDA and the frequency band of the signal.

upstage: how is it going to play its part of the detection system of not just a signal, but a random signal?

Signal noise at the output of PDA with integrating filter. There is practically no difference between the detection of random optical signals and the detection of deterministic signals. The quantum efficiency η may be regarded as the probability of one photon generating one photoelectron; then the mean number $\overline{N}$ of incident photons generates on the average $\overline{N_s} = \eta\overline{N}$ photoelectrons. This number is linked with the amplitudes of the optical signal $\overline{E_{sA}}$ and the photocurrent I_{sA} by obvious relations:

$$\overline{N_s} = \frac{I_{sA}T_s}{q} = S\overline{E_{sA}}\frac{T_s}{q} = \frac{\eta q}{h\nu}\overline{E_{sA}}\frac{T_s}{q} = \eta\frac{\overline{E_{sA}T_s}}{h\nu} = \eta\overline{N} \ . \quad (6.29)$$

The random number of photoelectrons N_s, generated by each optical pulse, also obeys the Poisson statistics [56].

It is more interesting to analyze the passage of the random signal through the filter-amplifier, to calculate the dispersion of the signal at its output. In this section we are going to consider a filter-integrator with a certain time of integration (storage) T_i (Fig. 6.2). The reason is that this example demonstrates in an especially clear way the physics of reception of signals split into individual quanta.

The filter under consideration integrates the short pulse of the input current caused by each separate photoelectron, so that the output signal is proportional to the charge of said input pulse (in this context, q). This signal is stored for the time $T_i = \tau_{eff}$. The exact value of the output signal caused by the one-electron input pulse is calculated on the basis of the pulse response of the filter. In case of filter-integrator the response if rectangular (Fig. 6.2a); its height on the interval τ_{eff} is $H = K_0/\tau_{eff}$ (4.7), (5.36). The output signal is determined by the area of the input current pulse q times the pulse response, $U_{sA} = qK_0/\tau_{eff}$ (4.10).

Now a few formal comments. Since the signal-to-noise ratio does not depend on the transfer coefficient K_0, in the preceding sections we assumed it to be equal to one, so as to simplify the notation. Further on we also assume that $K_0 = 1$. In our current-to-voltage converter the transfer coefficient K_0 is a dimensional quantity; accordingly, the assumption $K_0 = 1$ implies the transition to dimensionless (normalized) variables, which ought to be borne in mind.

So, when $K_0 = 1$ we have $H(t_m) = h(t_m) = 1/\tau_{eff}$, and the output signal is equal to $U_{sA} = q/\tau_{eff}$: the charge of the electron is as though smeared over the time interval τ_{eff}, producing the output signal (current pulse) q/τ_{eff}. Since the integrator remembers the signals from all photoelectrons N_τ, generated on the interval τ_{eff}, we must add up all these signals:

$$U_{sA} = qN_\tau/\tau_{eff} \ . \quad (6.30)$$

A simpler method of processing the input signal is hardly possible: the filter is nothing else but a counter which counts the photoelectrons N_τ generated on the interval τ_{eff}. Since the two random quantities U_{sA} and N_{s} are linked through the deterministic coefficient q/τ_{eff}, their mean value and dispersions, according to (2.35), (2.36), are also linked through this coefficient and its square:

$$\overline{U_{\text{sA}}} = \frac{q}{\tau_{\text{eff}}}\overline{N_\tau} \; ; \tag{6.31}$$

$$\overline{\Delta U_{\text{sA}}^2} = \frac{q^2}{\tau_{\text{eff}}^2}\overline{\Delta N_\tau^2} = \frac{q^2}{\tau_{\text{eff}}^2}\overline{N_\tau} \; . \tag{6.32}$$

The noise of the signal at the output of PDA is nonstationary. As long as there is no signal, there is naturally no noise. Let us consider the dispersion $\overline{\Delta U_{\text{s}}^2}$ for the time $t = t_{\text{m}}$, when the signal reaches its maximum. The calculation of $\overline{\Delta U_{\text{s}}^2}$ reduces to finding the number N_τ at the time t_{m}.

Inertial filter. If the effective time constant of the filter τ_{eff} is large enough — larger than the duration of the bell-shaped signal at its base,

$$\tau_{\text{eff}} \geq 2T_{\text{s}} \; , \tag{6.33}$$

then the integrator-counter will count (integrate) all the signal photoelectrons (Fig. 6.2b). Therefore, $N_\tau = N_{\text{s}}$, and the expressions for the mean output signal and its dispersion (6.31), (6.32) can be written as

$$\overline{N_\tau} = \overline{N} \; , \; \overline{U_{\text{s}}} = \frac{q}{\tau_{\text{eff}}}\overline{N_{\text{s}}} \; ; \tag{6.34}$$

$$\overline{\Delta U_{\text{s}}^2} = \frac{q^2}{\tau_{\text{eff}}^2}\overline{N_{\text{s}}} \; . \tag{6.35}$$

Expression (6.35) for the noise of the signal can be "disguised" as the Shottky formula: the term $q\overline{N_{\text{s}}}$ is related to the current I_{sA} (6.29), and the effective time constant is related to the noise band, $1/2\tau_{\text{eff}} = f_{\text{nw}}$ (Table 5.1):

$$\overline{\Delta U_{\text{s}}^2} = \frac{q^2}{\tau_{\text{eff}}^2}\overline{N_{\text{s}}} = 2q\frac{q\overline{N_{\text{s}}}}{T_{\text{s}}}\frac{T_{\text{s}}}{\tau_{\text{eff}}}\frac{1}{2\tau_{\text{eff}}} = 2qI_{\text{sA}}u_{\text{sA}}f_{\text{nw}} \; . \tag{6.36}$$

The transformation for the signal is similar:

$$\overline{U_{sA}} = \frac{q}{\tau_{eff}}\overline{N_s} = \frac{q\overline{N_s}}{T_s}\frac{T_s}{\tau_{eff}} = I_{sA}u_{sA} \ . \tag{6.37}$$

Equations (6.36), (6.37) involve the relative pulse sensitivity (relative amplitude) $T_s/\tau_{eff} = u_{sA}$. It turns out that such replacement is always possible (irrespective of the profiles of the signal and the pulse response), as long as the time constant of the filter is large, $\tau_{eff} >> T_s$. Indeed, in this case the input current pulse may be regarded as the delta-pulse with the area $I_{sA}T_s$, and then the output pulse reproduces the pulse response $\mathrm{H}(t_m) = \mathrm{h}(t_m) = 1/\tau_{eff}$:

$$\tau_{eff} >> T_s \ , \ u_{sA} = \frac{U_s(t_m)}{I_{sA}} = \frac{(I_{sA}T_s)\mathrm{h}(t_m)}{I_{sA}} = \frac{T_s}{\tau_{eff}} \ . \tag{6.38}$$

Low-inertia filter. When the time constant of the filter is very small ($\tau_{eff} << T_s$), the rate of generation on this time interval τ_{eff} may be considered constant (Fig. 6.2c). Its maximum value is given by the signal amplitude I_{sA}/q; the number of photoelectrons arriving at the input within the time τ_{eff} is $N_\tau = I_{sA}\tau_{eff}/q$. Therefore, for the output signal and its noise, in accordance with (6.31), (6.32), we get:

$$U_{sA} = \frac{q}{\tau_{eff}}\overline{N_\tau} = \frac{q}{\tau_{eff}}\frac{I_{sA}\tau_{eff}}{q} = I_{sA} \ , \tag{6.39}$$

$$\overline{\Delta U_s^2} = \frac{q^2}{\tau_{eff}^2}\overline{N_\tau} = 2q\frac{q\overline{N_\tau}}{T_s}\frac{1}{2\tau_{eff}} = 2qI_{sA}f_{nw} \ . \tag{6.40}$$

In fact, equations (6.39), (6.40) represent a particular case of (6.36), (6.37): the inertialess filter does not distort the signal, and therefore $u_{sA} = 1$. Since, however, the formulas (6.36), (6.37) hold in the limits of very large and very small time constant of the filter, it would seem natural to trust them at any intermediate value of τ_{eff}. And this is really so. The probability of one electron arriving within the interval dt is $e_s(t)dt/T_s$; accordingly, the probability of arriving within the arbitrary time interval τ_{eff} is given by the integral

$$\int_{-\tau_{eff}/2}^{\tau_{eff}/2}\frac{e_s(t)dt}{T_s} = \frac{\tau_{eff}}{T_s}\int_{-\tau_{eff}/2}^{\tau_{eff}/2} e_s(t)\frac{1}{\tau_{eff}}dt = \frac{\tau_{eff}}{T_s}\int_{-\infty}^{\infty} e_s(t)\mathrm{h}(t)dt = \frac{\tau_{eff}}{T_s}u_{sA} \ , \tag{6.41}$$

where

$$u_{sA} = \int_{-\infty}^{\infty} e_s(t)\mathrm{h}(t_m - t)dt \ .$$

Figure 6.2d helps to understand this calculation. The pulse response of the integrator is constant on the interval $[-\tau_{eff}/2, \tau_{eff}/2]$ and equal to $1/\tau_{eff}$, which allows us to change the limits of integration and introduce $h(t)$. The sought-for probability of finding one electron on the interval τ_{eff} is proportional to the area under the curve $e_s(t)$ on this interval. But then, it is the same area that defines the relative pulse response u_{sA}, the relative amplitude of the output signal (4.10). The mean number of the detected electrons $\overline{N_\tau}$ is equal to the mean number of the electrons arriving at the input times the probability (6.41) of each of them to arrive within the interval τ_{eff}:

$$\overline{N_\tau} = \overline{N_s}\left(\frac{\tau_{eff}}{T_s}u_{sA}\right) = \frac{\tau_{eff} q \overline{N_s}}{qT_s}u_{sA} = \frac{\tau_{eff}}{q}I_{sA}u_{sA} \,. \tag{6.42}$$

Substituting this value of $\overline{N_\tau}$ into the initial expressions for the signal and its noise (6.31), (6.32), we arrive at the formulas (6.37), (6.36):

$$\overline{U_{sA}} = \frac{q}{\tau_{eff}}\overline{N_\tau} = \frac{q}{\tau_{eff}}\left(\frac{\tau_{eff}}{q}I_{sA}u_{sA}\right) = I_{sA}u_{sA} \,, \tag{6.43}$$

$$\overline{\Delta U_s^2} = \frac{q^2}{\tau_{eff}^2}\left(\frac{\tau_{eff}}{q}I_{sA}u_{sA}\right) = 2qI_{sA}u_{sA}\frac{1}{2\tau_{eff}} = 2qI_{sA}u_{sA}f_{nw} \,. \tag{6.44}$$

These formulas for signal and noise are so natural that it seems we did not even need to derive them: they are nothing else but the well-familiar expressions for the shot noise of constant incident radiation, only this radiation now is the signal itself. This "replacement" of constant incident radiation with the signal is obvious when the time constant of the amplifier is small, $\tau_{eff} << T_s$. Indeed, during this short observation time the signal remains practically unchanged, and the PDA cannot distinguish it from the ordinary constant current. The latter is always accompanied by the white noise of spectral density $2qI_{sA}$ (at $t \approx t_m$), which acts in the frequency band of the amplifier f_{nw}, so that we may directly write the formula for the dispersion of the output current (6.44).

This reasoning can also be inverted: the constant background and dark current $I_0 = I_{pb} + I_d$ may be regarded as a "signal" of infinitely long duration T_s. Any amplifier having a finite time constant τ_{eff} will be inertialess for such a long signal, and so its noise will be given by (6.40), with $I_{sA} = I_0$, $u_{sA} = 1$. This brings us back to the formula for the shot noise derived earlier in Sect. 2.3.

In the general case the signal does not remain constant over the observation interval τ_{eff}. If, however, we spread evenly the signal charge arrived during this time over the interval τ_{eff}, we get the mean current $I_{sA}u_{sA}$ on this interval (6.43) — dashed line in Fig. 6.2d. This current is responsible for the noise, as described by the general expression (6.44). When $\tau_{eff} >> T_s$, the amplifier registers the entire charge of the signal $I_{sA}T_s$, and the mean current over the interval τ_{eff} is then $I_{sA}T_s/\tau_{eff}$ (Fig. 6.2b). This value of the current enters the special-case formula for the shot noise (6.36).

According to (6.36), (6.37), the ratio of the signal to its own noise is

$$\mathrm{SNR} = \frac{U_{\mathrm{sA}}}{\sqrt{\overline{\Delta U_{\mathrm{s}}^2}}} = \frac{q\overline{N_\tau}}{\tau_{\mathrm{eff}}} \frac{1}{\sqrt{\frac{q^2}{\tau_{\mathrm{eff}}^2}\overline{N_\tau}}} = \sqrt{\overline{N_\tau}} \ . \tag{6.45}$$

Comparing this with the SNR for the optical signal itself (6.26), we can easily see how and why the passband of the amplifier affects the signal-to-noise ratio. Optimal is the amplifier with a large time constant $\tau_{\mathrm{eff}} \geq 2T_{\mathrm{s}}$, which registers all the photoelectrons, $N_\tau = N_{\mathrm{s}}$. If this amplifier is noiseless, and the quantum yield is high, then there is practically no loss in the SNR, and $N_\tau = N_{\mathrm{s}} \approx N$. A fast amplifier fails to count out all the photoelectrons of the signal, $N_\tau < N_{\mathrm{s}}$; then, the smaller the time constant, the poorer the SNR.

In this way, our exemplary filter-integrator has helped us to understand the transformation of the signal noise. As demonstrated in Sect. 4.1, any filter in the first approximation may be regarded as integrator with a certain effective integration time τ_{eff}; because of this, our results should be universal. Still, it will be interesting to see how the particular frequency response (pulse response) affects the signal noise. Let us do the calculations once again, but this time for an arbitrary filter-amplifier with an arbitrary frequency response $k(f)$. For the sake of generality, we shall also include the multiplication.

Signal noise at the output of arbitrary filter-amplifier. Let us write the expression for the realization of the mixture of signal with the noise of the PDA at the output of this PDA:

$$U(t) = U_{\mathrm{n}}(t) + U_{\mathrm{s}}(t) \ . \tag{6.46}$$

Assume now that random is not only the noise of the PDA $U_{\mathrm{n}}(t)$, but also the signal itself $U_{\mathrm{s}}(t)$. Since we have agreed to define the threshold properties with respect to the mean value of the signal, let us separate in (6.46) the mean value of the signal $\overline{U_{\mathrm{s}}(t)}$ and its random component, the signal noise $\Delta U_{\mathrm{s}}(t) = U_{\mathrm{s}}(t) - \overline{U_{\mathrm{s}}(t)}$:

$$U(t) = U_{\mathrm{n}}(t) + \Delta U_{\mathrm{s}}(t) + \overline{U_{\mathrm{s}}(t)} \ . \tag{6.47}$$

The noise of the PDA $U_{\mathrm{n}}(t)$ is supplemented by the signal noise $\Delta U_{\mathrm{s}}(t)$ not correlated with it. This implies that the signal itself interferes with its own detection, and so, if the noise of the PDA is small, the designer of PDA must take into account not only the noise which is "his own", but also the noise of the signal.

Let us find the expression for the random signal $U_{\mathrm{s}}(t)$, and then calculate the mean value $\overline{U_s(t)}$ and the dispersion $\overline{\Delta U_s^2(t)}$. Each photoactive light-generated pair, splitting at the time t_i, induces in the external circuit a short current pulse carrying the charge qM (if there is no multiplication, $M = 1$). Accordingly, the pulse $qM\,\mathrm{H}(t - t_i)$ appears at the output of the PDA whose shape imitates the pulse response $\mathrm{H}(t - t_i)$. The output signal is determined by the superposition of such partial pulses caused by each split pair [91]:

$$U_s(t) = qM_1\,\mathrm{H}(t - t_1) + \ldots + qM_i\,\mathrm{H}(t - t_i) + \ldots + qM_{N_s}\,\mathrm{H}\left(t - t_{N_s}\right) . \tag{6.48}$$

Here, as in the optical signal (5.98), random are the times of arrival of the pulse t_i and the number of such pulses N_s. When the PDA is based on APD (and it is these photodetectors that are usually capable of dragging out the noise of the signal from the noise of the PDA), there is also a third random quantity, the multiplication M_i. In the calculation of the mean value of $U_s(t)$ we note that the random quantities t_i, M_i, N_s are independent, and therefore averaging can be carried out separately with respect to each of them in a row.

One quantum gives rise to the signal $U_s(t) = q\,\mathrm{H}(t - t_i)$ at the output, with the random time of arrival t_i. Since the probability of the instant t_i is given by $e_s(t)/T_s$, averaging with respect to t_i gives the following value of the mean signal caused by one quantum:

$$\overline{U_s(t)} = q \int_{-\infty}^{\infty} \frac{e_s(t_i)}{T_s}\,\mathrm{H}(t - t_i)\,dt_i \; . \tag{6.49}$$

When there is multiplication, the charge carried by each such pulse is multiplied, on the average, $\overline{M}$-fold. On the average, there are $\overline{N_s}$ such partial pulses at the output, and so for the mean signal we finally get:

$$\overline{U_s(t)} = q\overline{N_s}\,\overline{M}\left\{\frac{1}{T_s}\int_{-\infty}^{\infty} e_s(t_i)\,\mathrm{H}(t - t_i)\,dt_i\right\} . \tag{6.50}$$

At $t = t_m$, equation (6.50) gives the amplitude value, and the integral defines the relative amplitude u_{sA} (6.41). Carrying out the replacement $q\overline{N_s}/T_s = I_{sA} = SE_{sA}$ in accordance with (6.29), we get:

$$\overline{U_{sA}} = \frac{q\overline{N_s}}{T_s}\overline{M}\int_{-\infty}^{\infty} e_s(t_i)\,\mathrm{H}(t_m - t_i)\,dt_i = I_{sA}\overline{M}u_{sA} = SE_{sA}\overline{M}u_{sA} \; . \tag{6.51}$$

This is the standard expression for the signal at the output of PDA (4.10).

Now let us find the mean square value of the fluctuation of the signal $\overline{\Delta U_s^2}$. By contrast to the above calculation of the mean value $\overline{U_{sA}}$, we now have to average the square of the partial pulse $q^2 M_i^2 \mathrm{H}^2(t-t_i)$. Performing averaging in the same order as before, we get the expression which is similar to (6.50) but involves quadratic rather than linear terms q^2, $\overline{M_i^2} = \overline{M^2}$, H^2. Only the number $\overline{N_s}$ remains unsquared: all the partial pulses from each of the N_s photoelectrons are statistically independent, and so their dispersions are simply added together:

$$\overline{\Delta U_s^2(t_m)} = q^2 \overline{N_s}\, \overline{M^2} \left\{ \int_{-\infty}^{\infty} \frac{e_s(t_i)}{T_s} \mathrm{H}^2(t_m - t_i) dt_i \right\} . \tag{6.52}$$

Making once again our traditional replacement $q\overline{N_s}/T_s = I_{sA}$, we rewrite (6.52) as

$$\overline{\Delta U_s^2} = \overline{\Delta U_s^2(t_m)} = 2qI_{sA}\overline{M^2} \left\{ \frac{1}{2} \int_{-\infty}^{\infty} e_s(t_i) \mathrm{H}^2(t_m - t_i) dt_i \right\} . \tag{6.53}$$

Now the main thing is to work out the integral in braces in (6.52), (6.53). As always, let us try to find the key to the problem by considering extreme cases.

Inertial filter. When $\tau_{eff} >> T_s$, the pulse characteristic of the filter $\mathrm{H}(t_m - t_i)$ changes slowly compared to the fast signal $e_s(t_i)$; therefore, $\mathrm{H}^2(t_m - t_i)$ may be regarded as an almost constant quantity, which can be factored out of the integral in (6.52), (6.53):

$$\overline{\Delta U_s^2} = q^2 \overline{N_s}\, \overline{M^2}\, \mathrm{H}^2(t_m) \left\{ \frac{1}{T_s} \int_{-\infty}^{\infty} e_s(t_i) dt_i \right\} = \frac{q^2}{\tau_{eff}^2} \overline{N_s}\, \overline{M^2} , \tag{6.54}$$

$$\overline{\Delta U_s^2} = 2qI_{sA}\overline{M^2} \left\{ \mathrm{H}^2 \frac{1}{2} \int_{-\infty}^{\infty} e_s(t_i) dt_i \right\}$$

$$= 2qI_{sA}\overline{M^2} \left\{ \frac{T_s}{\tau_{eff}} \frac{1}{2\tau_{eff}} \right\} = 2qI_{sA}\overline{M^2} u_{sA} f_a . \tag{6.55}$$

Here we have used the fact that $\mathrm{H}(t_m) = \mathrm{h}(t_m) = 1/\tau_{eff}$; $1/2\tau_{eff} = f_a$ (Table 5.1), $u_{sA} = T_s/\tau_{eff}$ (6.38), and that the area under the curve $e_s(t_i)$ is the effective signal duration T_s. In exactly the same way, we get for the signal (6.49):

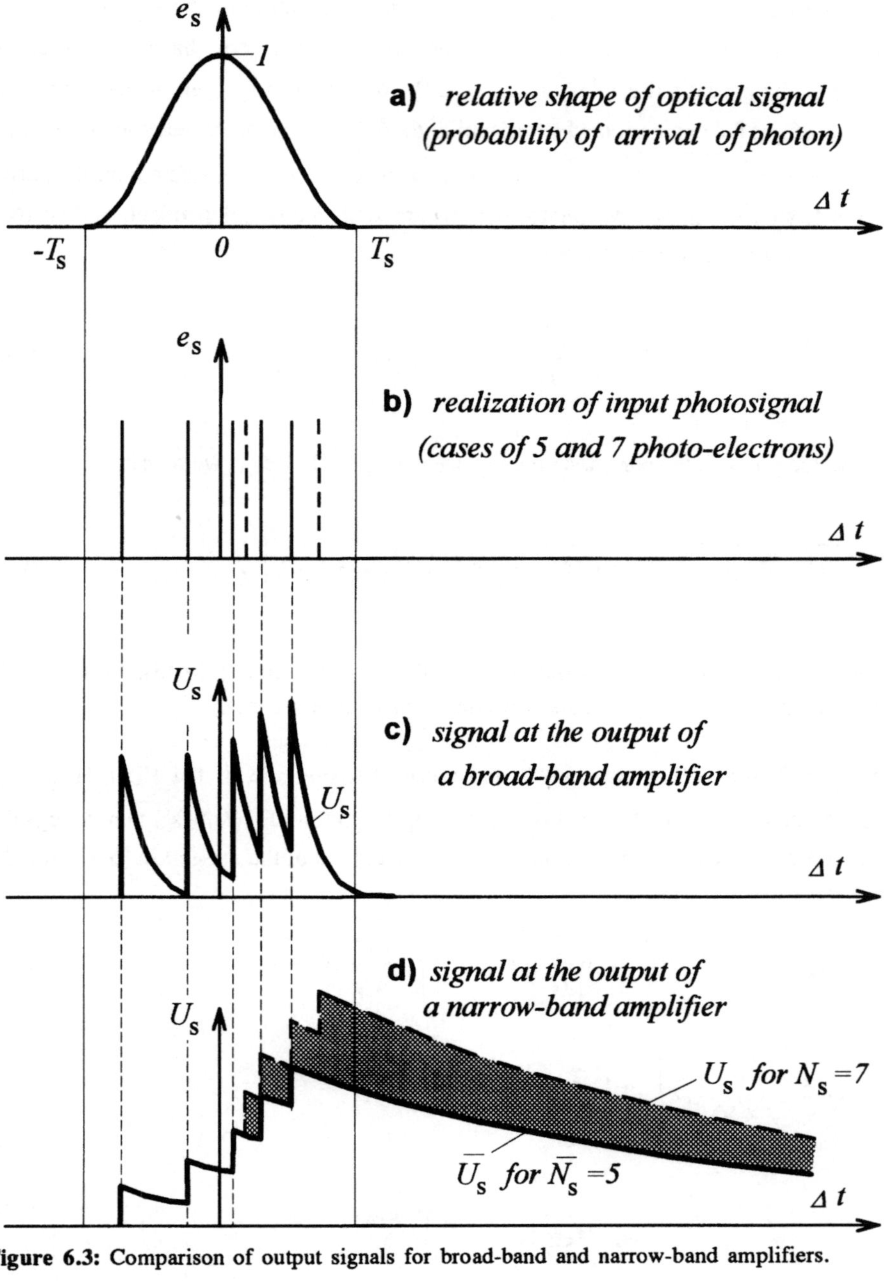

Figure 6.3: Comparison of output signals for broad-band and narrow-band amplifiers.

$$\overline{U_s(t_m)} = q\overline{N_s}\,\overline{M}\,\mathrm{H}(t_m)\left\{\frac{1}{T_s}\int_{-\infty}^{\infty} e_s(t_i)dt_i\right\} = \frac{q}{\tau_{\text{eff}}}\overline{N_s}\,\overline{M} \ . \tag{6.56}$$

This brings us back to our old expressions for the signal and its dispersion at the output of filter-amplifier (compare (6.56) and (6.34), and also (6.54) and (6.35)). The only novelty is the multiplication. There is nothing surprising in this similarity of the formulas: any filter is an integrator for a short pulse, since its response is proportional to the area of this short pulse. It turns out that any sufficiently inertial filter counts out all the photoelectrons of the signal N_s, and all the multiplied electrons MN_s when there is multiplication.

It is interesting to look at the shape of the signal in this case: the profile is deterministic and imitates the pulse response of the filter; random is only the amplitude which is defined by the random number of electrons MN_s at the input. Figure 6.3 clearly illustrates these arguments. Figure 6.3d shows two realizations of the output signal: the solid line corresponds to the case when the mean number of quanta arriving at the input is $\overline{N_s} = 5$, and the dashed line to $N_s = 7$. We see that both the mean signal $\overline{U_s(\Delta t)}$ and its noise $U_s(t) - \overline{U_s(t)}$ at $t > T_s$ exhibit the same dependence on the time, whereas the signal-to-noise ratio is defined by the SNR for the number of incident photons N (or, more precisely, for the number of photoelectrons $N_s = \eta N$). The latter is equal to $1/\sqrt{N_s}$ when $M = 1$. The same result follows from (6.55), (6.56):

$$\frac{\sqrt{\overline{\Delta U_s^2(t_m)}}}{\overline{U_s(t_m)}} = \frac{q\sqrt{\overline{N_s}\,\overline{M^2}}}{q\overline{N_s}\,\overline{M}} = \sqrt{\frac{\overline{M^2}}{\overline{N_s}\,\overline{M}^2}} = \sqrt{\frac{\mathcal{F}}{\overline{N_s}}} \; . \tag{6.57}$$

The deterioration by the factor of $\sqrt{\mathcal{F}}$ is caused by the fluctuation of the multiplication.

Fast amplifier. Let us return to our initial formula for the noise, and evaluate the integral in braces in (6.53) for the fast amplifier. At $\tau_{\text{eff}} << T_s$ the functions $e_s(t_i)$ and $\text{H}(t_m - t_i)$ exchange their roles: now the signal is almost constant, and can be factored out:

$$\left\{ e_s(t_m)\frac{1}{2}\int_{-\infty}^{\infty} \text{H}^2(t_m - t_i)dt_i \right\} = \left\{ \frac{1}{2}\int_{-\infty}^{\infty} \text{H}^2(t_m - t_i)dt_i \right\} \tag{6.58}$$

(recall that $e_s(t_m) = 1$). Now our integral resembles the convolution (4.10), used to find the output signal. More precisely, this integral coincides with the convolution when the input signal has the same shape as the pulse response — that is, when $E(t) = \text{H}(t_m - t_i)/2$:

$$U_{sA}(t_m) = \int_{-\infty}^{\infty} e_s(t_i)\text{H}(t_m - t_i)dt_i = \frac{1}{2}\int_{-\infty}^{\infty} \text{H}^2(t_m - t_i)dt_i \; . \tag{6.59}$$

But then, this is the case of the optimal filter!

In the space of frequencies the situation looks as follows: the amplitude spectrum of the signal applied to the input of the amplifier coincides with the relative frequency response of this amplifier, $\mathrm{E_s}(f) = \mathrm{e_s}(f) = k(f)$ (3.48). We have also noted that at low frequencies the spectral density of our signal is $\mathrm{H}(t_i)/2 = 1$:

$$E_{\mathrm{sA}} = \frac{\mathrm{H}(t_\mathrm{m})}{2} = \frac{1}{2\tau_\mathrm{eff}}\ , \quad T_\mathrm{s} = \tau_\mathrm{eff}\ , \quad \mathrm{E_s}(f) = 2T_\mathrm{s}E_\mathrm{sA} = 2\tau_\mathrm{eff}\left(\frac{1}{2\tau_\mathrm{eff}}\right) = 1\ . \tag{6.60}$$

Therefore, for the amplitude of the output signal we get:

$$U_\mathrm{sA}(t_\mathrm{m}) = \int_0^\infty \mathrm{E_s}(f)k(f)df = \int_0^\infty k^2(f)df\ . \tag{6.61}$$

This is a very useful result. The integral turns out to be equal to the amplitude of the signal at the output of the optimal filter U_sA, which, in turn, is equal to the noise band f_nw (this feature has already been pointed out in Sect. 3.3; see (3.54) with condition $\mathrm{H}(t_\mathrm{m})\tau_\mathrm{eff} = 1$). Thus, the formula for the signal noise (6.53) assumes the form which is quite natural for expressions of this kind:

$$\overline{\Delta U_\mathrm{s}^2} = 2qI_\mathrm{sA}\overline{M^2}f_\mathrm{nw}\ . \tag{6.62}$$

The shape of the output signal (one of the possible realizations) is shown in Fig. 6.3c. By contrast to the previously considered case of the inertial amplifier, now it is not only the amplitude that is random, but the entire profile $U_\mathrm{s}(t)$. It looks like a comb, in which each tooth is the response of the system to a single input electron arriving at a random time moment t_i.

The limiting expressions for the dispersion of the signal at the output of the inertial (6.55) and the noninertial (6.62) filters differ favorably from the general expression (6.53) because they are written in the canonical form: they involve the shot noise of the primary current $I_\mathrm{sA}u_\mathrm{sA}$, the multiplication term $\overline{M^2}$, and the frequency band f_a or f_nw. Unfortunately, the general expression (6.53) is not so transparent: instead of the frequency band and the relative amplitude u_sA it involves the integral in braces. This juxtaposition prompts us how to bring this expression to the canonical form: if we replace the integral

$$\left\{\frac{1}{2}\int_{-\infty}^{\infty} e_\mathrm{s}(t_i)\mathrm{H}^2(t_\mathrm{m} - t_i)dt_i\right\} \Rightarrow f'_\mathrm{nw}u_\mathrm{sA}\ ,$$

then we get

$$\overline{\Delta U_\mathrm{s}^2} = 2qI_\mathrm{sA}u_\mathrm{sA}\overline{M^2}f'_\mathrm{nw} = 2qI_\mathrm{sA}u_\mathrm{sA}\overline{M^2}\Gamma'_1 f_\mathrm{a}\ . \tag{6.63}$$

Here we have introduced the effective band for the signal noise f'_{nw} and, as always, its ratio Γ'_1 to the effective frequency band of the amplifier:

$$\Gamma'_1 = \frac{f'_{nw}}{f_a} = \frac{1}{u_{sA} f_a} \left\{ \frac{1}{2} \int_{-\infty}^{\infty} e_s(t_i) H^2(t_m - t_i) dt_i \right\} . \tag{6.64}$$

So, the PDA has played its role, and detected the random signal. The task is fulfilled: the mean value of the signal is transformed according to the same algorithm as the ordinary deterministic signal (6.51). During the effective observation time τ_{eff} the signal creates at the input the averaged current $I_{sA} u_{sA}$; this current is responsible for the noise of the signal. All peculiarity of the calculation reduces to the definition of the noise band f'_{nw} for the dispersion of the random signal. In the general case this band is defined not only by the amplifier, but also by the signal (its relative shape $e_s(t)$, and the relationship between the signal duration T_s and the effective time constant of the amplifier τ_{eff}). This fact, however, is of more theoretical than practical interest: the band f'_{nw} varies, depending on τ_{eff}/T_s, within sufficiently narrow limits from $f_{nw} = \Gamma_1 f_a$ to f_a; accordingly, the coefficient Γ'_1 varies from Γ_1 to 1 (see (6.55), (6.62)). In this book we consider four examples of filters-amplifiers (Table 5.1). For two of these (the integrator, already considered above, and the amplifier with the rectangular frequency response) the bands actually coincide: $f'_{nw} = f_{nw} = f_a$, $\Gamma'_1 = \Gamma_1 = 1$. It follows that this variation of the band f'_{nw} is only observed with RC-filters, and especially with the one-section filter, for which $\Gamma_1 = 0.5$. For calculating the noise dispersion (6.63), the following approximations for Γ'_1 may be suggested:

$$\Gamma'_{1app} = \frac{\Gamma_1 + 1}{2} , \tag{6.65}$$

$$\Gamma'_{1app} = \frac{\Gamma_1 + \left(\dfrac{\tau_{eff}}{T_s}\right)^2}{1 + \left(\dfrac{\tau_{eff}}{T_s}\right)^2} . \tag{6.66}$$

In (6.65) the coefficient Γ'_{1app} is assumed to be constant and equal to the mean value of $\Gamma_1(\tau_{eff})$. Even this rough approximation ensures that the error in the calculation of the noise $\sqrt{\overline{\Delta U_s^2}}$ for the one-section filter does not exceed the factor of $\sqrt{\Gamma'_{1app}/\Gamma_1} = \sqrt{0.75/0.5} = 1.22$ (and the factor of 1.1 for the two-section filter). A more demanding reader will appreciate the approximation (6.66): as shown in Fig. 6.4, it closely follows the actual variation of the noise band, and secures the accuracy of calculation of $\sqrt{\overline{\Delta U_s^2}}$ within 2...5% for the RC-filters under consideration.

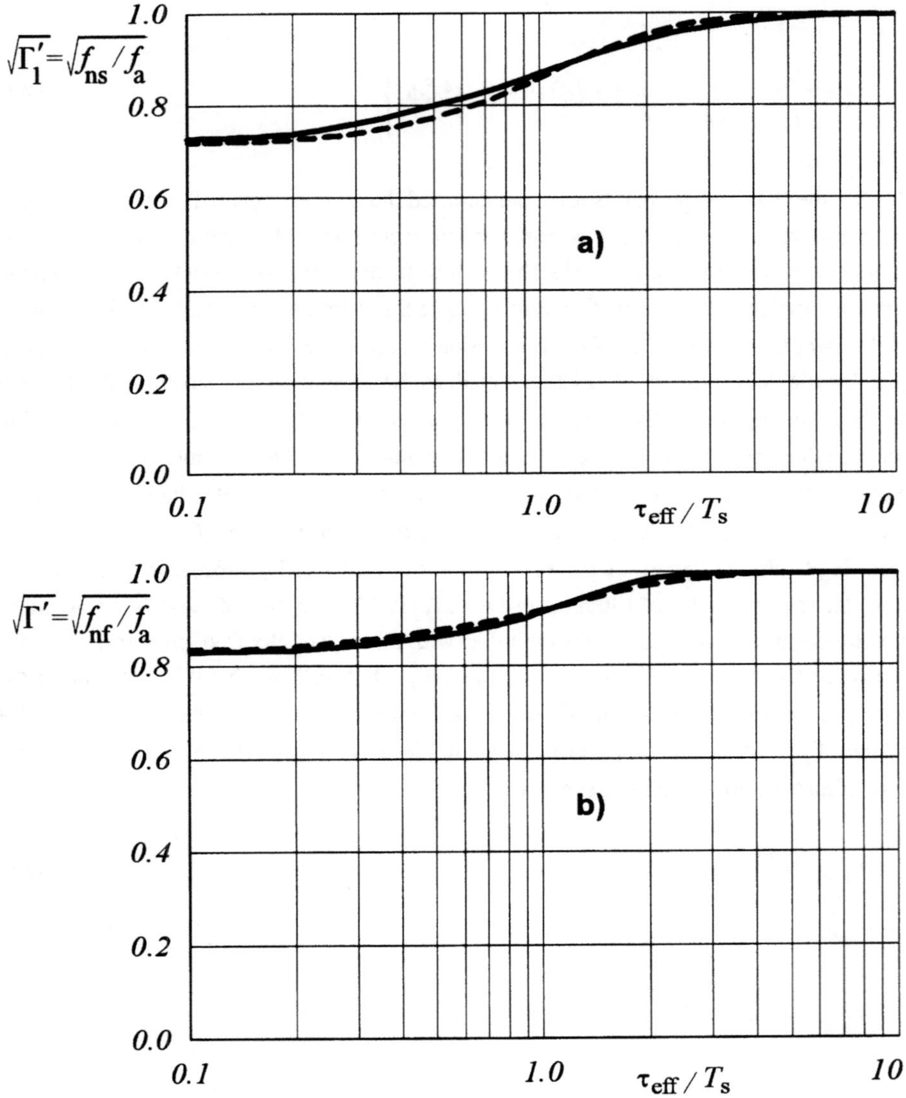

Figure 6.4: The peculiar feature of the effective band of amplifier for signal noise is that it exhibits dependence (although weak) on the length of the signal itself: (a) one-section filter; (b) two-section filter.

General case: the signal noise and the PDA noise are comparable. Noise number. When the signal is close to the quantum limit, the total noise of the PDA must be supplemented by yet another component, the signal noise (6.47), (6.63). This applies to all formulas for the SNR obtained in the previous sections. For example, (5.55) with the inclusion of I_n as defined by (6.53) takes on the form

$$\mathrm{SNR} = \frac{I_{\mathrm{sA}} u_{\mathrm{sA}} \overline{M}}{\sqrt{\frac{4\kappa T \Gamma_1 f_a}{R_{\mathrm{in}}} + 2q\left(I_{\mathrm{sA}} u_{\mathrm{sA}} \frac{\Gamma_1'}{\Gamma_1} + I_0\right) \mho \overline{M}^2 \Gamma_1 f_a + e_n^2 C_{\mathrm{in}}^2 (2\pi)^2 \Gamma_2 f_a^3}} . \tag{6.67}$$

The magnitudes of signals and their noises have been defined in terms of the number of photoelectrons $\overline{N_s}$, $\overline{N_\tau}$. It is convenient to express the noise of the PDA in terms of the so-called noise number $\overline{N_n}$. To go over from the current to the number of electrons we multiply the current by the observation time (the effective time constant τ_{eq} of the PDA), and find the registered charge. Dividing this by q, we find the number of photoelectrons (6.42).

Multiplying the numerator and the denominator of the expression for the SNR by τ_{eff}/q, we get the fraction whose numerator expresses the signal in terms of $\overline{N_\tau}$, and the denominator expresses the noise in terms of $\overline{N_n}$. To make the computation more comprehensible, let us retain only the shot noise component in the noise of the PDA:

$$\mathrm{SNR} = \frac{U_s}{U_n} = \frac{\frac{\tau_{\mathrm{eff}}}{q} I_{\mathrm{sA}} u_{\mathrm{sA}} \overline{M}}{\frac{\tau_{\mathrm{eff}}}{q}\sqrt{2qI_{\mathrm{sA}} u_{\mathrm{sA}} \mho \overline{M}^2 \Gamma_1' f_a + 2qI_0 \mho \overline{M}^2 \Gamma_1 f_a}} , \tag{6.68}$$

$$U_s = \frac{\tau_{\mathrm{eff}}}{q} I_{\mathrm{sA}} u_{\mathrm{sA}} \overline{M} = \overline{N_\tau}\,\overline{M} , \tag{6.69}$$

$$\overline{\Delta U_s^2} = \frac{\tau_{\mathrm{eff}}^2}{q^2} 2qI_{\mathrm{sA}} u_{\mathrm{sA}} \mho \overline{M}^2 \Gamma_1' f_a = \frac{\tau_{\mathrm{eff}} I_{\mathrm{sA}} u_{\mathrm{sA}}}{q} \overline{M}^2 \mho \Gamma_1' (2 f_a \tau_{\mathrm{eff}})$$

$$= \overline{N_\tau}\,\overline{M}^2 \mho \Gamma_1' , \tag{6.70}$$

$$\overline{U_{n\mathrm{PDA}}^2} = \frac{\tau_{\mathrm{eff}}^2}{q^2} 2qI_0 \mho \overline{M}^2 \Gamma_1 f_a = \frac{\tau_{\mathrm{eff}} I_0}{q} \overline{M}^2 \mho \Gamma_1 (2 f_a \tau_{\mathrm{eff}}) = \overline{N_0}\,\overline{M}^2 \mho \Gamma_1 . \tag{6.71}$$

Here we have separated the components of the noise of the signal and the noise of the PDA; to make the corresponding expressions symmetrical with respect to each other,

$$\overline{U_{n\mathrm{PDA}}^2} = N_n \overline{M}^2 \mho \Gamma_1' , \tag{6.72}$$

we must define the noise number as

$$N_n = \frac{\Gamma_1}{\Gamma_1'} \overline{N_0} . \tag{6.73}$$

In other words, the number N_n by its physical meaning is the number $\overline{N_0}$ of electrons of the dark (total) current as registered by the PDA over the time τ_{eff}. The shot noise of these primary electrons determines the noise of the PDA. The coefficient Γ_1/Γ_1' arises because the noise of the PDA which acts in the band $\Gamma_1 f_a$ is referred to the noise band of the signal $\Gamma_1' f_a$. In the broad-band amplifier and in the integrator this coefficient is not needed, since $\Gamma_1' = \Gamma_1$.

If the thermal noise of the PDA is also noticeable, we may replace it with the equivalent current generator I_n and define the equivalent numbers $\overline{N_0}$ and N_n. In this case in the expression for the noise of the PDA (6.71) we retain all the components included in (6.67). Now we only have to follow closely the rather cumbersome algebra of transition from $\overline{U^2_{nPDA}}$ to N_n:

$$\overline{U^2_{nPDA}} = \frac{\tau^2_{eff}}{q^2}\left[2qI_0\overline{M}^2\mathcal{F}\Gamma_1 f_a + \frac{4\kappa T}{R_{in}}\Gamma_1 f_a + (e_n C_{in} 2\pi f_a)^2 \Gamma_2 f_a\right]$$

$$= \frac{\tau_{eff} I_0}{q}\overline{M}^2\mathcal{F}\Gamma_1(2\tau_{eff} f_a) + \frac{\kappa T}{q}\frac{2\tau_{eff}}{qR_{in}}\Gamma_1(2\tau_{eff} f_a)$$

$$+\left(\frac{e_n C_{in}\pi}{q}\right)^2 \Gamma_2 (2\tau_{eff} f_a)^2 f_a$$

$$= \overline{N_0}\overline{M}^2\mathcal{F}\Gamma_1 + \frac{\varphi_t}{qR_{in} f_a}\Gamma_1 + \left(\frac{e_n C_{in}\pi}{q}\right)^2 f_a \Gamma_2 \ ;$$

$$N_n = \frac{\overline{U^2_{nPDA}}}{\overline{M}^2\mathcal{F}\Gamma_1'} = \frac{\Gamma_1}{\Gamma_1'}\left[\overline{N_0} + \frac{2\varphi_t \tau_{eff}}{2R_{in}\overline{M}^2\mathcal{F}} + \left(\frac{e_n C_{in}\pi}{q\overline{M}}\right)^2 \frac{f_a \Gamma_2}{\mathcal{F}\Gamma_1}\right] . \tag{6.74}$$

In terms of $\overline{N_\tau}$ and N_n, the signal-to-noise ratio (6.67) can be presented in a simple and conspicuous way:

$$\text{SNR} = \frac{U_s}{U_n} = \frac{\overline{N_\tau}\,\overline{M}}{\sqrt{\overline{N_\tau}\,\overline{M}^2\mathcal{F}\Gamma_1' + N_n \overline{M}^2\mathcal{F}\Gamma_1'}} = \frac{\overline{N_\tau}}{\sqrt{(\overline{N_\tau} + N_n)\mathcal{F}\Gamma_1'}} . \tag{6.75}$$

Lowest detectable intensity of signal. The lowest detectable power (energy) of the low-level signals cannot be calculated simply by multiplying the noise by the desired value of the SNR (see Sect. 3.1). The calculation is complicated by a number of circumstances.

In the first place, the noise itself is not stationary, and depends on the presence of the signal, so it is not known a priori which value of the noise should be taken into account: in the presence or in the absence of signal. Let us dwell awhile upon this issue.

1. There is no signal, and the total noise U_n is only determined by the noise of the PDA $\overline{M}^2 \mathfrak{F} \Gamma_1' N_n$ (6.74). The noise outbursts of the PDA may cause a false alarm: according to Sect. 3.1, its probability P_{fa} is defined by the ratio of the discrimination threshold U_t to the mean square noise U_n (3.3). We denote this partial ratio by n_{fa}, and represent it as

$$n_{fa} = \frac{U_t}{U_n} = \frac{\frac{\tau_{eff}}{q} U_t}{\frac{\tau_{eff}}{q} U_n} = \frac{\frac{\tau_{eff}}{q} U_t}{\overline{M}\sqrt{\mathfrak{F} \Gamma_1' N_n}} = \frac{N_t}{\sqrt{\mathfrak{F} \Gamma_1' N_n}} . \tag{6.76}$$

Since we are now using noise numbers, we have introduced the appropriate number for the discrimination threshold N_t. By analogy with its counterparts for the signal and the noise, this number is defined by multiplying the output value by τ_{eff}/q:

$$N_t = \frac{\tau_{eff}}{q} \frac{U_t}{\overline{M}} . \tag{6.77}$$

Recall that by convention we refer all the voltages from the output to the input of the amplifier stages (assuming $K_0 = 1$). In the APD there is yet another virtual stage with the gain $\overline{M}$, so it is logical to refer the value of N_t to the input of this stage by dividing it into the mean $\overline{M}$ (which means referring it to the primary number of carriers).

2. At the input of the amplifier there is a signal. Naturally, now both the PDA and the fluctuations of the signal itself (6.63) contribute to the total noise. It is in this situation that the target miss probability P_{tm} is calculated: when it is known in advance that the signal has arrived at the input of the PDA. According to (3.10), the probability P_{tm} (denoted below as n_{tm}) is determined in terms of the following ratio:

$$n_{tm} = \frac{U_{sA} - U_t}{U_n} = \frac{\frac{\tau_{eff}}{q} U_{sA} - \frac{\tau_{eff}}{q} U_t}{\frac{\tau_{eff}}{q} U_n} = \frac{\overline{N_\tau} - N_t}{\sqrt{\mathfrak{F} \Gamma_1' \left(\overline{N_\tau} + N_n\right)}} . \tag{6.78}$$

Now the SNR may be defined as the sum of the two partial ratios:

$$\text{SNR} = n_{fa} + n_{tm} . \tag{6.79}$$

According to (6.76), (6.78), (6.79), this definition of SNR takes care of the noise both in the presence and in the absence of the signal. When the fluctuations of the signal are negligibly small, equation (6.79) reduces to the familiar signal-to-noise ratio:

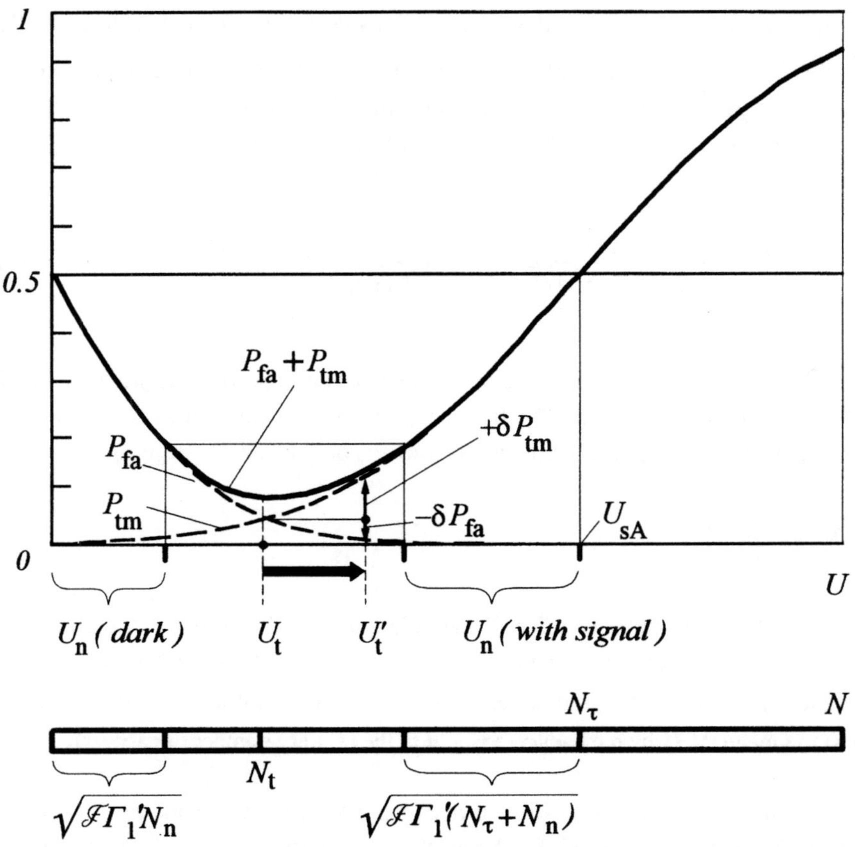

Figure 6.5: Discrimination threshold U_t, which secures minimum total error $P_{fa} + P_{tm}$ is selected as the intercept of curves $P_{fa}(U)$ and $P_{tm}(U)$.

$$\mathrm{SNR} = n_{fa} + n_{tm} = \frac{U_t}{U_n} + \frac{U_{sA} - U_t}{U_n} = \frac{U_{sA}}{U_n} \; . \tag{6.80}$$

Formulas for the partial ratios n_{fa} and n_{tm} (6.76), (6.78) allow us to write the equation for defining the lowest detectable signal intensity in terms of the number $\overline{N_\tau}$. Equation (6.76) is transformed to

$$N_t = n_{fa}\sqrt{\mathcal{F}\,\Gamma_1' N_n} \; , \tag{6.81}$$

and (6.78) to

$$\overline{N_\tau} - N_t = n_{tm}\sqrt{\mathcal{F}\,\Gamma_1'\left(\overline{N_\tau} + N_n\right)} \; . \tag{6.82}$$

Adding these together, we get:

$$\overline{N_\tau} = n_{\text{fa}}\sqrt{\mathcal{F}\,\Gamma_1'N_\text{n}} + n_{\text{tm}}\sqrt{\mathcal{F}\,\Gamma_1'\left(\overline{N_\tau}+N_\text{n}\right)}\ . \tag{6.83}$$

The parameters of the problem $\mathcal{F}$, Γ_1', n_{fa}, n_{tm} that enter this equation must be known, and the last two parameters (n_{fa}, n_{tm}) are uniquely defined by the preassigned values of P_{fa}, P_{tm} (3.3), (3.10). Then from (6.83) we may construct a quadratic equation in $\overline{N_\tau}$ — quadratic, because in the presence of the signal the noise in general depends on the signal itself (the number $\overline{N_\tau}$).

To simplify the calculations, we are going to consider first a particular case, when $n_{\text{fa}} = n_{\text{tm}}$, and therefore $P_{\text{fa}} = P_{\text{tm}}$ (3.3), (3.10). This is not a speculative but a very real situation, first of all for the lightwave communication systems, when the total error $P_{\text{fa}} + P_{\text{tm}}$ must be minimized (recall Sect. 3.1, Fig. 3.3). The diagram in Fig. 3.3 shows that as the threshold U_t is moved away to any side (for example, to the right) from the optimal point corresponding to $P_{\text{fa}} = P_{\text{tm}}$, one of the probabilities (target miss) grows faster than the other probability (false alarm) decreases. As long as we stay in the domain of the Gaussian law, the behavior of $P_{\text{fa}}(U_\text{t})$, $P_{\text{tm}}(U_\text{t})$ is not going to change dramatically. Previously, when the signal did not affect the noise, the distributions $P_{\text{fa}}(U_\text{t})$, $P_{\text{tm}}(U_\text{t})$ had the same shape, and the only difference was that one curve was shifted with respect to the other by the magnitude of the signal U_{sA}, Fig. 3.3. Now, owing to the increasing noise dispersion in the presence of the signal, the curve $P_{\text{tm}}(U_\text{t})$ has become slightly "fatter", Fig. 6.5; as before, however, the minimum total error corresponds to $n_{\text{fa}} = n_{\text{tm}}$, $P_{\text{fa}} = P_{\text{tm}}$. To be more precise, it is slightly displaced to the right of this point, because the slope of P_{tm} is slightly less than that of P_{fa} at $n_{\text{fa}} = n_{\text{tm}}$. This displacement, however, is not too important because the curves $P_{\text{fa}}(U_\text{t})$, $P_{\text{tm}}(U_\text{t})$ exhibit a fast change in this region.

So, we set $n_{\text{fa}} = n_{\text{tm}} = \text{SNR}/2$ and factor out the common multiplier in (6.83):

$$\overline{N_\tau} = \frac{\text{SNR}}{2}\sqrt{\mathcal{F}\,\Gamma_1'}\left(\sqrt{\overline{N_\tau}+N_\text{n}} + \sqrt{N_\text{n}}\right)\ . \tag{6.84}$$

Multiplying both sides of this equation by $\left(\sqrt{\overline{N_\tau}+N_\text{n}} - \sqrt{N_\text{n}}\right)$ and applying the textbook identity $(a+b)(a-b) \equiv a^2 - b^2$ to the right-hand side, we get:

$$\overline{N_\tau}\left(\sqrt{\overline{N_\tau}+N_\text{n}} - \sqrt{N_\text{n}}\right) = \frac{\text{SNR}}{2}\sqrt{\mathcal{F}\,\Gamma_1'}\ \overline{N_\tau}\ . \tag{6.85}$$

Now we cancel out $\overline{N_\tau}$ on both sides, carry over $\sqrt{N_\text{n}}$ from the left-hand to the right-hand side,

$$\sqrt{\overline{N_\tau}+N_\text{n}} = \frac{\text{SNR}}{2}\sqrt{\mathcal{F}\,\Gamma_1'} + \sqrt{N_\text{n}}\ , \tag{6.86}$$

and square the equation:

$$\overline{N_\tau} + N_n = \left(\frac{\mathrm{SNR}}{2}\right)^2 \mathcal{F}\Gamma_1' + \mathrm{SNR}\sqrt{\mathcal{F}\Gamma_1' N_n} + N_n \ . \tag{6.87}$$

Now we only have to cancel equal terms N_n on the two sides of the equation to get the final expression for the lowest detectable number of photoelectrons:

$$\overline{N_\tau} = \mathcal{F}\Gamma_1'\left(\frac{\mathrm{SNR}}{2}\right)^2 + \mathrm{SNR}\sqrt{\mathcal{F}\Gamma_1' N_n} \ . \tag{6.88}$$

Like most other main expressions of the detection theory, this formula is simple and clear.

When the noise of the amplifier is negligibly small, the discrimination threshold may be set low, $N_t << \overline{N_\tau}$ (recall that noise outbursts in the dark are very rare). Then, according to (6.78), the signal intensity is determined from considerations of the allowable target miss:

$$n_{tm} = \frac{\overline{N_\tau} - N_t}{\sqrt{\mathcal{F}\Gamma_1' \overline{N_\tau}}} \approx \frac{\overline{N_\tau}}{\sqrt{\mathcal{F}\Gamma_1' \overline{N_\tau}}} = \sqrt{\frac{\overline{N_\tau}}{\mathcal{F}\Gamma_1'}} \ ,$$

$$\overline{N_\tau} = \mathcal{F}\Gamma_1' n_{tm}^2 = \mathcal{F}\Gamma_1'\left(\frac{\mathrm{SNR}}{2}\right)^2 . \tag{6.89}$$

In the other extreme, when the noise of the PDA prevails, we get a trivial dependence $\overline{N_\tau} = \mathrm{SNR}\sqrt{\mathcal{F}\Gamma_1' N_n}$. Formula (6.88) sums up these two values! In other words, the lowest detectable signal intensity is additive: its one part ($\mathcal{F}\Gamma_1'(\mathrm{SNR}/2)^2$) is determined only by the noise of the signal, and the other ($\mathrm{SNR}\sqrt{\mathcal{F}\Gamma_1' N_n}$) only by the noise of the PDA.

Equally simple is the expression for the lowest detectable number of photoelectrons when $P_{fa} \neq P_{tm}$ and $n_{fa} \neq n_{tm}$. From (6.83) we get a quadratic equation in $\overline{N_\tau}$, whose solution is

$$\overline{N_\tau} = \mathcal{F}\Gamma_1' \frac{n_{tm}^2}{2} + n_{fa}\sqrt{\mathcal{F}\Gamma_1' N_n}$$
$$+ \sqrt{\left(\mathcal{F}\Gamma_1' \frac{n_{tm}^2}{2}\right)^2 + (\mathcal{F}\Gamma_1')^{3/2} n_{tm}^2 n_{fa}\sqrt{N_n} + \mathcal{F}\Gamma_1' N_n n_{tm}^2} \ . \tag{6.90}$$

Expression (6.90) is cumbersome; in the middle term under the root, however, we may replace $n_{fa} \Rightarrow n_{tm}$, which practically brings us back to (6.88):

$$\overline{N_\tau} = \mathcal{F}\Gamma_1' n_{\rm tm}^2 + \mathrm{SNR}\sqrt{\mathcal{F}\Gamma_1' N_{\rm n}} \ . \tag{6.91}$$

Even in the most unfavorable situation, with $\mathcal{F}\Gamma_1' N_{\rm n} \approx 1$ and asymmetrical values of $n_{\rm tm} = 3$, $n_{\rm fa} = 6$ (see Sect. 3.1), the error of this approximation is within 6...9%.

What kind of PDA is capable of detecting fluctuations of the signal? The expressions obtained so far are sufficient for going over to numerical evaluations. Namely, we are going to determine the level of the PDA noise which does not yet jam completely the noise of the signal. For the ultimately admissible level we choose such value of the PDA noise $[N_{\rm n}]$ for which the lowest detectable signal N_τ increases two-fold. Then, according to (6.88),

$$\mathcal{F}\Gamma_1'\left(\frac{\mathrm{SNR}}{2}\right)^2 \equiv \mathrm{SNR}\sqrt{\mathcal{F}\Gamma_1'[N_{\rm n}]} \ ,$$

$$N_{\rm n} < [N_{\rm n}] = \frac{1}{4}\left(\frac{\mathrm{SNR}}{2}\right)^2 \mathcal{F}\Gamma_1' = \frac{1}{4}\cdot(3...6)^2\cdot 4\cdot 0.68 \approx 6...24 \ . \tag{6.92}$$

Here and further we use the example of the broad-band PDA with the two-section filter and silicon APD ($\mathcal{F} = 4$, $\Gamma_1 = \Gamma_1' = 0.68$, $\mathrm{SNR} = 6...12$). We have obtained a very rigorous condition: it is required that the noise power of the PDA should be one-fourth of the power of the signal noise $\overline{\Delta N_\tau^2} \approx N_\tau = \mathcal{F}\Gamma_1'(\mathrm{SNR}/2)^2$ (6.89). Of course, this low value of $[N_{\rm n}]$ cannot affect considerably the total noise $\overline{N_\tau} + [N_{\rm n}]$ in the presence of the signal (6.77). The reason behind this rigorous requirement is different: it is the false alarm. The noise $[N_{\rm n}]$ gives rise to noise outbursts in the absence of the signal; because of this, we must raise the threshold, and, as a consequence, the power of the signal.

Let us evaluate each component of the PDA noise, starting with the main of these, the high-frequency component. Retaining it in the expression for $N_{\rm n}$ (6.74), we get:

$$N_{\rm n} = \left(\frac{e_{\rm n} C_{\rm in}\pi}{q\overline{M}}\right)^2 \frac{f_{\rm a}\Gamma_2}{\mathcal{F}\Gamma_1'} < [N_{\rm n}] = \left(\frac{\mathrm{SNR}}{2}\right)^2 \frac{\mathcal{F}\Gamma_1'}{4} \ . \tag{6.93}$$

Naturally, the contribution of the high-frequency noise decreases as the frequency band is made narrower; accordingly, for a given set of $e_{\rm n}$, $C_{\rm in}$, $\overline{M}$ there always exists a characteristic bandwidth $[f_{\rm a}]$ below which the signal noise prevails:

$$f < [f_{\rm a}] = \left(\frac{\mathcal{F}\Gamma_1' q\overline{M}\,\mathrm{SNR}}{2\pi e_{\rm n} C_{\rm in}}\right)^2 \frac{1}{4\Gamma_2} =$$

$$= \left(\frac{4 \cdot 0.68 \cdot 1.6 \cdot 10^{-19} \cdot 100 \cdot (6...12)}{2\pi \cdot 2 \cdot 10^{-9} \cdot 5 \cdot 10^{-12}} \right)^2 \frac{1}{4 \cdot 0.51} \approx (0.8...3) \cdot 10^7 \text{ Hz} . \quad (6.94)$$

The accepted values of e_n, C_{in}, $\overline{M}$ (far not the record-breaking) are clear from the calculation. We see that in amplifiers with a very broad band (about 10^7 Hz) it is possible to achieve the value of $[N_n] = 6...24$ noise electrons. To a large extent, of course, this is the merit of the APD: such number $[N_n]$ is obtained with reference to the conventional input of the PDA before the multiplication stage, with respect to the primary photocurrent. The power at the input of the amplifier itself is greater than that given by (6.92) by the factor of $\overline{M}^2 = 10^4$, and is of the order of $(6...24) \cdot 10^4$ noise electrons.

Now we know the band f_a, and hence the effective time constant $\tau_{eff} = 1/2f_a$, which means that we also know the number of dark electrons registered within this time: $N_0 = I_0 \tau_{eff}/q = I_0/2qf_a$. Therefore, we can define the restriction imposed on the dark (total) current I_0:

$$I_0 = 2qf_a N_0 < 2qf_a [N_n] = 2 \cdot 1.6 \cdot 10^{-19} (0.8...3) \cdot 10^7 \cdot (6...24)$$

$$\approx (1.5...25) \cdot 10^{-11} \text{ A} . \quad (6.95)$$

The minimum load is, in turn, determined by the obtained value of the PD current (2.183):

$$R_{in} \le \frac{2\varphi_t}{I_0 \overline{M}^2 \mathcal{F}} = \frac{2 \cdot 0.026}{(1.5...25) \cdot 10^{-11} \cdot 10^4 \cdot 4} \approx 0.5 \cdot 10^4 ... 10^5 \, \Omega . \quad (6.96)$$

These estimates indicate that the analysis of signal fluctuations is not of purely theoretical interest: the requirements imposed on the PDA are realistic, and the noise of the signal may be comparable to or even greater than the noise of the PDA.

Optimal filtering with due account for the signal noise. So, we have obtained the expressions for the signal-to-noise ratio with due account for the noise of the signal, and found the lowest detectable number of photoelectrons $\overline{N_\tau}$. These expressions can be used for calculating SNR and $\overline{N_\tau}$ given the passband (time constant) of the filter $f_a = 1/2\tau_{eff}$. In the broad-band systems the band is determined by the allowable distortion of the signal profile (Sect. 5.3), and may be regarded as fixed. In case of optimal filtering, however, the passband f_a itself (and the frequency response profile) are chosen in such a way as to maximize the SNR. Optimal for the noiseless PDA is the large time constant, which allows the filter to count out all the photoelectrons of the signal; accordingly, for the bell-shaped signal we have $\tau_{eff} \ge 2T_s$ (6.33). When the white noise of the amplifier dominates, the maximum SNR is achieved with the limited time constant of the filter, equal to the signal duration, $\tau_{eff} = T_s$ (see Sect. 3.2). Thus, we are faced with another problem: the optimal filter should be adaptive, and its frequency response has to accommodate itself to the relationship between the noise of the signal and the noise of the PDA. In other words, we have just formulated one of the sophisticated nonlinear problems of the detection theory. Fortunately, it is possible to get away from solving this problem:

1. In the classical case of rectangular signal and white noise of the PDA, the filter-integrator with the time constant $\tau_{eff} = T_s$ does not need to be tuned up. Such filter is optimal whether the signal noise or the amplifier noise dominates.

2. There is also no need to tune up the filter when the high-frequency noise of the PDA prevails. Optimal for such noise spectrum is a large time constant $\tau_{eff} \approx 1/\pi f_{high}$ (5.57), which is sufficient for the integration of the short signal, $\tau_{eff} >> T_s$.

3. Finally, the most unfavorable situation occurs when the signal is bell-shaped, and the noise of the PDA is white. In defiance of the theory, let us try to get by without retuning the filter in this case as well, in the hope that the flat tuning curve will come to our rescue. Once again we use the example of the two-section RC-filter (Ch. 5), but this time with a certain medium-valued time constant $\tau_{eff} = e\tau_f = \sqrt{3}T_s$. This is somewhat larger than the time constant of quasi-optimal filter in case of the white noise ($\tau_{eff} = T_s$), but less than the value required for the detection of the signal when only the signal fluctuations are taken into account ($\tau_{eff} \geq 2T_s$). Of course, this detuning will result in a certain loss in the SNR. We already have the necessary tools for evaluating this loss: they are the formulas (5.29), (5.66), (6.45), (6.42).
The loss coefficient in case of the prevalent white noise of the PDA is

$$\mathcal{L} \approx \sqrt{\frac{\dfrac{2\tau_{eff}}{T_s}}{1+\left(\dfrac{\tau_{eff}}{T_s}\right)^2}} = \sqrt{\frac{2\sqrt{3}}{1+\left(\sqrt{3}\right)^2}} = 0.93\ , \tag{6.97}$$

and the same in case of the prevalent signal noise is

$$\mathcal{L} \approx \sqrt{\frac{N_\tau}{N_s}} = \sqrt{\frac{\tau_{eff}}{T_s}u_{sA}} = \sqrt{\frac{\tau_{eff}}{T_s}\frac{1}{\sqrt{1+\left(\dfrac{\tau_{eff}}{T_s}\right)^2}}} = \sqrt{\frac{\sqrt{3}}{\sqrt{1+\left(\sqrt{3}\right)^2}}} = 0.93\ . \tag{6.98}$$

It is hardly worthwhile to develop the theory of nonlinear filtering and synthesize sophisticated adaptive filters all because of the mere 7%.

Inclusion of the noise distribution law. We are not yet through with the calculation of the lowest detectable number of signal quanta $\overline{N_\tau}$. So far we have only considered two factors: the nonstationary nature of the noise (which increases when the signal arrives at the input), and the optimization of the passband. The most difficult problem associated with the detection of quantized signals, however, is the third one: the distribution of the noise at the output of the PDA may be non-Gaussian. So far we have been calculating both the error probabilities P_{fa}, P_{tm}, and the lowest detectable signal $\overline{N_\tau}$ under the assumption that the distribution is Gaussian. Have we been sweating in vain?

As indicated in Sect. 2.2 and 3.1, the universality of the Gaussian (normal) law is explained by the fact that this is the law of large numbers. In our context the number is the mean number of electrons $N_0 + N_\tau$, registered over the observation time τ_{eff}. For most PDAs this number is large. For example, for the PDAs of spectral ranges II – V, in accordance with Table 2.1 and (6.95), and with the typical values of $\tau_{eff} \approx 10^{-6}...10^{-3}$ s, we get:

$$u_{sA} = 1 \; ,$$

$$\overline{N_0} = \frac{I_0 \tau_{eff}}{q} = \frac{\left(10^{-9}...2\cdot 10^{-4}\right)\left(10^{-6}...10^{-3}\right)}{1.6\cdot 10^{-19}} \approx 6\cdot 10^{3}...10^{12} \; . \tag{6.99}$$

From this estimate it follows that the response at the output of such PDAs is a superposition of a very large number of pulses, $N_0 \approx \overline{N_0}$. Because of the fluctuation of N_0, it is practically any value of $U_n(t_m)$ at the time t_m that is feasible at the output of the PDA: at least, the probability of any value of U_n within the margin of several values of the dispersion is noticeable. The discreteness of the possible values is very small, $\delta U_n / U_n \approx 1/\sqrt{N_0}$. All this has allowed us to regard the noise as a random arbitrary function, and the large number N_0 resulted in the Gaussian distribution.

The situation is entirely different when on the average it is one electron or less that falls within the observation time τ_{eff}:

$$I_0 \le \frac{q}{\tau_{eff}} = \frac{1.6\cdot 10^{-19}}{10^{-6}...10^{-9}} = 1.6\cdot\left(10^{-13}...10^{-10}\right) \text{A} \; . \tag{6.100}$$

Such currents are achievable with silicon PDs. The values of $\tau_{eff} = 10^{-6}...10^{-9}$ s correspond to the actual response time of PDAs based on such PDs. The noise at the output of the PDA, when it is determined by the fluctuation of such small currents, cannot be any arbitrary random function: now only those values are possible which correspond to the two most probable events: the electron either has or has not arrived at the input during the observation time τ_{eff}. The noise distribution at the output of the PDA cannot be Gaussian (at least because the Gaussian law implies a broad spectrum of values of U_n). The distribution function is found from the general representation of the output response as a pulsed random sequence — the set of pulses $M_i \mathrm{H}(t_m - t_i)$, with the random time of arrival of the primary electron t_i, and random amplitude M_i when there is multiplication. Strictly speaking, the distribution of the output noise (the amplitude of the response $U_n(t_m)$) differs in general from the Poisson distribution of the input noise (the number of electrons $N_0 + N_\tau$). This is clear from the following argument. Assume that one event has occurred at the input: one primary electron has arrived. At the output, at the time t_m, we have a whole spectrum of the possible events: depending on the time of arrival of the primary electron t_i, and on the value of the multiplication coefficient M_i, the output voltage may take on not one but many different values $M_i \mathrm{H}(t_m - t_i)$. In this way, we have two more "large numbers": the number of possible times of arrival t_i, and the number of possible values of the multiplication coefficient M_i.

Let us leave these factors out of consideration, and confine ourselves to the classical case of optimal reception of rectangular signal against the background of the shot noise of the dark current of PD. In this way we shall fully elucidate the "abnormality" of the Poisson distribution, and its effects on the characteristics of detection. Optimal for the case in question is the well familiar filter-integrator with the effective integration time $T_i = \tau_{eff} = T_s$, for which $\Gamma_i' = 1$ (Table 5.1). We also assume that $\mathcal{F} = 1$, $N_n = \overline{N_0}$ — in other words, the noise of multiplication and the noise of the amplifier are not taken into account (the feasibility of such regime is considered in Sect. 6.3 and 7.1). As indicated above, the output voltage of filter-integrator is proportional to the number of signal electrons and dark electrons $N_\tau + N_0$, arriving within the interval τ_{eff} (Fig. 6.2); therefore, its fluctuation repeats the Poisson fluctuation of this number.

The calculation algorithm for the lowest detectable signal is the same for any distribution function of the output noise. Like in case of the Gaussian distribution, we first choose the discrimination threshold N_t. A false alarm will occur in the absence of the signal if the number N_0 of dark carriers generated over the time τ_{eff} exceeds the discrimination threshold N_t. Naturally, the aggregate probability of all realizations $N_0 > N_t$ must not exceed the allowable false alarm probability P_{fa}. For the Poisson distribution we get:

$$P(N_0 \geq N_t) = \sum_{N_0 = N_t}^{\infty} e^{-\overline{N_0}} \frac{\overline{N_0}^{N_0}}{N_0!} \leq P_{fa} . \tag{6.101}$$

Then, using the selected value of N_t, we find the mean value of the signal $\overline{N_\tau}$ such that the overall probability of all realizations $N_0 + N_\tau < N_t$ does not exceed the preassigned target miss probability P_{tm}:

$$P(N < N_t) = \sum_{N=0}^{N_t - 1} e^{-\overline{N}} \frac{\overline{N}^N}{N!} \leq P_{tm} , \quad N = N_0 + N_\tau . \tag{6.102}$$

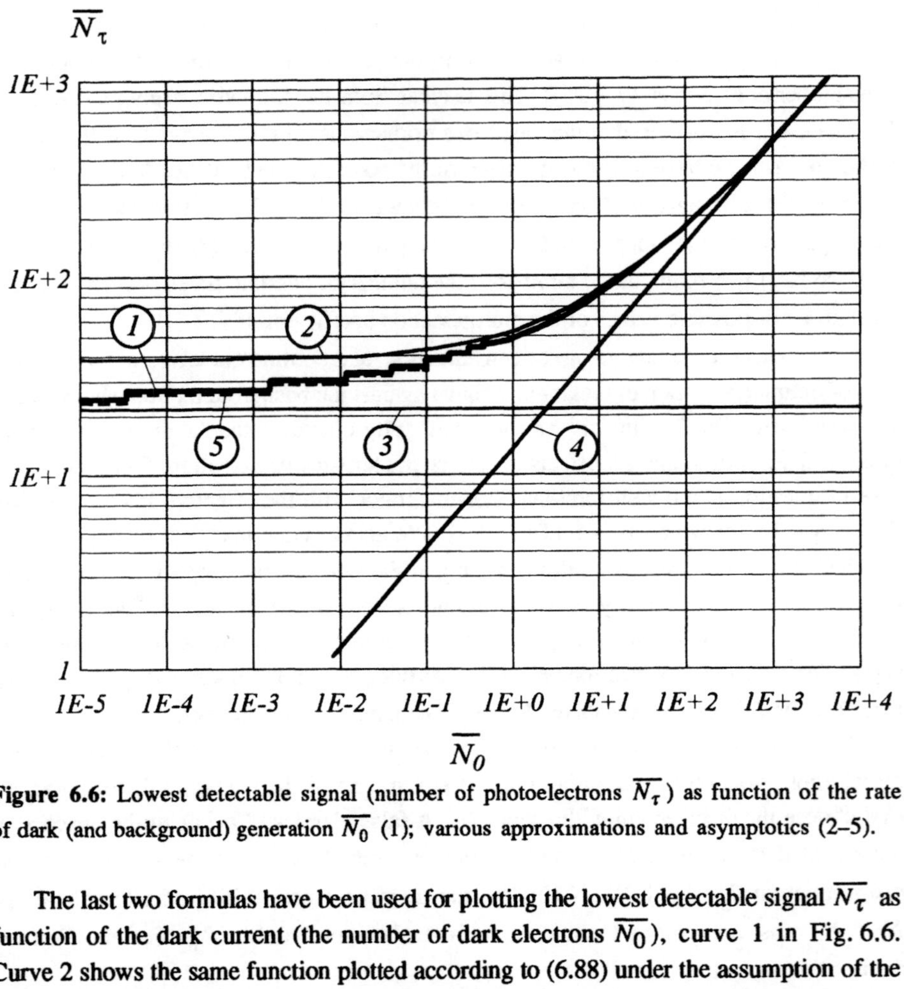

Figure 6.6: Lowest detectable signal (number of photoelectrons $\overline{N_\tau}$) as function of the rate of dark (and background) generation $\overline{N_0}$ (1); various approximations and asymptotics (2–5).

The last two formulas have been used for plotting the lowest detectable signal $\overline{N_\tau}$ as function of the dark current (the number of dark electrons $\overline{N_0}$), curve 1 in Fig. 6.6. Curve 2 shows the same function plotted according to (6.88) under the assumption of the Gaussian distribution of the signal noise and the dark current noise. And now for the limiting curves: in the absence of dark generation ($\overline{N_0} = 0$) we set the threshold at $N_t = 1$, and then the lowest detectable signal was $\overline{N_\tau} = \ln(1/P_{tm})$ (6.24); straight line 3 in Fig. 6.6. In the other extreme, when the dark current is very large, the shot noise of the latter dominates, and delimits the lowest detectable signal by the value of $\overline{N_\tau} = \mathrm{SNR}\sqrt{\overline{N_0}}$ (straight line 4).

Now in Fig. 6.6 we have a full display of all the particulars in the reception of quantized signals. The fluctuations of the dark current become definitive when $\overline{N_0} \geq 200...1000$: in this region curves 1 and 4 are indistinguishable, and the lowest detectable signal may be calculated by the trivial formula $\overline{N_\tau} = \mathrm{SNR}\sqrt{\overline{N_0}}$. When the rate of dark generation is lower, the effects of the signal noise become more and more noticeable. In this current range we get a pleasant surprise: curves 1 and 2 for the Poisson and the Gaussian distributions practically coincide down to $\overline{N_0} \sim 1$. The number of dark and signal carriers in this range is still relatively large, and so the Poisson distribution stays close to the Gaussian one.

The number $\overline{N_0} \leq 1$, however, is by no means large, and it is not surprising that the difference between curves 1 and 2 becomes considerable. The PDA reacts keenly even to a very small number of dark electrons $\overline{N_0} << 1$, which seemingly could have been disregarded. Of course, the noise of this small number of electrons ($\sqrt{\overline{N_0}} << 1$) is less than the noise of the signal electrons ($\sqrt{\overline{N_\tau}} = \sqrt{7...21} \approx 3...5$), and may be neglected in the calculation of target miss probability, but not in the calculation of false alarms. Even this very low rate of generation gives rise to noticeable noise outbursts, which makes it necessary to raise the discrimination threshold N_t, and this, in turn, increases the lowest detectable value of the signal $\overline{N_\tau}$.

This may be supported by the following calculations. When the rate of dark generation is low, $\overline{N_0} << 1$, equation (6.101), which relates this rate to the required threshold N_t, becomes simpler. Now only the pulses with $N_0 = N_t$ are responsible for the false alarms, since the probability of occurrence of pulses with a larger number of electrons $N_0 > N_t$ (6.22) is negligible. This allows us to retain in (6.101) only one term with $N_0 = N_t$:

$$P\left(N_0 \geq N_t\right) = \sum_{N_0=N_t}^{\infty} e^{-\overline{N_0}} \frac{\overline{N_0}^{N_0}}{N_0!} \approx e^{-\overline{N_0}} \frac{\overline{N_0}^{N_t}}{N_t!} \approx \frac{\overline{N_0}^{N_t}}{N_t!} . \tag{6.103}$$

The highest allowable rate of dark generation $\overline{N_0}$ for the selected discrimination threshold N_t may be written explicitly as

$$\overline{N_0} \leq \sqrt[N_t]{N_t! P_{fa}} . \tag{6.104}$$

Assume, for example, that $P_{fa} = 5 \cdot 10^{-10}$. Then the PDA pays no attention to the dark current as long as $\overline{N_0} \leq P_{fa} = 5 \cdot 10^{-10}$. Indeed, according to (6.19), the probability of one-electron pulses is $p(1) = \overline{N_0} < 5 \cdot 10^{-10}$, and so, retaining the initial value of the threshold $N_t = 1$, we get a quite acceptable probability of false alarms $P_{fa} = p(1) = \overline{N_0} < 5 \cdot 10^{-10}$. When the rate of dark generation is higher, the discrimination threshold must be raised to $N_t = 2$. This value is sufficient as long as $\overline{N_0} < \sqrt[2]{1 \cdot 2 \cdot \left(5 \cdot 10^{-10}\right)} \approx 3.1 \cdot 10^{-5}$. When $\overline{N_0} > 3.1 \cdot 10^{-5}$, we must raise the threshold again, setting $N_t = 3$, etc. Since N_0 and N_t are integers, they change in steps, each time by one. Because of this, the lowest detectable signal $\overline{N_\tau}$ also changes stepwise, which becomes noticeable when the dark current is extremely low.

Having selected N_t, for calculating $\overline{N_\tau}$ we again use the "kinship" between the Poisson and the Gaussian distributions: the number of photoelectrons in our signals is not too small ($\overline{N_\tau} \gtrsim 7...21$), so we may hope that the distributions will be similar. So as to ensure the desired probability of target miss P_{tm}, we must maintain the same gap between the signal amplitude and the threshold — or, to be more precise, maintain the ratio of this gap to the noise (6.77):

$$n_{tm} = \frac{U_s - U_t}{U_n} \approx \frac{\overline{N_\tau} - N_t}{\sqrt{\overline{N_\tau}}} = \sqrt{\overline{N_\tau}} - \frac{N_t}{\sqrt{\overline{N_\tau}}} = \text{const} . \tag{6.105}$$

The derivative of a constant must be equal to zero,

$$\left(\frac{1}{2\sqrt{\overline{N_\tau}}} + \frac{N_t}{2\overline{N_\tau}\sqrt{\overline{N_\tau}}}\right)\delta\overline{N_\tau} - \frac{\delta N_t}{\sqrt{\overline{N_\tau}}} = 0 \ ; \ \delta\overline{N_\tau} = \frac{2\delta N_t}{1 + \dfrac{N_t}{\overline{N_\tau}}} \approx 2\delta N_t . \tag{6.106}$$

Numerical analysis for the Poisson distribution (6.102) reveals that the formula obtained is an approximation for the increments $\delta\overline{N_\tau} = \overline{N_\tau} - \overline{N_{\tau 0}}$, $\delta N_t = N_t - 1$. Only when the false alarm probability is reduced from $5 \cdot 10^{-4}$ to $5 \cdot 10^{-10}$, the numerical coefficient increases from 2 to 2.6. The increments are counted from the initial point, when there are no dark carriers, $N_t = 1$, and $\overline{N_{\tau 0}} = \ln(1/P_{tm})$ (6.24). Hence,

$$\overline{N_\tau} \approx \overline{N_{\tau 0}} + (2...2.6)(N_t - 1) = \ln\frac{1}{P_{tm}} + (2...2.6)(N_t - 1) . \tag{6.107}$$

Curve 5 in Fig. 6.6 is plotted from analytical approximations (6.104), (6.107).

So, we have successfully dealt with the last and most difficult problem, and took the real Poisson distribution into account. It turns out that with $\overline{N_0} \gtrsim 1$ we may calculate the lowest detectable signal using our old analytical approximations (6.88), (6.91), obtained under assumption of the Gaussian nature of the PDA and signal noise. After all, one dark electron per signal length T_s is a very low rate of dark generation. The dark current of the modern silicon-PD-based PDAs is usually not lower than $10^{-13}...10^{-10}$ A, which means that the dark generation rate is not lower than the value of $\overline{N_0} \sim 1$ as indicated above (see (6.101)). Of course, with special designs, and especially with cooling, it is possible to beat the records and make the dark current as small as $\sim 10^{-16}...10^{-14}$ A, which means that $\overline{N_0} << 1$ (Sect. 6.3). The particulars of the Poisson distribution only come up when the rate of dark generation is as low as that. This circumstance should not be overlooked: the calculation under the assumption of Gaussian distribution may exaggerate the lowest detectable signal power up to 1.8-fold (cf. (6.23) and (6.24)). Moreover, for the Poisson distribution we have at our disposal convenient analytical expressions (6.104), (6.107).

Summary. *The quantum nature of radiation comes upstage in the detection of ultimately low optical signals. Because of the random losses on the way from the optical* *source to the PDA the number of photons in the incident signal fluctuates. It is not possible to detect signals which contain less than 7...21 photons — otherwise the probability of realization of pulses with the zero number of photons will exceed the allowable probability of target miss ($5 \cdot 10^{-4} \ldots 5 \cdot 10^{-10}$). In the general case, the noise of the signal must be taken into account along with the noise of the PDA. This gives rise to a number of problems.*

1. Since the signal noise is observed only when there is a signal at the input, the total noise is nonstationary.

2. The optimal filter becomes adaptive. When the white noise of the PDA dominates, the effective time constant of the filter τ_{eff} is made equal to the effective signal duration T_s. When the signal noise dominates, the filter must be capable of registering all the photoelectrons, and so its time constant must be greater than the signal duration at its base, and so $\tau_{eff} \geq 2T_s$ for the bell-shaped signal.

3. The signal noise at the input obeys the Poisson rather than the Gaussian distribution as has been assumed so far in this book, and the distribution of the signal noise at the output may be quite diverse.

The theory supplies solutions to these problems.

1. The nonstationarity of the noise is easily taken into account: the false alarm probability is calculated, as always, for the intrinsic (dark-current) noise of the PDA, and the signal noise is added to the latter in the calculation of the target miss probability. The signal noise corresponds to the shot noise of the signal current averaged over the observation time τ_{eff}.

2. The flat tuning curve allows tuning the filter to a certain midway setting, without making it adaptive. The loss in the SNR *will not exceed 7% at worst.*

3. The Poisson nature of the fluctuations tells only in case of the extremely low dark current, when no more than one pair of carriers is generated during the time of observation τ_{eff}. When the dark current is larger, the lowest detectable signal may be calculated under the assumption of the Gaussian distribution of the noise. Previously we did this by multiplying the PDA noise $\sqrt{\mathfrak{F}\,\Gamma\acute{\imath} N_n}$ by the desired value of the SNR. *Now we must include the noise of the signal itself, which is very easy to do: the term* $\mathrm{SNR}\sqrt{\mathfrak{F}\,\Gamma\acute{\imath} N_n}$ *is supplemented by the term $n_{tm}^2 \mathfrak{F}\,\Gamma\acute{\imath}$, which depends on the preassigned target miss probability.*

6.3. Photon Counting Mode of Avalanche Photodiode

As indicated in Sect. 6.2, in case of a very low generation rate and a low-inertia amplifier, when the period between two consecutively arriving electrons T_{per} is greater than the effective time constant of the amplifier τ_{eff}, the current as though splits into separate non-overlapping one-electron pulses. Because of this, it becomes possible to register each separate electron. The photodetector can fully reveal the capabilities of the intrinsic photoeffect and literally become a counter of the incident photons, like a photomultiplier tube. The photon count mode may be realized with the aid of APD, because the avalanche multiplication helps to suppress the noise of the amplifier.

Avalanche photodiode as photon counter. Let us try to count photons with an APD employed in the ordinary linear regime, when the working voltage U_{w} is less than the breakdown voltage U_{B}, and the avalanche is essentially a linear preamplification stage with the gain equal to the multiplication coefficient $\overline{M}$.

Photon counting with linear multiplication. The decision that the signal (one primary electron in the multiplication layer) has arrived at the input is taken when the amplitude of the output voltage exceeds the preset threshold

$$U_{\text{t}} = n_{\text{fa}} U_{\text{n}} = n_{\text{fa}} e_{\text{n}} C_{\text{in}} 2\pi \sqrt{\Gamma_2 f_{\text{a}}^3} \; . \tag{6.108}$$

The threshold voltage U_{t} is chosen by the factor of $n_{\text{fa}} = 3...6$ higher than the noise U_{n}; otherwise the frequency of false alarms because of the noise outbursts will exceed the allowable level (Sect. 3.1). For the noise in (6.108) we only retain the basically unremovable high-frequency component (5.54). The multiplication coefficient is selected in such a way that the signal N_{s} caused by one primary electron q should be sufficient for triggering the discriminator. This value of the multiplication coefficient M_{t} will be called the threshold value, and so

$$U_{\text{s}} = M_{\text{t}} I_{\text{s}} = \frac{M_{\text{t}} q}{\tau_{\text{eff}}} = M_{\text{t}} q 2 f_{\text{a}} \geq U_{\text{t}} = n_{\text{fa}} e_{\text{n}} C_{\text{in}} 2\pi \sqrt{\Gamma_2 f_{\text{a}}^3} \; . \tag{6.109}$$

In this section we respect our tradition proclaimed in Sect. 3.3, and assume that $K_0 = 1$; therefore, the output voltage must be regarded as a dimensionless quantity, or the value referred to the input.

From inequality (6.109) and equation (6.104) it follows that

$$M_{\text{t}} \geq \frac{n_{\text{fa}} e_{\text{n}} C_{\text{in}} \pi \sqrt{\Gamma_2 f_{\text{a}}}}{q} \; . \tag{6.110}$$

This inequality defines the required value of the multiplication coefficient. The value of the multiplication coefficient reflects the peculiarities of the photon count mode, due to the random nature of the avalanche. In this book we actually consider three situations corresponding to three levels of signals.

We began with the ordinary signals, with $N_\tau \sim 10^2 \ldots 10^3$ photoelectrons and more (see Sect. 1.2, 6.1). The avalanche multiplication of this packet of photoelectrons was described by a deterministic quantity, the mean multiplication coefficient $\overline{M}$. Indeed, each act of detection of such a relatively large signal packet may be regarded as a straightforward experimental assessment of the mean value $\overline{M}$. Of course, the multiplication of each separate electron from this packet is a random process: the first electron is multiplied M_1 times, the second M_2 times, the ith M_i times. The amplifier is an integrator with the effective integration time τ_{eff}. It adds up all the secondary electrons generated during this time, $U_s \sim \sum_{i=1}^{N_\tau} M_i$. The multiplication coefficient is defined as the ratio of the total of all secondary electrons to the total of all primary electrons, $M \sim \sum_{i=1}^{N_\tau} M_i \big/ N_\tau$. But this is the standard algorithm for finding the mean! The greater the number of samples is (which in our case is the number of electrons in the packet N_τ, the closer the measured value of M is to its mean value $\overline{M}$. This is why we assume that $M \approx \overline{M}$ when the signal is relatively strong ($N_\tau > 100 \ldots 1000$). The random nature of multiplication in this case comes up in the calculation not of the signal but of the noise (multiplication of the dark carriers).

The second case refers to the weaker signals with $N_\tau \sim 10 \ldots 100$ (see Sect. 6.2). For calculating the mean value of the output signal we have also used the mean value of the multiplication coefficient $\overline{M}$. However, because the value of N_τ is relatively small, the fluctuations of M from one pulse to another become noticeable, so that the target miss probability cannot be computed without taking into account the vacillation of the output signal about its mean value. We had to take care of the random nature of the avalanche not only with respect to the thermal noise, but also with respect to the noise of the signal, where it was reflected by the noise factor $\mathcal{F}$.

And now we encounter the third situation, which is the most complicated of all: the multiplication of just a single electron. The law of large numbers does not help us in any way, since $N_\tau = 1$. The random nature of the output signal qM/τ_{eff} entirely repeats the random nature of the multiplication coefficient M, and so further analysis is not possible without knowing the distribution of this coefficient. The probability $p(M)$ that one electron will be multiplied M times is calculated in [67, 93]. This distribution $p(M)$ for one electron is much "worse" than that for the packet consisting of many electrons. The probable values of M in the latter case concentrate around the mean value $\overline{M}$, whereas the distribution function for one electron is much more diffuse. The most probable are the small values of M, $M = 1,2,3,\ldots$. The larger M, the less likely it is to occur. Consequently, it is not possible to secure the desired value of M_t for each separately taken carrier. This brings us to a situation very typical of the detection theory (see

Sect. 3.1): there is a finite and quite considerable probability of realizations for which $M < M_t$. In such cases the discriminator will not be triggered, which will result in further reduction of the quantum efficiency in the photon count mode. Accordingly, the quantum efficiency of the APD is now $\eta\eta_c'$, where η_c' is the photon count efficiency which is equal to the sum of probabilities of all realizations with $M \geq M_t$. A simple analytical expression can be obtained for such integral probability in case of the one-carrier process, when multiplied is only one type of carriers [67]:

$$\alpha >> \beta \ , \quad \eta_c' = P(M \geq M_c) = \exp\left(-\frac{M_t - 1}{\overline{M}}\right) . \tag{6.111}$$

From (6.111) it follows that the desired counting efficiency η_c' (for example, 0.8...0.9) will be ensured if the mean multiplication coefficient $\overline{M}$ is chosen so as to exceed the threshold value M_t required for each realization by the factor of

$$\overline{M} >> 1 \ ; \quad 1 - \eta_c' << 1 \ , \quad \ln \eta_c' = -\frac{M_t}{\overline{M}} - \frac{1}{\overline{M}} \approx -\frac{M_t}{\overline{M}} \ ,$$

$$\gamma = \frac{\overline{M}}{M_t} \approx \frac{1}{-\ln \eta_c'} = \frac{1}{-\ln\left[1 - (1 - \eta_c')\right]} \approx \frac{1}{1 - \eta_c'} = \frac{1}{1 - (0.8...0.9)} = 5...10 \ . \tag{6.112}$$

This estimate reveals the unwelcome feature of the photon count mode: the additional loss in the quantum efficiency η_c' because of the fluctuation of the multiplication coefficient (the diffuse function $P(M_t)$). In order to make this loss acceptable ($\eta_c' = 0.8...0.9$), the mean value $\overline{M}$ has to be made quite large: 5 to 10 times as large as the value of M_t (in case of one-carrier process).

Amplifier passband. Condition (6.110) must be satisfied in order that the noise of the amplifier should not interfere with the registration of the single electron. To make a numerical estimate, we assume that $\gamma = 5...10$, $n_{fa} = 3...6$, $e_n = 2\,\mathrm{nV/Hz^{1/2}}$, $C_{in} = 5\,\mathrm{pF}$, $\overline{M} = 200$. This is a rather high, though not a record-breaking performance. Now we only have to choose the amplifier passband f_a. The narrower this passband, the less the amplifier noise interferes with the count of electrons. According to (6.110), (6.112), the maximum allowable passband f_a (other parameters being fixed) is

$$f_a \leq \frac{1}{\Gamma_2}\left(\frac{q\overline{M}}{\gamma n_{fa} e_n C_{in} \pi}\right)^2 = \frac{1}{0.5}\left[\frac{1.6 \cdot 10^{-19} \cdot 200}{(5...10) \cdot (3...6) \cdot 2 \cdot 10^{-9} \cdot 5 \cdot 10^{-12} \cdot \pi}\right]^2$$

$$\approx 0.6 \cdot 10^3 ... 10^4\ \mathrm{Hz} \ . \tag{6.113}$$

The capability of the APD in the ordinary linear regime turns out to be more than modest: it is only able to drag out the one-electron signal from the noise of a very narrow-band amplifier. So, in principle, the APD counts electrons, but does this very slowly. If we take the "evolutionary" approach, and try to improve upon the values of $\overline{M}$, e_n, C_{in}, it may be possible to expand the band by one or two orders of magnitude, but this will still be not sufficient for fast systems. Because of this, there had to be a revolution: if the broad band calls for the practically infinite multiplication $\overline{M}$ (6.110), let it be so. And so, we go over to the breakdown regime.

Photon counting in the breakdown regime. The working voltage applied to the APD is now $U_w > U_B$. The electron falling into the strong-field region triggers the avalanche, and the current starts to increase practically infinitely. In practice, the current is delimited by nonlinear effects: the field is shielded by the multiplied carriers, and becomes not strong enough for impact ionization. The maximum current density in the APD, which effectively shields the field, is of the order of $10...100\,\mathrm{A/mm^2}$ [94], which translates into the current pulse amplitude $I_A \approx 1...40\,\mathrm{mA}$ in the APD with the diameter of photosensitive element about $0.1...0.2\,\mathrm{mm}$. The operating regime with $U_w > U_B$ may be called the trigger mode, the switching (the "ignition" of the current pulse) being triggered by the primary electron. Now the APD has come into "the line of fire": when the current amplitude is as large as that, we need not bother about the amplifier noise, and may use a low-resistance input load and broad passband. For example, with the input resistance of the amplifier $R_{in} \approx 50...100\Omega$, connected in series with the APD, the pulse voltage is

$$U_A = R_{in} I_A = (50...100)\left(10^{-3}...4 \cdot 10^{-2}\right) = 0.05...4\,\mathrm{V}\ . \tag{6.114}$$

The amplifier noise is much less than that, even if the equivalent noise resistance is $R_n \approx 1\,\mathrm{k}\Omega$, and the passband is $f_a = 1\,\mathrm{GHz}$:

$$U_n = \sqrt{4\kappa T R_n \Gamma_1 f_a} = \sqrt{1.6 \cdot 10^{-20} \cdot 10^3 \cdot 0.68 \cdot 10^9} \approx 100\,\mu\mathrm{V}\ . \tag{6.115}$$

Current pulses in the trigger mode (in the breakdown regime). In the linear regime of the APD, the current pulse induced by one primary electron has a deterministic shape, and its length is determined by the time constant of the APD (when we consider the current pulse of the APD itself), or by the time constant of the PDA (when we consider the output pulse of the PDA). Random is the amplitude of the pulse: it is proportional to the random-valued coefficient M. In the breakdown regime the situation is reversed: the amplitude is deterministic (being defined by the saturation current), and the pulse duration is random. There is a very small but finite probability that all the carriers will traverse the avalanche region without being engaged in the ionizing collision, and then the avalanche will spontaneously stop. In the early APDs, whose photosensitive element was not quite homogeneous, local breakdowns occurred in the form of microplasmas (local regions with premature breakdown), and the corresponding pulses are often referred to as the microplasma pulses [58].

The random duration of the pulse does not fit in with the photon counting mode: as long as the avalanche continues (the switch is closed), the APD is not able to count the next quantum. Because of this, it was necessary to design measures for forced termination of the pulse, either passive or active [95, 96]. In the passive circuit, a large ballast R_L is connected in series with the bias voltage source U_w. When the APD carries the current pulse I_A, the voltage drop across this resistance is $I_A R_L$. Adjusting the ballast in such a way that $U_w - I_A R_L < U_B$, the voltage across the APD is made less than the breakdown voltage, so that the avalanche is quenched. Then the duration of the pulse is determined by the RC-constant of the circuit. The active quenching circuit operates in a different way: the pulse at the output of the PDA controls the voltage applied to the APD, and at the appropriate instant kills the working voltage. This is the fastest and the most reliable solution for the counting mode.

Quantum efficiency of counting in the breakdown regime. Unfortunately, even in the breakdown regime the counting efficiency is less than one. There is a finite probability $p(M=1,\ W)$ that the primary carrier will traverse the entire avalanche layer of thickness W without causing the avalanche. This probability is easy to calculate. By $p(1,\ x+dx)$ we denote the probability that the primary electron will pass part of the avalanche layer (from its boundary $x=0$ to point $x+dx$) without collision. This event may be split into two consecutive phases: first, the electron must reach without collision point x (the corresponding probability being $p(1,x)$), and then cover the distance dx, again without collision (the probability of this event being $p(1,dx)$). Naturally, $p(1,x+dx)=p(1,x)p(1,dx)$. By definition of the coefficient of impact ionization α, the probability of ionizing collision on the path dx is αdx, and hence the probability of non-collision is $1-\alpha dx = p(1,dx)$. This brings us to a simple differential equation:

$$p(1,x+dx)=p(1,x)p(1,dx)=p(1,x)(1-\alpha dx)\ ;$$

$$\frac{p(1,x+dx)-p(1,x)}{dx}+\alpha p(1,x)=0\ ;$$

$$\frac{dp(1,x)}{dx}+\alpha p(1,x)=0\ . \tag{6.116}$$

Solving this at the initial condition $p(1,0)=1$, we get

$$p(1,x)=\exp(-\alpha x)\ ;\ \ \eta_c' < 1-p(1,W)=1-\exp(-\alpha W)\ . \tag{6.117}$$

In principle, the secondary carriers n, generated after several rounds of collisions, may fly out of the avalanche layer without further multiplication. This will also reduce the quantum efficiency of counting, and hence the "less" sign in (6.117). However, the greater the number of primary collisions, the larger n is, and the smaller the probability that all

n carriers will leave the avalanche layer without being engaged in ionizing collisions. Accordingly, there are two most probable states of the avalanche: either it is not triggered at all (with the probability given by (6.117)), or it increases up to saturation.

From (6.117) it follows that the greater the overvoltage $\Delta U = U_w - U_B$ and the coefficient α are, the higher the counting quantum efficiency. An especially large amount of overvoltage ΔU is required for *npip*-structure reach-through APDs, and also for the APDs based on materials with closely-values coefficients of impact ionizations for electrons and holes, $\alpha \approx \beta$. In the first case this is due to the large thickness W_i of the intrinsic layer, which calls for the additional (super-breakdown) amount of field strength $\Delta\mathcal{E}$ to produce the additional amount of overvoltage $\Delta\mathcal{E}W_i$. In the second case the reason is different: the value of αW, corresponding to the breakdown voltage, is too small (of the order of unity). At the same time, in accordance with (6.117) its value must be larger ($\alpha W \approx 2$) to secure high quantum efficiency — that is, the added field itself must be large. In practice, the overvoltage is usually $\Delta U = 10...40$ V [96, 97]. The breakdown regime with the overvoltage as large as that requires specially designed APDs with extremely homogeneous properties throughout the entire photosensitive element. The tiniest variations in the depth and amount of doping, minor structural defects may result in microplasmas, only local breakdowns in the junction. Another nuisance is the high concentration of free carriers in the avalanche region, which increases the probability of the carriers being caught in the traps.

Dark count rate and the decision concerning the presence of signal. Along with the photoelectrons, the APD also counts the dark electrons. The frequency of arrival of dark electrons, with due account for the quantum efficiency η_c', determines the dark count rate

$$f_c = \frac{\eta_c'}{T_{per}} = \frac{\eta_c' I_0}{q} . \tag{6.118}$$

It must be emphasized that the current I_0 here is only the component which goes to the multiplication layer (the non-multiplied leakage currents are not taken into account). The dark count rate uniquely defines the rate of false alarms, $f_{fa} = f_c$. This brings us to yet another rigorous requirement imposed on the APD operating in the counting mode: the requirement of extremely low currents. Accordingly, record-breaking values are achieved in such APDs: $\overline{I_0} = \overline{I_d} \approx 10^{-14}$ A at room temperature, and $\overline{I_d} \approx 10^{-16}$ A with cooling. This corresponds to the dark count rate of

$$f_c = \frac{(0.8...0.9)\cdot\left(10^{-16}...10^{-14}\right)}{1.6\cdot 10^{-19}} \approx 5\cdot 10^2...5\cdot 10^4 \text{ Hz} . \tag{6.119}$$

In some systems (for example, in reflectometers, used for checking the performance

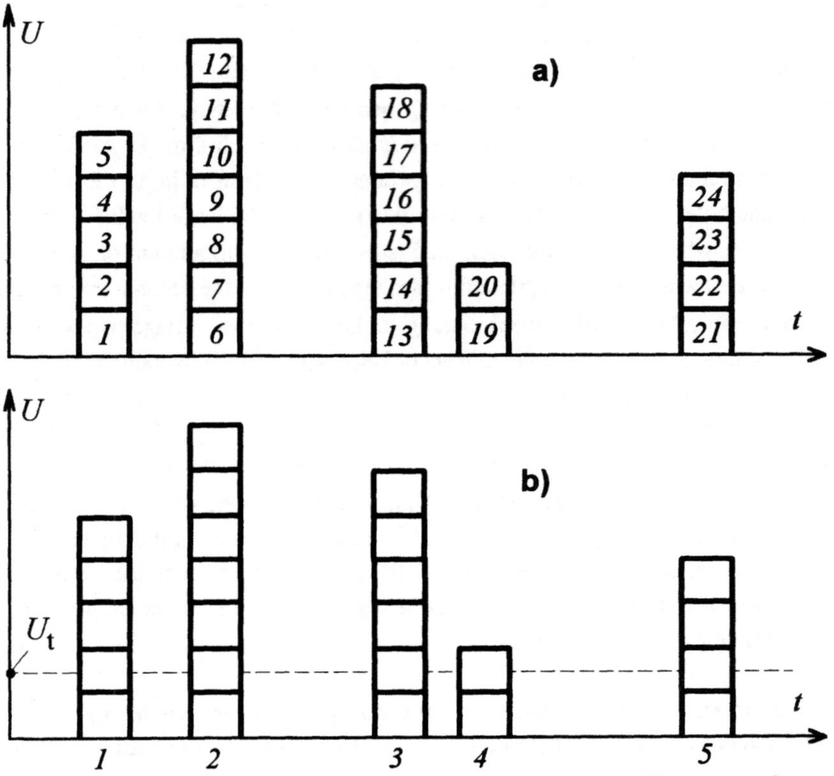

Figure 6.7: Filter-integrator counts (a) all secondary electrons of APD in the analog mode; (b) the number of elementary pulses (primary electrons) in the photon counting mode.

of fiber optical lightguides) this rate may be acceptable. In most optoelectronic systems (detection, guidance, communications), however, such rate cannot be tolerated. Accordingly, the decision-making procedure must use the standard algorithm (see Sect. 3.1 and 6.2) the device will count the electrons on the interval τ_{eff}, equal to the signal duration T_s, and take the decision that the signal is there when the counted number of electrons N_τ exceeds the preset threshold N_t. The greater the number of dark electrons $\overline{N_0}$, the higher the threshold N_t (6.101), (6.111), (6.75) must be raised. This means that prior to the reception of the signal we must measure the value of $\overline{N_0}$. To improve the precision, the time of measurement must be made large enough; to avoid the possible temperature instability and time drift of $\overline{N_0}$ the method of synchronous counting is used: the electrons are counted alternately on the intervals when there is no signal, and on the intervals when the signal is likely to arrive [98]. This complicates the circuit: recall that in case of the analog linear regime we only had to filter off the permanent component.

All these snags of the photon counting mode have to be overcome in the name of the main target: the improvement of the threshold characteristics.

Threshold characteristics in photon counting mode. Now what is the advantage of the photon count mode in terms of the sensitivity threshold. This inquiry may at first seem impertinent: after all, we have already stated in Sect. 3.2 that the attempts to improve upon the performance of the optimal filter are the same thing as trying to build a perpetuum mobile. The threshold characteristics of the optimal filter cannot be surpassed. As a matter of fact, any filter-amplifier may be regarded as a kind of effective accumulator, integrator, counter (Sect. 4.1). Recall the reception of rectangular signal against the background of the shot noise of the current. In this classical case the optimal filter faithfully counts the electrons: it adds up the responses caused by all electrons arriving at the input over the interval $T_i = \tau_{eff} = T_s$, and the output signal is proportional to the number of input electrons N_τ (Sect. 6.2). What is then the peculiarity of the counting mode currently under consideration?

The filter-integrator counts the *total number of electrons* that have arrived at the input over the time τ_{eff}. The photon counter with the threshold device counts the *number of arrivals of elementary pulses* over the same time, whereas the amplitude of the pulse has no effect on the result of count. In the absence of multiplication these procedures are the same: each input pulse consists of one electron. Multiplication, however, changes the situation dramatically: each elementary ith pulse contains now M_i secondary (multiplied) electrons. Accordingly, the output signal of filter-integrator is proportional to the *total number of secondary electrons* $\sum_{i=1}^{N_\tau} M_i$, whereas the counter circuit, as before, counts the acts of arrival, which is the *number of primary electrons* N_τ (Fig. 6.7).

We see that there is count and count. In the filter-integrator the fluctuations of the number of input electrons N_τ are supplemented by the fluctuations of multiplication. Owing to the presence of the threshold device, the value of the multiplication coefficient in the counter circuits is not critical, as long as it exceeds the threshold M_t. *The advantage of the counting mode consists in the suppression of the multiplication noise.* Until the advent of "noiseless" APDs having arbitrarily large multiplication coefficient M, it will be worthwhile to continue using the counting mode so as to eliminate the noise of the avalanche.

The magnitude of the lowest detectable signal in the counting mode has actually been calculated in Sect. 6.2 (Fig. 6.6, curve 1). It is the counting mode that allows suppressing the amplifier noise without introducing the multiplication noise, and to register the ultimately low signals against the background of the shot noise of the dark (and thermal background) current of the APD.

How much does the multiplication degrade the threshold characteristics? The rigorous calculation must include the statistics of the multiplication. In Sect. 6.2, however, we demonstrated that the calculations may be based on the Gaussian distribution down to extremely low currents, such as correspond to the occurrence of just one electron (on the average) during the time of observation. The current in question corresponds to about 60 electrons in the signal pulse (curves 1, 2 in Fig. 6.6, point $\overline{N_0} = 1$). This number is large enough for assuming that the multiplication obeys the Gaussian law [93, 99].

Because of this, we may compare the ordinary linear avalanche regime with the counting mode on the basis of equations (6.88), (6.91), derived in the Gaussian approximation. From these formulas it follows that, owing to elimination of the multiplication noise in the counting mode, the threshold characteristics are improved by the factor of $\sqrt{\mathcal{F}}\ldots\mathcal{F}$, where $\mathcal{F}$ is the noise factor of the avalanche. Even for the lowest-noise silicon APDs with $\mathcal{F}=4$, the gain will be at least $\sqrt{\mathcal{F}}=2$, which is quite considerable. In each particular case this numerical estimate must be revised: the gain may be somewhat smaller (because of the counting loss $\eta_c'\eta_c''$), ore somewhat larger (when the avalanche is more noisy, or the dark current is very low, $\overline{N_0} << 1$).

The counting mode is not an alternative to the optimal filter, but rather a way of its realization, which permits counting the primary electrons. The counter circuit may be designed as a digital filter: the threshold device (with the threshold M_t) is follows by the counter, which counts the number of incoming electrons N_τ, and the decision-making device with the threshold N_t. If $N_\tau \geq N_t$, the device decides that the signal is there. The detection errors (false alarm and target miss) are determined by the levels of the dark noise $\sqrt{\overline{N_0}}$, the threshold N_t, and the signal $\overline{N_\tau}$ in the usual way. In principle, one can also realize the desired optimal (quasi-optimal) pulse response $\mathrm{H}(t_m - t)$ by making the counter more sophisticated, so as to add up the pulses from the first threshold device with the weight $\mathrm{H}(t_m - t_i)$, depending on the time of arrival of this pulse t_i.

Time resolution. Following the arrival of the first electron, a pulse of a certain effective duration T_{eff} arises at the output of the PDA amplifier. In the broad-band amplifier and overvoltage regime of the APD the time T_{eff} is determined by the time of quenching the avalanche (the RC-constant of the input circuit in case of passive quenching, and the feedback time constant in case of passive quenching). In the narrow-band amplifier the time T_{eff} may be defined by the amplifier itself: $T_{eff} = \tau_{eff} = 1/2f_a$. The threshold device is switched on for the time T_{eff}. If a second electron arrives within the same time interval, it will not change the state of the threshold device, and will not be registered. Because of this, two primary electrons will be counted only if they are separated by the time interval greater than T_{eff}. The time resolution T_{eff} is yet another parameter peculiar to the counting mode.

What is the required time resolution? It all depends on the number of dark electrons generated within the signal duration T_s, which dictates the threshold N_t, and hence the number of electrons which have to be registered, $N_\tau \geq N_t$. For definiteness, assume that the dark current is $I_0 \approx 2\cdot 10^{-15}$ A, and the error probabilities are $P_{fa} = P_{tm} = 5\cdot 10^{-10}$. This current corresponds to the period of succession of dark electrons $T_{per} = q/I_0 \approx 10^{-4}$ s. The number of thermal electrons generated over the typical signal duration in the high-frequency systems ($T_s = 10^{-9}\ldots10^{-6}$ s) is

$$\overline{N_0} = \frac{I_0 T_s}{q} = \frac{T_s}{T_{per}} = \frac{10^{-9}...10^{-6}}{10^{-4}} = 10^{-5}...10^{-2} . \qquad (6.120)$$

According to (6.103), the discrimination threshold must be set at $N_t = 2...5$. To take the decision about the presence of signal, the device must register not less than N_t signal electrons. By the time the last of these arrives, its $N_t - 1$ predecessors must be registered, which will take $(N_t - 1)T_{eff}$ seconds. If the last electron is not lucky, and falls into the same time interval, it will not be counted. The device will detect only the happy electron which falls into the "unoccupied" time interval $T_s - (N_t - 1)T_{eff}$. This is yet another reason of the reduced quantum efficiency in the counting mode, which is due not to the fluctuations of multiplication, but to the finite counting time:

$$T_{per} >> T_s ;$$

$$\eta_c'' = \frac{T_s - (N_t - 1)T_{eff}}{T_s} = 1 - \frac{(N_t - 1)T_{eff}}{T_s} . \qquad (6.121)$$

Assuming again that the reasonable value of the quantum efficiency is $\eta_c'' \gtrsim 0.9$, we get the following estimate for the required time resolution:

$$1 - \eta_c'' = \frac{(N_t - 1)T_{eff}}{T_s} ; \; T_{eff} \le \frac{(1 - \eta_c'')T_s}{N_t - 1} = \frac{0.1 T_s}{1...4} = \frac{T_s}{10...40} . \qquad (6.122)$$

Now let us consider a long signal pulse $T_s >> T_{per}$, during which a large number of dark electrons are generated, $\overline{N_0} = T_s / T_{per} >> 1$. Evaluation of T_{eff} is simplified when the $\overline{N_0}$ is so large that it exceeds the number of signal electrons $\overline{N_\tau}$. This is ensured by condition

$$\frac{\overline{N_\tau}}{\overline{N_0}} = \frac{\text{SNR}\sqrt{\overline{N_0}}}{\overline{N_0}} = \frac{\text{SNR}}{\sqrt{\overline{N_0}}} << 1 , \; \overline{N_0} >> \text{SNR}^2 \approx (12)^2 \approx 150 . \qquad (6.123)$$

Here we have taken into account that fluctuations are normalized when $\overline{N_0} \gtrsim 100$; recall Fig. 4.12.

If $\overline{N_0} >> \overline{N_\tau}$, then the miss of any signal electron (from the first, number 1, to the last, number $\overline{N_\tau}$) is equiprobable. Indeed, in this case the loss in count will be determined by the dead time $\overline{N_0} T_{eff}$, spent on counting the dark electrons (because the time taken to count the signal electrons $T_{eff} ... \overline{N_\tau} T_{eff}$ is small compared to $\overline{N_0} T_{eff}$). The device will register the "lucky" electron that arrives during the "free" time interval $T_s - \overline{N_0} T_{eff}$. By analogy with (6.122) we get

$$\eta_c'' = \frac{T_s - \overline{N_0} T_{eff}}{T_s} = 1 - \frac{\overline{N_0}}{T_s} T_{eff} = 1 - \frac{T_{eff}}{T_{per}} ,$$

$$T_{eff} = (1 - \eta_c'') T_{per} = 0.1 T_{per} , \qquad (6.124)$$

where the numerical coefficient again corresponds to $\eta_c'' = 0.9$. Now the fast action of the counter is compared not with the signal duration, but rather to the rate of dark pulses, compare (6.122) and (6.124).

Observe that both equations (6.122), (6.124) give only a qualitative estimate of the required time resolution T_{eff}. The exact model calls for calculating the probability $P(N_t || \overline{N_0} + \overline{N_\tau})$, which is the probability that, given that the mean number of the incoming electrons is $\overline{N_0} + \overline{N_\tau}$, the number of separate (non-overlapping) pulses will be not less than N_t. As T_{eff} increases, the number of overlapping pulses becomes greater, and the probability $P(N_t || \overline{N_0} + \overline{N_\tau})$ decreases. To maintain its value above $1 - P_{tm}$, one must increase the number of input (signal) electrons. In principle, this allows one to find the precise dependence of additional loss of signal electrons (quantum efficiency η_c'') on T_{eff}, and to determine the desired value of T_{eff} in terms of η_c''. This statistical problem takes some elbow grease: its solution requires sorting item-by-item a large number of combinations (different superpositions of pulses). Because of this, here we confined ourselves to the simplest qualitative estimate.

Requirements to the amplifier passband. Equations (6.122), (6.124) for the time resolution establish the following requirements to the passband of the amplifier:

$$T_{per} >> T_s \ ; \ f_a \geq \frac{1}{2T_{eff}} = \frac{10...40}{2T_s} = \frac{10...40}{2 \cdot (10^{-9}...10^{-6})} = 5 \cdot 10^6 ... 2 \cdot 10^{10} \text{ Hz} \ ;$$

$$(6.125)$$

$$T_{per} << T_s \ ; \ f_a \geq \frac{10...20}{2T_{per}} = \frac{10...20}{2 \cdot 10^{-4}} = (5...10) \cdot 10^4 \text{ Hz} \ .$$

It follows that the passband of the PDA in the counting mode must be much broader than in the traditional analog mode because it is necessary to resolve each electron. While the band f_a in the analog mode is of the same order of magnitude as the bandwidth of the signal $1/2T_s$ (and considerably narrower than in case of optimal or quasi-optimal filtering and noticeable high-frequency noise), in the counting mode the passband f_a must be broader than $1/2T_s$ by an order of magnitude or more. When the pulse is very long, the passband f_a is determined not by the signal duration ($1/2T_s$), but by the much shorter period of succession of the dark pulses ($1/2T_{per}$). Since the requirements to the time resolution are especially tough in the detection of short pulses, the passband of the amplifier must be extremely broad — in fact, we need an UHF amplifier. Of course, equation (6.125) also defines the requirements to the time response of the APD. The experiments indicate, however, that the main problems in the way of improvements on the time resolution are associated not with the use of HF or UHF technologies, and not with the frequency limitations of the APDs, but with the trapping of carriers.

Trapping of carriers. Having solved one problem (high amplification), we created new difficulties. Overvoltage produces a very high concentration of carriers in the avalanche region, high enough for shielding the field. When the concentration is that high, the carriers are caught on the traps in SCR. When the avalanche is quenched, all mobile carriers are expelled by the field from the SCR. As the overvoltage is reapplied to the APD, the trapped carrier after some time τ_{trap} may be ejected into the band, triggering a new avalanche and causing a false alarm (increasing the dark count rate). Elimination of traps and making τ_{trap} as small as possible are specific engineering tasks in the development of APDs operating in the counting mode. Active quenching circuits are especially useful for reducing the probability of false alarm caused by carriers released from the traps. The length of the quenching pulse must be made greater than the trapping time constant, $T_{eff} = (3...10)\tau_{trap}$. Since cooling increases the trapping time constant τ_{trap}, there is a certain optimal operating temperature for the APD: τ_{trap} is too large below, and the dark current $\overline{I_d}$ is too large above. This optimal temperature depends on the type of the APD (concentration and type of traps and levels of generation), and the requirements imposed by the system on the dark count rate and the time resolution. The optimal working temperature for silicon APDs studied in [97] was found to be 200 K.

The achievable time resolution determines also the dynamic range of the photon counting circuit. Since the count of each electron takes the time T_{eff}, the number of electrons counted over the time T_s is not greater than $N_{max} = T_s/T_{eff}$. If $\tau_{trap} \approx 10^{-8}$ s, $T_{eff} = 5 \cdot 10^{-8}$ s, then $N_{max} = 20$ even when $T_s = 10^{-6}$ s. In case of short signals the limitation starts practically from the threshold values of N_τ. Because of this, one has to use simultaneously both the counting mode and the analog mode if it is necessary not only to detect the signal but also to measure its magnitude.

Summary. *When the rate of generation of carriers is low, the current as though splits into separate non-superimposed pulses, each caused by a single carrier, which makes it possible to count each separate electron. Specially designed regimes and circuits are used for realizing the counting mode. For example, overvoltage is applied to the APD, and the avalanche triggered by the primary electron is forcibly quenched (by killing the voltage across the APD); broad-band amplifiers are employed (which almost exactly reproduce the shape of the avalanche pulse); a threshold device is used for registering the pulse triggered by one electron. This method of registration is not sensitive to the value of the multiplication coefficient and its fluctuations, and suppresses the noise of the avalanche. This constitutes the advantage of the counting mode over the traditional analog mode. The counting mode is characterized by specific parameters: the dark count rate (frequency) caused by the dark current, the time resolution as determined by the time* *taken to register one electron (which depends on the time of avalanche quenching, the passband of the amplifier, the time of release of carriers captured on the traps in the course of breakdown), the quantum counting efficiency (the additional loss — omission of photons — is due to the overlapping of pulses, and to the possibility that a carrier will traverse the avalanche layer without experiencing an ionizing collision, which is associated with a small but finite probability).*

It is only the most advanced APDs that can operate in the counting mode: such as exhibit a perfectly homogeneous breakdown across the entire element, a low concentration of traps, an extremely low dark current.

Chapter 7

DETECTION AND PROCESSING OF OPTICAL SIGNALS IN MATRICES

The task of most diverse optoelectronic systems consists in the definition of coordinates of optical spot. Systems with mechanical scanning may use single-element photodetectors and PDAs. Using multielement devices and increasing the number of elements it is possible to improve resolution and to simplify the mechanical scanning system, which may be completely replaced with the electronic scanning. For the target, however, which has been our concern throughout this book, multielement devices offer a new possibility of improving the detective properties of the system. Obviously, M_e elements working together will do a better job than one element, and will be able to detect a weaker optical signal. How do the detective properties depend on the number of elements M_e? What are the specific features of the detection of low-level signals by multielement PDAs, including focal plane arrays? How is the signal processed in the integration cell, which is the basic element of focal plane array, and to what extent this cell is inferior to the optimal filter? Such are the questions that constitute the topic of this chapter.

7.1. Integration Cell As Quasi-Optimal Filter

Optimal filtering is nothing else but integration of the signal. We demonstrated this in Sect. 4.2 in the analysis of the optimal filter, when we considered convolutions (4.9), (4.10), and summed up the signal and noise samples. Throughout the entire book we have been dealing with integration, accumulation, summation. In Sect. 4.1 we showed that any filter-amplifier, in accordance with the convolution, sums up the samples with the weights corresponding to the pulse response. In other words, we regarded it as a device with the effective observation (integration) time $1/2f_a$, where f_a is the passband of the amplifier. Optimal filter for the rectangular pulse of length T_s (Fig. 5.1) directly contains an integrator, and, using the delay and subtraction circuits, performs integration over the time T_s. Quasi-optimal filter, based on one-section RC-network (Fig. 7.1), also carries out integration. The circuit contains just two components, each performing its own function. The capacitance C is the memory of the system; it stores the charge (the integral of the charging current). The resistance R performs the opposite function: it erases the information after the time $\tau_f = RC$. The dialectical unity of these two

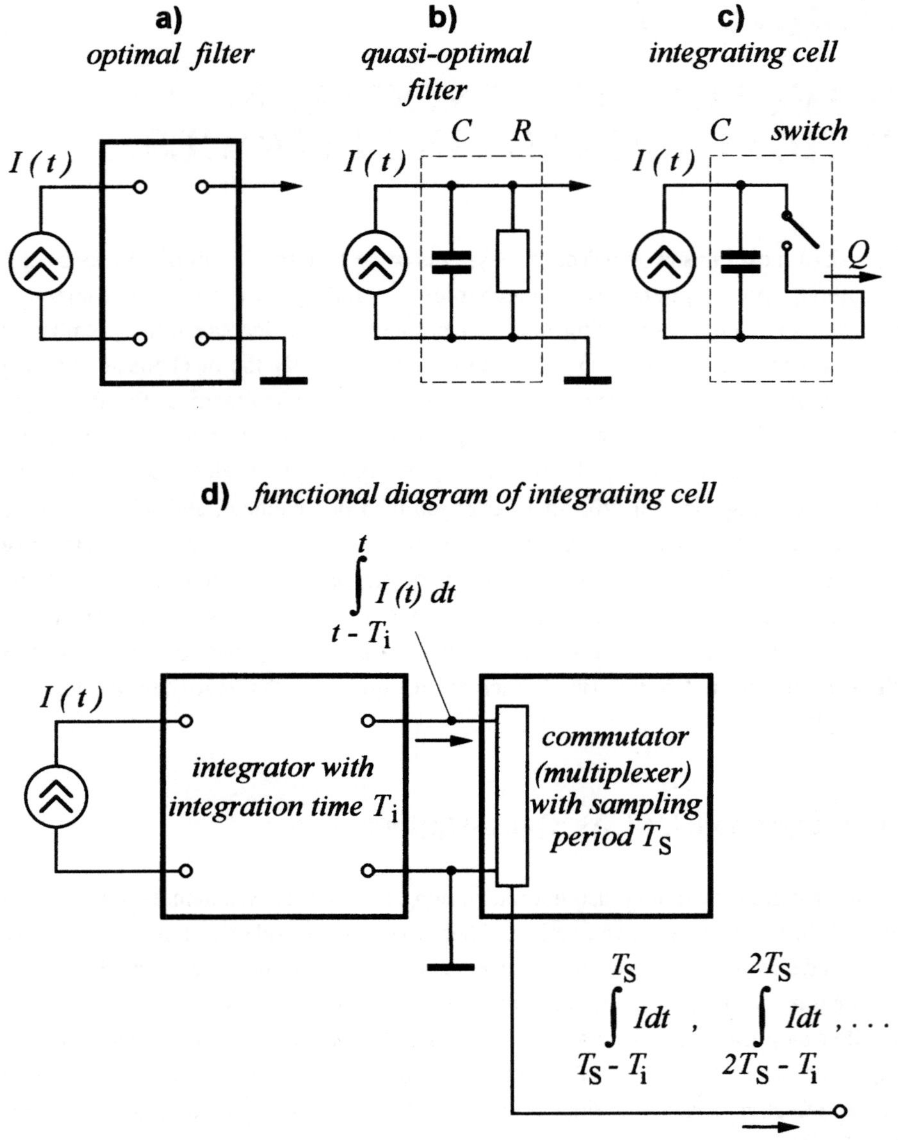

Figure 7.1: From optimal filter to cell with integration: (a) optimal filter; (b) quasi-optimal filter; (c) integration cell; (d) functional diagram of integration cell.

counteracting elements brings the desired result: integration of charge over the prescribed observation time, $\tau_f = T_s$.

The capacitor can be discharged, however, not only across a resistance: the resistor may be replaced with a switch, which periodically closes after the time interval T_i (Fig. 7.1). Then we get yet another embodiment of optimal filtering: the cell with capacitance and switch. Its simplicity permits solving one of the main tasks of optoelectronics: the development of multielement matrix PDAs with near-optimal signal processing. It would be simply not possible to squeeze a sophisticated filter with the optimal frequency response into each of the $10^3 \ldots 10^6$ channels, with the available space as small as 100 by 100 or even 30 by 30 micrometers. The simplest and the most universal circuitry used in almost all matrix PDAs is the cell with three obligatory components: photodetector, capacitance, and switch. It is not important whether it is the capacitance of the photodiode or of the CCD cell, whether the switch is the FET of the adjacent CCD element. It is not important how the accumulated charge is read out: by measuring the voltage across the capacitance, like in photodiode matrices with MIS switch, or it is conveyed to the output. The principle of signal processing in the cell is universal and common for all one and two-dimensional CCD and CID matrices, bucket-brigade devices, lattices, and even in some one-channel PDAs. The use of the capacitance-switch circuit is reported for silicon PD with low dark current [100]. In some devices the circuitry of the cell may be modified. In hybrid CCDs, for example, the input units may contain differentiating circuits [23, 24, 101]. In photoresistor the charge carriers are stored directly in the bulk of semiconductor (in the valence band or conduction band), and the erasing is accomplished through recombination. The equivalent diagram of this cell is shown in Fig. 7.1b; observe that $\tau_f = \tau_{life}$, where τ_{life} is the lifetime. Other mechanisms may be employed, but the optimal signal processing always involves accumulation and erasing, which seems a perfectly obvious, self-evident thing. In spite of this, however, some devices are still being designed without the integrating element — for example, some hybrid matrices [102]. Of course, it is technically difficult to realize accumulation in the cell because of the limited capacitance (see Sect. 1.1); such matrix, however, will not be a quasi-optimal device. The information about the signal is lost, since its energy is only used at the time of interrogation of the illuminated cell.

From arguments developed above it follows that the analysis of signal processing in the photodetector – capacitance – switch cell is actually the analysis of signal reception as it occurs in practically all kinds of multielement PDAs with integration. It is also clear that the capabilities of such cell are limited: its relative frequency response and pulse response are predetermined, so the cell cannot be tuned to the signal spectrum of arbitrary form. Our next assignment is also clear: we must evaluate the deterioration of the signal-to-noise ratio of such cell as compared with the optimal filter.

Capacitance – switch cell as optimal filter. Let us find the charge Q which is accumulated by the capacitance C (Fig. 7.1c). Time after time, with the period T_i, the switch is closed, and the charge accumulated by the capacitance is erased. Accordingly, T_i is the time of integration. The charge is dumped to the output to be registered. In this interrogation procedure the sampling instant coincides with the instant the charge is erased, and the period between the samples coincides with the integration time T_i:

$$Q = \int_0^{T_i} I(t)dt \ , \tag{7.1}$$

where $I(t)$ is the realization of the photodetector current.

Now let us rewrite (7.1) as

$$Q = \int_{-\infty}^{\infty} \mathrm{H}(t) I(t) dt \ , \tag{7.2}$$

where

$$\mathrm{H}(t) = \begin{cases} 1 \text{ at } 0 < t < T_{\mathrm{i}}; \\ 0 \text{ at } t < 0, \ t > T_{\mathrm{i}}. \end{cases} \tag{7.3}$$

From this representation it immediately becomes clear that the pulse response of the capacitance – switch cell is a unit-valued rectangular function of length T_{i}. Hence, according to Sect. 4.2, it follows that the cell is the optimal filter under the following three conditions:

— the shape of the input signal is also rectangular;
— the sampling instants coincide with the beginning (and the end) of the input pulse, and the integration time is equal to the length of the pulse, $T_{\mathrm{i}} = T_{\mathrm{s}}$;
— the noise is white.

We encountered this kind of pulse response more than once in Sect. 4.1, 5.2, 6.2, where it pertained to the current-to-voltage converter, and its dimension was Ω/s. Our current cell with integration is a current-to-charge converter, whose transfer coefficient K_0 has the dimension of seconds. Accordingly, the pulse response is dimensionless:

$$\dim\mathrm{H} = \frac{\dim K}{\mathrm{s}} = \frac{\mathrm{s}}{\mathrm{s}} = [\text{dimensionless}] \ .$$

We used to start the analysis from the calculation of the signal-to-noise ratio for such classical case (see Sect. 3.3, equation (3.57)). The same equation can be obtained by considering directly the charge accumulated by the capacitance. The fluctuation of the dark current (and of the background current, if any) $I_0(t)$ results in the fluctuation of the charges $Q_1, Q_2, \ldots, Q_n$, accumulated during different frames — that is, on the sequence of the time intervals $0 \ldots T_{\mathrm{i}}; \ T_{\mathrm{i}} \ldots 2T_{\mathrm{i}}; \ \ldots; \ (n-1)T_{\mathrm{i}} \ldots nT_{\mathrm{i}}$. The mean values of the charge and the number of electrons over the sequence of frames are:

$$\overline{Q_0} = \overline{I_0} T_{\mathrm{i}} \ , \quad \overline{N_0} = \frac{\overline{I_0} T_{\mathrm{i}}}{q} \ . \tag{7.4}$$

Since the fluctuations obey the Poisson statistics, by analogy with (4.32) we have:

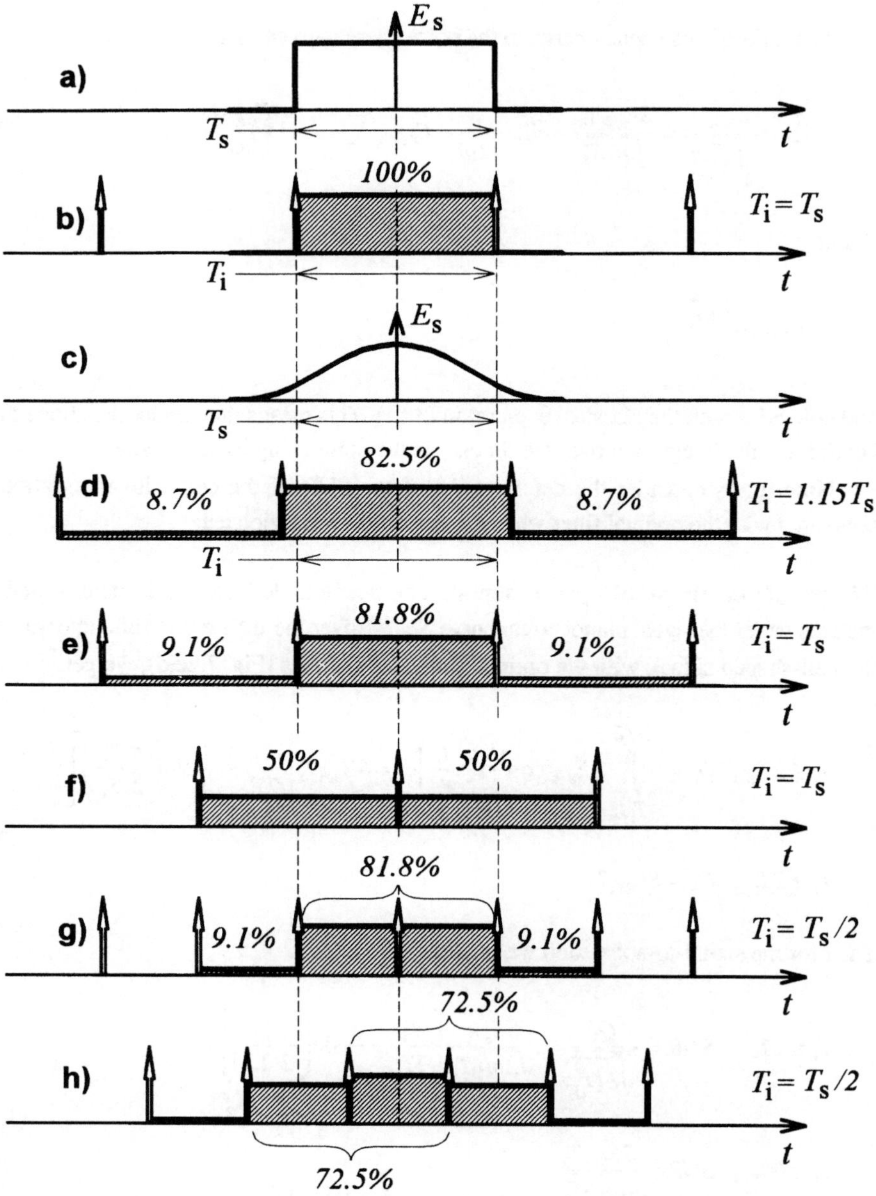

Figure 7.2: Accumulated charge depends on the relative position of signal and samples: (a,c) shapes of input optical pulses; (b,d–h) arrangements of sampling instants.

$$\overline{\Delta N_0^2} = \overline{N_0}\ , \quad \overline{\Delta Q_0^2} = q^2 \overline{\Delta N_0^2} = q^2 \overline{N_0} = q\overline{Q_0} = q\overline{I_0} T_i \ . \tag{7.5}$$

The optical signal E_s causes additional signal photocurrent SE_{sA}, which adds the signal charge to the capacitance (Fig. 7.2a,b):

$$Q_s = SE_{sA}T_s \ . \tag{7.6}$$

The ratio of this signal charge to the noise charge, given that $T_i = T_s$, is

$$\text{SNR} = \frac{Q_s}{\sqrt{\Delta Q_0^2}} = \frac{SE_{sA}T_s}{\sqrt{qI_0T_s}} = \sqrt{2}\frac{S}{\sqrt{2qI_0}}E_{sA}\sqrt{T_s} = 1.41\frac{E_{sA}\sqrt{T_s}}{\text{NEP}} \ , \tag{7.7}$$

where

$$\text{NEP} = \frac{\sqrt{2qI_0}}{S} \ , \tag{7.8}$$

and coincides with the classical expression (3.61). This means that, under the above three conditions, the integration over the time T_s is the optimal signal processing.

Now let us consider the deterioration of the SNR of the cell with integration as compared with the optimal filter when these conditions are violated.

Effects of the shape of optical signal. The accumulated charge is determined by the area under the signal photocurrent curve $SE_s(t)$ over the integration time interval. For the bell-shaped signal, with the optimal timing of samples (Fig. 7.2c,d), we get

$$\begin{aligned} &T_i \le 2T_s \ , \ Q_s = \int_{-T_i/2}^{T_i/2} SE_{sA}\cos^2\left(\frac{\pi}{2}\frac{t}{T_s}\right)dt = SE_{sA}T_s\left[\frac{T_i}{2T_s} + \frac{1}{\pi}\sin\left(\frac{\pi}{2}\frac{T_i}{T_s}\right)\right] ; \\ &T_i \ge 2T_s \ , \ Q_s = SE_{sA}T_s \ . \end{aligned} \tag{7.9}$$

Then for the signal-to-noise ratio we get:

$$\begin{aligned} &T_i \le 2T_s \ , \ \text{SNR} = \frac{Q_s}{\sqrt{\Delta Q_0^2}} = \frac{SE_{sA}}{\sqrt{2qI_0T_i}}\left[\frac{T_i}{2T_s} + \frac{1}{\pi}\sin\left(\frac{\pi}{2}\frac{T_i}{T_s}\right)\right] ; \\ &T_i \ge 2T_s \ , \ \text{SNR} = \frac{SE_{sA}T_s}{\sqrt{qI_0T_i}} \ . \end{aligned} \tag{7.10}$$

As follows from (7.9), (7.10), SNR increases with T_i when T_i is small: $\text{SNR} \propto \sqrt{T_i}$; when T_i is large, the signal-to-noise ratio starts to fall ($\text{SNR} \propto 1/\sqrt{T_i}$) because all of the signal is accumulated over the time $T_i \ge 2T_s$, and integration over a longer period only adds to the noise. The maximum SNR corresponds to a certain optimal time $T_{i\,opt}$. Setting the derivative of (7.10) equal to zero, we find:

$$T_{\text{i opt}} = 1.149 T_{\text{s}} \ , \tag{7.11}$$

$$\text{SNR}\left(T_{\text{i opt}}\right) = 1.167 \frac{E_{\text{sA}}\sqrt{T_{\text{s}}}}{\text{NEP}} \ . \tag{7.12}$$

From Sect. 5.2 we know that the tuning curves (which portray the SNR as a function of the passband, and hence of the time T_{i}) are flat, and there is no need to be too pedantic about fulfilling the condition (7.11). In practice it convenient to make the integration time equal to the effective signal duration, just like in case of the rectangular signal. The additional amount of loss in the SNR is as small as 1% (Fig. 7.2e):

$$T_{\text{i}} = T_{\text{s}} \ , \ \ \text{SNR} = 1.156 \frac{E_{\text{sA}}\sqrt{T_{\text{s}}}}{\text{NEP}} \ . \tag{7.13}$$

From Fig. 7.2c-e, in case of the bell-shaped signal it is only possible to accumulate 82% of the signal charge (in case of the rectangular signal the charge is stored completely, Fig. 7.2b). As a matter of fact, however, in case of optimal filtering the SNR for the bell-shaped signal is also 13.5% less than that for the rectangular signal (Table 3.2). Accordingly, the deterioration of the SNR for the bell-shaped signal in the cell with integration is just 5% as compared with the optimal filter, which is even less than the same for the one-section quasi-optimal RC-filter (5.62), (5.68).

Effects of stagger-timed sampling. Integration interval and period of sampling. So far we have been considering the case when the interrogation (sampling) times were precisely synchronized with the location of the signal on the time axis, and were positioned in the best possible way — that is, the maximum charge was accumulated in one of the frames (Fig. 7.2a–d). In most optoelectronic systems, however, the time of arrival of the signal is not known in advance, and so the relative disposition of the signal and the samples is random. Figure 7.2f illustrates the worst possible case when the incoming signal is split in halves between two adjacent frames. Then the cell with integration loses rather considerably to the optimal filter: 50% in case of the rectangular signal, and 42% in case of the bell-shaped signal:

$$\text{SNR} = 0.5 \frac{S E_{\text{sA}} T_{\text{s}}}{\sqrt{q I_0 T_{\text{s}}}} = 0.707 \frac{E_{\text{sA}}\sqrt{T_{\text{s}}}}{\text{NEP}} \ . \tag{7.14}$$

This is the price that has to be paid for the introduction of nonstationary element (switch) into the cell, whose resistance changes with the time. The circuit has become parametric, and so we may assign to it a phase-frequency characteristic which depends on the time (the instant the switch is closed). When this characteristic of the cell is not phased with the signal spectrum, we lose in the SNR. By contrast, the optimal filter (Fig. 7.1a; (4.1)) is stationary. Its memory (the delay line) ensures that the amplitude signal value U_{sA} is obtained at the output as soon as a certain predetermined time has

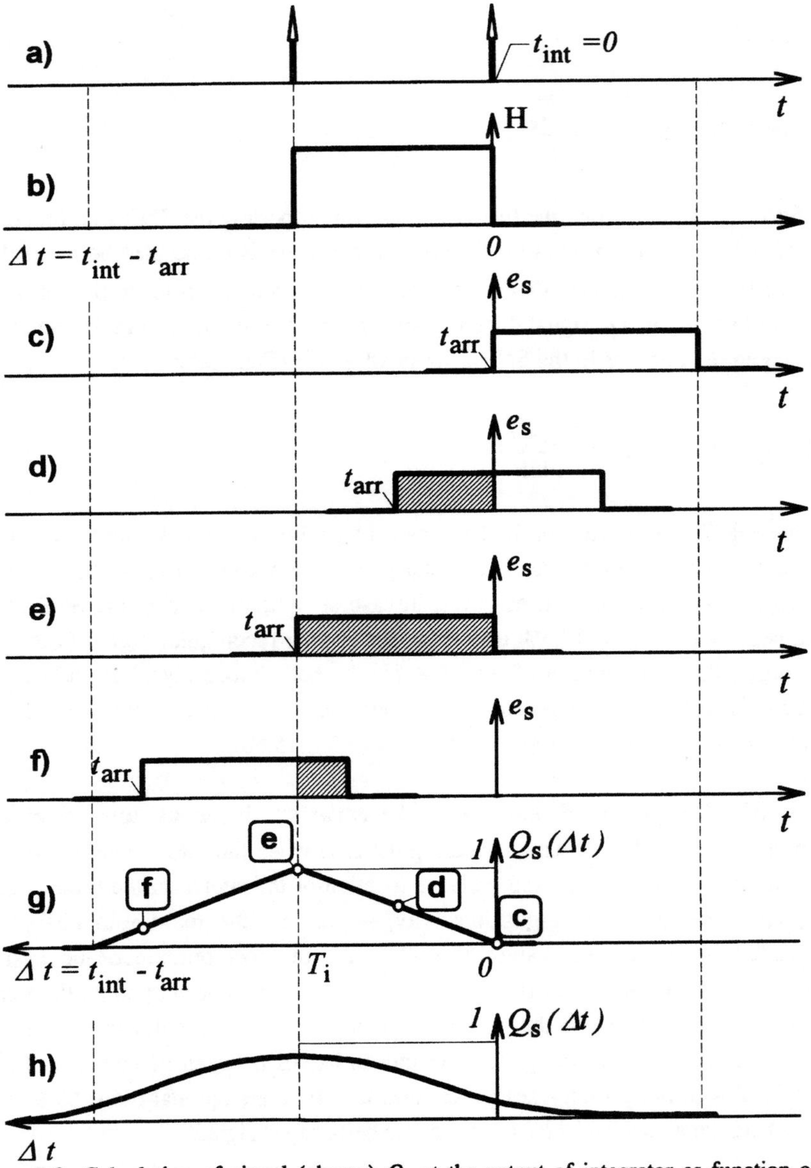

Figure 7.3: Calculation of signal (charge) Q_s at the output of integrator as function of real time.

elapsed after arrival of the signal at the input. Because of this, the value of SNR in invariant with respect to the time of arrival of the signal.

So, we have considered the extreme cases of the best (Fig. 7.2a–e) and the worst (Fig. 7.2f) arrangements of samples. Now let us analyze a situation in between (Fig. 7.3). The sample (interrogation time) is aligned with time zero, $t_{int} = 0$ (Fig. 7.3a). The range of integration is indicated in Fig. 7.3b, where the rectangular function $H(\Delta t)$ is the pulse response of the integrator at $t_{int} = 0$. By moving the signal along the time axis, it is easy to find the accumulated charge for any arbitrary arrangement of the signal (Fig. 7.3c–g). This charge is proportional to the area of overlap of the signal with the rectangular pulse response H(Δt). Obviously, when the signal is rectangular, the charge Q_s is a linear function of the time $\Delta t = t_{int} - t_{arr}$, where t_{arr} is the time of arrival of the signal (the front edge). Now, both Fig. 7.3 and the method of calculation of the output signal of the cell (the accumulated charge $Q_s(\Delta t)$) are quite familiar to us! It was in exactly the same way that we calculated the signal shape at the output of the optimal filter (Fig. 4.7): at any arbitrary time the signal is also equal to the integral of the input current over the precedent period T_i.

This analogy is very important. It allows us to add the final touch to the formal transformation of the equivalent diagram, and to replace the capacitance and the switch by two functional units (Fig. 7.1c,d). The first unit is the stationary (non-parametric) integrator with the integration time T_i, entirely similar to the optimal filter for the rectangular signal shown in Fig. 5.1. Its output is a continuous signal $Q_s(\Delta t)$. The second functional unit is concerned with the multiplexing (commutation) with the period $T_S = T_i$, and takes samples of this signal $Q_s(\Delta t)$. In this way we have obtained the functional diagram of the multielement PDA, with which we actually started this book (see Sect. 1.1, Fig. 1.1a). The diagram shown in Fig. 7.1d is good for any shape of the input signal. This technique can be used for calculating the signal $Q_s(\Delta t)$ in case of the bell-shaped input pulse (Fig. 7.3h, where t_{arr} is measured at half-magnitude of the pulse edge).

Now that we have the functional diagram of the cell, we may identify the root of our trouble. The integrator itself is practically perfect: as shown above, there is no loss when the signal is rectangular, and when the signal is bell-shaped the loss is just 5% as compared with the optimal filter. The problems are associated with the commutation. In the one-channel PDA the commutator is not needed, so we may continuously monitor the output signal and will always be able to register its maximum. In the multielement PDA we have to take discrete samples from each channel. Of course, the period of sampling must be made small enough, so as not to miss the maximum of the signal. In the traditional PDA with amplifier in each channel and subsequent commutator the integration time T_i (which essentially is the effective time constant of the amplifier τ_{eff}) is not matched with the sampling period. In devices of this kind the time T_S is obviously limited too, and so the loss may occur because of the staggered sampling; it is possible, however, to minimize this loss by making the sampling time small, $T_S << T_i$. The disadvantage of the cell with integration comes from the fact that the sampling period is quite large, $T_S = T_i$, and is comparable with the time constant of the output signal T_s. The samples may miss the maximum of $Q_s(\Delta t)$, which results in losing as much as 42...50% of the signal charge (Fig. 7.2f).

So, the signal-to-noise ratio may vary within relatively broad limits (0.5...1 in case of the rectangular pulse, and 0.58...1 in case of the bell-shaped pulse), because the samples are rare and are not phased with the signal. These values of SNR were obtained when we optimized the integration time T_i in a favorable situation: the samples were

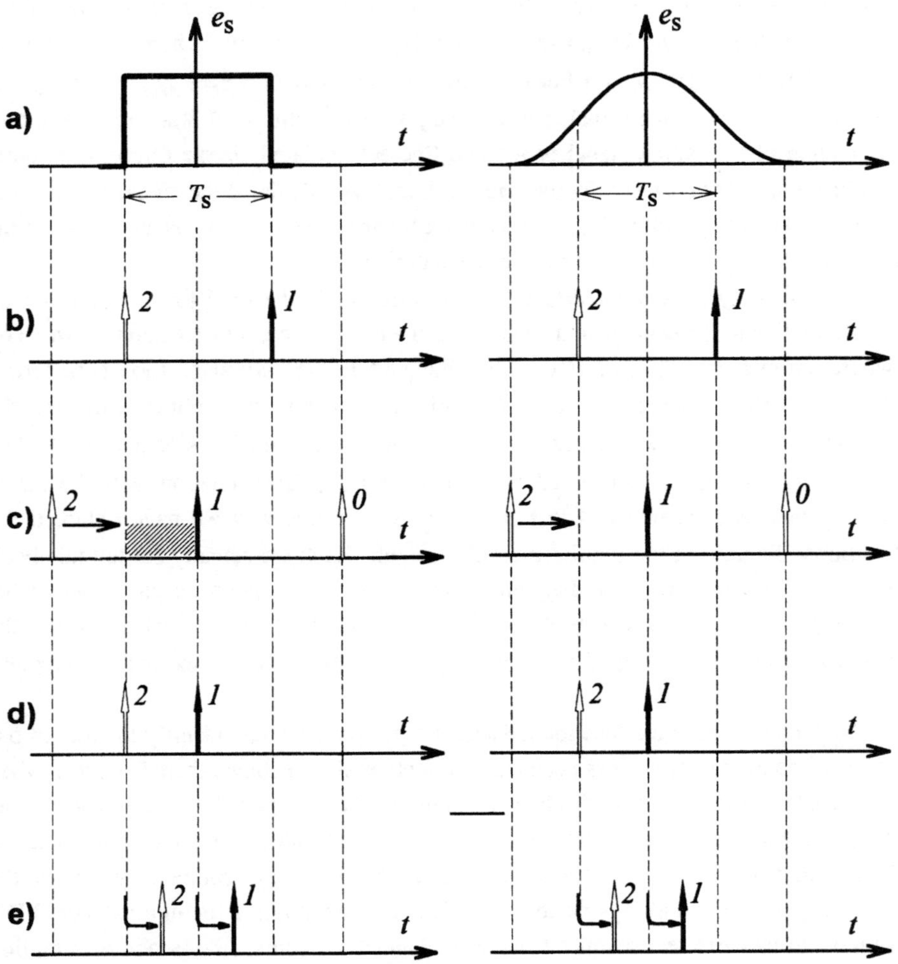

Figure 7.4: Optimization of integration time for unfavorable arrangement of samples.

arranged symmetrically with respect to the maximum (center) of the signal, and most of the charge fell into one of the frames (samples 2...1; Fig. 7.4a,b). Now let us do the same thing for the worst possible situation. Namely, we fix one of the samples (sample 1) in the maximum (center) of the signal (Fig. 7.4c); then each of the adjacent frames (2—1, 1—0) will get no more than half of the signal charge. From Fig. 7.4c it is quite clear that the integration period $T_i = T_s$ is too long for the rectangular signal. The signal is accumulated for only one half of this time, whereas the second half is dedicated to the accumulation of the noise. It follows that for the rectangular pulse we should set $T_i = T_s/2$: then the signal will remain the same, $Q_s = 0.5SE_{sA}T_s$, and the noise will be reduced by the factor of $\sqrt{2}$:

$$\mathrm{SNR} = \frac{Q_\mathrm{s}}{\sqrt{\Delta Q_0^2}} = \frac{1}{2}\frac{SE_{\mathrm{sA}}T_\mathrm{s}}{\sqrt{qI_0\frac{T_\mathrm{s}}{2}}} = E_{\mathrm{sA}}\sqrt{T_\mathrm{s}}\frac{S}{\sqrt{2qI_0}} = \frac{E_{\mathrm{sA}}\sqrt{T_\mathrm{s}}}{\mathrm{NEP}} . \tag{7.15}$$

The same integration time $T_\mathrm{i} = T_\mathrm{s}/2$ — or, more precisely, the time $T_\mathrm{i} = T_{\mathrm{iopt}}/2 = 0.507T_\mathrm{s}$ — must be used for the bell-shaped signal. If this values is surpassed, the noise grows faster than the signal. We have proved this earlier for the best disposition of samples (Fig. 7.2d). In our current case the signal is simply divided in two (frame 1, Fig. 7.4d), $Q_\mathrm{s} = 0.818SE_{\mathrm{sA}}T_\mathrm{s}/2$, and we get:

$$\mathrm{SNR} = \frac{Q_\mathrm{s}}{\sqrt{\Delta Q_0^2}} = 0.409\frac{SE_{\mathrm{sA}}T_\mathrm{s}}{\sqrt{qI_0\frac{T_\mathrm{s}}{2}}} = 0.818\frac{S}{\sqrt{2qI_0}}E_{\mathrm{sA}}\sqrt{T_\mathrm{s}} = 0.818\frac{E_{\mathrm{sA}}\sqrt{T_\mathrm{s}}}{\mathrm{NEP}} . \tag{7.16}$$

After such reduction of the integration time, the value of SNR ceases to depend on the timing of the sample in case of the rectangular pulse, and exhibits a very weak dependence in case of the bell-shaped pulse (Fig. 7.4d,e):

$$\mathrm{SNR} = \frac{Q_\mathrm{s}}{\sqrt{\Delta Q_0^2}} = (0.409...0.475)\frac{SE_{\mathrm{sA}}T_\mathrm{s}}{\sqrt{qI_0\frac{T_\mathrm{s}}{2}}} = (0.818...0.95)\frac{E_{\mathrm{sA}}\sqrt{T_\mathrm{s}}}{\mathrm{NEP}} . \tag{7.17}$$

Although we have not reached the maximum SNR as given by (7.8), (7.13), neither have we fallen down to the minimum of (7.14). For arbitrary timing of the sample we get a certain intermediate value of the signal and the signal-to-noise ratio (7.16), (7.17). It it good or bad? From Sect. 3.1 we know that the detection error is by no means a linear function of SNR: deterioration of SNR causes the error to increase dramatically. Because of this, it is important not to let the signal-to-noise ratio fall. This can be achieved by cutting the integration time by half, $T_\mathrm{i} = T_\mathrm{s}/2$.

There still exists the possibility, however, to improve SNR by about 30...40% (compare (7.15), (7.16) with (7.8), (7.13)). To make use of this margin, one must "disengage" the times T_S and T_i. The interrogation technique based on the independent times T_S and T_i will be discussed somewhat later; presently, following the order of the day, we are going to consider the violation of the third and last condition: the condition that the noise is white.

Excess 1/*f* noise [103]. In practice, the noise may be not white because of the presence of the excess $1/f$ noise. An example of matrix with this kind of noise is the hybrid IR CCD, where the excess noise likes to make its appearance both in the photodetector (PbS photoresistor or MCT photodiode, especially with bias), and in the CCD input device (especially when the input current is considerable) [104-106].

Let us try to find the fluctuation of the charge in the presence of excess noise. We regard the integrator as a simple linear converter whose input is the current. In the calculation of the excess noise we specify the spectrum of the excess input current (2.165) – (2.169):

$$\overline{J^2(f)} = \frac{2qI_0 f_{\text{low}}}{f} \ . \tag{7.18}$$

At the output we register the charge, the calculation of which requires knowing the transfer coefficient of the current-to-charge converter — the frequency response of the integrator. Since the pulse response of the integrator is known (it is rectangular, (7.3)), its spectrum

$$K_{\text{int}}(f) = \frac{\sin \pi f T_{\text{i}}}{\pi f} = K_0 k(f) \ ; \ \ K_0 = T_{\text{i}} \ , \ \ k(f) = \frac{\sin \pi f T_{\text{i}}}{\pi f T_{\text{i}}} \tag{7.19}$$

is the sought-for frequency response of the integrator (see Sect. 2.1 and Table 5.1). From Table 5.1 we see that the effective band of the integrator is $1/2T_{\text{i}}$ both for the signal and for the white noise. Using equations (7.18), (7.19) we may also find the fluctuation of the charge caused by the excess noise:

$$\overline{\Delta Q_0^2} = \int_0^\infty K_{\text{int}}^2(f)\overline{J^2(f)}df = 2qI_0 f_{\text{low}} T_{\text{i}}^2 \int_0^\infty \left(\frac{\sin \pi f T_{\text{i}}}{\pi f T_{\text{i}}}\right)^2 \frac{df}{f} \ . \tag{7.20}$$

In the neighborhood of the zero frequency, where the relative frequency response of the integrator is $k(f) = 1$, equation (7.20) is simplified:

$$\overline{\Delta Q_0^2} \propto \int_0^{\Delta f} df / f \ .$$

This integral diverges when $f \to 0$. Owing to the fact that the integrator transmits all low frequencies at which the power of the excess noise is theoretically infinite, the dispersion of the charge is infinite too. Does this mean that the integrator is in principle not suitable for the detection of low-level signals in the presence of excess noise?

The straightforward one-frame integration is not suitable indeed. If, however, we throw interframe processing into the bargain — for example, by subtracting the charges Q_1 and Q_2 accumulated during two subsequent frames — this nuisance will be removed, and the dispersion will no longer be infinite:

$$\Delta Q_{12} = Q_1 - Q_2 = \int_0^{T_{\text{i}}} I(t)dt - \int_{-T_{\text{i}}}^0 I(t)dt \ . \tag{7.21}$$

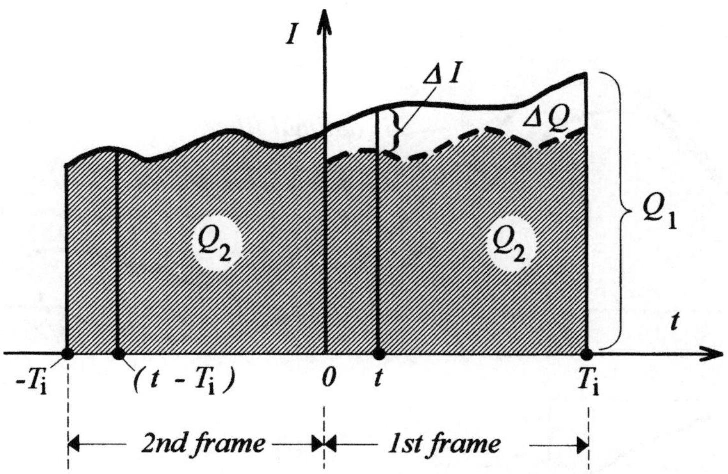

Figure 7.5: Subtraction of adjacent frames is equivalent to integration of differential current $\Delta I(t) = I(t) - I(t - T_i)$.

As a matter of fact, the low-frequency fluctuation is a slow fluctuation. Accordingly, the current $I(t)$ varies very slowly during the frame time T_i, and the charges Q_1 and Q_2 accumulated during two subsequent frames are almost equal to each other. Therefore, there difference (the noise after subtraction) is very small (Fig. 7.5). Now let us transform equation (7.21) to

$$\Delta Q_{12} = \int_0^{T_i} \left[I(t) - I(t - T_i) \right] dt \ . \tag{7.22}$$

In (7.22) the order of linear transformations is reversed: first we find the increment of the current $\Delta I(t)$ after the time T_i, and then integrate this increment (the differential current). Figure 7.5 graphically illustrates the feasibility of such reversal. In this way, two linear transformations are now applied to the input current: time delay by T_i and subtraction, and then the already known operation of integration of the difference (the increment $\Delta I(t)$). Accordingly, yet another factor must be introduced into (7.20), the frequency response of the subtractor. This characteristic is very easy to find. For the sine-wave input signal $I_{in} = I_S \sin 2\pi f t$, at the output of the subtractor with time delay T_i we get:

$$I_{out}(t) = I_{in}(t) - I_{in}(t - T_i) = I_S \sin 2\pi f t - I_S \sin 2\pi f (t - T_i)$$

$$= 2 \sin \pi f T_i I_S \sin 2\pi f \left(t - \frac{T_i}{2} \right) . \tag{7.23}$$

The additional linear multiplier is nothing else but the amplitude-frequency characteristic of the subtractor:

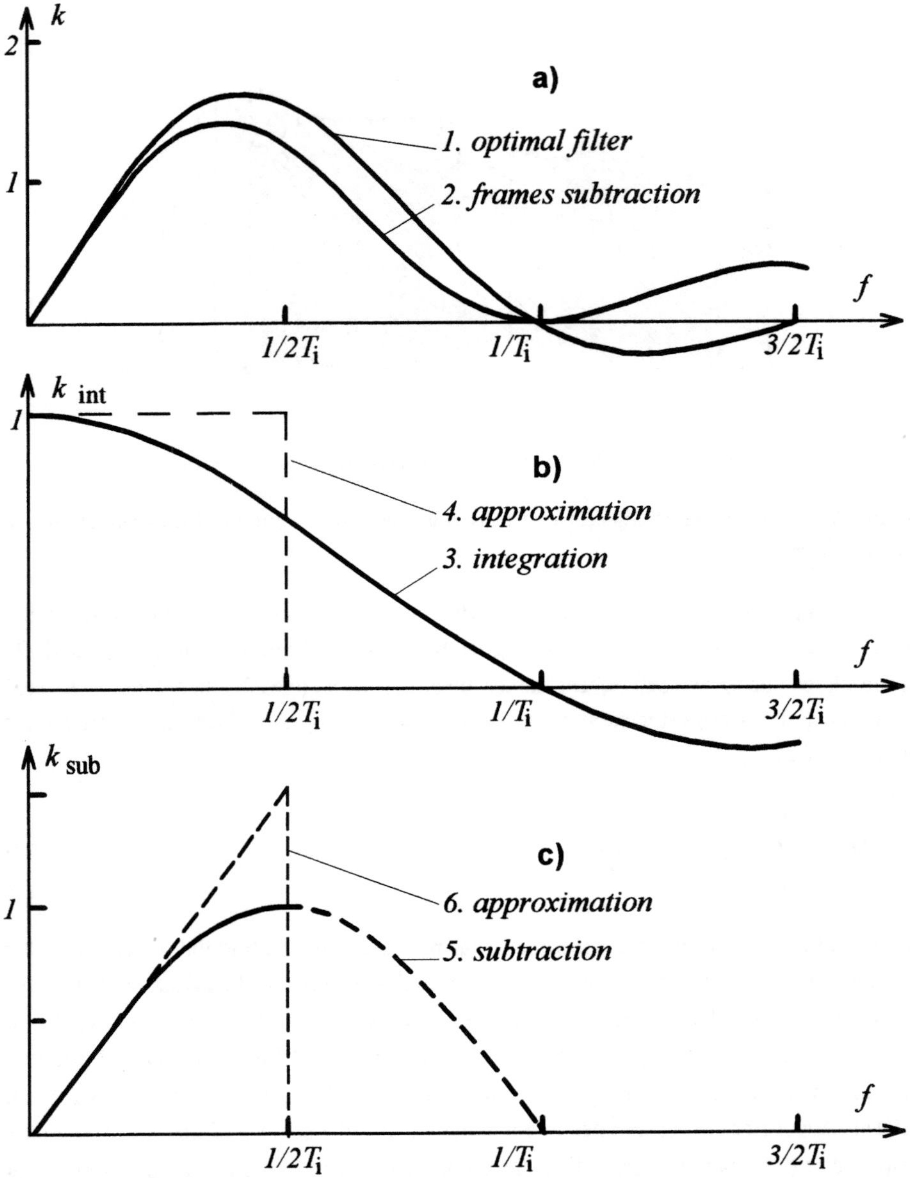

Figure 7.6: Frequency response curve in frame subtraction technique.

$$K_{sub} = 2\sin \pi f T_i \ . \tag{7.24}$$

The interpretation of this characteristic is also simple. The constant component and the low-frequency harmonics, whose period $1/f$ is very large compared to T_i, are eliminated by subtraction, and $K_{sub} \approx 0$. By contrast, the harmonics delayed by half-period add up, and hence the factor 2.

Now let us include K_{sub} into (7.20) for the charge noise:

$$\overline{\Delta Q_{12}^2} = \int_0^\infty K_{sub}^2 K_{int}^2 \overline{J^2} df = \int_0^\infty \left(2\sin \pi f T_i \frac{\sin \pi f T_i}{\pi f} \right)^2 2qI_0 \frac{f_{low}}{f} df \ . \tag{7.25}$$

Equation (7.25) may be rewritten as

$$\overline{\Delta Q_{12}^2} = (2qI_0 T_i)(f_{low} T_i)\,\mathrm{Jn} = (2qI_0 T_i)(f_{low} T_i)\cdot 2.77 \ , \tag{7.26}$$

where

$$\mathrm{Jn} = \int_0^\infty K_{sub}^2 k_{int}^2 \frac{df}{f} = \int_0^\infty \left(2\sin \pi f T_i \frac{\sin \pi f T_i}{\pi f T_i} \right)^2 \frac{df}{f} = 4\ln 2 = 2.77 \ . \tag{7.27}$$

The frequency response of integrator with subtraction of frames is shown in Fig. 7.6a; the diagrams in Fig. 7.6b,c show each of its two components: the frequency curves of, respectively, the integrator and the subtractor. As usual, simple approximations will be useful for understanding the physics and evaluating the integral (7.27). Approximation for the frequency response of the integrator $k_{int}(f)$ is standard: we assume that k_{int} = const in the effective band $1/2T_i$ (curve 4). In the subtractor the main feature is the roll-off at low frequencies; for $f \ll 1/T_i$ we get:

$$K_{sub} = 2\sin \pi f T_i \approx 2\pi f T_i \ . \tag{7.28}$$

The simplest thing is to use this linear approximation for $K_{sub}(f)$ (curve 6); then the overall characteristic takes on the form

$$f \le 1/2T_i \ , \quad K = k_{int} K_{sub} \approx 2\pi f T_i \ . \tag{7.29}$$

Each of the operations secures the required cutoff: subtraction does this at low frequencies (arresting the excess noise), and integration establishes the high-frequency limit (corresponding to the signal bandwidth $1/2T_s$. The excess noise integral in (7.27) in this approximation is equal to

$$\mathrm{Jn} = \int_0^\infty K_{sub}^2 k_{int}^2 \frac{df}{f} \approx \int_0^{1/2T_i} (2\pi f T_i)^2 \frac{df}{f} = (2\pi T_i)^2 \int_0^{1/2T_i} f df = \frac{(2\pi)^2}{8} \approx 4.9 \ . \tag{7.30}$$

Although this value is not quite the same as that given by the exact expression (7.27), it is good by the order of magnitude.

This is as far as the noise is concerned. Now, what does subtraction do to the signal?

The signal may be somewhat reduced, because the signal charge is accumulated in both the first (Q_{s1}) and in the second (Q_{s2}) frames. The largest signal ΔQ_s is accumulated with the optimal arrangement of the samples (Fig. 7.2e): then

$$\Delta Q_s = Q_{s1} - Q_{s2} = 0.82 Q_s - 0.091 Q_s = 0.72 S E_{sA} T_s \ . \tag{7.31}$$

In the other extreme, with the worst arrangement of samples with respect to the signal (Fig. 7.2f), when the main charge (pertaining to the first frame) is small, the signal charge is fortunately not accumulated during the second frame at all, so that subtraction does not reduce the signal any further:

$$\Delta Q_s = Q_{s1} - Q_{s2} = 0.5 Q_s - 0 \cdot Q_s = 0.5 S E_{sA} T_s \ . \tag{7.32}$$

Having found the signal and the noise, we can finally find their ratio:

$$\mathrm{SNR} = \frac{\Delta Q_s}{\sqrt{\Delta Q_0^2}} = \frac{(0.5 \ldots 0.72) S E_{sA} T_s}{\sqrt{2 q I_0 T_s \cdot 2.77 \cdot f_{\mathrm{low}} T_s}} = (0.3 \ldots 0.43) \frac{E_{sA}}{\mathrm{NEP}\sqrt{f_{\mathrm{low}}}} \ . \tag{7.33}$$

Let us compare (7.33) with the corresponding formula for the optimal filter (3.84). The loss of the cell with integration is relatively low, and may be as small as 26% (at worst, it may increase to 49% because of the non-synchronized sampling). It is the subtraction that has allowed us to finalize the shaping of the frequency response of the cell, bringing it close to the optimum (compare curves 1, 2 in Fig. 7.6). The optimal curve 1 is plotted according to (3.49), (2.14), and $e_s(f)/\Theta \propto f e_s(f)$.

Such a good agreement between the frequency curves 1, 2 is not surprising: in Sect. 4.1 we indicated that the optimal filter carries out subtraction. The approximation of the pulse response of optimal filter in Fig. 3.9 allows us to regard its signal processing algorithm as subtraction, which involves, however, not two but three frames: half of the charge accumulated in each of the adjacent frames 2, 3 is deduced from the charge of the main frame 1. Accordingly, the above technique of calculating the SNR may also be used for approximate evaluation of SNR of the optimal filter.

Superposition of frames. The drawbacks of the one-frame integration are the following: the pulse response (and, respectively, the frequency response) are strictly predetermined; predetermined is also the sampling interval which cannot be made shorter than the integration interval. As shown above, this may result in considerable deterioration; in the presence of excess noise the one-frame integration is not suitable at all, and we have to go over to the simplest version of the frame-to-frame processing: the charge of the preceding subframe Q_2 is subtracted from the accumulated charge Q_1. In principle, however, it is possible to construct a linear superposition of n subframes with arbitrary coefficients a_i:

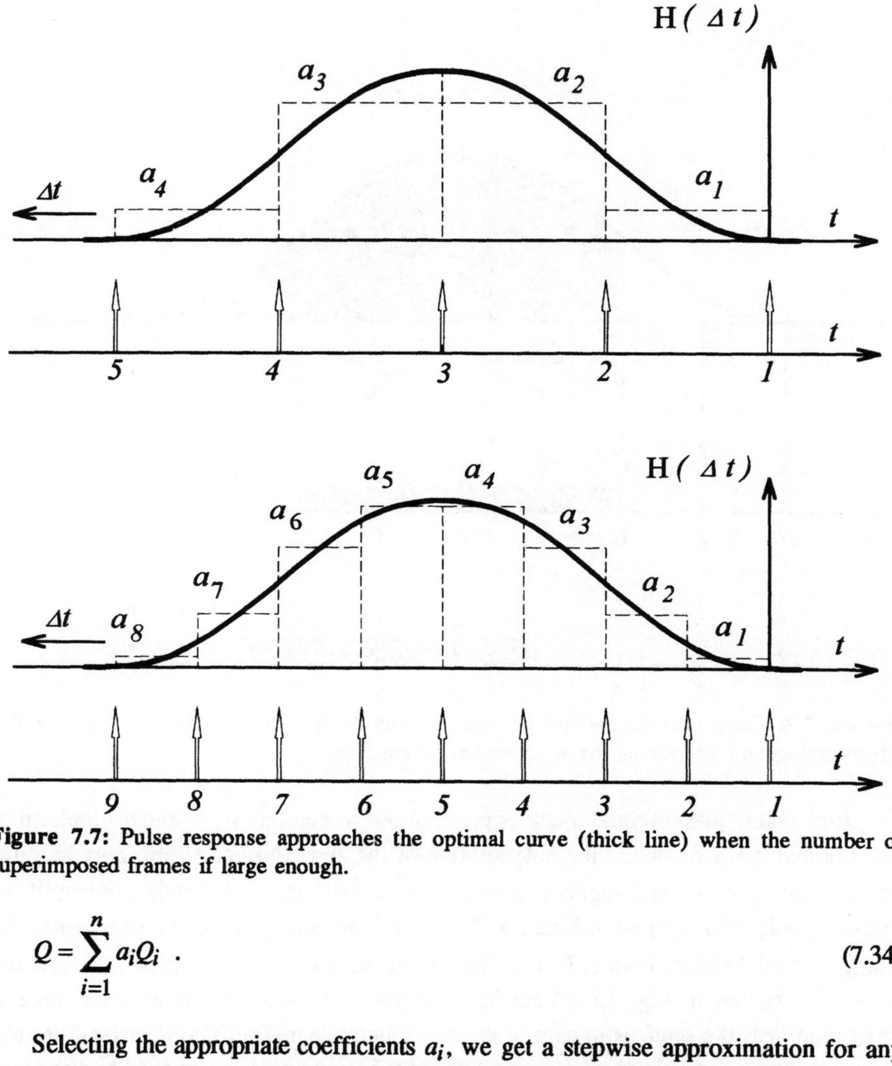

Figure 7.7: Pulse response approaches the optimal curve (thick line) when the number of superimposed frames if large enough.

$$Q = \sum_{i=1}^{n} a_i Q_i \ . \tag{7.34}$$

Selecting the appropriate coefficients a_i, we get a stepwise approximation for any desired pulse response; the greater the number of subframes n, the more accurate the approximation (Fig. 7.7). When n is large, however, we encounter fundamental hardware-associated problems:

— the charge of each subframe is reduced n-fold, and so the contribution of the downstream noises of charge transfer and output devices increases;

— the frequency of interrogation becomes n times as high, and may exceed the technical limits of the matrix; additional distortion of the signal may occur;

— the required capacity of external memory is n times larger, the processing algorithm becomes more sophisticated, and the number of operations performed by external devices increases.

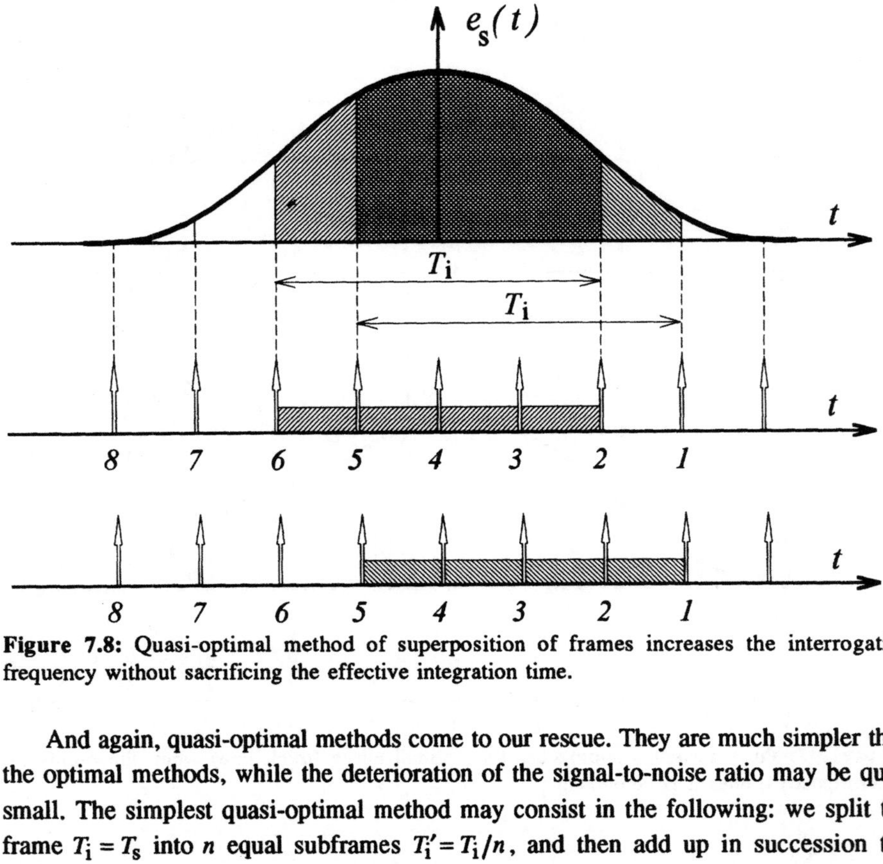

Figure 7.8: Quasi-optimal method of superposition of frames increases the interrogation frequency without sacrificing the effective integration time.

And again, quasi-optimal methods come to our rescue. They are much simpler than the optimal methods, while the deterioration of the signal-to-noise ratio may be quite small. The simplest quasi-optimal method may consist in the following: we split the frame $T_i = T_s$ into n equal subframes $T_i' = T_i/n$, and then add up in succession the charges of all these subframes. First we add up the samples 5, 4, 3, 2, then 4, 3, 2, 1, and so on, as shown in Fig. 7.8. What has changed? The effective integration time is preserved: it is the same as in case of the one-frame integration. The interrogation rate, however, has become n times faster, $T_S = T_i'$. In this way we have managed to disengage the integration time and the sampling time, and to reduce T_S: after all, it was large T_S that caused the loss of signal. The functional diagram in Fig. 7.1d is now entirely similar to that of the traditional PDA. The latter also includes a commutator; the sampling time in such PDA does not depend on the effective band of the amplifier, and the same can be said now about our matrix with integration cell. The loss of signal because of the commutator in the matrix with accumulation now are the same as those in the traditional PDA with amplifiers.

The method of calculation of the magnitude of the signal as function of the sampling period $T_S = T_i/n$ has been developed earlier (Fig. 7.3). Curve $Q_s(\Delta t)$ in Fig. 7.3 depicts the charge at the output of the integrator, and allows us to find, setting $\Delta t = T_S$, the signal $Q_s(T_S) = Q_s(T_s/n)$, and then the signal-to-noise ratio (Fig. 7.9a). In practice,

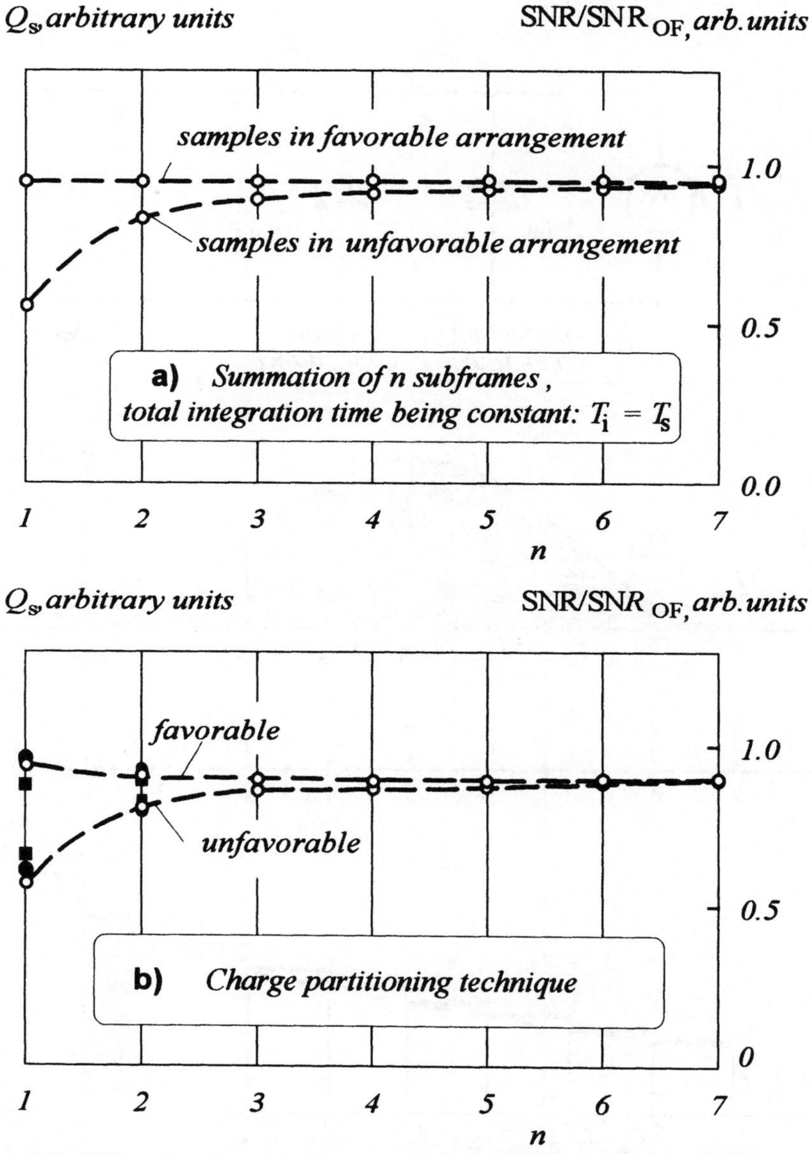

Figure 7.9: SNR as function of n for two quasi-optimal methods of superposition of frames (the charge is referred to the total signal charge $SE_{sA}T_s$; the input pulse is square-cosine; $n = T_s/T_S$).

three to five subframes (three to five samples over the effective signal duration T_S) are sufficient for not missing the signal maximum.

The described technique simplifies the processing algorithm, but it still requires storing the signals from all n subframes in the external memory. The integration cell itself may be entrusted with memorizing the charges from the preceding subframes. In the act of interrogation we only have to "spill" and read out the kth portion of the accumulated charge; then the remaining $(1-k)$th part will carry information about the

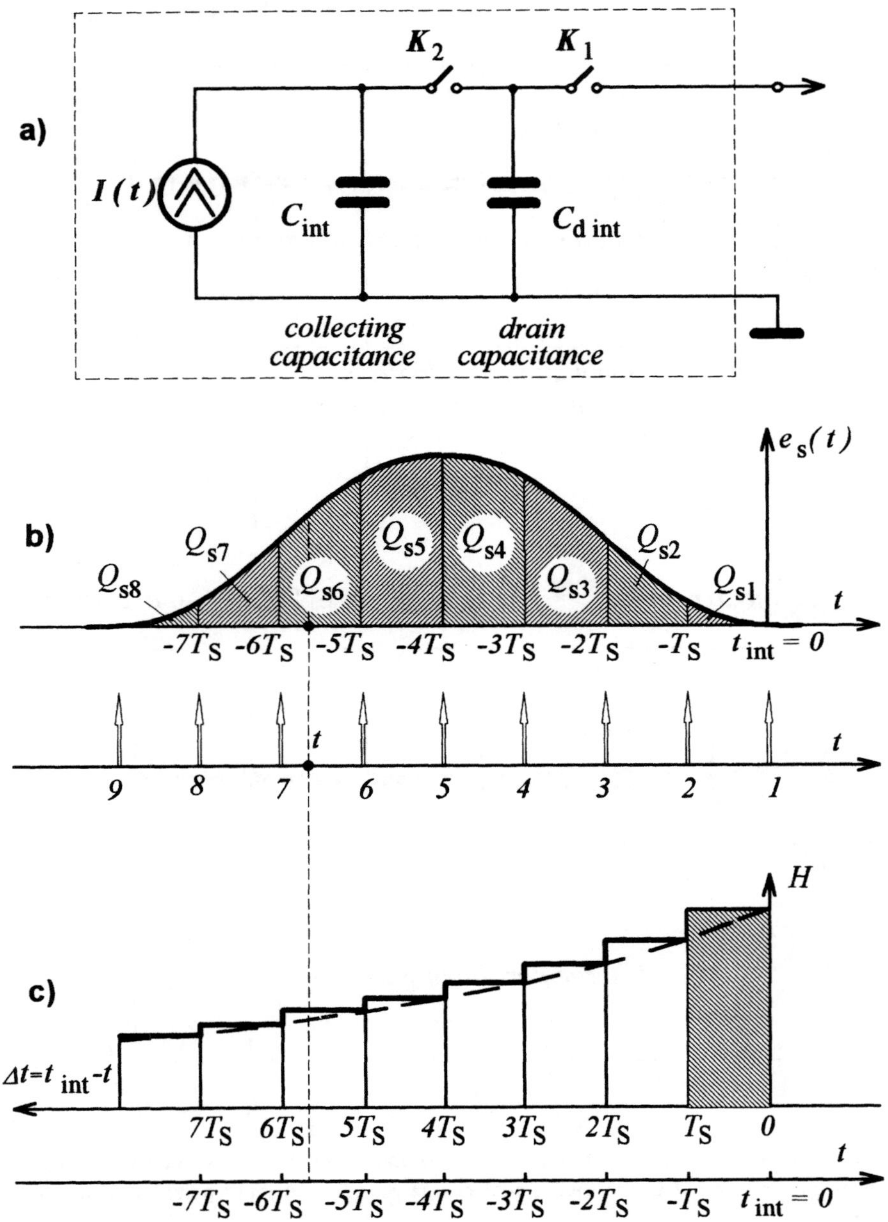

Figure 7.10: Charge partitioning technique.

preceding frames. This is implemented by introducing the additional receiving (drain) capacitance $C_{d\,int}$, and the second switch K_2 between the main integrating capacitance C_{int} and the drain capacitance $C_{d\,int}$ (Fig. 7.10a). Now, when the switch K_2 is closed, the following part of the accumulated charge Q_{int} spills from C_{int} into $C_{d\,int}$:

$$Q_{\mathrm{d\,int}} = \frac{C_{\mathrm{d\,int}} Q_{\mathrm{int}}}{C_{\mathrm{d\,int}} + C_{\mathrm{int}}} = kQ_{\mathrm{int}} \;, \quad k = \frac{C_{\mathrm{d\,int}}}{C_{\mathrm{d\,int}} + C_{\mathrm{int}}} \;. \tag{7.35}$$

Then the switch K_1, like in the ordinary cell, transmits all the charge $Q_{\mathrm{d\,int}}$ to the output to be registered. The switches may be either MIS transistors, or CCD gates between the CCD cells (integration cell capacitance C_{int} and drain cell capacitance $C_{\mathrm{d\,int}}$). The pulse response of the cell is shown in Fig. 7.10c. If the charge had been erased completely, it would have been a rectangle of length T_S (hatched area). Since the charge is not erased completely, however, the integrating capacitance retains some charge from the preceding frame — but only its $(1-k)$th part, because its kth part has been removed at the time of sampling $t=-T_S$ (Fig. 7.10b). Accordingly, the second integration interval $-2T_S...-T_S$ is also depicted by a similar but smaller rectangle, to reflect the coefficient $(1-k)$. Two samplings have taken place after the integration over the third interval $-3T_S...-2T_S$, and the height of the corresponding rectangle is $(1-k)^2$. Accordingly, for any time interval $\Delta t = t_{\mathrm{int}} - t$, where t_{int} is the time of interrogation and t is the time of arrival of the signal, we get:

$$(n-1)T_S < \Delta t < nT_S \;,$$

$$\mathrm{H}(\Delta t) = (1-k)^{(n-1)} = \exp[(n-1)\ln(1-k)] = \exp\left[-(n-1)\frac{T_S}{\tau_{\mathrm{eff}}}\right] \;; \tag{7.36}$$

$$\tau_{\mathrm{eff}} = \frac{T_S}{-\ln(1-k)} \;. \tag{7.37}$$

Expression (7.37) defines the equivalent time constant of discharge of the integrating capacitance. When the coefficient of charge partitioning k is small, this expression takes on a very simple form:

$$k \ll 1 \;, \quad \tau_{\mathrm{eff}} = \frac{T_S}{-\ln(1-k)} \approx \frac{T_S}{k} = T_S \frac{C_{\mathrm{d\,int}} + C_{\mathrm{int}}}{C_{\mathrm{d\,int}}} \approx T_S \frac{C_{\mathrm{int}}}{C_{\mathrm{d\,int}}} \;. \tag{7.38}$$

When the charge is spilled frequently by small amounts ($k \ll 1$, $T_S \ll \tau_{\mathrm{eff}}$), the pulse response of the integrator becomes exponential, like that of the ordinary RC-circuit. The physics of this phenomenon is extremely simple. The discharge δQ of the capacitance C across the resistance R over a small time interval δt is proportional to the total charge on this capacitance:

$$\delta Q = \frac{U\delta t}{R} = \frac{Q\delta t}{CR} = \frac{Q\delta t}{\tau_{\mathrm{eff}}} \;. \tag{7.39}$$

In exactly the same way the integrating capacitance discharges over the small time interval $\delta t \approx T_S << \tau_{eff}$: from (7.35) it follows that

$$\delta Q = Q_{d\,int} = kQ_{int} = \frac{Q_{int}T_S}{\frac{T_S}{k}} \ . \tag{7.40}$$

Comparing (7.39) and (7.40), we arrive at the same value of the equivalent time constant of discharge of the integrating capacitance, $\tau_{eff} = T_S/k$ (7.38).

Since the pulse response $H(\Delta t)$ is now known, further procedure of calculation of signal and noise hardly calls for any comment. From Fig. 7.10b,c we see that by the time of interrogation (sampling) $t_{int} = 0$ the capacitance C_{int} will preserve $\left((1-k)^{n-1}\right)$th part of each charge Q_{sn} accumulated over the nth subframe. The signal charge $Q_s(t_{int})$ is found as the superposition of the hatched areas in Fig. 7.10b with the weight $H(\Delta t) = (1-k)^{n-1}$. The dark (and background) charges accumulated over each subframe are the same: $Q_0 = I_0 T_S$. Their dispersions $\overline{\Delta Q_d^2} = qI_d T_S$ add up with the weight $H^2(\Delta t) = (1-k)^{2(n-1)}$, and so for the noise we get:

$$\overline{\Delta Q_d^2} = \sum_{n=1}^{\infty} (1-k)^{2(n-1)} qI_0 T_S = \frac{qI_0 T_S}{1-(1-k)^2} = \frac{qI_0 T_S}{k(2-k)} \ . \tag{7.41}$$

Here we have duly noted that this is the sum of geometric progression with the denominator $(1-k)^2$; the first term is equal to one, and so the sum of all terms is $1/\left(1-(1-k)^2\right)$.

This technique has been used for calculating the signal and the noise for the bell-shaped input signal; the resulting SNR curves are plotted in Fig. 7.9b. The parameter here is the number $n = T_s/T_S$, which is already familiar to us from the first method of superposition of subframes. Now, however, there is yet another degree of freedom, the coefficient of partitioning k. When k is small, the cell is identical to the one-section RC-filter. The signal-to-noise ratio for such filter attains its maximum when the effective time constant of the filter τ_{eff} is equal (almost equal) to the signal duration T_s (see Sect. 5.2). According to (7.38), this condition is satisfied if k is the reciprocal of the number of samples n:

$$k = \frac{1}{n} = \frac{T_S}{T_s} \ , \quad \tau_{eff} = \frac{T_S}{k} \equiv \frac{T_S}{T_S/T_s} \equiv T_s \ . \tag{7.42}$$

Figure 7.9b completely confirms these arguments: the best possible SNR in this technique is the same as that for the one-section RC-filter even with the non-synchronized sampling, and is achieved with as few as 3...5 samples within the time T_s.

A similar number of samples was found to be sufficient in the method of superposition of frames. Figure 7.9b also shows how the function SNR(n) is affected by the choice of charge partitioning coefficient $k(n)$. When n is small ($n = 1,2,3$), the stepwise shape of the pulse response is still noticeable, and therefore the accepted value of $k = 1/n$, strictly speaking, may be not optimal. Accordingly, we also selected such values of k which optimize the signal-to-noise ratio for the least favorable (squares) and the most favorable (black dots) arrangements of samples. Variations in the value of k bring about rather small (1...15%) variations of SNR, and those only for $n = 1,2$. This is not surprising: once again we reconfirm our conclusion that SNR is not critical to the tuning of the system.

Thus, the proposed method of charge partitioning allows us to improve the signal-to-noise ratio by the factor of 1.5 as compared to the ordinary one-frame integration (compare points $n = 1$ and $n = 5$ in Fig. 7.9b for the unfavorable arrangement of samples). The gain is almost the same as that obtained with the method of superposition of frames. Which of the two methods is better? The answer depends on the system: if it is not desirable to complicate the matrix, one should use the method of superposition of frames; if it is not desirable to complicate the external processing, one should use the method of charge partitioning.

Yet another advantage of the methods discussed above is that the frequency of interrogation is increased by the factor of 3...5 as compared with the straightforward integration, which is important for systems which require precise definition of the coordinates of the radiation source.

Accumulation of the dc component. The blocking capacitor in the traditional PDA arrests the direct-current component of the photodetector current to the output of the amplifier. The integrator stores indiscriminately the charge carriers generated by the signal and by thermal and background radiation. The large dc component I_0 may give rise to two problems.

The first problem is associated with the fact that the low-level signal charge has to be separated from its mixture with a very large constant charge Q_0. The ratio of the constant charge to the noise charge (7.5)

$$\frac{Q_0}{\sqrt{\overline{\Delta Q_0^2}}} = \frac{Q_0}{\sqrt{qQ_0}} = \sqrt{\frac{Q_0}{q}} = \sqrt{N_0} \qquad (7.43)$$

is proportional to square root of the number of accumulated electrons (see also Sect. 6.2, equation (6.45)). For the typical values of specific capacitance $(3\ldots5)\cdot10^{-8}\ \mathrm{F/cm^2}$, PSE area $A_e = 30\,\mu\mathrm{m}\times30\,\mu\mathrm{m} = 9\cdot10^{-6}\ \mathrm{cm^2}$, $\Delta U = 5\,\mathrm{V}$, we get the following numerical estimates:

$$C_{int} = (3\ldots5)\cdot10^{-8}\cdot9\cdot10^{-6} = (2.7\ldots4.5)\cdot10^{-13}\ \mathrm{F}\ , \qquad (7.44)$$

$$Q_0 = C_{int}\Delta U = (2.7...4.5)\cdot 10^{-13}\cdot 5 = (1.35...2.25)\cdot 10^{-12}\,\mathrm{C}\ , \tag{7.45}$$

$$N_0 = \frac{Q_0}{q} = \frac{(1.35...2.25)\cdot 10^{-12}}{1.6\cdot 10^{-19}} \approx 10^7\ ,\ \sqrt{N_0} \approx 3000\ . \tag{7.46}$$

This is the rule of three thousands: when the typical integrating capacitance is charged, the ratio of the dc component to the noise and the noise itself (the number of noise electrons) are quite large, and equal to about 3000.

The second problem is the limited time of integration. If the voltage across the integrating capacitance C_{int} may vary within $|\Delta U|$, then the largest allowable integration time is

$$T_{i\,max} = \frac{C_{int}|\Delta U|}{I_0} = \frac{\frac{C_{int}}{A_e}|\Delta U|}{j_0}\ . \tag{7.47}$$

If the integrating capacitance is the capacitance of the photodetector, the time $T_{i\,max}$ does not depend on the area of the photosensitive element A_e. The calculated values of $T_{i\,max}$ for IR PDAs of spectral ranges III – V are shown in Table 7.1.

Table 7.1. **Maximum integration time**

Parameter	Spectral range			
	III	IV	V	
Wavelength, μm	2...3	3...5	8...10	8...12
j_{db}, $\mathrm{A/cm^2}$	~10^{-6}	$2.5\cdot 10^{-4}$	~$8\cdot 10^{-3}$	$2.4\cdot 10^{-2}$
$T_{i\,max}$, s	0.3	~10^{-3}	$4\cdot 10^{-5}$	$1.2\cdot 10^{-5}$

Calculations have been carried out for the field of view ±27° and the typical values of parameters $\eta = 0.8$; $\Delta U = 5...10\,\mathrm{V}$; $C_{int}/A_e = (3...5)\cdot 10^{-8}\,\mathrm{F/cm^2}$; $C_{int}\Delta U/A_e \approx 3\cdot 10^{-7}\,\mathrm{C/cm^2}$. The reduction of the allowable integration time because of the thermal radiation current is noticeable already in the spectral range 3...5 μm, and becomes quite considerable in the range 8...12 μm. When the time $T_{i\,max}$ is exceeded, the cell is “dazzled”, and image blooming occurs in the CCD: the charge spills over into the adjacent cells. This unwanted phenomenon is eliminated by introducing specially designed anti-blooming charge drains.

The time $T_{i\max}$ can be increased by reducing the current I_0 — that is, by implementing the same measures as used for improving the radiation threshold (see Sect. 1.2 and 2.5). When such measures (being limited by the general requirements to the optoelectronic system) are exhausted, further increase in $T_{i\max}$ can be achieved by using special devices for charge division and subtraction. Division of the charge in CCDs is accomplished by introducing, in addition to the main capacitance C_{int}, the capacitance $C_{\text{d int}} = kC_{\text{int}}$, like in the cell with charge partitioning (Fig. 7.10a), and a third capacitance, the end storage capacitor $C_{\text{int}2} \approx C_{\text{int}}$. From time to time, with the period of $<< T_{i\max}$, the maximum charge Q_{int} accumulated by the main capacitance is split, and its kth part is first transferred to the capacitance $C_{\text{d int}}$ (7.40), and then to the end capacitance $C_{\text{int}2}$. The main capacitance C_{int} is cleared (as different from the method of charge partitioning considered earlier) and is ready to receive the charge from the next subframe. As a result, the charge $kQ_0 = kI_0T_i$ is stored in the end capacitance, so the time $T_{i\max}$ is increased by the factor of $1/k$. If the voltage applied to the main capacitance C_{int} exceeds the voltage across the drain capacitance $C_{\text{d int}}$ by $|\Delta U|$, the main capacitance will retain the charge $Q_{\text{const}} = C_{\text{int}}|\Delta U|$. Thus, the subtraction is done: to the drain goes the charge $Q - Q_{\text{const}}$ if $k = 1$, or the part of the charge $k(Q - Q_{\text{const}})$ if $k < 1$ (in the latter case the subtraction is combined with division).

Unfortunately, the introduction of anti-blooming, division and subtraction deprives the cell of one of its most attractive features, its simplicity. Anti-blooming requires additional space, and so the photosensitive element in two-dimensional matrices has to shrink considerably. Because of the reduced useful area the sensitivity falls, and often the required space is simply not available. On top of that, there is the noise of division, the KTC-noise, the adverse effects of the spread of surface potential. All these phenomena limit the utility of the technical solutions discussed above.

Commutation noise. The charge accumulated in one cell of the CCD is drained practically completely to the adjacent cell. In other cases (in the floating diffusion CCD region, in the photodetector-switch cell) the capacitance is not discharged completely: the switch is used for establishing the initial state of the capacitance: the potential U_{ini} and the charge $Q_{\text{ini}} = C_{\text{int}}U_{\text{ini}}$ (Fig. 7.11a). In the absence of the photodetector current (the dark frame) this state must remain unchanged. In the course of integration, the photodetector current discharges the capacitance by the amount of $(I_0 + I_s)T_i$ (cf. Fig. 7.11c,e). Accordingly, the informative charge $Q_0 = (I_0 + I_s)T_i$ has to be registered above the constant component $C_{\text{int}}U_{\text{ini}}$. Unfortunately, this charge is not strictly constant, and changes after each commutation of the switch, so that instead of the idealized graph of the potential in Fig. 7.11c we get a graph with the potential steps as shown in Fig. 7.11d (the real graph is even worse because of the transient effects which will be discussed later on). The potential jumps (fluctuations $\overline{\Delta U_{\text{ini}}^2}$) are explained by the fact that the switch connects the capacitance C_{int} not only to the constant voltage

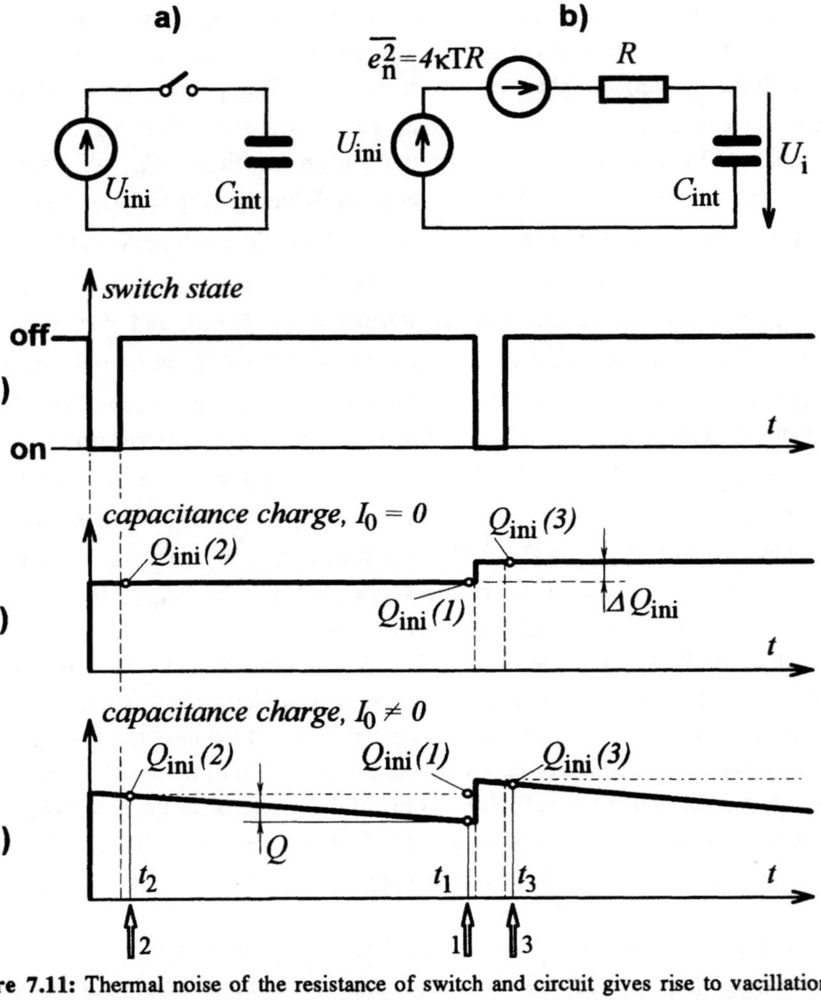

Figure 7.11: Thermal noise of the resistance of switch and circuit gives rise to vacillation of voltage and charge on the capacitance (KTC-noise): (a) voltage setting with the switch; (b) equivalent diagram with switch closed; (c) on-off switching; (d) charge on the capacitance ($I_0 = 0$); (e) charge on the capacitance ($I_0 \neq 0$).

generator U_{ini}, but also to the thermal noise emf generator $\overline{e_n^2} = 4\kappa TR$, where R is the resistance of the switch and the current-carrying circuit (Fig. 7.11b). At low frequencies the noise emf $\overline{e_n^2}$ drops completely across the capacitance, and at high frequencies it drops across the resistance R. The characteristic noise frequency band for this one-section filter with the time constant $\tau_f = RC_{int}$ is $1/4\tau_f$ (Table 5.1), and the dispersion of the noise is therefore

$$\overline{\Delta U_{ini}^2} = \overline{e_n^2} f_{nw} = 4\kappa TR \frac{1}{4RC_{int}} = \frac{\kappa T}{C_{int}}, \tag{7.48}$$

$$\overline{\Delta Q_{ini}^2} = C_{int}^2 \overline{\Delta U_{ini}^2} = \kappa \mathrm{T} C_{int} \; . \tag{7.49}$$

This noise caused by the instability of the initial potential (and initial charge) is known as the KTC-noise. These expressions have been derived for the steady state model; they also hold for the dynamic model — provided, of course, that the capacitance is connected to the voltage source for the time long enough as compared with the time constant $\tau_f = RC_{int}$. It is interesting that the dispersions $\overline{\Delta U_{ini}^2}$, $\overline{\Delta Q_{ini}^2}$ do not depend on the resistance, but the noise spectrum does. As R increases, the spectral density $\overline{e_n^2} \propto R$ increases at low frequencies, but the band $f_{nw} \propto 1/R$ decreases, and so the noise power $\overline{e_n^2} f_{nw}$ remains the same. When the switch is open, its resistance goes to infinity, the noise power is concentrated near the zero frequency, and for the time T_i the state of the integrating capacitance does not fluctuate.

When the background noise is low, the KTC-noise may delimit the lowest detectable magnitude of the signal. For the typical values of $C_{int} = 3 \cdot 10^{-13}$ F (7.44) and the working temperatures 77...300 K, the KTC-noise will prevent us from detecting fewer electrons than

$$N_n = \frac{\sqrt{\overline{\Delta Q_{ini}^2}}}{q} = \frac{\sqrt{\kappa \mathrm{T} C_{int}}}{q} = \sqrt{\frac{\kappa \mathrm{T}}{q} \cdot \frac{C_{int}}{q}}$$

$$= \sqrt{(0.006...0.026) \frac{3 \cdot 10^{-13}}{1.6 \cdot 10^{-19}}} \approx 100...200 \text{ electrons.} \tag{7.50}$$

Is it possible to eliminate the KTC-noise? It all depends on the method employed for taking and processing the readings (Fig. 7.11d,e). The diagrams show three samples: at the beginning of the frame $Q(1) = Q_{ini}(1)$, at the end of the frame $Q(2)$, and at the beginning of the next frame $Q(3) = Q_{ini}(3)$. Manipulating with these three samples, we may find the sought-for informative charge Q_0 in at least three ways.

The first and the most straightforward method consists in using only one sample at the end of the frame $Q(2)$. To find the informative charge Q_0, the external device must subtract the mean steady charge $\overline{Q_{ini}}$:

$$Q(2) - \overline{Q_{ini}} = [Q_{ini}(2) - Q_0] - \overline{Q_{ini}} = [Q_{ini}(2) - \overline{Q_{ini}}] - Q_0 \; . \tag{7.51}$$

The drawbacks of this method are obvious: the external device must establish the values of $\overline{Q_{ini}}$ for each cell, averaging Q_{ini} over many samples in the absence of the signal, and then remember all the values of $\overline{Q_{ini}}$ (which may be different for each cell in multielement PDAs). From (7.51) it is clear that this method does not do away with the fluctuations of $Q_{ini}(2)$ with respect to its mean (that is, the KTC-noise).

The second method is the subtraction of the two adjacent samples $Q(2)$ and $Q(3)$:

$$Q(3)-Q(2)=Q_{\text{ini}}(3)-\left[Q_{\text{ini}}(2)-Q_0\right]=\left[Q_{\text{ini}}(3)-Q_{\text{ini}}(2)\right]+Q_0 \ . \tag{7.52}$$

The job of the external device has been made easier: it does not need to establish the values of $\overline{Q_{\text{ini}}}$, and, which is the most important, there is no need to remember the entire matrix of these values. Now, however, we are registering fluctuations of the two samples $Q_{\text{ini}}(3)$ and $Q_{\text{ini}}(2)$, and the power of the KTC-noise is doubled.

The third method, which is eventually capable of eliminating the KTC-noise, is the subtraction of the samples 1 and 2, between which the switch does not change its state, and only the charge accumulation (or transfer through charge coupling) takes place. This method is known as correlated double sampling (CDS): double because there are two samples, correlated because the initial charge Q_{ini} between these samples does not change, and so the samples in the dark are entirely correlated, $Q_{\text{ini}}(1)=Q_{\text{ini}}(2)$ (Fig. 7.11d).

When the background charge is large, the KTC noise is drowned in the shot noise of this background. Equations (7.5), (7.49) offer an easy way for finding this regime from condition of charging-discharging of the integrator because of this background, $\Delta U = I_0 T_{\text{i}}/C_{\text{int}}$:

$$\overline{\Delta Q_0^2} > \overline{\Delta Q_{\text{ini}}^2} \ , \ \overline{\Delta Q_0^2} = qQ_0 = qC_{\text{int}}\Delta U > \kappa\, T C_{\text{int}} \ ; \ \Delta U > \frac{\kappa\, T}{q} = \varphi_{\text{th}} \ . \tag{7.53}$$

It follows that if the integrator is (dis)charged by the current and the voltage change exceeds a few thermal potentials, which is observed in IR PDAs when the integration time reaches $T_{\text{i}} \approx 0.1 T_{\text{i max}}$, the KTC-noise may be totally disregarded.

More difficult it is to combat the interference from the control pulses. Commutation may induce stray currents through the integrating capacitance and the signal buses caused by the recharging currents of the stray capacitance of the switch C_{sw} (7.11a); spurious pulses may arise because of the stray capacitances. The control pulse voltage is of the same order of magnitude or slightly higher than the voltage of the capacitance charging in case of integration of the background currents (about 2...20 V); and thus exceeds the threshold-level signals by the factor of 3000 and more. When the background is low, this figure is even higher. Obviously, the requirements to the suppression of the commutation noise have to be extremely rigorous if we want to detect the signal on the level of the intrinsic noise. The methods used for reducing the noise level include arrangement of screening electrodes between clock and signal buses; the correlated samples are taken at the moments when the potential of the control bus is the same; the edges of the clock pulses are smoothed, and their amplitude is limited; the time of interrogation is increased so that the samples could be taken when steady-state values are established. If the switch is closed between the consecutive interrogations of the photodetector-switch cell (samples 1, 3 in Fig. 7.1e), the pulses of recharging of the stray capacitance of the switch are eliminated by including the integrator into the output device. Since the charging and the discharging currents of the switch capacitance carry equal charges and flow in opposite directions, the charges in the integrator cancel out.

We see that the processing and registration of weak signals becomes more complicated when commutation is involved. Although the commutation noise can in principle be eliminated, it may spoil considerably the detectivity of the real systems, and should always be treated with due diligence and consideration.

Summary. *The integration cell is comprised of the lowest number of functional elements: photodetector – capacitance – switch.*

If the input signal is rectangular, the noise is white, the integration time is $T_\mathrm{i} = T_\mathrm{s}$, *and the samples are synchronized with the beginning and the end of the pulse, this cell is the optimal filter. Even when these conditions are not satisfied, the loss in the signal-to-noise ratio is relatively small:* 5% *when the signal is bell-shaped, about* 26% *when the noise is not white* ($1/f$*-noise), and the subtraction technique is used. The loss caused by non-synchronized sampling is considerable and, at worst, may be as large as* 42...50% *(when the signal charge is split in half between two consecutive frames). Improved signal processing allows reducing the loss to* 5...10%, *increasing the interrogation rate while preserving the desired integration time (which also improves the accuracy of definition of the time of arrival of the signal). One such method (superposition of subframes) consists in splitting the desired integration time* $T_\mathrm{i} = T_\mathrm{s}$ *into* n *subframes,* $T_\mathrm{i}' = T_\mathrm{i}/n$, *and the number of subframes need not be too large:* $n \approx 3...5$. *Then the external processor computes the sum of charges accumulated in each of the* n *subframes. In another technique of charge partitioning only* $(1/n)$th *part of the accumulated charge is drained from the capacitance and registered at the output. The removal of charge is performed often enough, with the same period* T_i/n.

It follows that the integration cell in the general case is a quasi-optimal filter, and the signal processing is similar to that performed in the traditional PDA with amplifier and subsequent commutator. It is this fascinating property of the integration cell to perform nearly optimal signal processing together with its utmost simplicity that have made it the main building block of practically all kinds of modern imaging PDAs; there is practically no alternative to this type of cell for two-dimensional imaging focal plane arrays with more than $10^3...10^4$ *pixels.*

7.2. Multielement Traditional-Type PDAs and Focal Plane Arrays

SNR for one channel. We begin by writing the expression for the signal-to-noise ratio for PDA based on the optimal filter. This expression will become a benchmark for calculating SNR in real PDAs, both traditional (with an amplifier in each channel), and those with charge integration. The signal is assumed to be bell-shaped. The approximation for SNR in case of optimal filtering has been obtained in Sect. 3.3 (see (3.94)); now we rewrite it as

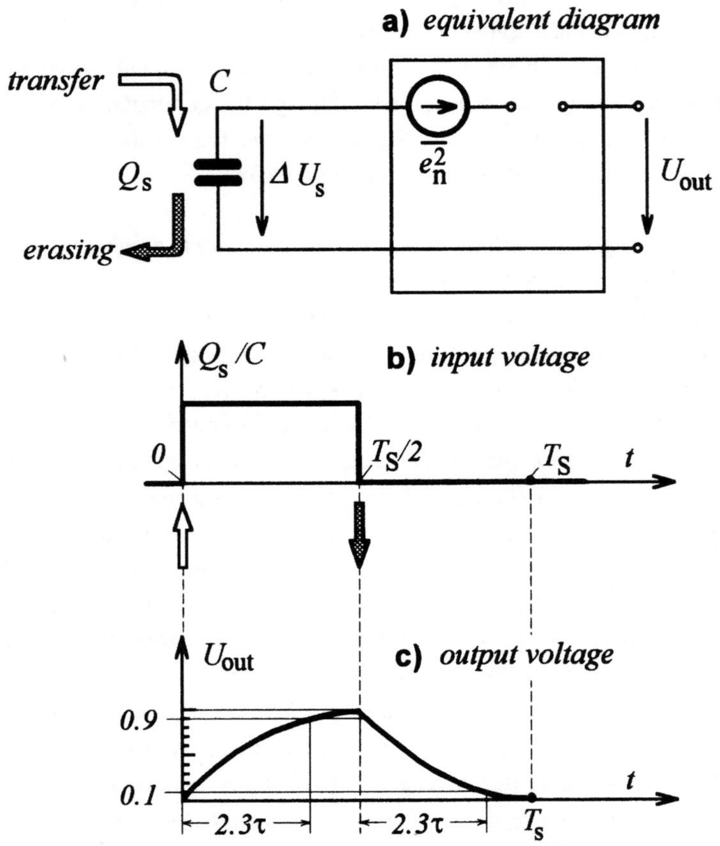

Figure 7.12: (a) Equivalent diagram of output device (charge measuring device); idealized graphs of (b) input, and (c) output voltage.

$$\mathrm{SNR} = \frac{0.865 S E_{sA} T_s}{\sqrt{q I_0 T_s (4.2 f_s T_s) + q I_0 T_s + \dfrac{0.24 q I_0}{f_{high}}}} . \tag{7.54}$$

The results of Sect. 7.1 can be used for calculating the value of SNR in multielement PDAs with integration. We shall write the sought-for expression for the subtraction technique, since this will allow us to consider the general case, with the inclusion of the excess noise. The dispersion of the charge accumulated in the cell is given by (7.26) in case of excess noise, and by (7.5) in case of white noise. We only have to include yet another source, the emf noise of the amplifier (the commutation noise is excluded from consideration as being, in principle, eradicable). The signal charge Q_s accumulated in the cell is transferred to the capacitance C, causing the potential jump $\Delta U_s = Q_s / C$ (Fig. 7.12a). Its readout takes a certain limited interrogation time T_{int}, after which the charge Q_s is erased from the capacitance C, which becomes ready for the readout of the charge from the next cell. In the idealized case the task reduces to detecting a rectangular pulse with the amplitude ΔU_s and length T_{int} against the background of white noise $\overline{e_n^2}$. Making use of equation (3.61) for optimal filtering with the replacements $SE_{sA} \Rightarrow \Delta U_s$, $2qI_n \Rightarrow \overline{e_n^2}$, $T_s \Rightarrow T_{int}$, we get:

$$\mathrm{SNR} = \frac{\sqrt{2}\Delta U_s \sqrt{T_{int}}}{\sqrt{\overline{e_n^2}}} = \frac{\sqrt{2} Q_s \sqrt{T_{int}}}{C\sqrt{\overline{e_n^2}}} . \tag{7.55}$$

Hence we may find the lowest detectable signal charge Q_{st} from condition SNR = 1 (that is, setting the signal equal to the noise):

$$\overline{Q_{st}^2} = a_t \frac{C^2 \overline{e_n^2}}{T_{int}} , \text{ where } a_t = 0.5 . \tag{7.56}$$

So far so good — but unfortunately, owing to its limited passband, the optimal filter protracts the signal, doubling its length at the base (see Sect. 4.2). The amplifier is common for all cells (groups of cells), and this protraction leads to intercoupling of channels. Because of this, it is usually required that all the transient processes should be finished within the allocated time T_{int}, which means delimiting the time during which the charge occupies the capacitance. Assume, for example, that this time is equal to $0.5T_{int}$. If the passband of the amplifier is shaped by a one-section RC-filter, the length of the pulse edge is about $2.3\tau_f$, and the time constant of the filter must be selected from condition $\tau_f = 0.5T_{int}/2.3 \approx 0.2T_{int}$. With τ_f as small as that, the time T_{int} will be enough for the pulse to reach its amplitude value and then relax almost to zero (Fig. 7.12b,c). For this interrogation mode we get from (5.5) and Table 5.1:

$$\mathrm{SNR} = \frac{\Delta U_s}{\sqrt{\overline{e_n^2}\Gamma_1 f_a}} = \frac{Q_s}{C\sqrt{\overline{e_n^2}\frac{\Gamma_1}{2\tau_f}}} ;$$

$$\overline{Q_{st}^2} = \frac{\Gamma_1 C^2 \overline{e_n^2}}{2\tau_f} = \frac{\Gamma_1 C^2 \overline{e_n^2}}{2(0.2T_{int})} = a_t \frac{C^2 \overline{e_n^2}}{T_{int}} , \quad a_t = \frac{\Gamma_1}{0.4} = \frac{0.5}{0.4} = 1.25 . \tag{7.57}$$

In real output amplifiers the frequency response and the signal shape $\Delta U_s(t)$ may differ. The above examples outline the method of calculation of the noise of such amplifier and indicate the range of typical values of constant $a_t \approx 0.5...1.25$.

The quantity $\overline{Q_{st}^2} = a_t C^2 \overline{e_n^2} / T_{int}$ restricts the lowest detectable signal charge in exactly the same way as the dispersion of the charge of the cell itself (7.5), (7.26). Accordingly, all these components of the noise power must simply be added together. The signal charge is given by (7.31), (7.32), and so its ratio to the total equivalent noise is

$$\mathrm{SNR} = \frac{(0.5...0.72) S E_{\mathrm{sA}} T_{\mathrm{i}}}{\sqrt{2 q I_0 T_{\mathrm{i}} (2.77 f_{\mathrm{low}} T_{\mathrm{i}}) + 2 q I_0 T_{\mathrm{i}} + \dfrac{2 \overline{e_{\mathrm{n}}^2} C^2 a_{\mathrm{t}}}{T_{\mathrm{int}}}}} . \tag{7.58}$$

The structural similarity between (7.54) for PDA with optimal filter and (7.58) for PDA with integration is obvious. The signal here is expressed in terms of the accumulated charge, which is justified not only for the cell with integration, but also for the amplifier (whether optimal of quasi-optimal), which may be regarded as an integrator with the effective integration time $\tau_{\mathrm{eff}} \approx T_{\mathrm{i}}$ (see Sect. 4.1). The components of the low-frequency noise and the white noise are also the same (we discussed this in detail in Sect. 7.1). These formulas, however, are different in the following two respects.

The first distinction concerns the numerical coefficients at the low-frequency and the white-noise components. In Sect. 7.1 we discussed why and how much these coefficients differ. When the noise is white, the loss in the SNR, according to (7.58), is increased by the factor of $\sqrt{2}$, because the noise power of the two frames is combined in the subtraction technique. This does not mean, however, that the difference in the coefficients reflects any inherent flaw in PDAs with integration, that, in terms of the SNR, they are inferior by the factor of 1.4...2.4 to the optimal PDAs. The loss can be reduced dramatically by using more sophisticated methods of superposition of frames, and then PDAs with integration will be almost as good as the optimal PDAs in terms of the signal-to-noise ratio. The same can be said about traditional PDAs with amplifier in each channel. The loss in the amplifier in case of high-quality (quasi-optimal) signal processing constitute 3...10% (see Sect. 5.2), and they increase further (by 3...30%) because of discrete sampling by the commutator (see Sect. 7.1). From these arguments it is clear that the spread of the coefficients in (7.54), (7.58) simply reflects the range of SNR values in real systems. When advanced signal processing techniques are used, the value of SNR in PDAs with amplifiers and in PDAs with integration may be close to the maximum attainable value as given by (7.54), falling short of the latter by just 3.5...5.5% (in case of the white noise and two-section RC-filter, or multi-frame integration). When the signal processing method consists in simple superposition of frames, the loss in the SNR may be as large as 2.5-fold.

The second difference in the formulas (7.54), (7.58) is found in the component of the amplifier emf noise; this is related to the difference in the effective frequency bands of the PDA with optimal filter and PDA with integration. We shall discuss this feature later, in connection with the analysis of this noise component.

The next due step consists in relating the obtained value of SNR for each channel to the number of elements (channels) in the multielement PDA. For this we need to get acquainted with the design principles of optico-mechanical system which detects the point optical target with the aid of PDA.

Linkage between the parameters of PDA and optico-mechanical system. The image formed in the focal plane of the optical system is the frame area $A = L \times L$, where L is the side of the frame. Assume that this focal plane image contains the light spot (target image) with the diameter d and the power E_{sA}. We denote by $m = L/d$ the number of lines, and by $\mathcal{M} = A/d^2 = m^2$ the number of image decomposition elements. Our task consists in detecting the spot and finding its coordinates. For this purpose we shall employ one-element PDA, PDA based on the linear array of photodetectors (linear array), and PDA based on the two-dimensional array (matrix). The photosensitive element area in all three types of PDAs is assumed to be the same, and large enough to nest exactly the optical spot:

$$A_e = d^2 = \frac{A}{\mathcal{M}} . \tag{7.59}$$

Readout of information about all the elements of the image with the aid of one-element PDA requires fast line-by-line scanning of the entire frame with this PDA (Fig. 7.13a). Scanning the image with the linear array gives the signals from all lines in one go (Fig. 7.13b) (provided that the number of elements in the array is equal to the number of lines in the image, $M_e = m = \sqrt{\mathcal{M}}$). The matrix FPA built up of $M_e = \mathcal{M}$ elements occupies the entire area of the frame, and eliminates the need for mechanical scanning (Fig. 7.13c). The coordinate of the spot is determined by the serial number of the element which has detected the spot (in the linear array and the matrix), and by the time of arrival of the signal (in the linear array and one-element PDA).

Defined for the optico-mechanical system is also the frame time T_f, which is allocated for the readout of information about one frame. After the time T_f the position of the spot may change, which will be reflected in the information of the next frame. As follows from Fig. 7.13a–c, the length of the optical pulse is different for each of the three types of the PDAs. One-element PDA has to scan very fast, because it must take a look at each of the $\mathcal{M}$ image elements within the time T_f. Scanning with the linear array is slower: each of its photodetectors has to scan one line of m image elements, whereas each element in the matrix FPA looks at one image element for the entire frame time T_f. Accordingly, we have

$$T_s = \frac{T_f}{\mathcal{M}} \tag{7.60}$$

for one-element PDA,

$$T_s = \frac{T_f}{m} = \frac{T_f}{\sqrt{\mathcal{M}}} \tag{7.61}$$

for the linear array, and

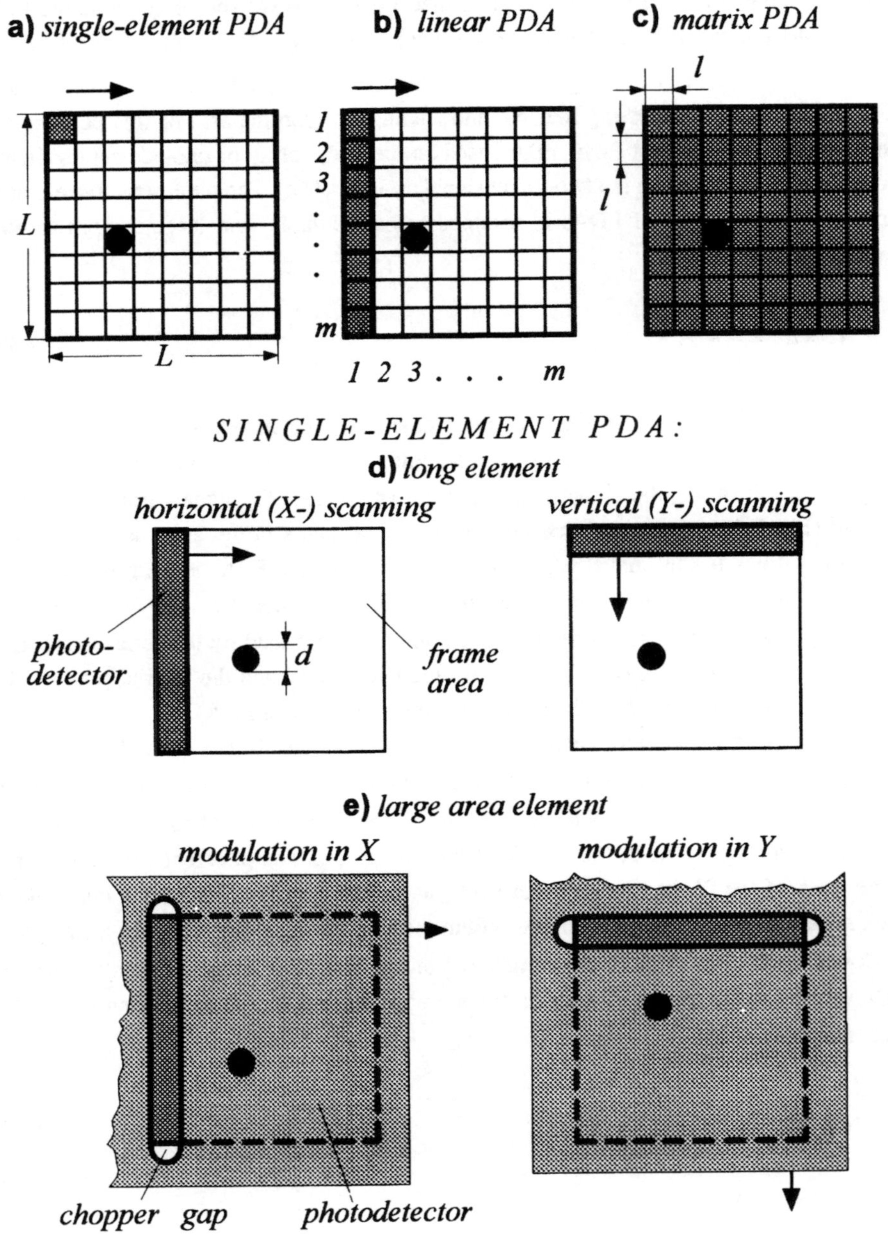

Figure 7.13: Principles of scanning and modulation in optoelectronic systems for detection of point sources using PDAs with different topologies of photodetectors.

$$T_s = T_f \tag{7.62}$$

for the matrix FPA.

The general expression for all three cases is

$$T_s = \frac{T_f M_e}{\mathcal{M}} . \tag{7.63}$$

As far as the improvement of the detective properties is concerned, the simple formulas (7.59) — (7.63) pinpoint the very essence of the multielement approach. Two factors are at work when the number of elements is increased. First, the area of each photosensitive element is reduced (see (7.59)), and so are its current, intrinsic noise, capacitance. Secondly, the duration of the signal is increased (see (7.60) — (7.63)), which makes it possible to carry out integration for a longer time. We must always distinguish two methods of increasing the number of elements. In fact, it is the first method that is illustrated in Fig. 7.13a–c. Making the area of the photosensitive element A_e constant, we change the configuration of the photodetector itself: adding new elements, we go over from one-element PDA to a linear array (possibly comprised of several rows), from linear array to a matrix FPA. As a consequence, the signal duration increases. The other approach to increasing the number of elements consists in reducing the diameter of the optical spot d (of course, if this is feasible), and the area of each photosensitive element $A_e = d^2$. Then, given that the frame area remains the same, the resolution $\mathcal{M}$ is improved. Accordingly, the number of elements M_e in the multielement PDA will increase: recall that $M_e = m = \sqrt{\mathcal{M}}$ in the linear array, and $M_e = \mathcal{M}$ in the matrix. In this case the improvement is mainly associated with the reduction of the photosensitive element area (7.59) and the photodetector noise. At the same time, the signal duration falls as $\mathcal{M}$ increases (see (7.61), (7.62)), which has an adverse effect on the detection of the signal.

The simple formulas (7.59) — (7.63) highlight the direction of further development of modern optoelectronics: if each photosensitive element has practically reached its theoretical quantum limit, if the electronic signal processing is close to optimal, then the only way to improve the detective properties of the system consists in using multielement assemblies. Accordingly, the number of elements M_e becomes yet another parameter which characterizes the detection. Tell me the number of elements in your PDA, and I will tell you what threshold you are able to achieve. The trick is very simple: just substitute the area (7.59) and the signal length (7.60) — (7.63) into the expressions for SNR (7.54) — (7.58). As usual, we shall analyze the signal-to-noise ratio for each of the three cases, corresponding to the predominance of white, low-frequency and high-frequency noise components.

White noise. This classical case is observed primarily in IR PDAs, sensitive to thermal radiation (see Sect. 1.2 and 2.5), when the shot noise component caused by thermal radiation exceeds all the other noise components, and the radiation limit is

Table 7.2. **Signal-to noise ratio**

PDA	Scanning system	Times		LF-noise	White noise	Noise emf of PDA with: amplifier	integrator
		$T_s = T_{int}$	T_{int}	$\frac{B_L E_{sA}}{NEP^* \sqrt{f_{low} A}} \times$	$\frac{B_W E_{sA} \sqrt{T_f}}{NEP^* \sqrt{A}} \times$	$\frac{B_H E_{sA} T_f^{3/2}}{e_n C_{in}/S} \times$	
Single-element	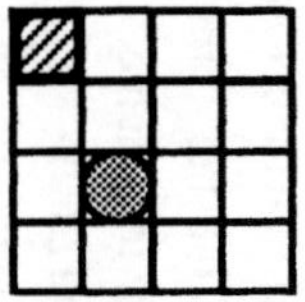	$\frac{T_f}{\mathcal{M}}$	$\frac{T_f}{\mathcal{M}}$	$\times\sqrt{\mathcal{M}}$	×1 (no effect)	$\times\frac{1}{\mathcal{M}^{3/2}}$	$\times\frac{1}{\mathcal{M}^{3/2}}$
Linear array	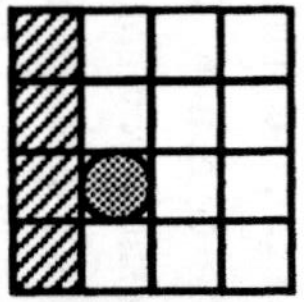	$\frac{T}{m}$	$\frac{T_f}{\mathcal{M}}$	$\times\sqrt{\mathcal{M}}$	$\times\sqrt{m}$	$\times\frac{1}{m^{3/2}}$	$\times\frac{1}{\mathcal{M}}$
2D matrix	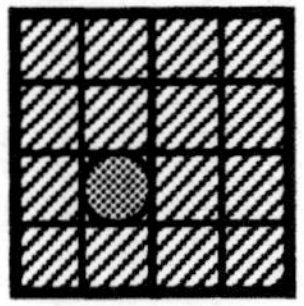	T_f	$\frac{T_f}{\mathcal{M}}$	$\times\sqrt{\mathcal{M}}$	$\times\sqrt{\mathcal{M}}$	×1 (no effect)	$\times\frac{1}{\sqrt{\mathcal{M}}}$
Increase in number of elements	Transition from single-element PDAs to linear arrays, matrices (A_e = const)			No effect	Favorable	Favorable	Favorable
	Increased resolution (reduced A_e)			Favorable	Favorable	Non-Favorable	Non-Favorable

attained. Then equations (7.54), (7.58) take on the known structure

$$\mathrm{SNR} = \frac{B_{\mathrm{W}} S E_{\mathrm{sA}} \sqrt{T_{\mathrm{s}}}}{\sqrt{2qI_0}} = \frac{B_{\mathrm{W}} S E_{\mathrm{sA}} \sqrt{T_{\mathrm{s}}}}{\sqrt{2qj_0 A_{\mathrm{e}}}} = \frac{B_{\mathrm{W}} E_{\mathrm{sA}} \sqrt{\dfrac{\mathcal{M} T_{\mathrm{s}}}{A}}}{\mathrm{NEP}^*} , \tag{7.64}$$

where

$$B_{\mathrm{W}} = 0.5 \ldots 1.22 \ .$$

In (7.64) we have used the specific parameters for the current ($j_0 = I_0/A_{\mathrm{e}}$) and the specific noise equivalent power ($\mathrm{NEP}^* = \mathrm{NEP}/\sqrt{A_{\mathrm{e}}}$), and the frame area A. Further on we assume these parameters to be constant. Substituting here T_{s} from (7.60) — (7.63), we obtain the final expressions for SNR in terms of the parameters of the system A, T_{f}, and, most importantly, the number of elements M_{e} (or $\mathcal{M}$). For quick reference these formulas are displayed in Table 7.2.

As we go over from one-element PDA to linear array, from linear array to matrix FPA, the signal duration increases in proportion to the number of elements. Since SNR is, in turn, proportional to square root of the signal duration, we arrive at the main law of the multielement PDAs [1]: *when the noise is white, the signal-to-noise ratio is proportional to square root of the number of elements* $\sqrt{M_{\mathrm{e}}}$. Accordingly, the threshold illumination E_{sAt} is reduced by the factor of $\sqrt{M_{\mathrm{e}}}$ while the signal-to-noise ratio in each channel remains the same. Observe that, as the number of image elements $\mathcal{M}$ increases while the frame area A remains constant, the signal-to-noise ratio for the matrix also increases as $\sqrt{\mathcal{M}}$, which is due not to the increasing integration time, but to the decreasing area of the photosensitive element. In the linear arrays this dependence is weaker, $\sim \sqrt{m} = \sqrt[4]{\mathcal{M}}$, because the increase in the SNR, owing to the decreasing area and intrinsic noise of the photosensitive element, is hampered by the decreasing integration time. In the one-element PDA these two counteracting factors balance out completely, and the signal-to-noise ratio becomes independent of the number of image elements (provided that $A = \mathrm{const}$ and $T_{\mathrm{f}} = \mathrm{const}$).

Low-frequency noise. As we go over to the linear arrays (especially those consisting of several rows) and the matrices, the effective frequency band of the signal $1/2T_{\mathrm{s}}$ narrows down, and the excess $1/f$ noise may manifest itself at low frequencies. When this type of noise dominates, equations (7.54) and (7.58) assume the form

$$\mathrm{SNR} = \frac{B_{\mathrm{L}} S E_{\mathrm{sA}}}{\sqrt{2qI_0 f_{\mathrm{low}}}} = \frac{B_{\mathrm{L}} S E_{\mathrm{sA}}}{\sqrt{2qj_0 A_{\mathrm{e}} f_{\mathrm{low}}}} = \frac{B_{\mathrm{L}} E_{\mathrm{sA}}}{\mathrm{NEP}^*} \sqrt{\frac{\mathcal{M}}{A f_{\mathrm{low}}}} \ . \tag{7.65}$$

This result is known to us from Sect. 3.3: in case of the excess noise the signal-to-noise ratio does not at all depend on the duration of the signal. Although the required frequency band is narrower when the signal is longer, the noise dispersion in the signal band does not decrease, since the noise density at low frequencies is proportional to f_{low}/f. As a result, the increase in the number of elements M_e (from one-element PDA to the linear array, from the linear array to the matrix FPA) does not tell on the threshold characteristics: the resulting increase in the signal duration T_s does not improve the signal-to-noise ratio. This ratio is the same for all topologies of the PDAs under consideration (Table 7.2). As the number of image elements $\mathcal{M}$ increases, however, the second of the two factors named above still works: as the area of the photosensitive element A_e becomes smaller, its current and its noise decrease (recall that we are comparing devices having the same specific parameters). This is why the signal-to-noise ratio for all three types of PDAs increases proportionally to $\sqrt{\mathcal{M}}$ (not to be confused with the number of photosensitive elements M_e).

The excess noise restricts the improvement of the detective properties of the system. And, even though we may increase the integration time (the signal duration T_s), increase the number of elements (using linear arrays and matrices), this is not sufficient for the desired improvement of detection parameters: one has to eliminate the excess noise component at the frequencies comparable with the effective signal band $1/2T_s$.

High-frequency noise. If the current caused by the background radiation is low (as is the case in silicon-based PDAs or in the PDAs for the middle and far IR ranges operating at reduced background temperatures and limited field of view), if the time of interrogation is small, then the high-frequency noise may become predominant: the emf noise of the amplifier. In case of optimal filtering (7.54), the signal-to-noise ratio is proportional to the duration of the signal — and hence to the number of elements M_e (7.60) – (7.63). It is hard to imagine, however, that any designer will risk to implement optimal filtering in multi-element PDA of traditional type. Indeed, the effective passband of the optimal filter f_{high} is constant, and is narrower than the signal band $1/2T_s$. This protracts the signal, and prevents discrimination of close-by point targets. Accordingly, as indicated in Sect. 5.3, the passband of the amplifier is made equal to or greater than the bandwidth of the signal, $f_a \approx 0.9...1.8 f_s$ (Fig. 5.19). As follows from (5.106), the signal-to-noise ratio in this case exhibits an even stronger dependence on the signal length:

$$\text{SNR} = \frac{SE_{\text{sA}}u_{\text{sA}}}{\sqrt{(2\pi e_n C)^2 \Gamma_2 f_a^3}} = \frac{u_{\text{sA}} SE_{\text{sA}} T_s^{3/2}}{2\pi\sqrt{\Gamma_2 f_a^3 T_s^3}\, e_n C} = B_H \frac{E_{\text{sA}} T_s^{3/2}}{\frac{e_n C}{S}}, \tag{7.66}$$

$$B_H = \frac{u_{\text{sA}}}{2\pi\sqrt{\Gamma_2 f_a^3 T_s^3}} = \frac{0.65...0.9}{2\pi\sqrt{0.51\cdot(0.45...0.9)^3}} \approx 0.23...0.48 .$$

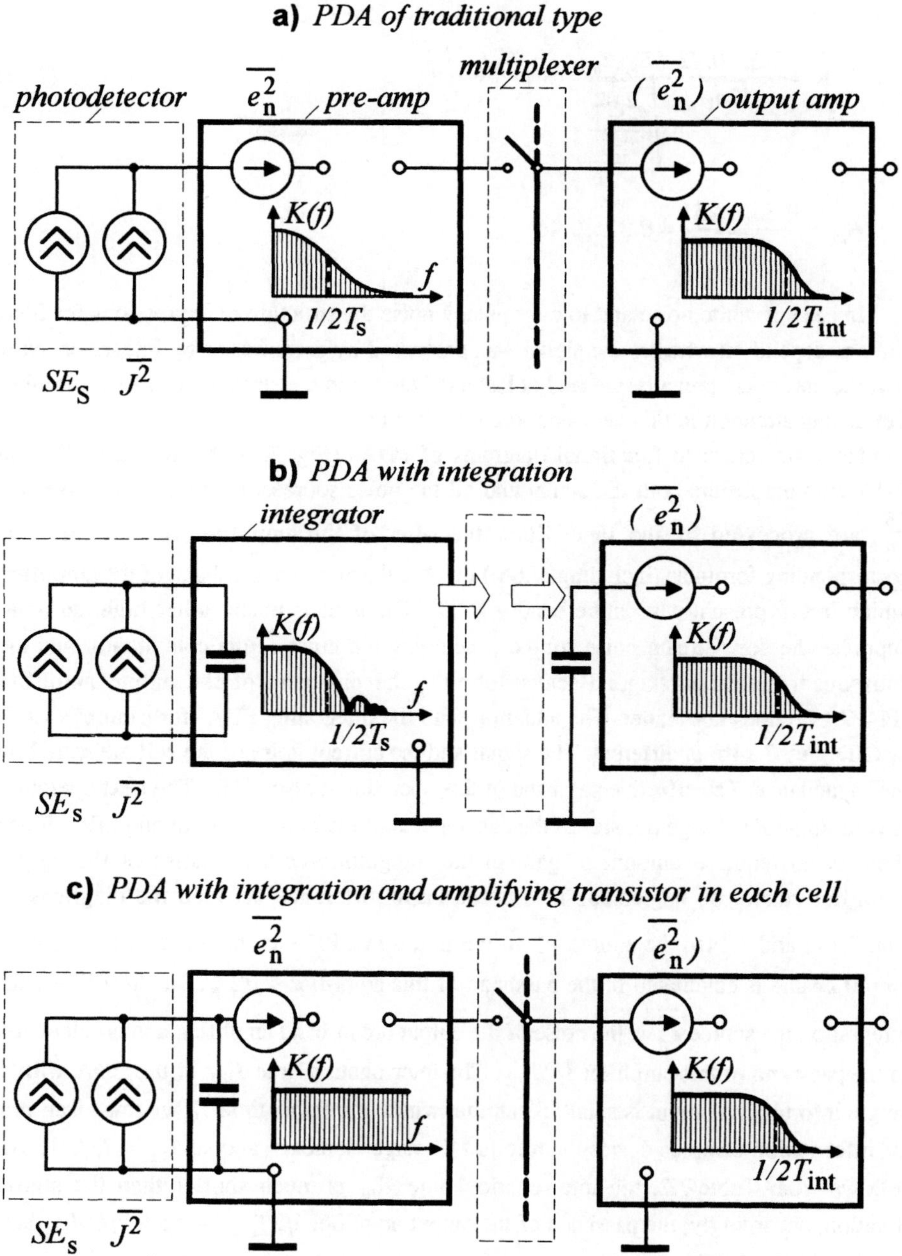

Figure 7.14: Noise emf in PDA with preamplifiers acts in the narrow band of the preamplifier; in FPA with integration it acts in the broad band of the input amplifier.

The numerical coefficient B_H here is evaluated for the two-section RC-filter having the specified passband, which corresponds to the protraction of the length of the bell-shaped signal just by 6...16% (Fig. 5.19).

Expression for the signal-to-noise ratio for PDA with integration dominated by the emf noise follows from (7.58):

$$\mathrm{SNR} = \frac{0.5...0.72}{\sqrt{2a_t}} \frac{SE_{sA}T_s}{\sqrt{\frac{\overline{e_n^2}C^2}{T_{int}}}} = B_H \frac{E_{sA}T_sT_{int}^{1/2}}{\frac{e_nC}{S}} , \tag{7.67}$$

$$B_H = \frac{0.35...0.51}{\sqrt{a_t}} = 0.31...0.46 .$$

In case of white noise and low-frequency noise the structure of expressions for SNR did not depend on whether the signal was processed by amplifier or by integrator. Now that the emf noise prevails, we find at last a difference in the structure of these formulas. Let us pay attention to this very important distinction.

Now we return to functional diagrams of PDAs (Fig. 7.14a,b). In the traditional PDA with amplifiers both the signal and all the noise sources, including the noise emf $\overline{e_n^2}$, are processed by the same filter, the filter of the amplifier. Accordingly, the corresponding formulas (including (7.66)) involve the effective passband of the amplifier, which in our present case is the band $\sim 1/2T_s$. The gain is made rather high, so as to suppress the downstream noise sources, including the noise of the commutator and the subsequent amplifier stage. Because of this, the passband of the output amplifier (Fig. 7.14a) does not matter. The situation with the integrating PDA (for example, CCD or CID, Fig. 7.14b) is different. The signal and the current noise of the cell are stored by the capacitance. The effective passband of this operation is also $1/2T_s$. The next operation is commutation: charge transfer to the capacitance of the output readout amplifier. Since there is no required amount of gain in the integration cell, the noise of the output amplifier cannot be neglected. This is the basic difference between the diagrams in Fig. 7.14a and 7.14b: the source $\overline{e_n^2}$ in the traditional PDA is the noise of the channel amplifier and is calculated in the passband of this amplifier $\sim 1/2T_s$. In the PDA with integration the source $\overline{e_n^2}$ is the noise of the output broad-band amplifier, and is calculated in the passband of this amplifier $1/2T_{int}$. The interrogation time T_{int} in the PDA with a single information output is small: recall that within the frame time T_f we must transmit $\mathcal{M}$ information samples corresponding to $\mathcal{M}$ image elements, and so $T_{int} = T_f/\mathcal{M}$. As follows from Table 7.2, the interrogation time T_{int} is much shorter than the signal duration; accordingly, the passband of the output amplifier $1/2T_{int}$ is much broader than the signal band $1/2T_s$.

We see that the specific feature (and not a good one) of the PDA with integration is the existence of two characteristic frequency bands: the band of the integrator for the signal and the intrinsic noise of the cell, $1/2T_i = 1/2T_s$, and the broader band of the output amplifier $1/2T_{int}$ for its source of the emf noise. One may often hear the question which of the two passbands (integration band or interrogation band) should be taken into account in calculating the threshold characteristics of PDA with integration. Now the answer is clear: both, each for its own noise component.

Observe that two characteristic frequency bands are peculiar also to the lattice-type matrices with photodetector, switch, MIS FET amplifier (Fig. 7.14c). Although there is an amplifier in each cell, there is no filtering, because the passband of such amplifier is not limited from above. Accordingly, the noise $\overline{e_n^2}$ of the transistor of each cell acts in the broad band $1/2T_{int}$ of the same readout amplifier.

Substituting the characteristic times T_s and T_{int} into (7.66), (7.67), we find SNR as a function of the topology and the number of elements (see Table 7.2). It is only in the one-element PDA that the signal-to-noise ratio is practically the same for the PDA with amplifier and PDA with integrator, since $T_{int} \cong T_s$. In linear arrays and matrix FPAs the integrator is worse (for the first time in our experience!), and the reason is now clear. The transition from one-element PDAs to linear arrays and matrix FPAs is now especially advantageous: the signal duration factor is active, and the value of SNR exhibits an especially strong dependence on the signal length T_s (see (7.66), (7.67)). The alternative method of increasing the number of elements — by reducing the PSE area while retaining the linear or matrix arrangement — is not advantageous at all; in fact, its effect is adverse. So, trying to improve the detective properties by using multielement arrays, we may arrive at the opposite result! The reduction of the PSE area in this case does not play any role, since the noise of the PSE is assumed to be small anyway — provided, of course, that the capacitance of the PSE does not affect the capacitance of the output readout amplifier (the output device). Therefore, the result of the increased resolution (number of image elements) $\mathcal{M}$ is only negative: the reduced interrogation time T_{int} in all PDAs, and the reduced signal duration T_s in one-element PDAs and linear arrays. As a consequence, the signal-to-noise ratio deteriorates, especially in the PDAs with integration, since they are sensitive to the reduction of each of the two characteristic times. The only exception is the matrix with amplifiers: here each of the elements, irrespective of their number, looks at its own pixel for the entire frame time, $T_s = T_f$, so that the noise in the band $1/2T_f$ does not depend on the number of elements $\mathcal{M} = M_e$. This situation is, however, only of academical interest, since a PDA with two-dimensional photodetector and amplifier in each channel is technically not feasible when the number of channels is large.

In this way, while the increased resolution improved the detective properties when the white noise prevailed, the effect is opposite when the emf noise is at play, and the price of the high resolution is the deterioration of the threshold parameters.

Much attention must be paid to the reduction of this noise component. Unfortunately, the possibilities for reducing $\overline{e_n^2}$ are limited. The output transistor in CCDs and CIDs usually has to be a MIS FET, as dictated by the technological compatibility with the manufacture of photosensitive cells. At the same time, the noise $\overline{e_n^2}$ in MIS FETs is quite considerable ($50..200\,\mathrm{nV/Hz^{-1/2}}$). Therefore, the primary attention is paid to the reduction of the capacitance of the output device (amplifier) in CCDs. Very low values (0.1pF) have been achieved. In terms of this crucial parameter, CID is much inferior to CCD: such a low capacitance is in principle not feasible because of the long vertical column bus (see Sect. 1.1). This bus is charge-coupled with all the cells in the column, which one by one transfer their charge to this bus in the course of the readout. Because of this, a CID cannot be used as a threshold-level device in the circumstances of low background; also, the length of the column is limited, and so is the number of elements in the column (the usual number is 32...64).

To reduce the noise of the output device, it was proposed to use n identical amplifiers with charge-coupled capacitances. This device is known as distributed floating gate amplifier (DFGA) [107]. Although the DFGA, because of its complexity, has not become popular, its principle of signal processing deserves attention. The readout charge is supplied in turn to the inputs of all n amplifiers, so that at their outputs we have the same signal U_s and noise U_n. The outputs of all amplifiers are added together, and the resulting signal is thus nU_s. Since the noises of the amplifiers are not correlated, the noise resulting after summation is $\sqrt{n\overline{U_n^2}}$. Then the ratio of the total signal to the total noise is

$$\mathrm{SNR} = \frac{nU_s}{\sqrt{n\overline{U_n^2}}} = \sqrt{n}\,\frac{U_s}{U_n}\ , \tag{7.68}$$

that is, n amplifiers, like n photosensitive elements, improve the signal-to-noise ratio by the factor of $\sqrt{n}$.

The operating principle of the DFGA can also be explained in a different way: each amplifier interrogates (reads out) the charge for the time T_{int}; accordingly, all n amplifiers do this for the time nT_{int}. According to (7.67), the n-fold increase of the interrogation time corresponds to the gain of $\sqrt{n}$. This prompts yet another method of suppression of the high-frequency noise: the transition from serial readout from all M_e channels via a single information output to the parallel-serial readout via n information outputs. Then the time available for readout from each channel will be increased n-fold, giving the same gain of $\sqrt{n}$ in the signal-to-noise ratio. If the number of output leads has to be minimized, a second commutator may be installed after the n intermediate amplifiers. Thanks to the intermediate amplifiers (having the narrow band $1/2nT_{int}$), the noise of the common output amplifier (with the broad band $\approx 1/2T_{int}$) will have no effect. This technique is usually employed for signal processing in the CIDs. The intermediate amplifiers are installed at the output of each column, followed by the common commutator. Since the buses from each column are connected each to a separate amplifier (and not all to a common one), the added bonus consists in the reduced capacitance, which, unfortunately, still remains large in the CID.

Theoretically, there also is a possibility (not yet realized in practice) to multiply the carriers in the strong field of the output pn junction prior to readout in exactly the same way as the photocarriers are multiplied in the avalanche photodiode.

Options, options, options... The examples considered above do not exhaust the diversity of the possible situations. Options start from the very beginning: from the choice of the scanning system and the topology of the PDA. To simplify the analysis, we considered a one-element PDA having small PSE area A_e — the same as that in the multielement PDAs. When, however, a one-element PDA is chosen, the scanning system can be made simpler if the photosensitive element spans the entire frame length, $A_e = L \times d$, as shown in Fig. 7.13d. If we select a large-area photodetector (whose are covers the entire frame, $A_e = A$, scanning can be replaced with modulation. The choice of the modulation technique is a separate task in itself [1, 6]. The simplest method is illustrated in Fig. 7.13e: scanning with a slit of the size $L \times d$ first in the horizontal, and then in the vertical direction (similarly to the procedure shown in Fig. 7.13d). In these examples the signal length is just one-half that in the linear array, $T_s = T_f/2m$. Nevertheless, the replacement of such one-element PDAs by linear arrays is profitable: although the integration time does not increase considerably, the other factor (the decrease of the area A_e and the reduction of the intrinsic noise of the channel) is important.

In our analysis of the white and the low-frequency noise power components we assumed that they originate only in the photosensitive element, and are proportional to its area A_e. It is also possible, however, that these noise components depend on the perimeter of the PDA; they may also be to some extent generated in the amplifier. Then the dependence on the area A_e and on the number of image elements $\mathcal{M}$ becomes weaker. By contrast, for the emf noise component, for which the effects of reduced area A_e have been neglected, this factor may be of some importance — for example, when the capacitance of the PSE is comparable to the other capacitances of the input circuit, or in the CID, where the column length determines the capacitance of the readout (output) bus.

The contribution to the total noise may come not from one, but from several noise components at once — for example, the white noise and the high-frequency noise. Then, theoretically, there must be a certain optimal number of pixels $\mathcal{M}_{opt}$, below which the signal-to-noise ratio deteriorates because of the white noise, and above which the signal-to-noise ratio is spoiled by the high-frequency noise. In the output device of the CCD, in the cell with a switch the KTC noise (7.53), (7.49) may be important.

We may also consider a system with the rectangular (non-square) frame, take into account the possible variations in T_f, A_f, S. All these arguments imply that the detailed account for different factors in the calculation of the detective properties is the task of the designers of the particular optoelectronic system and the particular PDA. Such calculation may refine the particular dependence of SNR on $\mathcal{M}$ (or M_e); nevertheless, the relations displayed in Table 7.2 give a comprehensive account of all the basic trends, and the results will be explained by the interplay of the two factors described above: the signal length T_s and the PSE area A_e.

Inhomogeneity of the elements. So far we have been dealing with the idealized system of homogeneous elements. The reality is, of course, not perfect: the greater the number of elements, the greater the spread of their parameters, the higher the probability of channel failures during both manufacture and operation (failed elements in CCDs are referred to as "dark points" and "bright points").

To secure the preassigned false alarm probability P_{fa} for the PDA as a whole, we demand that the false alarm probability for each channel should not exceed $\tilde{P}_{fa} = P_{fa}/M_e$. Then, one and the same ratio of the threshold voltage U_{ti} to the noise must be established for all channels: $n_{fa} = U_{ti}/U_{ni} = \text{const}$. The numerical value of n_{fa} is calculated according to (3.3) and (3.6) for the desired $\tilde{P}_{fa}$, and is usually $n_{fa} = 3\ldots6$ (see Sect. 3.1).

The target miss probability P_{tm} for the PDA as a whole is determined as the sum of conditional probabilities of target miss by each element $\tilde{P}_{tmi}$ (the condition being that the target falls on the given ith element). Accordingly, $\tilde{P}_{tmi}$ must be multiplied by the probability of this event $p(i)$:

$$P_{tm} = \sum_{i=1}^{M_e} p(i)\tilde{P}_{tmi} = \frac{1}{M_e}\sum_{i=1}^{M_e} \tilde{P}_{tmi}$$

$$\approx \frac{1}{M_e\sqrt{2\pi}}\sum_{i=1}^{M_e} \frac{U_{ni}}{U_{si}-U_{ti}} \exp\left[-\frac{(U_{si}-U_{ti})^2}{2U_{ni}^2}\right]. \tag{7.69}$$

Subscript i here denotes the parameters of the ith element (channel); the probability that the light spot will fall on any element is assumed to be equal for all elements, $p(i) = 1/M_e$; for the probability $\tilde{P}_{tmi}$ we use the approximate expression (3.11).

Now let us try to express (7.69) in terms of the parameters of PDA averaged with respect to all its channels. We introduce the mean signal-to-noise ratio

$$\overline{\text{SNR}} = \frac{\sum_{i=1}^{M_e} \text{SNR}_i}{M_e},$$

the deviation from the mean $\Delta\text{SNR}_i = \text{SNR}_i - \overline{\text{SNR}}$, and transform the exponent as follows:

$$\frac{(U_{si}-U_{ti})^2}{2U_{ni}^2} = \frac{(\text{SNR}_i - n_{fa})^2}{2} = \frac{(\overline{\text{SNR}} - n_{fa} + \Delta\text{SNR}_i)^2}{2}$$

$$= \frac{(\overline{\text{SNR}} - n_{fa})^2}{2} + (\overline{\text{SNR}} - n_{fa})\Delta\text{SNR}_i + \frac{\Delta\text{SNR}_i^2}{2}. \tag{7.70}$$

When the deviation is very small,

$$\Delta\,\mathrm{SNR}_i << \frac{1}{\overline{\mathrm{SNR}} - n_{\mathrm{fa}}} \,, \tag{7.71}$$

the expression for P_{tm} is simplified. First of all, we may neglect the quadratic term $\Delta\,\mathrm{SNR}_i^2$ in the exponent (7.70), and then (7.71) takes on the form

$$P_{\mathrm{tm}} \approx \frac{1}{M_{\mathrm{e}}\sqrt{2\pi}} \sum_{i=1}^{M_{\mathrm{e}}} \frac{\exp\left[-\frac{1}{2}\left(\overline{\mathrm{SNR}} - n_{\mathrm{fa}}\right)^2 - \left(\overline{\mathrm{SNR}} - n_{\mathrm{fa}}\right)\Delta\,\mathrm{SNR}_i\right]}{\left(\overline{\mathrm{SNR}} - n_{\mathrm{fa}}\right) + \Delta\,\mathrm{SNR}_i}$$

$$= \frac{1}{M_{\mathrm{e}}\sqrt{2\pi}} \frac{\exp\left[-\frac{1}{2}\left(\overline{\mathrm{SNR}} - n_{\mathrm{fa}}\right)^2\right]}{\overline{\mathrm{SNR}} - n_{\mathrm{fa}}} \sum_{i=1}^{M_{\mathrm{e}}} \frac{\exp\left[-\left(\overline{\mathrm{SNR}} - n_{\mathrm{fa}}\right)\Delta\,\mathrm{SNR}_i\right]}{1 + \dfrac{\Delta\,\mathrm{SNR}_i}{\overline{\mathrm{SNR}} - n_{\mathrm{fa}}}} \,. \tag{7.72}$$

Next we can compute the sum in (7.72), again assuming that the deviations $\Delta\,\mathrm{SNR}_i$ (7.71) are small, and their sum is equal to zero:

$$\sum_{i=1}^{M_{\mathrm{e}}} \frac{\exp\left[-\left(\overline{\mathrm{SNR}} - n_{\mathrm{fa}}\right)\Delta\,\mathrm{SNR}_i\right]}{1 + \dfrac{\Delta\,\mathrm{SNR}_i}{\overline{\mathrm{SNR}} - n_{\mathrm{fa}}}} = \sum_{i=1}^{M_{\mathrm{e}}} \frac{1 - \left(\overline{\mathrm{SNR}} - n_{\mathrm{fa}}\right)\Delta\,\mathrm{SNR}_i}{1 + \dfrac{\Delta\,\mathrm{SNR}_i}{\overline{\mathrm{SNR}} - n_{\mathrm{fa}}}}$$

$$\approx \sum_{i=1}^{M_{\mathrm{e}}} \left[1 - \left(\overline{\mathrm{SNR}} - n_{\mathrm{fa}}\right)\Delta\,\mathrm{SNR}_i - \frac{\Delta\,\mathrm{SNR}_i}{\overline{\mathrm{SNR}} - n_{\mathrm{fa}}}\right] = M_{\mathrm{e}} \,. \tag{7.73}$$

Now for the sought-for probability P_{tm} we finally get

$$P_{\mathrm{tm}} = \frac{\exp\left[-\frac{1}{2}\left(\overline{\mathrm{SNR}} - n_{\mathrm{fa}}\right)^2\right]}{\sqrt{2\pi}\left(\overline{\mathrm{SNR}} - n_{\mathrm{fa}}\right)} \,. \tag{7.74}$$

From these calculations we see that the target miss probability is determined by the mean signal-to-noise ratio $\overline{\mathrm{SNR}}$ only when the homogeneity of the channels is very high; for the typical values of $\overline{\mathrm{SNR}} - n_{\mathrm{fa}} \approx 3 \ldots 6$, $\overline{\mathrm{SNR}} \approx 6 \ldots 12$ (see Sect. 3.1) it must be not worse than (see (7.71))

$$\Delta\,\mathrm{SNR}_i < \frac{1}{\overline{\mathrm{SNR}} - n_{\mathrm{fa}}} = \frac{1}{3 \ldots 6} \approx 0.15 \ldots 0.3 \,;$$

$$\frac{\Delta\,\mathrm{SNR}_i}{\overline{\mathrm{SNR}}} = \frac{0.15 \ldots 0.3}{6 \ldots 12} = 0.025 \,. \tag{7.75}$$

Such rigorous requirements to the homogeneity are satisfied very rarely: one can name CCDs based on Shottky barriers, and PDAs operating in the time delay and integration (TDI) mode. As a rule, the spread of parameters of PDA is much greater than the required 2.5% [108, 109]. Then the largest contribution to the sum in (7.69) for the target miss probability P_{tm} comes from the channels with the low value of SNR_i, because of the strong nonlinear dependence of $\tilde{P}_{tm\,i}$ on SNR_i. The increase in the overall P_{tm} because of such bad channels is not compensated by the high $\tilde{P}_{tm\,i}$ in good channels with the high value of SNR_i. It follows that the portrait of the multielement PDA is given not by its statistical-average channel, but rather by its worst representatives. Because of this, one must specify not the mean value of $\overline{SNR}$ (or the mean detective properties), but the least allowable values of SNR, which are actually definitive for the detection errors P_{fa}, P_{tm}. This is the reason why in Sect. 7.1, in our analysis of the deterioration of SNR caused by non-synchronized sampling, we did not average the SNR with respect to all possible dispositions of samples, paying instead special attention to the least advantageous arrangements (the lowest values of SNR), and devising frame superposition techniques and charge partitioning aimed at reducing the loss caused by non-synchronized sampling.

When the number of elements is large, we must also envisage a certain number of defective channels m_d. Since they are unable to detect the target, for each of such channels we have $\tilde{P}_{tm\,i} = 1$, and, according to (7.69), the contribution of these m_d channels to the overall target miss probability is m_d/M_e. Assume that the limit of allowable error $[P_{tm}]$ is exhausted by the defective elements, whereas all the remaining $M_e - m_d$ operate error-free. Then the allowable value of $[m_d]$ is found from the simple relation

$$\frac{[m_d]}{M_e} = [P_{tm}]\,, \quad [m_d] = M_e[P_{tm}]\,. \tag{7.76}$$

So, even a relatively mild requirement $[P_{tm}] = 10^{-2}...10^{-3}$ implies that we are allowed only one defective channel per hundred or thousand. To ensure the low probability of error in operation (even when the number of failed channels is rather large, $[m_d] > M_e[P_{tm}]$), it is necessary to provide some redundancy, backing up the defective elements with good ones. In the matrix one can use additional frames: the decision is taken on the basis of two (or more) subframes, each time moving the matrix by step d. The linear array may have built-in redundancy, being made up of two or more rows of elements. Signal processing in case of the linear array consisting of several rows of elements has some peculiar features.

Time delay and integration mode (TDI). Consider a linear array of n rows. Such arrangement may as well be regarded as an $m \times n$ matrix comprised of $M_e = m \times n$ elements. According to our theory, the n–fold increase in the number of elements

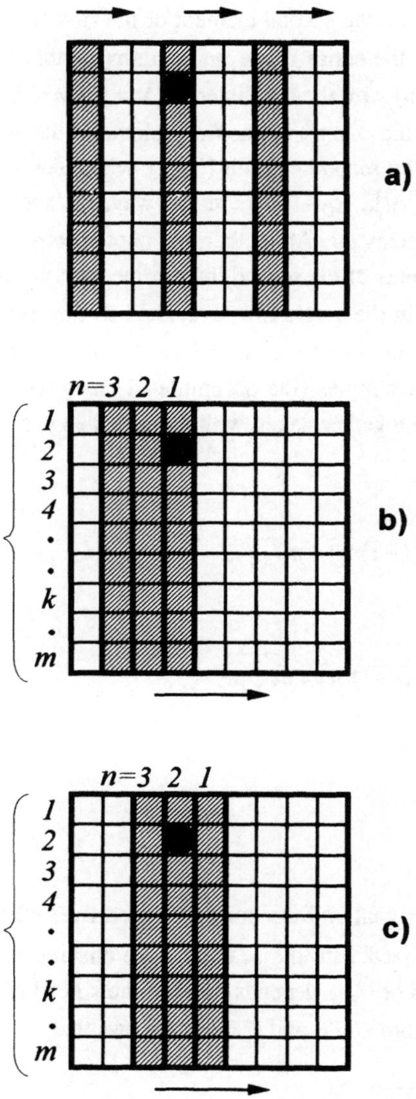

Figure 7.15: Principle of signal processing in time delay and integration mode.

improves the detective properties by the factor of $\sqrt{n}$ (given that the noise is white). If we had followed our former procedure, and entrusted each element with scanning its own part of the frame, we should have split the frame into n sections, scanning each section with a separate linear array as shown in Fig. 7.15a. Then the scanning rate would have been n times slower as compared with that for the single-row array, while the signal length would have increased n-fold, $T_s = (n/m)T_f$, and the gain in the SNR would have been $\propto \sqrt{T_s} \propto \sqrt{n}$. In practice, however, a different technique is used, known as the time delay and integration mode. Its principle is illustrated in Fig. 7.15b,c: by contrast to the method just described, each row scans the entire area of the frame. Assume that at the time t_0 the light spot falls on the second element of the first column, Fig. 7.15b. Since the linear array must scan the entire frame (m columns) within the time T_f, it has to move one step (one column) with the time interval $\Delta t = T_f/m = T_s$. From t_0 to $t_0 + \Delta t$ the spot stays (building up the charge) on the first column of the array, and after the time Δt the spot will move to the second column ($i = 2$), where the integration will proceed over the time interval $(t_0 + \Delta t)\ldots(t_0 + 2\Delta t)$. In this way, the spot will visit each of the n elements with the time delay Δt. Altogether, the photoelectrons will be generated by the light spot in all n elements of the second line for the time period of $n\Delta t = T_f n/m$ — that is, n times as long as in the single-row array, and so the signal-to-noise ratio must be improved by the same factor of $\sqrt{n}$ as before. Further on the ideology of signal processing is obvious: the samples (the accumulated charges) from each ith element ($i = 1,\ldots,n$) must be added together, taken with the time delay Δt between two adjacent elements:

$$U_s(k) = \sum_{i=1}^{n} U_{ski}\left[t_0 + (i-1)\Delta t\right] = n\overline{U_{sk}} \ , \tag{7.77}$$

$$\overline{U_n^2(k)} = \sum_{i=1}^{n} \overline{U_{nki}^2}\left[t_0 + (i-1)\Delta t\right] = n\overline{U_{nk}^2} \ ;$$

$$\overline{U_{nk}^2} = \frac{1}{n}\sum_{i=1}^{n} \overline{U_{nki}^2} \ , \tag{7.78}$$

where $\overline{U_{sk}}$, $\overline{U_{nk}^2}$ are the signal and the noise averaged over all n elements of the kth line. Time delay synchronized with the scanning rate ensured phased sampling of all channels, so that the spread of U_{ski} depends only on the spread of the sensitivity of the elements of the kth line. From (7.77) and (7.78) it follows that

$$\mathrm{SNR} = \frac{U_s(k)}{\sqrt{\overline{U_n^2(k)}}} = \frac{n\overline{U_{sk}}}{\sqrt{n\overline{U_{nk}^2}}} = \sqrt{n}\,\overline{\mathrm{SNR}_k} \ ; \tag{7.79}$$

$$\overline{\mathrm{SNR}_k} = \frac{\overline{U_{sk}}}{\sqrt{\overline{U_{nk}^2}}} \ .$$

So, we have got what we ought to: the gain of $\sqrt{n}$ in the signal-to-noise ratio. Of course, when the noise is white, the TDI mode cannot beat the conventional regime, the imaginary technique illustrated in Fig. 7.15a. Collective action, however, (the detection of the point target by all n line elements of multilinear array) has important advantages.

First, equation (7.77) has been derived without reference to a particular noise spectrum; accordingly, the gain of $\sqrt{n}$ pertains to any type of spectrum, including the excess noise. This is understandable: the noises of different channels are not correlated, and therefor add up geometrically. When the integration time is increased n-fold in one channel, however, it is the same as if n samples of channel's own noise were added together; in case of the $1/f$ noise these samples are correlated, and they add arithmetically rather than geometrically:

$$U_{\rm n}(k)=\sum_{i=1}^{n}U_{{\rm n}ki}=nU_{{\rm n}k}\ ,\quad \overline{U_{\rm n}^2(k)}=n^2\overline{U_{{\rm n}k}^2}\ , \tag{7.80}$$

so that there is no gain in the SNR. It follows that in the presence of the excess noise the $m\times n$ matrix operated in the TDI mode is "$\sqrt{n}$ times better" than the single-row linear array, and than the prototype $m\times n$ matrix operated as illustrated in Fig. 7.15a, and even better than the full-frame $m\times m$ matrix in the conventional mode.

Secondly, in the prototype $m\times n$ matrix of Fig. 7.15a the signal is read out independently from each of the $M_{\rm e}=m\times n$ channels. The parameters of detection are thus described by the $m\times n$ matrix $[{\rm SNR}_{ki}]$, $k=1,\ldots,m$; $i=1,\ldots,n$. The homogeneity of the matrix is characterized by the spread of its terms — that is, by their mean square deviation $\sqrt{\overline{\Delta\,{\rm SNR}_{mn}^2}}$. In the TDI mode the signals are taken from m lines, the values of ${\rm SNR}_{ki}$ are averaged over the n elements of the line, so that the matrix of detection parameters contains now only m numbers $[{\rm SNR}_k]$, $k=1,\ldots,m$, and each of the numbers ${\rm SNR}_k$ becomes closer to the mean value of $\overline{\rm SNR}$: the mean square deviation $\sqrt{\overline{\Delta\,{\rm SNR}_m^2}}$ is reduced by the factor of $\sqrt{n}$:

$$\sqrt{\overline{\Delta\,{\rm SNR}_m^2}}=\frac{\sqrt{\overline{\Delta\,{\rm SNR}_{mn}^2}}}{\sqrt{n}}\ . \tag{7.81}$$

The improved homogeneity is an important advantage of the TDI mode, and facilitates signal processing by external devices.

The third advantage is the consequence of the second one. Assume, for example, that the initial spread is

$$\frac{\sqrt{\overline{\Delta\,{\rm SNR}_{mn}^2}}}{\overline{\rm SNR}}=\frac{\sqrt{\overline{\Delta S^2}}}{\overline{S}}=10\ldots20\%\ . \tag{7.82}$$

If the number of rows is large enough, $n=25\ldots100$, from (7.81) it follows that the spread of parameters of the matrix in the TDI mode is reduced to 2%, and condition (7.75) is satisfied: the target miss probability is given by (7.76). Accordingly, the matrix as a whole is characterized by the mean value of $\overline{SNR}$ rather than by the lowest signal-to noise ratios as before.

If there are m_d defective elements in any line, the gain in the SNR in the TDI mode will be $\sqrt{n-m_d}$ instead of $\sqrt{n}$ (see (7.79)). The deterioration as compared with the completely operating line will be expressed by the factor of $\sqrt{(n-m_d)/n}$. So, even if one element out of ten fails, the loss will be as small as 5%! In this way the TDI mode is a solution to the problem of securing the desired target miss probability P_{tm} through redundancy.

The TDI mode may be realized both by external devices, and within the PDA itself. A CCD is as though specially invented for implementation of the TDI mode. It is not surprising that this mode is often realized in CCDs: the cells of one line (Fig. 7.15b,c) are charge coupled, and the accumulated charges are moved leftwards at the same rate as the image is moved with respect to the CCD — that is, with the clock frequency T_f/m. Then the signal charge is accumulated in the cell on which the light spot falls (cell i-1 in Fig. 7.15b, cell i-2 in Fig. 7.15c, etc.). When the light spot leaves the matrix (cell i-n passes the spot), the integration is over, and the charge goes to the output to be registered.

The TDI mode can also be realized in the photoresistor: the accumulation of the package of photocarriers takes place in the bulk of the semiconductor (during the lifetime of carriers), and the displacement of the package (synchronized with scanning) is accomplished through drift in the strong field. The total time of drift must be less than the time of diffusion spreading of the package. This device (see Sect. 1.1) is known as SPRITE (for Signal PRocessing In The Element). Both m and n in SPRITE are small (ten or so), but it was possible to manufacture it on the MCT-basis for the spectral range 8...12 μm, where the number of elements is generally limited.

In some systems (for example, orbital spectrophotometers), the linewise scanning is performed by the movement of the spacecraft itself. It seems that the TDI mode is then the only possibility for increasing the number of elements by replacing the linear array with the matrix. In silicon-based CCDs the number of elements in the line may be of the same order as the number of lines, $n \approx m$, which may be as large as several hundreds [110].

Pattern noise. As shown in Sect. 7.1, the signal in PDA with integration comes to the output mixed with the constant component U_{0i}, which, in case of the strong background, may exceed the noise by the factor of 3000. The constant component U_{0i} varies (vacillates) from one element to another even when the background is constant. This is caused by all kinds of nonuniformities: the spread of sensitivity, transfer coefficient, integrating capacitances. Fluctuation of the constant component U_{0i} is known as "pattern noise", because it masks the spatial (geometrical) distribution of the power of optical signal. Even when the parameters of PDA elements are highly uniform (as reported in [108, 109], the spread is 10...20%, and can be made less than that only in PDAs based on Shottky barriers), the ratio of the pattern noise to the intrinsic noise of the cell is very large, about $(0.1...0.2)\cdot 3000...6000 \approx 2^8...2^{10}$. When the background itself is not homogeneous, and varies from the maximum value to the dark level, this ratio may be as high as $\sim 3000 \approx 2^{11}...2^{12}$ (see Sect. 7.1). In the lattice-type matrices the pattern noise is supplemented by the inhomogeneity of erasing (the initial setting of the potential U_{ini}). The above estimate reveals that the registration of threshold-level signals requires using an 8–12 bit analog-digital converter (ADC), memorizing the constant component of each channel, carrying out the channel-by-channel subtraction of the constant component — in short, digital signal processing.

The simplest subtraction technique consists in subtracting the charges accumulated in the presence and in the absence of the signal (the charges of the signal frame and the dark frame). In this technique, however, the noise power is doubled (see (7.60)). If there is no excess noise, it is expedient to calculate the constant $\overline{U_{0i}}$ by averaging its value over n dark frames. The dispersion will then be n times less (see (7.81)): $\overline{\Delta U_{0i}^2(n)} = \overline{\Delta U_{0i}^2}/n$, so that the noise will only increase by the factor of $\sqrt{1+1/n}$. To eliminate the effects of the long-term drift, such measurements must be renewed from time to time.

Well, all this may be a blessing in disguise. If a microprocessor is necessary for dealing with the pattern noise, it can obviously be used for implementing more sophisticated signal processing algorithms.

Maximum number of elements and detection. Since the modern development of optoelectronics is mainly directed towards increasing the number of elements M_e, it would be natural to ask what is the maximum achievable value of M_e. Of course, this limit is to a large extent determined by the available technologies. The advances in silicon technology, the successful development of CCDs over the past twenty-five years, have made it possible to produce silicon CCD-based linear arrays and matrices with a very large number of elements: $(1...4)\cdot 10^3$ in linear arrays, $2.5\cdot 10^5$ in matrices (sometimes as many as 10^6 and more) [15, 111]. With silicon sensitized in the medium IR range (creating Shottky barriers), matrices for this range can be made of the same format as matrices for the visible range [27, 112]. In hybrid CCDs and CCDs based on the narrow-band semiconductors, the number of elements used to be limited for technological reasons, especially at the early stage. In recent years, however, full-format matrices have been successfully realized [113].

In this book we are more interested, however, in the fundamental restrictions imposed on the number of elements from the side of low-level signal processing. The maximum allowable number of channels M_e connected to one information output can be defined from the following simple consideration:

$$M_e = \frac{T_f}{T_{\text{int}}} . \tag{7.83}$$

Theoretically, the charge integration devices may register signals on the level of their own fluctuations, in the same manner as APD in the photon counting mode (Sect. 6.2, 6.3).

For such ultimately low signal charges the noise of the output device (7.56) must be less than the noise of the signal (6.32):

$$\overline{\Delta Q_{\text{ini}}^2} < \overline{\Delta Q_0^2} = \overline{\Delta Q_{\text{s}}^2} \; ; \quad \frac{a_{\text{t}} C^2 \overline{e_{\text{n}}^2}}{T_{\text{int}}} < q^2 \overline{N_\tau} \; . \tag{7.84}$$

Hence follows the restriction on the interrogation time,

$$T_{\text{int}} > \frac{a_{\text{t}}}{\overline{N_\tau}} \left(\frac{C e_{\text{n}}}{q} \right)^2 = \frac{0.5...1.25}{\overline{N_\tau}} \left[\frac{10^{-13} \cdot (50...200) \cdot 10^{-9}}{1.6 \cdot 10^{-19}} \right]^2$$

$$= \frac{5 \cdot 10^{-4} ... 2 \cdot 10^{-2}}{\overline{N_\tau}} \; . \tag{7.85}$$

This numerical estimate is based on the typical values of the capacitance of the CCD output device [114, 115], and the noise emf (Table 2.2).

The interrogation time is found to be inversely proportional to the number of registered photoelectrons. When $\overline{N_\tau} = 20...1000$, the interrogation time varies between $5 \cdot 10^{-7} ... 10^{-3}$ s, so that the TV standard ($T_{\text{f}} = 0.02$ s) in the threshold-level mode restricts the number of elements to $M_{\text{e}} = 20...4 \cdot 10^4$ per each information output.

If the number of photoelectrons is large, the value of T_{int} falls, and when the cell is filled to capacity ($\overline{N_\tau} \sim 10^7$), its shot noise is so large that the amplifier noise only becomes noticeable with extremely small interrogation times (7.85) (less than $5 \cdot 10^{-11} ... 2 \cdot 10^{-9}$ s). In this case the interrogation time T_{int} is limited by the operating rate of the matrix and the signal processing system. The values of transfer time and corresponding T_{int} achieved in silicon CCDs are about 10^{-7} s. This is not yet the theoretical limit: the transfer time of 10^{-9} s has been reportedly obtained in specially designed peristaltic CCDs. Such low values of T_{int} permit obtaining a quite large number of elements in TV-standard CCD matrices, $M_{\text{e}} = 0.02/10^{-7} = 2 \cdot 10^5$. Lattice matrices, CCDs based on narrow-band materials, and hybrid CCDs are slower than silicon-based CCDs. In case of the real-time signal processing, the time T_{int} is limited by the speed of both the processor itself, and the signal converters. As indicated above, an 8...12 bit ADC with the response time of $10^{-7} ... 5 \cdot 10^{-6}$ s is required for signal processing when the background charge is large.

It is especially difficult to increase the number of elements in IR PDAs for spectral ranges 3...5 and 8...14 μm in case of the strong thermal background. On the one hand, because of the strong background the integrator fills up very soon, and the accumulated charge has to be dumped quite often: $T_{\text{i max}} \approx 10^{-5} ... 10^{-3}$ s (Table 7.1). On the other hand, separation of the weak signal from the background (1:3000) requires a multi-bit

ADC, which delimits the speed of information readout: $T_{int} \approx 10^{-7}...5 \cdot 10^{-6}$ s. Therefore, one information output in such threshold-level PDAs can, according to (7.83), serve only a limited number of elements,

$$M_e = \frac{10^{-5}...10^{-3}}{10^{-7}...5\cdot 10^{-6}} \approx 10...10^4.$$

However, IR CCDs based on doped silicon or on Shottky barriers have a large number of elements, $M_e = 10^4...2\cdot 10^5$. How can this be done? The answer is simple: the quantum yield η in such CCDs is small, about 20...1000 times less than that in the photodetectors based on intrinsic materials. The current I_0 decreases proportionally, and $T_{i\,max}$ and the number of elements increase. This, however, is a Pyrrhic victory: there is no gain of $\sqrt{M_e}$ in the signal-to-noise ratio. This conclusion follows directly from (7.64), which may be rewritten as

$$\mathrm{SNR} = B_W \frac{SE_{sA}}{\sqrt{2qi_0}}\sqrt{T_s\frac{\mathcal{M}}{A}} = B_W \frac{SE_{sA}}{\sqrt{2qSE_b}}\sqrt{\frac{T_f}{\mathcal{M}}M_e\frac{\mathcal{M}}{A}} = B_W E_{sA}\frac{\sqrt{SM_eT_f}}{\sqrt{2qE_bA}}, \quad (7.86)$$

where E_b is the density of effective flux of the background (thermal) radiation (see Sect. 1.2); E_{sA} is the signal. Both E_b and E_{sA} are referred to the maximum sensitivity wavelength λ_{max}, and S is the sensitivity at this wavelength (Sect. 1.2). The product SM_eT_f is a generalized criterion of quality of PDA as a radiation-limited threshold-level device. The value of M_e by itself is not yet a figure of merit of the PDA, if it is increased at the price of reduced sensitivity S (as in the previous example), or the time T_f. Unfortunately, it is usually possible to maximize only two of these parameters. To clarify this point, let us consider examples of PDAs intended for the optoelectronic system with the following parameters:

Spectral range, μm	3...5
Frame format, pixels	100×100
Frame time, s	10^{-2}
Solid angle of view, sterad	2π

In principle, the following parameters can be achieved in the PDA for such system:

Sensitivity ($\eta = 0.8$, $\lambda_{max} = 5$μm), A/W	3.2
Maximum integration time $T_{i\,max}$, s	10^{-4}
Interrogation time T_{int}, s	10^{-6}

The first PDA type is the matrix with a large number of elements and low sensitivity. Assume that in this matrix $M_e = 100 \times 100 = 10^4$, $\eta = 0.008$,

$S = 3.2 \cdot 10^{-2}$ A/W. Because of the hundred-fold decrease in the sensitivity and the background current, the allowable integration time is increased by the same factor, and becomes equal to 10^{-2} s. Then the integration may be carried out throughout the entire frame time, $T_i = T_f = 10^{-2}$ s, routing the signals from all elements to the same information output: $M_e = T_f / T_{int} = 10^{-2}/10^{-6} = 10^4$. The figure of merit for such matrix is

$$SM_eT_f = \left(3.2 \cdot 10^{-2}\right) \cdot 10^4 \cdot 10^{-2} = 3.2 \ . \tag{7.87}$$

The second PDA type is the linear array with $M_e = 1 \times 100 = 10^2$. Technologically it is possible to achieve the photodetector sensitivity of $S = 3.2$ A/W. The length of the signal pulse $T_s = T_f / m = 10^{-2}/10^2 = 10^{-4}$ s does not exceed the time $T_{i\max}$. For this linear array we have

$$SM_eT_f = 3.2 \cdot 10^2 \cdot 10^{-2} = 3.2 \ . \tag{7.88}$$

The third PDA type is again the matrix with the same number of elements as before, $M_e = 100 \times 100 = 10^4$. This time we assume, however, that we have been able to improve on our first matrix, and achieved the sensitivity of $S = 3.2$ A/W. If we use only one information output, each element can only be interrogated once in a frame. Then the signal can be integrated over the time $T_i \approx T_{i\max} \approx 10^{-4}$ s — otherwise the matrix will be dazzled (the time T_i can be delimited with mechanical or electronic shutters). For such matrix we find again that

$$SM_eT_f = 3.2 \cdot 10^4 \cdot 10^{-4} = 3.2 \ . \tag{7.89}$$

These three examples illustrate the directions of the modern research and development in the field of IR PDAs for systems with high-level background. All three types of PDAs are entirely identical in terms of their threshold characteristics, and the choice between them will depend on the particular design requirements and operating conditions. In technical literature it is often said that the low sensitivity of the first matrix is compensated by the large integration time. It would be equally right to argue, however, that the high sensitivity compensates for the low number of the elements in the linear array, or for the short integration time in our last example. But, alas, the loss in any of the three parameters S, M_e, T_f cannot be compensated. In fact, the comparison must be drawn with the ideal PDA, in which these parameters are close to their best achievable values:

$$SM_eT_f = 3.2 \cdot 10^4 \cdot 10^{-2} = 320 \ . \tag{7.90}$$

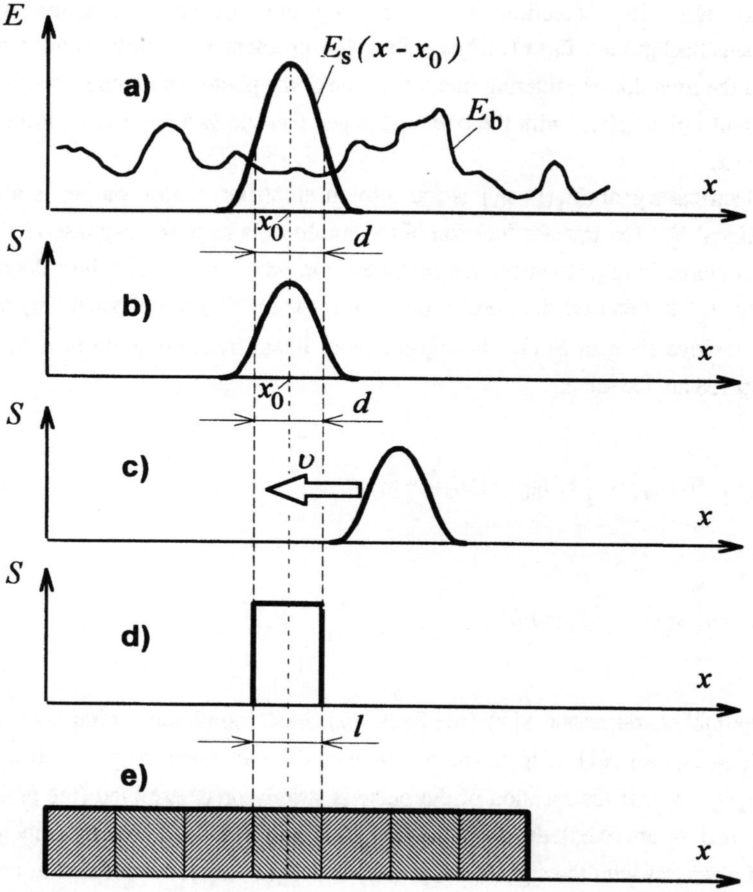

Figure 7.16: Spatial filtering.

This ideal PDA is better than our three exemplary PDAs by two orders of magnitude in terms of the threshold characteristics. The path towards its technical realization lies through the improvement of the PDA of the third type: parallel output of information, increased speed of output, charge partitioning and scimming in the cell, delimited useful information readout.

Such are the main limitations which restrict the number of elements in the threshold-level matrices, and the ways to overcome these limitations.

Spatial filtering. In the beginning of this section in connection with the analysis of scanning systems we assumed that the size of the element l is equal to the diameter of the light spot d. Let us now formulate the problem in general.

Optimal spatial filter. We start our analysis with the simple one-dimensional case: let the illumination within the light spot obey a certain distribution law $E_s(x - x_0)$, where x_0 is the center of the spot (corresponding to the maximum of illumination; Fig. 7.16a). Now the detection is hindered by the Poisson fluctuation of the homogeneous background $E_b(x)$. Thus stated, the problem is entirely similar to the problem of the time-domain filtering (the separation of the photocurrent pulse $SE_s(t-t_0)$ from the shot noise $i_n(t)$), with the only difference that the variable t is replaced with the variable x.

The electrical signal $SE_s(t-t_0)$ is fed into an amplifier, whose output is also an electrical signal U_s. The transfer function of the amplifier is its pulse response $\mathrm{H}(t'-t)$. The optical signal is registered by the photodetector, which converts it into the signal photocurrent I_p. In this case the transfer function is the distribution of sensitivity across the photosensitive element $S(x)$. The algorithms of linear transformation in both cases being compared are the same:

$$t' = t_m \ , \quad U_s(t_m) = \int_{-\infty}^{\infty} \mathrm{H}(t_m - t) SE_s(t - t_0)\, dt \ , \tag{7.91}$$

$$I_s(x_0) = \int_{-\infty}^{\infty} S(x - x_0) E_s(x)\, dx \ . \tag{7.92}$$

The optimal characteristic $S(x)$ (see Sect. 4.1) must completely imitate the signal profile, $S(x-x_0) \propto E_s(x)$ (Fig. 7.16a,b), in exactly the same way as $\mathrm{H}(t_m - t_i)$ imitates $E_s(t_i - t_0)$. If the location of the pulse is strictly predetermined (the points of maxima t_0 and x_0 are fixed), the functions $\mathrm{H}(t_m - t)$ and $S(x-x_0)$ exactly coincide on the axis of abscissas with the signals $SE_s(t-t_0)$ and $E_s(x-x_0)$. Recall that in case of the time-domain filtering this is ensured by the correct placement of the origin. If the position of the pulse $SE_s(t-t_0)$ on the time axis is not known (that is, not known is the time t_0), we had to watch the output signal continuously (take samples) throughout the entire time interval T on which the signal may be expected to arrive. If the spatial position of the optical spot $E_s(x-x_0)$ within the frame (of length L) is not known, we must get a continuous sequence of samples $I_s(x_0)$ (7.92) by moving the sensitivity maximum $S(x-x_0)$ along the x–axis. Ergo, it is necessary to move the photodetector with respect to the image with a certain velocity v. The coordinate x is linearly transformed into the coordinate $t = x/v$ (Fig. 7.16c).

The filter-amplifier in both cases is tuned to the signal photocurrent pulse $I_s(t)$. In the first case, when the spatial position of the spot is known, and scanning is not necessary, the photocurrent is constant throughout the frame time T_f, and there the integration is optimal. In the second case the photocurrent $I_s(t)$ in the course of scanning imitates the correlation function for the bell-shaped pulse (Fig. 4.8g). It is to this pulse that the filter must be tuned.

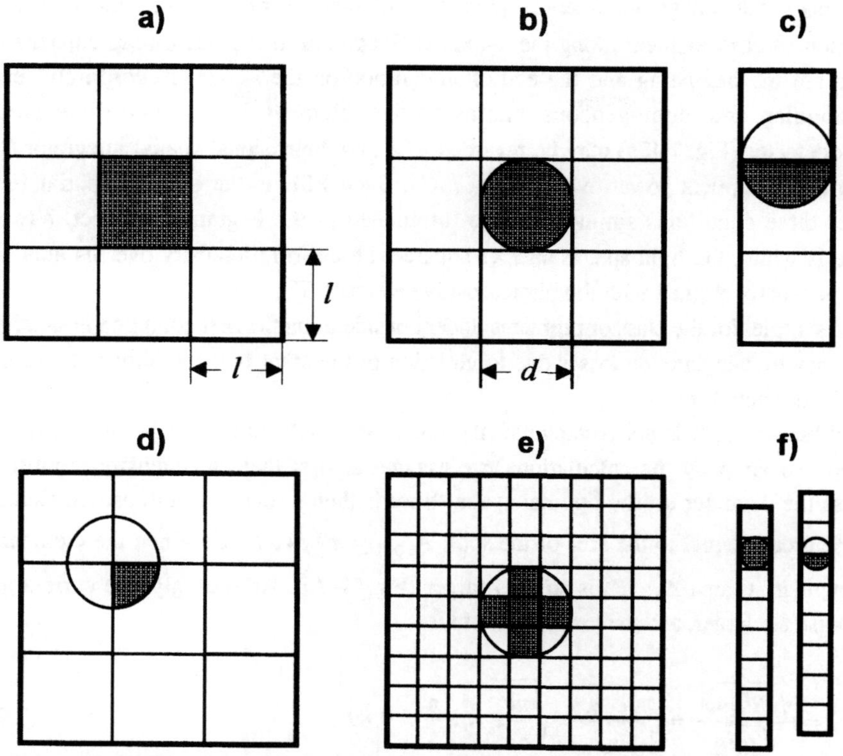

Figure 7.17: Matrix may be regarded as quasi-optimal spatial filter: each photosensitive element integrates the incident optical power.

Generalization to the two-dimensional situation appears to be a formal operation. The distribution function of illumination $E_0(x-x_0,\ y-y_0)$ uniquely defines the optimal distribution of sensitivity across the photosensitive element $S(x-x_0,\ y-y_0) \propto E_s(x-x_0,\ y-y_0)$, and when position of the center of the spot (x_0, y_0) is fixed, the sensitivity maximum is aligned with this point. When the position of the spot is not known, however, the situation is more complicated. By analogy with the one-dimensional case, the optimal filtering calls for the replacement of variables $x \Rightarrow t,\ y \Rightarrow t'$. But there is only one time axis t — the time axis t' does not exist, and so there can be no one-to-one correspondence between the two-dimensional space $(0,x)$, $(0,y)$ and the one-dimensional space $(0,t)$.

Matrix of elements as quasi-optimal spatial filter. Nevertheless, there is no reason for despair: quasi-optimal methods come again to our rescue. We start with the simpler one-dimensional problem, and replace the PSE with a rather sophisticated bell-shaped distribution $S(x-x_0)$ by a PSE with uniform sensitivity and length l (Fig. 7.16b,d). Once again we get a rectangular characteristic $S(x-x_0)$, exactly the same as the response of the integrating capacitance $\mathrm{H}(t-t_0)$. The PSE integrates the radiation over its entire length l, in the same manner as the capacitance stores up the charge carriers over the integration interval T_i. In the multielement linear array (Fig. 7.16d) the position of each element along the x–axis is fixed, and in the integrating capacitance fixed are the beginning and the end of integration on the t–axis. Accordingly, each rectangular and homogeneous photosensitive element in the two-dimensional photodetector (Fig. 7.17a) may be regarded as a two-dimensional spatial integrator (the integrator of optical power over the area $l \times l$). Such PSE is the optimal spatial filter under three conditions similar to those formulated in the beginning of Sect. 7.1: the noise is white, the light spot is an $l \times l$ square with uniform intensity over its area, the spot is exactly aligned with the photosensitive element (Fig. 7.17a).

As a rule, for the class of problems under consideration the first condition is satisfied. It is easy to calculate the losses due to violation of the other two conditions; the results will be not new for us.

When the spot is not rectangular, the losses are small. In reality the optical spot is round. To simplify the calculations, we assume at first that the intensity is uniform across the diameter d. The optimal spatial filter is then a round photosensitive element whose area is equal to the area of the spot, $A_{e\,\mathrm{opt}} = \pi l^2/4$, $l = d$. When the element is rectangular, its area $A_e = l^2$ is slightly larger (Fig. 7.17b). Accordingly, the current and the noise are larger, and the loss coefficient is

$$\mathcal{L} = \frac{\sqrt{2qI_{0\,\mathrm{opt}}}}{\sqrt{2qI_0}} = \sqrt{\frac{A_{e\,\mathrm{opt}}}{A_e}} = \sqrt{\frac{\pi l^2}{4l^2}} = \sqrt{\frac{\pi}{4}} = 0.89 \; . \tag{7.93}$$

Interestingly enough, the deterioration is the same 11% when the energy distribution across the spot is Gaussian, if the side of the element l is made equal to the effective diameter of the spot.

The loss may be quite considerable when the centers of the spot and the photosensitive element are misaligned. In the worst case, when the energy of the spot is equally shared between two adjacent elements, the signal is reduced by half when the scanning is done with the linear array (Fig. 7.17c, where we observe complete analogy with the non-synchronized sampling of the time integrator; see Sect. 7.1). In the matrix the signal may be reduced to one-fourth (Fig. 7.17d).

The ways to combat this problem were outlined in Sect. 7.1: the interval of summation is divided into 3...5 subintervals, and the readings from these subintervals are added together one by one. Then the step between consecutive readings along each of the axes becomes 3...5 times smaller. This means that it is necessary to increase the number of elements in the matrix by the factor of $3^2 \ldots 5^2 = 9 \ldots 25$ (Fig. 7.17e). In the matrix shown in the diagram the spot covers 3×3 elements. When the signal from the five hatched elements are added together, the misalignment of the centers of the spot and the PSE results in the reduction of the signal by the factor of 1.3 (and not 4, as in our former matrix with $l = d$). The largest loss in SNR related to the optimal space filter is by the factor of 1.45 (as compared with 4.5 in the matrix with $l = d$). If we increase the number of elements in the matrix further, so that the spot covers 5×5 elements, we practically come close to the optimal spatial filter: depending on the position of the center of the spot, the signal loss will vary within 2...16%, which translates into the loss in the SNR of 5...19% (when the signals from the elements hatched in Fig. 7.17e are added together). All these numerical estimates have been obtained for a round light spot of uniform intensity.

So, increasing the number of elements is good in terms of both time-domain and spatial filtering. The two advantages indicated above (increased integration time and reduced noise with increasing M_e and $\mathcal{M}$) are supplemented by a third one: the registered energy of the spot becomes practically independent of its alignment, and the precision of measurement of the coordinate is improved.

This is theory. Practice imposes certain restrictions on the maximum number of elements M_e. As a rule, the pitch l in matrices can be made not less than the effective spot diameter d, as we had assumed. This does not mean, however, that one has to put up with a four-fold energy loss. If the frame time in the matrix is large enough, one can split it into four subframes, $T_f' = T_f/4$, and take four readings, moving the matrix each time by the step $l/2$ along the axes x, y. In one of these four subframes the spot will be almost completely captured by one of the elements. The four-fold reduction of the integration time causes deterioration in the signal-to-noise ratio by the factor of $\sqrt{4} = 2$ (and not 4, as in case of one-frame integration). Using the linear array, it is sufficient to split the frame into two subframes, and the loss as compared with the optimal spatial filter is by the factor of $\sqrt{2}$. One can also use a two-column array, with the chessboard arrangement of the two columns (Fig. 7.17f).

Scattering function and modulation transfer function. The analogy between time-domain and spatial filtering may be carried further. Let us return to the linear array (Fig. 7.16e), and illuminate the central element with the infinitesimally narrow spot. Because of the finite size of the element, the output image is perceived as a spot of size l (Fig. 7.16d). Such (in this case rectangular) response is the spatial pulse characteristic of our device (see Sect. 4.1). In the image receivers this characteristic is known as the scattering function. The spectrum of the pulse response of amplifier is its frequency response, and the spectrum of the scattering function is called the modulation transfer function (MTF), which describes the transmission of the spatial frequencies. The effective passband of the amplifier is related to its integration time as $f_a = 1/2T_i$; in the same way, the effective passband for the spatial frequencies is, ideally, $1/2l$, where l is the side of the element. In real devices the scattering function broadens because of intercoupling between the elements, and the passband for the spatial frequency narrows down.

Space-time filtering. Finally, the problem in its general statement consists in describing the source of radiation in the three-dimensional space (x, y, t). If the three-dimensional illumination function $E_s(x, y, t)$ is known, we face the necessity to construct a three-dimensional filter $S(x, y, t)$. In the class of detection problems under consideration the intensity distribution in the spot is known, and so is the function $S(x,y) \propto E_s(x,y)$. The path of the spot, however, is usually not known in advance, and the parameters of the filter $S(x,y,t)$ cannot be defined a priori. So, the task consists in finding this path (the time dependence of the position of maximum $x_0(t)$, $y_0(t)$). This, however, is the subject of the more general theory of reception and measurement of signals.

The theory of reception proves that for the threshold-level signals, whose intensity is close to that of the noise, the optimal detection system is also the optimal measuring instrument (or, more precisely, its constituent). In our case, unknown is one of the parameters of the signal, the velocity of the spot v. To measure this velocity, we must at first interrogate all elements (at $t = t_0$), managing to do this within the time $T_f \approx d/v$. Having found the coordinates of the targets $x_{0i}(t_0)$, $y_{0i}(t_0)$, we may then interrogate the limited number of elements in the d–neighborhood of the current coordinates x_{0i}, y_{0i}.

Summary. *If the capabilities of the photosensitive element are exhausted (the quantum yield is close to 100%), and the electronics has done its best (the filtering is close to optimal), further improvement on the detection parameters is only possible with multielement arrays. As the number of elements M_e increases, increased is the time of integration (each element has more time to look at its part of the image), while the area of the element becomes smaller, which reduces the noise. Owing to these two factors, the signal-to-noise ratio in the classical case of the white noise is improved by the factor of $\sqrt{M_e}$. In case of the high-frequency and the low-frequency noises this rule does not hold: for the low-frequency noise the longer integration time does not improve the signal-to-noise ratio, and increasing the number of elements M_e is of no avail if the low-frequency noise cannot be eliminated. When the high-frequency noise dominates (caused by the emf noise of the amplifier), the increased resolution results in the increased interrogation frequency; accordingly, the passband and the noise of the amplifier both increase, so that the signal-to-noise ratio may even become worse.*

The quality of the PDA is determined by its worst elements in terms of the signal-to-noise ratio, rather than by the statistical-average element. The time delay and integration (TDI) mode is used to improve the uniformity and to back up the defective elements with good ones (thus reducing the target miss probability). In this mode the scene is scanned with the n-column array ($m \times n$ matrix), and the signals from all n columns corresponding to one and the same point of the scene are added together.

Spatial filtering is similar to the time-domain filtering. Each element may be regarded as a quasi-optimal spatial filter-integrator, which integrates the incident optical power. If we manage to squeeze 3...5 elements on the length equal to the diameter of the spot, the loss is 5...31% as compared to the optimal spatial filter; if the side of the element is equal to the diameter of the spot, the deterioration may be as large as by the factor of 4...4.5 (in the worst situation, when the spot falls on to four adjacent elements, so that its energy is shared equally among them).

Chapter 8

DETECTION OF WEAK OPTICAL SIGNALS IN SPECIAL-PURPOSE OPTOELECTRONIC SYSTEMS

The basics of the theory of detection of low-level optical signals are necessary not only for the development of optoelectronic systems that only have to detect optical signals. This theory is actually applicable to any kind of optoelectronic system: range finders (which not only detect the optical pulse, but also measure the time of its arrival), lightwave communication systems (which register pulse sequences rather than stand-alone pulses), IR imagers (which detect the temperature difference of heated bodies). Whatever is the particular task of the optoelectronic system, first of all it has to discern weak optical signals against the noise. The better we can do this (the higher is the signal-to-noise ratio we are able to achieve), the more successfully we shall deal with our main task: fewer errors in the information exchanged via optical communications, higher accuracy and longer range in optical range finders, smaller detectable temperature difference ΔT in IR imagers.

Some particular features of detection in lightwave communications (broad-band amplification, APD operating regimes), and in lidars (the use of multielement PDAs) were discussed in the previous chapter. Now we shall supplement this discussion with two important applications of optimal detection systems: timing of arrival of optical pulse (Sect. 8.1), and IR imagers (Sect. 8.2).

8.1. Timing of Arrival of Optical Pulse

Circuit diagram. In some optical systems — such as range finders, lidars based on linear photodetector arrays, reflectometers for locating defects in fibers by timing the arrival of the reflected pulse — it is necessary to measure the time of arrival of the signal. This is often done with conventional detection systems, and the time of arrival is identified with the instant at which the discriminator is set off. When the signal is weak, and its amplitude is comparable with the threshold setting U_t, the discriminator is triggered when the output signal is more or less at its maximum, $t \approx t_m$; when the signal is strong, this happens as soon as the signal arrives. Accordingly, the timing error is of the same order as the duration of the front edge of the signal pulse, which for the bell-shaped signal and broad-band amplifier is about equal to the duration T_s of the optical signal itself (cf. Fig. 5.18).

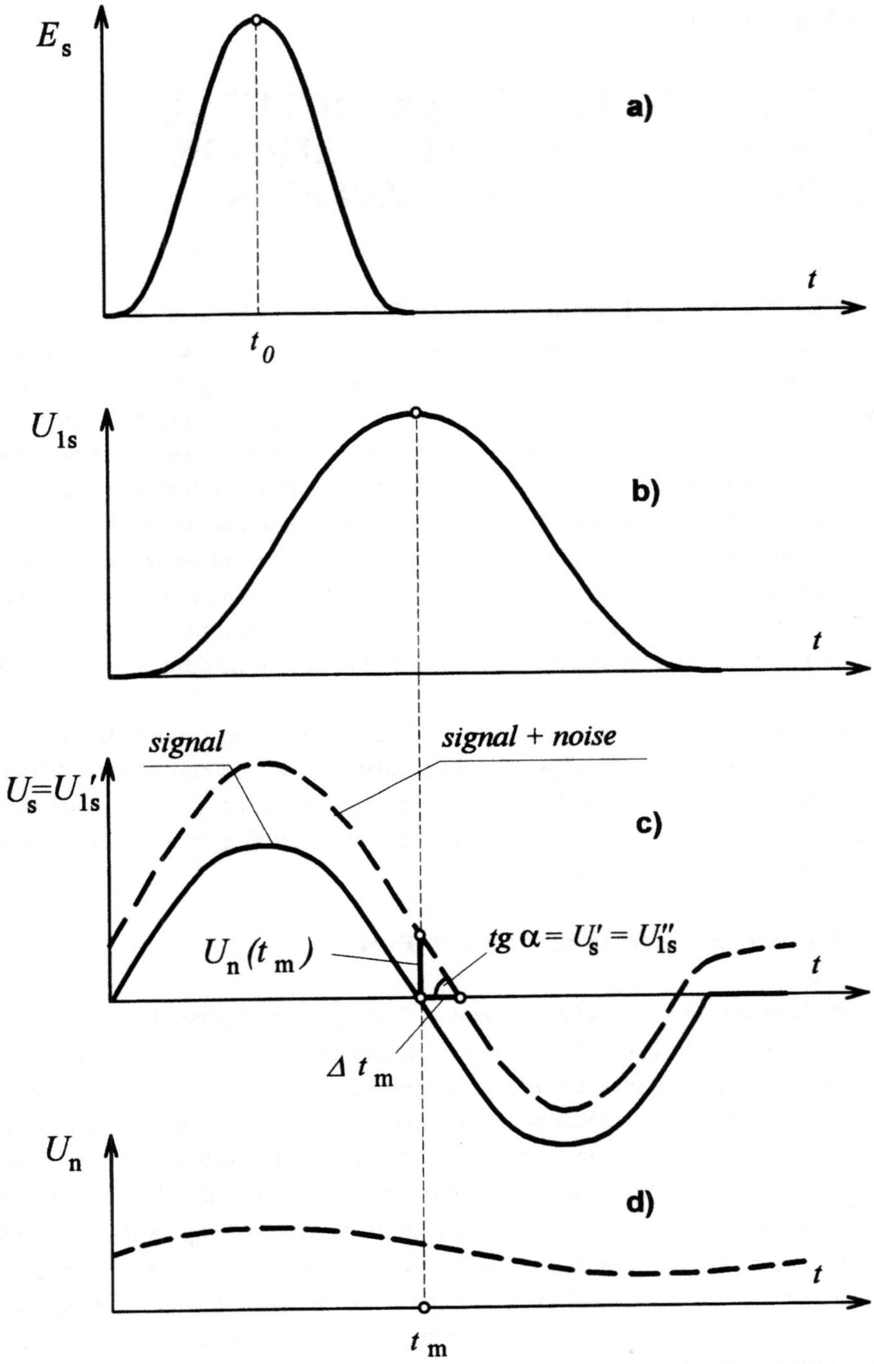

Figure 8.1: Noise superimposed on the signal gives rise to error in the measurement of t_m (a–c).

To improve the accuracy of measurements, we must refer to a certain characteristic point of the output pulse. Such point is obviously the maximum of the pulse ($t = t_m$, Fig. 8.1a). The shape of the pulsed signal and the time delay of the system are assumed to be known, which means that the time of arrival of the optical pulse is uniquely determined by the time t_m (Fig. 8.1a). At maximum the derivative is zero; therefore, we must first find the derivative, and then the time when this derivative becomes equal to zero (Fig. 8.1b).

So, in order to implement this algorithm, we modify the detection system as follows (see Sect. 3.1): a differentiating stage is placed between the amplifier and the decision-making device, and set the discriminator threshold to $U = 0$ (recall that formerly we set the threshold in the detection system at $U_t \approx 3...6U_n$; see Sect. 3.1). This setup is shown in Fig. 8.2. Of course, so far it is only good for a theoretician: it allows analyzing the signal and the noise, calculating the error in the measurement of t_m, and is thus sufficient for the purposes of this section. The diagram in Fig. 8.2, however, will not satisfy a practical engineer: it will respond to the maxima of not only the signal, but also of all the noise spikes. Because of this, it must also include nonlinear selection stages. For example, a discriminator with the threshold U_t may be placed between the amplifier and the differentiating stage; this discriminator will block the noise pulses whose amplitude is less than U_t. It is also possible to combine two devices, the detection system and the measuring device. The decision-making device of the detection system generates its signal if $U \geq U_t$, and the discriminator of the measuring device is set to $U = 0$. The signals generated by the two devices must have the duration of about T_s. The decision concerning the arrival of the signal and the location of its maximum is taken by an additional AND circuit (when the two signals arrive simultaneously at its inputs).

Of course, a large-amplitude stray noise pulse will trigger also the last two circuits. Such false alarms, however, are inevitable, and their frequency is controlled by setting the appropriate level U_t, in the same way as in the conventional detection system (see Sect. 3.1).

Error of measurement of the time of arrival of the signal. The noise superimposed on the signal may displace the latter, causing an error in the measurement of t_m. The calculation of this error becomes especially instructive if we assume that the noise realization $U_n(t)$ is almost constant in the neighborhood of t_m, as shown in Fig. 8.1b,c. This assumption is physically justified, because

— the amplifier stage 1 with the effective time constant τ_{eff} equally protracts both the signal pulse and the noise pulse, and so their durations T_{05} and τ_n are of the same order (if $\tau_{eff} \gtrsim T_s$; see Sect. 2.2, 5.1);

— the correlation time τ_{cor} coincides with the duration of the noise pulse (Sect. 2.2);

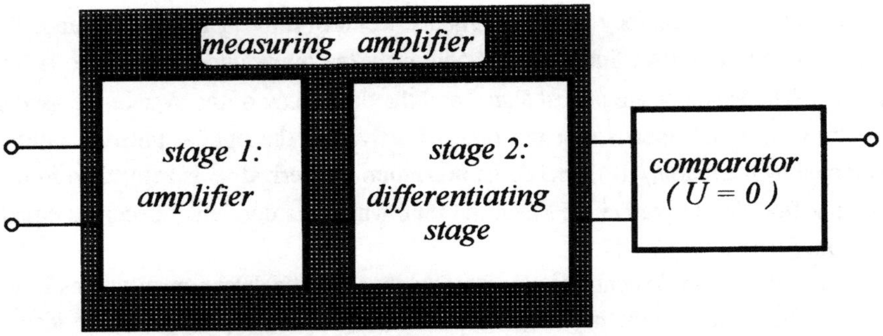

Figure 8.2: Structural diagram of device for measuring the location of signal maximum.

— T_{05}, τ_{eff} and τ_{cor} are of the same order of magnitude;

— the interval in the neighborhood of t_m, within which the point $U = 0$ is displaced (Fig. 8.1b), is small compared to T_{05}; therefore, it is small compared to τ_{cor}.

In case of the broad-band amplification, τ_n and τ_{cor} are, generally speaking, less than $T_{05} \approx T_s$, but the interval of interest in the neighborhood of t_m is even smaller.

So, we have arrived to the simplest model shown in Fig. 8.1b. The signal in the neighborhood of t_m is a straight line with the slope U_s'; the noise is a small constant addition U_n which shifts the straight line so that its intercept with the axis of abscissas is displaced through Δt_m. The magnitude of Δt_m is found by solving the triangle in Fig. 8.1b:

$$\tan\alpha = U_s' = \frac{U_n(t_m)}{\Delta t_m}\ , \quad U_n(t_m) = U_s'\Delta t_m\ . \tag{8.1}$$

The noise sample $U_n(t_m)$ is a random quantity. Therefore, by averaging over all possible values of $U_n(t_m)$ we find the rms error $\sqrt{\overline{\Delta t_m^2}}$ (the rms displacement of the point $U_s + U_n = 0$) [92]:

$$\left[U_s'(t_m)\right]^2 \overline{\Delta t_m^2} = \overline{U_n^2(t_m)} = \overline{U_n^2}\ , \quad \sqrt{\overline{\Delta t_m^2}} = \frac{U_n}{U_s'(t_m)}\ . \tag{8.2}$$

Further calculations are standard. Let k_1 be the frequency response of the amplifier stage 1, k_2 the frequency response of the differentiating stage 2, $k = k_1k_2$ the frequency response of the overall measuring amplifier (Fig. 8.2). We also denote by U_{1s}, U_{1n} the signal and the noise at the output of the amplifier stage 1, and by U_s, U_n the signal and the noise at the output of the overall measuring amplifier. The signal and the noise at the output of the amplifier with arbitrary frequency response have been calculated in Sect. 3.2:

$$U_{\mathrm{s}}(t_{\mathrm{m}}) = SE_{\mathrm{sA}} u_{\mathrm{s}}(t_{\mathrm{m}})\ ;\quad u_{\mathrm{s}}(t_{\mathrm{m}}) = 2T_{\mathrm{s}} \int_0^{\infty} k\, e_{\mathrm{s}} \cos(\omega t_{\mathrm{m}} + \psi_{\mathrm{s}} + \psi_{\mathrm{a}})\, df\ , \tag{8.3}$$

$$U_{\mathrm{n}} = \sqrt{2qI_{\mathrm{n}}}\, u_{\mathrm{n}}\ ,\quad u_{\mathrm{n}} = \sqrt{\int_0^{\infty} \Theta k^2\, df}\ . \tag{8.4}$$

Accordingly, the values of $U_{1\mathrm{s}}(t_{\mathrm{m}})$, $U_{1\mathrm{n}}$ can be found by replacing $k \Rightarrow k_1$ in these formulas.

To calculate the rms error $\sqrt{\Delta t_{\mathrm{m}}^2}$, we must first learn how to find the derivative of the signal. We take one signal harmonic $u_{\mathrm{s}} \propto \cos(\omega t + \psi_{\mathrm{s}} + \psi_{\mathrm{a}})$ and find its derivative:

$$\frac{d}{dt} u_{\mathrm{s}} \propto \frac{d}{dt} \cos(\omega t + \psi_{\mathrm{s}} + \psi_{\mathrm{a}}) = -\omega \sin(\omega t + \psi_{\mathrm{s}} + \psi_{\mathrm{a}})$$

$$= \omega \cos\left(\omega t + \psi_{\mathrm{s}} + \psi_{\mathrm{a}} + \frac{\pi}{2} \right)\ . \tag{8.5}$$

We see that to differentiate the harmonic we must multiply it by the cyclic frequency ω and shift the phase by $\pi/2$. Using this rule, we find the derivative of the signal (8.3):

$$U'_{\mathrm{s}}(t_{\mathrm{m}}) = SE_{\mathrm{sA}} u'_{\mathrm{s}}(t_{\mathrm{m}})\ ;$$

$$u'_{\mathrm{s}}(t_{\mathrm{m}}) = 2T_{\mathrm{s}} \int_0^{\infty} k e_{\mathrm{s}} \omega \cos\left(\omega t_{\mathrm{m}} + \psi_{\mathrm{s}} + \psi_{\mathrm{a}} + \frac{\pi}{2} \right) df\ , \tag{8.6}$$

Now we substitute (8.4) and (8.6) into (8.2) to get the expression for $\sqrt{\Delta t_{\mathrm{m}}^2}$:

$$\sqrt{\Delta t_{\mathrm{m}}^2} = \frac{U_{\mathrm{n}}}{U'_{\mathrm{s}}(t_{\mathrm{m}})} = \frac{\sqrt{2qI_{\mathrm{n}}}}{SE_{\mathrm{sA}}} \frac{\sqrt{\int_0^{\infty} \Theta k^2\, df}}{2T_{\mathrm{s}} \int_0^{\infty} k e_{\mathrm{s}} \omega \cos\left(\omega t_{\mathrm{m}} + \psi_{\mathrm{s}} + \psi_{\mathrm{a}} + \frac{\pi}{2} \right) df}\ . \tag{8.7}$$

This is the most general representation of the rms error $\sqrt{\Delta t_{\mathrm{m}}^2}$. It applies to any configuration of the measuring device with the arbitrary frequency response $k(f)$. It is worth the effort to particularize this formula using our knowledge about the circuit of the measuring device. First of all, we know that the second stage is differentiating. This allows us to express the parameters of the signal and the noise at the output of the device in terms of those at the output of the amplifier stage 1:

$$u_s(t) = u'_{1s}(t)\ ,\quad u'_s(t) = u''_{1s}(t)\ ;$$

$$U_s(t) = SE_{sA}u_s(t)\ ,\quad U'_s(t_m) = SE_{sA}u''_{1s}(t_m)\ ; \tag{8.8}$$

$$k_2 = \omega\ ,\quad k = k_1 k_2 = \omega k_1\ .$$

Using these prefabricated parts, we transform (8.2) into

$$\sqrt{\overline{\Delta t_m^2}} = \frac{U_n}{U'_s(t_m)} = \frac{\sqrt{2qI_n}}{SE_{sA}} \frac{\sqrt{\int_0^\infty \omega^2 k_1^2 \Theta\, df}}{u''_{1s}(t_m)}$$

$$= \frac{\sqrt{2qI_n \int_0^\infty k_1^2 \Theta\, df}}{SE_{sA}u_{1s}(t_m)} \times \frac{u_{1s}(t_m)}{u''_{1s}(t_m)} \times \sqrt{\frac{\int_0^\infty \omega^2 k_1^2 \Theta\, df}{\int_0^\infty k_1^2 \Theta\, df}} \tag{8.9}$$

The first and the third terms on the right-hand side hardly call for any comments. The first term

$$\frac{\sqrt{2qI_n \int_0^\infty k_1^2 \Theta\, df}}{SE_{sA}u_{1s}(t_m)} = \frac{1}{\mathrm{SNR}} \tag{8.10}$$

is the reciprocal of the SNR; we used to deal with it all the way throughout the book. The third term is the characteristic frequency of positive noise spikes ω_{char}. This parameter came up not that often (see Sect. 3.1, 5.1, equations (3.25), (5.90)):

$$\omega_{char} = 2\pi f_{char}\ ;\quad f_{char} = \sqrt{\frac{\int_0^\infty f^2 k_1^2 \Theta\, df}{\int_0^\infty k_1^2 \Theta\, df}} \tag{8.11}$$

When the noise is white, the frequency f_{char} is equal (up to a coefficient) to the effective passband of the amplifier for the signal f_a. This follows from (5.39), (5.44), (5.47), (5.52):

$$f_{char} = \sqrt{\frac{\int_0^\infty f^2 k_1^2 \Theta \, df}{\int_0^\infty k_1^2 \Theta \, df}} = \sqrt{\frac{f_a^2 f_{nh}}{f_{nw}}} = \sqrt{\frac{f_a^3 \Gamma_2}{f_a \Gamma_1}} = f_a \sqrt{\frac{\Gamma_2}{\Gamma_1}} . \qquad (8.12)$$

New for us is only the second term in (8.9), the ratio $u_{1s}(t_m)/u''_{1s}(t_m)$. This ratio we shall also evaluate for a particular typical situation when the optical pulse has a standard bell-shaped profile, and the amplifier stage 1 is broad-band. Then the output signal U_{1s} is also bell-shaped, and in the neighborhood of the maximum it can, as a rule, be approximated with the square-cosine function. This can be seen in Fig. 5.7, 5.9, 5.12 (where the passband of the amplifier is delimited by the optimal filter, and by one-section and two-section RC-filters, respectively). In the square-cosine approximation the derivative is

$$\left|\frac{d^2}{dt^2} u_{1s}(t)\right\|_{t=t_m} \approx \left|\frac{d^2}{dt^2} u_{1s}(t_m) \cos^2\left(\frac{\pi}{2}\frac{\Delta t}{T_{05}}\right)\right\|_{\Delta t=0}$$

$$= \frac{u_{1s}(t_m)}{2}\left|\frac{d^2}{dt^2}\left(1+\cos\frac{\pi \Delta t}{T_{05}}\right)\right\|_{\Delta t=0} = \frac{u_{1s}(t_m)}{2}\left(\frac{\pi}{T_{05}}\right)^2 . \qquad (8.13)$$

If you have made sure that the square-cosine approximation of the output signal in the neighborhood of t_m is valid in your particular case, you can calculate $\sqrt{\Delta t_m^2}$ by the formula which follows from (8.9) – (8.13):

$$\sqrt{\Delta t_m^2} \approx \frac{1}{SNR_1}\frac{2T_{05}^2}{\pi^2}\omega_{char} = \frac{2}{\pi}\frac{T_s}{SNR_1}\left(\frac{T_{05}}{T_s}\right)^2 (f_{char} 2T_s) . \qquad (8.14)$$

Each of the expressions for $\sqrt{\Delta t_m^2}$ (8.7), (8.14) has its own niche. The generalized expression (8.7) will be used for solving the problem which arises in connection with the development of any system: the search for the limiting values of the parameters (the best possible accuracy of measurement of t_m). Expression (8.14) has a more pronounced physical appearance: it involves the familiar parameters of the detection theory SNR_1, T_{05}, f_{char}. This formula is convenient for analyzing the sources of errors in real devices for measuring the time position of the signal maximum.

Optimal measuring device. Let us find the frequency response $k(f)$ of the optimal measuring device from the criterion of minimum error $\sqrt{\overline{\Delta t_{\mathrm{m}}^2}}$. At first sight, the search for the extremum of (8.7) seems a formidable task. The theory of detection, however, helps us out again. As a matter of fact, we have already found the extremum of a similar expression: it was the extremum of the signal-to-noise ratio (see Sect. 3.2). Let us write the ratio of the signal (8.3) to the noise (8.4) at the output of the amplifier:

$$\mathrm{SNR}_1 = \frac{SE_{\mathrm{sA}} 2T_{\mathrm{s}}}{\sqrt{2qI_{\mathrm{n}}}} \frac{\int\limits_0^\infty k_1 \mathrm{e}_{\mathrm{s}} \cos(\omega t_{\mathrm{m}} + \psi_{\mathrm{s}} + \psi_{\mathrm{a}}) df}{\sqrt{\int\limits_0^\infty \Theta k_1^2 \, df}} . \tag{8.15}$$

To facilitate further analysis, we have made the replacement $k \Rightarrow k_1$.

The extrema of expressions (8.15), (8.7) are attained with the following phase-frequency characteristics of the amplifier:

$$\omega t_{\mathrm{m}} + \psi_{\mathrm{s}} + \psi_{\mathrm{a}} = 0 \,, \quad \psi_{\mathrm{a}} = -\omega t_{\mathrm{m}} - \psi_{\mathrm{s}} \tag{8.16}$$

for the optimal detection system, and

$$\omega t_{\mathrm{m}} + \psi_{\mathrm{s}} + \psi_{\mathrm{a}} + \frac{\pi}{2} = 0 \,, \quad \psi_{\mathrm{a}} = -\omega t_{\mathrm{m}} - \psi_{\mathrm{s}} - \frac{\pi}{2} \tag{8.17}$$

for the optimal measuring device. Then equations (8.15) and (8.7) become, respectively,

$$\mathrm{SNR}_1 = \frac{SE_{\mathrm{sA}} 2T_{\mathrm{s}}}{\sqrt{2qI_{\mathrm{n}}}} \frac{\int\limits_0^\infty k_1 \mathrm{e}_{\mathrm{s}} \, df}{\sqrt{\int\limits_0^\infty \Theta k_1^2 \, df}} , \tag{8.18}$$

$$\sqrt{\overline{\Delta t_{\mathrm{m}}^2}} = \frac{\sqrt{2qI_{\mathrm{n}}}}{SE_{\mathrm{sA}} 2T_{\mathrm{s}}} \frac{\sqrt{\int\limits_0^\infty \Theta k^2 \, df}}{\int\limits_0^\infty k(\omega \mathrm{e}_{\mathrm{s}}) \, df} . \tag{8.19}$$

We see that SNR_1 and $\sqrt{\overline{\Delta t_{\mathrm{m}}^2}}$ are expressed as the ratios of the same integrals, only the spectrum e_s in the formula for $\sqrt{\overline{\Delta t_m^2}}$ is replaced with the spectrum ωe_s. The fact that the numerator and the denominator change places does not matter for the search of the extremum. As shown in Sect. 3.2, the extremum (maximum) of SNR_1 is achieved with the optimal frequency response

$$k_1 = \frac{e_s}{\Theta} . \tag{8.20}$$

Since the "spectrum" in the expression for $\sqrt{\overline{\Delta t_m^2}}$ is $e_s\omega$, with the replacement $e_s \Rightarrow e_s\omega$ we get

$$k = \frac{e_s}{\Theta}\omega = k_1 k_2 . \tag{8.21}$$

Substituting (8.20) into (8.18), we arrive at the known expression for the signal-to-noise ratio in the optimal detection system,

$$SNR_1 = \frac{SE_{sA}2T_s}{\sqrt{2qI_n}} \frac{\int_0^\infty \left(\frac{e_s}{\Theta}\right) e_s\, df}{\sqrt{\int_0^\infty \Theta \left(\frac{e_s}{\Theta}\right)^2 df}} = \frac{SE_{sA}2T_s}{\sqrt{2qI_n}} \sqrt{\int_0^\infty \frac{e_s^2}{\Theta} df} . \tag{8.22}$$

And, substituting (8.21) into (8.19), we find the minimum error of the optimal measuring device:

$$\sqrt{\overline{\Delta t_m^2}} = \frac{\sqrt{2qI_n}}{SE_{sA}2T_s} \frac{\sqrt{\int_0^\infty \Theta \left(\frac{e_s^2}{\Theta^2}\right) \omega^2\, df}}{\int_0^\infty \left(\frac{\omega e_s}{\Theta}\right)(\omega e_s)\, df} = \frac{\sqrt{2qI_n}}{SE_{sA}2T_s} \frac{1}{\sqrt{\int_0^\infty \omega^2 \left(\frac{e_s^2}{\Theta}\right) df}}$$

$$= \frac{\sqrt{2qI_n}}{SE_{sA}2T_s \sqrt{\int_0^\infty \left(\frac{e_s^2}{\Theta}\right) df}} \times \sqrt{\frac{\int_0^\infty \left(\frac{e_s^2}{\Theta}\right) df}{\int_0^\infty \omega^2 \left(\frac{e_s^2}{\Theta}\right) df}} . \tag{8.23}$$

The first term on the right-hand side of (8.23) is the reciprocal of the signal-to-noise ratio at the output of the optimal filter (8.22), and the second term is the reciprocal of the characteristic frequency of this filter (cf. (8.11)):

$$\sqrt{\frac{\int_0^\infty \left(\frac{e_s^2}{\Theta}\right) df}{\int_0^\infty \omega^2 \left(\frac{e_s^2}{\Theta}\right) df}} = \sqrt{\frac{\int_0^\infty \Theta\left(\frac{e_s}{\Theta}\right)^2 df}{\int_0^\infty \omega^2 \Theta\left(\frac{e_s}{\Theta}\right)^2 df}} = \frac{1}{\omega_{char}} . \tag{8.24}$$

Finally, for the lowest achievable error of the ideal measuring device we get [92]:

$$\sqrt{\overline{\Delta t_m^2}} = \frac{1}{SNR_{1OF}\, \omega_{char}} . \tag{8.25}$$

Equations (8.21) and (8.25) are the high point of this section, and confirm the statements proclaimed in the beginning of this chapter.

1. *To solve the problem of the optimal measurement of t_m (with the highest possible accuracy) one must first solve the problem of detection.* Therefore, the optimal measuring device must include the optimal filter (the first term $k_1 = e_s/\Theta$ in (8.21)). Its output is connected to the differentiating stage (the second term $k_2 = \omega$ in (8.21)). Equation (8.21) proves that our structural diagram of the measuring device (Fig. 8.2) is correct, and that this setup is optimal if the optimal filter is used for the amplifier stage 1.

In practice, the complex representation of the transfer coefficient $\mathbf{K}(\omega)$ is often used. The complex transfer coefficient of the optimal filter $\mathbf{K}_1$ was defined by (3.50). The phase shift through $\pi/2$ by the differentiating stage (8.16), (8.17) is equivalent to multiplication by the imaginary unit j. Thus,

$$\mathbf{K}_1 = \frac{\exp(-j\omega t_m) e_s^*}{\Theta} , \quad \mathbf{K}_2 = j\omega , \quad \mathbf{K} = \mathbf{K}_1 \mathbf{K}_2 = \frac{j\omega \exp(-j\omega t_m) e_s^*}{\Theta} . \tag{8.26}$$

2. *The better we handle the detection problem (achieving a high signal-to-noise ratio), the better we cope with our main task (measuring t_m with higher accuracy).* This statement applies to both optimal and quasi-optimal measuring devices (at least when the noise is white): according to (8.14) and (8.25), we have $\sqrt{\overline{\Delta t_m^2}} \propto 1/SNR_1$.

Let us calculate the minimum error $\sqrt{\overline{\Delta t_m^2}}$ for two standard cases, when either the white noise or the high-frequency noise components prevail.

White noise. The calculation of $\sqrt{\overline{\Delta t_m^2}}$, according to (8.25), amounts to calculating the characteristic frequency. Assuming that $\Theta = 1$, $k_1 = e_s$, for the bell-shaped signal spectrum (2.14) we get:

$$\omega_{\text{char}} = \sqrt{\frac{\int_0^\infty \omega^2 e_s^2\, df}{\int_0^\infty e_s^2\, df}} = \sqrt{\frac{\int_0^\infty \frac{\omega^2 \sin^2 \omega T_s\, df}{\omega^2 T_s^2\left[1-\left(\frac{\omega T_s}{\pi}\right)^2\right]^2}}{\int_0^\infty \frac{\sin^2 \omega T_s\, df}{\omega^2 T_s^2\left[1-\left(\frac{\omega T_s}{\pi}\right)^2\right]^2}}} = \sqrt{\frac{\frac{\pi^2}{8T_s^3}}{\frac{3}{8T_s}}} = \frac{\pi}{\sqrt{3}}\frac{1}{T_s}. \tag{8.27}$$

So, according to (8.25) and (8.27), the minimum rms error of measurement is

$$\sqrt{\overline{\Delta t_m^2}} = \frac{\sqrt{3}}{\pi}\frac{T_s}{\text{SNR}_1} = \frac{T_s}{1.81\,\text{SNR}_1}. \tag{8.28}$$

We see that theoretically it is possible to discern the displacement of the pulse by about $1/2\,\text{SNR}_1$-th part of its duration. For the ultimately low signals we have $\text{SNR}_1 \approx 6\ldots12$ (see Sect. 3.1), and the smallest detectable displacement is therefore about $0.05\ldots0.1T_s$. It follows that the optimal measuring device improves the accuracy of measurement by an order of magnitude (as compared with the method of Sect. 5.2) even when the signal is extremely low.

High-frequency noise. The characteristic frequency (8.24) must now be calculated for $\Theta = 1 + \omega^2/\omega_{\text{high}}^2$:

$$\omega_{\text{char}} = \sqrt{\frac{\int_0^\infty \omega^2 \frac{e_s^2}{\Theta}\, df}{\int_0^\infty \frac{e_s^2}{\Theta}\, df}} = \sqrt{\frac{\int_0^\infty \frac{\omega^2 e_s^2\, df}{1+\left(\frac{\omega}{\omega_{\text{high}}}\right)^2}}{\int_0^\infty \frac{e_s^2\, df}{1+\left(\frac{\omega}{\omega_{\text{high}}}\right)^2}}} \approx \sqrt{\frac{\omega_{\text{high}}^2 \int_0^\infty e_s^2\, df}{\int_0^\infty \frac{df}{1+\left(\frac{f}{f_{\text{high}}}\right)^2}}}$$

$$= \sqrt{\frac{\omega_{\text{high}}^2 \frac{3}{8T_s}}{f_{\text{high}}\frac{\pi}{2}}} = \sqrt{\frac{3}{2}\frac{1}{\tau_{\text{high}} T_s}}. \tag{8.29}$$

In computing the integrals we have noted that the high-frequency noise prevails practically in the entire signal band, $f_s >> f_{high}$.

Substituting the characteristic frequency (8.29) into (8.25), we get:

$$\sqrt{\overline{\Delta t_m^2}} = \frac{\sqrt{\tau_{high} T_s}}{1.22\,\mathrm{SNR}_1}\,, \quad \tau_{high} = \frac{1}{\omega_{high}} = \frac{1}{2\pi f_{high}}\,. \tag{8.30}$$

From comparison between (8.28) and (8.30) we see that in case of the high-frequency noise the rms error $\sqrt{\overline{\Delta t_m^2}}$ is much greater (by the factor of about $1.5\sqrt{\tau_{high}/T_s}$) than that in case of the white noise, even if the value of SNR_1 is the same. This is due to the fact that the passband of the optimal filter for the high-frequency noise is narrow (of the order of f_{high}), and the output signal is protracted considerably (its length is about $1.4\tau_{high}$, see Sect. 4.2).

Quasi-optimal measuring device. The measuring device differs from the detection system in that it includes an additional differentiating stage with the frequency response $k = j\omega$. The circuit of this stage is simple: it is a one-section differentiating filter having a large time constant. The main difficulty consists in synthesizing the first amplifier stage with the optimal response e_s/Θ. And again we are rescued by the theory of detection. In Sect. 5.1, 5.2 we laid out the guidelines for the design of quasi-optimal detection systems whose SNR is almost as good as that of the optimal ones. The same principles can certainly be applied to the development of quasi-optimal measuring devices.

It would be a mistake, however, to apply these principles in a straightforward manner. There are additional restrictions, as can be seen from the expressions for the rms error (8.14). The three terms in this expression are three sources of error.

The first term is inversely proportional to SNR_1. The parameters of quasi-optimal filter (the time constants τ_f of its sections) are chosen so as to maximize the value of SNR_1. This suits our purpose: the lower the noise $U_n(t_m)$, the smaller the displacement of the signal $U_s(t)$ (recall Fig. 8.1b), and the smaller the error Δt_m.

The second term $(T_{05}/T_s)^2$ indicates that the error of measurement increases because of the protraction of the output signal. This is also clear from Fig. 8.1a,b: curve $U_s(t)$ is flattened, which results in larger Δt_m. The effective passbands of the optimal and quasi-optimal filters for the signal f_a are practically the same, and so these filters protract the signal by the same (almost the same) amount (the particular features of the case of high-frequency noise will be discussed later).

The specific features of the design of measuring device are associated with the third term. The higher the characteristic frequency f_{char} is with respect to the signal band $f_s = 1/2T_s$, the greater the error of measurement. According to definition (8.11), the high ratio f_{char}/f_s points to the noticeable high-frequency noise. In Fig. 8.1b,c the noise realization was assumed to be almost constant in the neighborhood of t_m. When, however, the high frequencies are present in the noise spectrum, the realization $U_n(t)$ changes faster, and noise spikes occur even in the small neighborhood of t_m (recall that f_{char} is the rate of positive noise spikes). The noise spikes superimposed on the flat maximum of the signal $U_s(t)$ may sneak to the input of the differentiating stage. Then the measuring device will be betrayed, and will measure the position of the maximum of not only the signal but also of the noise spike.

In the synthesis of quasi-optimal filter (see Sect. 5.1), the number of sections and their time constants were chosen without regard for f_{char}. If f_{char} was greater by an order of magnitude than that in the optimal filter, the desired frequency of false alarms was ensured by a small increase in the discrimination threshold U_t (by 4...20%). Equation (8.19) indicates that it will be not that easy to make up for the increase in f_{char}, and it is not yet sufficient to maximize SNR_1 by making the passband of quasi-optimal filter f_a equal to the passband of the optimal filter. As follows from definition of f_{char} (8.11), one also has to ensure a rapid fall-off of $k_1(f)$ at the high frequencies $f > f_a$: $k \le 1/f^2$ in case of the white noise, and $k \le 1/f^3$ in case of the high-frequency noise.

White noise. A one-section filter does not provide the required high-frequency fall-off: for such filter we have $k \propto 1/f$ (at $f >> f_s$), and $f_{char} \to \infty$ (see (5.91)). Therefore, the quasi-optimal filter in the measuring device must be two-sectional, with $\tau_{eff} = e\tau_f = T_s$ (see Sect. 5.1). Let us evaluate the rms error $\sqrt{\Delta t_m^2}$ for the measuring device based on such quasi-optimal filter using the results of Sect. 5.1.

The first source of error is the loss in the SNR_1. According to (5.69),

$$\mathcal{L} = \frac{(SNR_1)_{qua}}{SNR_{OF}} = 0.965 \ . \tag{8.31}$$

The second source of error is the protraction of the output pulse, which in the quasi-optimal filter is somewhat greater than in the optimal filter (recall Sect. 4.2, 5.3, and equation (5.99)):

$$\left(\frac{\tau_{05\,qua}}{\tau_{05\,opt}}\right)^2 = \left(\frac{1.35T_s}{1.28T_s}\right)^2 = 1.11 \ . \tag{8.32}$$

The third source of error is the change in the characteristic frequency. According to (8.12) and Table 5.1, the frequency f_{char} for the selected quasi-optimal filter is

$$f_{char\,qua} = f_a\sqrt{\frac{\Gamma_2}{\Gamma_1}} = f_a\sqrt{\frac{e^2}{\pi^2}} = \frac{1}{2T_s}\frac{e}{\pi} \ . \tag{8.33}$$

So, the characteristic frequency increases as compared to its value for the optimal filter (8.27) by the factor of

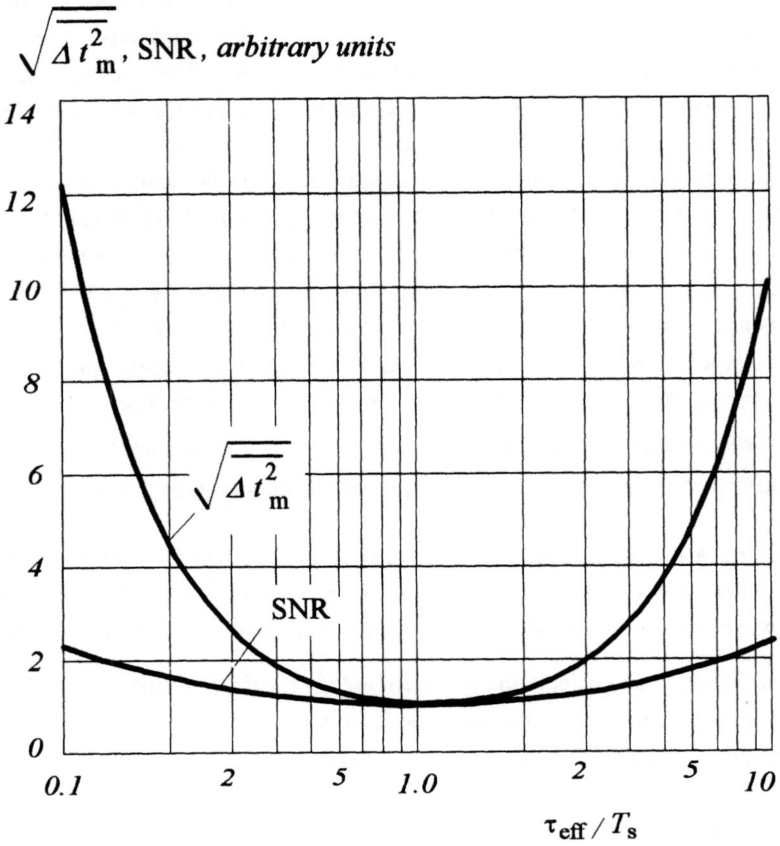

Figure 8.3: Mean square error $\sqrt{\overline{\Delta t_m^2}}$ is more sensitive than SNR to the tuning of filter.

$$\frac{f_{\text{char qua}}}{f_{\text{char opt}}} = \left(\frac{e}{2\pi}\right)\left(\frac{1}{T_s}\right)\left(2\sqrt{3}T_s\right) = \frac{e\sqrt{3}}{\pi} = 1.5 \ . \tag{8.34}$$

All these losses combined increase the rms error for quasi-optimal two-section filter by the factor of

$$\frac{\left(\sqrt{\overline{\Delta t_m^2}}\right)_{\text{qua}}}{\left(\sqrt{\overline{\Delta t_m^2}}\right)_{\text{opt}}} \approx \frac{1}{0.965} \cdot 1.11 \cdot 1.5 = 1.7 \ . \tag{8.35}$$

We see that the main contribution to the error comes from the high characteristic frequency.

Now let us find the tuning curve — the value of $\sqrt{\overline{\Delta t_{\mathrm{m}}^2}}$ as function of the effective time constant of the filter $\tau_{\mathrm{eff}} = e\tau_{\mathrm{f}}$. The necessary approximations for $\mathrm{SNR}_1(\tau_{\mathrm{eff}})$, $T_{05}(\tau_{\mathrm{eff}})$ can be found in Sect. 5.1. Using (5.66), (5.30), we get:

$$\frac{\sqrt{\overline{\Delta t_{\mathrm{m}}^2}}}{\left(\sqrt{\overline{\Delta t_{\mathrm{m}}^2}}\right)_{\mathrm{qua}}} = \frac{\mathrm{SNR}_1}{\mathrm{SNR}_{1\mathrm{qua}}} \frac{T_{05}^2}{T_{05\mathrm{qua}}^2} \frac{f_{\mathrm{a}}}{f_{\mathrm{a\,qua}}} \approx \sqrt{\frac{\dfrac{T_{\mathrm{s}}}{\tau_{\mathrm{eff}}} + \dfrac{\tau_{\mathrm{eff}}}{T_{\mathrm{s}}}}{2} \frac{T_{\mathrm{s}}^2 + 0.81\tau_{\mathrm{eff}}^2}{T_{\mathrm{s}}^2 + 0.81 T_{\mathrm{s}}^2} \frac{T_{\mathrm{s}}}{\tau_{\mathrm{eff}}}}$$

$$= \sqrt{\frac{\dfrac{T_{\mathrm{s}}}{\tau_{\mathrm{eff}}} + \dfrac{\tau_{\mathrm{eff}}}{T_{\mathrm{s}}}}{2} \frac{\dfrac{T_{\mathrm{s}}}{\tau_{\mathrm{eff}}} + 0.81\dfrac{\tau_{\mathrm{eff}}}{T_{\mathrm{s}}}}{1.81}} \,. \tag{8.36}$$

From comparison of (5.66) and (8.36) and from Fig. 8.3 it is clear that the rms error is more sensitive to the tuning of the filter than the value of SNR_1. In the narrow-band filter ($\tau_{\mathrm{eff}} >> T_{\mathrm{s}}$) the error $\sqrt{\overline{\Delta t_{\mathrm{m}}^2}}$ increases because of the protraction of the output signal, and in the broad-band filter ($\tau_{\mathrm{eff}} << T_{\mathrm{s}}$) because of the increase in the characteristic frequency f_{char}.

High-frequency noise. When the high-frequency noise prevails, a two-section filter is not suitable:

$$k \propto \frac{1}{f^2}\,, \quad \Theta \propto f^2\,;$$

therefore, the integral (8.11) diverges, $f_{\mathrm{char}} \to \infty$. As indicated in Sect. 5.1, a finite value of f_{char} can only be secured with an RC-filter of three or more sections. If the effective passband of the three-section filter is made about equal to that of the optimal filter, it is possible to achieve near-optimal values of f_{char} and SNR_1. The loss in the accuracy of measurement with the three-section filter will still be large because of the low value of the derivative $u_{1\mathrm{s}}''(t_{\mathrm{m}})$. This is explained by the fact that the phase-frequency characteristics of the two filters under consideration are different: the output signal of the three-section filter is flat in the neighborhood of the maximum, whereas the maximum of the signal from the output of the optimal filter is sharp, see Fig. 4.9 (approximation (8.13) for the optimal filter does not work in case of the high-frequency noise). As it is, the three-section filter ought to be tuned to the signal band f_{s} rather than to the band f_{high}: the loss in the SNR_1 is overcompensated by the increase in $u_{1\mathrm{s}}''(t_{\mathrm{m}})$. Even with such tuning, however, the three-section filter is inferior to the optimal filter as far as the accuracy of measurement is concerned. More sophisticated filters can be proposed, but the radical solution consists in using APDs in measuring devices (see Sect. 5.3).

Summary. *The optimal measuring device must include the optimal filter which maximizes the signal-to-noise ratio. A differentiating stage is connected to the output of this filter, and the optimal measuring device (its linear channel) is ready. The differentiating stage takes the derivative of the output signal of the optimal filter, and a subsequent comparator pinpoints the instant at which this derivative becomes equal to zero. Such is the optimal algorithm for measuring the time when the signal reaches its maximum. The optimal measuring device is able to discern the displacement of the bell-shaped pulse by about* 1/2 SNR *-th part of its duration.*

When the noise is white, one can replace the optimal filter in real measuring devices by a two-section filter tuned so as to maximize the signal-to-noise ratio, and use a differentiating RC-section with a large time constant. The error of measurement of such quasi-optimal measuring device is greater than that of the optimal device by the factor of about 1.7.

8.2. IR Imager

IR imager is a device for conversion of thermal images, and detects the temperature difference ΔT of heated bodies. In this book we are interested with its temperature sensitivity: the smallest temperature difference the device is able to discern. This threshold temperature difference causes a signal equal to the noise; accordingly, it is commonly referred to as the "noise equivalent temperature difference", abbreviated (and symbolized) NETD or NEΔT [9, 114].

In the previous sections we have always assumed that the optical signal of the photodetector is predetermined. Now predetermined is the temperature difference of the objects viewed by the IR imager. Therefore, our analysis must start with the calculation of optical signal — the flux of thermal radiation received by the photodetector. Before embarking upon this task, the reader is advised to refresh the formulas obtained in Sect. 1.2, 2.3, 3.4 for parameters associated with Planck's distribution: the effective flux of photons n_{eff} in the spectral range of photodetector $\lambda'...\lambda''$, the relative temperature coefficient for this flux $\xi = (1/n_{\text{eff}})(dn_{\text{eff}}/d\mathrm{T})$, the background noise $\sqrt{2q^2 n_{\text{eff}}\mathcal{F}}$, and the background-limited specific detectivity D^*_{BLIP}. We shall need these formulas for calculating the characteristics of the IR imager.

Radiation of a body slightly heated with respect to the background. Consider an idealized model: the background is homogeneous and emits blackbody radiation with the temperature T. The spectral density of this radiation $n(\lambda)$ is given by Planck's law (1.30). The optical signal of the photodetector is the small addition to the background radiation coming from the slightly heated object:

$$\delta n(\lambda)=\frac{dn(\lambda)}{d\mathrm{T}}\Delta\mathrm{T}=\frac{d}{d\mathrm{T}}\frac{2\pi c}{\lambda^4\left(\exp\frac{\mathrm{h}c}{\kappa\mathrm{T}\lambda}-1\right)}\Delta\mathrm{T}$$

$$=\frac{2\pi c\exp\frac{\mathrm{h}c}{\kappa\mathrm{T}\lambda}}{\lambda^4\left(\exp\frac{\mathrm{h}c}{\kappa\mathrm{T}\lambda}-1\right)^2}\frac{\mathrm{h}c}{\kappa\mathrm{T}^2\lambda}\Delta\mathrm{T}=\frac{n(\lambda)}{\lambda}\frac{\mathrm{h}c}{\kappa\mathrm{T}^2}\frac{\exp\frac{\mathrm{h}c}{\kappa\mathrm{T}\lambda}}{\exp\frac{\mathrm{h}c}{\kappa\mathrm{T}\lambda}-1}\Delta\mathrm{T}$$

$$\approx\frac{n(\lambda)}{\lambda}\frac{\mathrm{h}c}{\kappa\mathrm{T}^2}\Delta\mathrm{T}. \tag{8.37}$$

Recall that Wien's approximation holds in the spectral range of interest ($\lambda''<14\mu\text{m}$): the exponential term is much greater than one ($\exp(\mathrm{h}c/\kappa\mathrm{T}\lambda)\gtrsim 33.4$). This allows us to drop "minus one" in the denominator. Naturally, when the temperature difference $\Delta\mathrm{T}$ is small, the added signal radiation $\delta n(\lambda)$ is proportional to this difference; therefore, the registered value of $\delta n(\lambda)$ may be uniquely translated into the magnitude of $\Delta\mathrm{T}$.

For future analysis we shall need the derivative $dn/d\mathrm{T}$. It may be treated as the unit signal coming from the body overheated by $\Delta\mathrm{T}=1\mathrm{K}$. From (8.37) we get:

$$\delta n(\lambda)\big|_{\Delta\mathrm{T}=1\mathrm{K}}=\frac{dn(\lambda)}{d\mathrm{T}}=n(\lambda)\cdot\frac{\mathrm{h}c}{\lambda}\cdot\frac{1}{\kappa\mathrm{T}^2}. \tag{8.38}$$

In this equation we distinguish three terms; each of them will play its role in the intrigue of this section. These three terms correspond to three important features: the unit signal of selective photodetector, which registers monochromatic radiation at wavelength λ, is

— proportional to the background radiation $n(\lambda)$ at this wavelength,
— inversely proportional to the wavelength λ itself,
— inversely proportional to $\kappa\mathrm{T}^2$.

Equation (8.37) indicates that these properties are a consequence of Planck's distribution.

As it is, these features with certain reservations remain valid also for photodetectors with a broad spectral range (if $\lambda''<14\mu\text{m}$). This is because the ideal photodetector is not sensitive to radiation with $\lambda>\lambda''$, whereas at $\lambda<\lambda''$ the function $n(\lambda)$ falls off very rapidly (especially when $\lambda<9\mu\text{m}$). Accordingly, the signal of photodetector is actually determined by the quasi-monochromatic radiation in the neighborhood of $\lambda\approx\lambda''$, and it was for monochromatic radiation that equation (8.38) was obtained.

These arguments can be translated into the language of integrals:

$$\delta n_{\text{eff}} = \int_{\lambda'}^{\lambda''} \delta n(\lambda) d\lambda = \frac{\text{h} c \Delta \text{T}}{\kappa \text{T}^2} \int_{\lambda'}^{\lambda''} \frac{n(\lambda)}{\lambda} d\lambda \approx \frac{\text{h} c \Delta \text{T}}{\kappa \text{T}^2 \lambda''} \int_{\lambda'}^{\lambda''} n(\lambda) d\lambda$$

$$= \frac{\text{h} c}{\kappa \text{T}^2 \lambda''} n_{\text{eff}} \Delta \text{T} . \tag{8.39}$$

Here we have factored out the function $1/\lambda \approx 1/\lambda''$, because it varies slowly compared to $n(\lambda)$ in the neighborhood of its maximum $\lambda = \lambda''$.

In the range $\lambda = 9...14\mu\text{m}$ the rates of variation of λ and $n(\lambda)$ are comparable, and the evaluations of the integral in (8.39) must be made for both the largest ($\lambda = \lambda''$) and the smallest ($\lambda = \lambda'$) values of the denominator:

$$\delta n_{\text{eff}} = \left(\frac{\lambda''}{\lambda''} ... \frac{\lambda''}{\lambda'}\right) \frac{\text{h} c n_{\text{eff}}}{\kappa \text{T}^2 \lambda''} \Delta \text{T} = \left(1 ... \frac{14}{9}\right) \frac{\text{h} c n_{\text{eff}}}{\kappa \text{T}^2 \lambda''} \Delta \text{T} . \tag{8.40}$$

We see that in the spectral range of interest the regularities of (8.38) still hold in general, only at $\lambda'' \to 14\ \mu\text{m}$ we get a certain numerical coefficient greater than one. The exact expression for δn_{eff} can be obtained using the relative temperature coefficient of the radiation flux ξ:

$$\delta n_{\text{eff}} = \frac{dn_{\text{eff}}}{d\text{T}} \Delta \text{T} = n_{\text{eff}} \left(\frac{1}{n_{\text{eff}}} \frac{dn_{\text{eff}}}{d\text{T}}\right) \Delta \text{T} = n_{\text{eff}} \xi \Delta \text{T} , \tag{8.41}$$

where ξ is given by (1.61). In Sect. 1.2 we also quoted three different approximations for ξ: (1.64), (1.65), (1.62). The first of these is sufficient for our current purpose; substituting it into (8.41), we get:

$$\delta n_{\text{eff}} = n_{\text{eff}} \cdot \frac{\text{h} c}{\kappa \text{T}^2 \lambda''} \left(1 + \frac{1}{x''}\right) \cdot \Delta \text{T} , \tag{8.42}$$

and, for the unit signal at $\Delta \text{T} = 1\text{K}$,

$$\delta n_{\text{eff}} \big|_{\Delta \text{T} = 1\text{K}} = \frac{dn_{\text{eff}}}{d\text{T}} = n_{\text{eff}} \cdot \frac{\text{h} c}{\kappa \text{T}^2 \lambda''} \left(1 + \frac{1}{x''}\right) . \tag{8.43}$$

There is practically nothing new for us in this expression: the additional term (compare with (8.38) for the signal of selective photodetector) does not differ much from unity:

$$\lambda'' \le 14\mu\text{m} , \quad x'' = \frac{\text{h} c}{\kappa \text{T} \lambda''} \ge 3.51 , \quad 1 + \frac{1}{x''} \le 1 + \frac{1}{3.51} = 1.28 . \tag{8.44}$$

So, we have calculated the number of quanta emitted from the unit surface area of

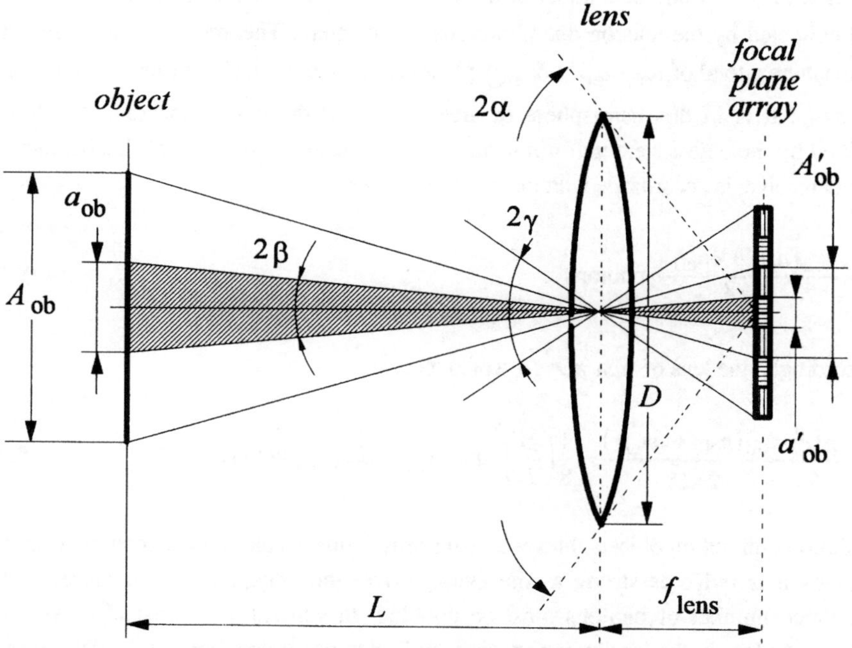

Figure 8.4: To the calculation of optical signal in IR imager.

overheated body. Now we have to find what proportion of these quanta reaches the photosensitive element to cause the photosignal, what is the accompanying noise, and what is the signal-to-noise ratio.

For this we must turn to the optical part of the device.

Signal-to-noise ratio. The lens of the IR imager throws the image of the object on to the focal plane (on the photodetector) with the reduction factor of f_{lens}/L, where f_{lens} is the focal length, and L is the distance between the object and the lens (Fig. 8.4a; here we assume that $L >> f_{lens}$). Given that $L_{ob} \times L_{ob} = A_{ob}$, where L_{ob} is the side of the object, and A_{ob} is its area, for the focal plane image of the object we get

$$L'_{ob} = \frac{f_{lens}}{L} L_{ob} \;, \quad A'_{ob} = L'_{ob} \times L'_{ob} = \left(\frac{f_{lens}}{L} \right)^2 A_{ob} \,. \tag{8.45}$$

Assume that the image of the object covers several elements of the photodetector. One element with the area A_e receives radiation coming from a part of the object area a_{ob}:

$$A_e = a'_{ob} = \left(\frac{f_{lens}}{L} \right)^2 a_{ob} \,. \tag{8.46}$$

It is easy to count the number of those photons which are emitted from the area a_{ob} and collected by the lens on one photosensitive element. The area a_{ob} emits into the hemisphere a total of $a_{ob}(n_{eff} + \delta n_{eff})$ photons. We are interested in the hemisphere of radius L: it is in this hemisphere of area $2\pi L^2$ that the lens is located. If radiation emitted by the object were uniform in all directions, then the unit area of this hemisphere would receive, in accordance with the conservation law,

$$\frac{a_{ob}(n_{eff} + \delta n_{eff})}{2\pi L^2} \text{ photons} . \tag{8.47}$$

Accordingly, the lens of area $\pi D^2/4$ would receive

$$\frac{\pi D^2}{4} \frac{a_{ob}(n_{eff} + \delta n_{eff})}{2\pi L^2} = \frac{1}{8}\left(\frac{D}{L}\right)^2 a_{ob}(n_{eff} + \delta n_{eff}) \text{ photons} . \tag{8.48}$$

Radiation emitted by objects, however, is usually Lambertian: in the direction normal to the lens it is twice as strong as the average over the hemisphere, and therefore the calculated number of photons must be doubled. In general, one must also take into account the loss in the lens by introducing the lens transmission factor τ_{lens} (the loss in the scanning system can be accounted for in a similar way). So, for the number of photons $N_{eff} + \delta N_{eff}$, projected by the lens from the object on to one photosensitive element, we get

$$N_{eff} + \delta N_{eff} = \tau_{lens} \frac{D^2}{4} \frac{a_{ob}}{L^2} (n_{eff} + \delta n_{eff}) . \tag{8.49}$$

Now let us eliminate inconvenient and inconsistent parameters L, a_{ob}. Equation (8.46) will help us to replace them with the known parameters of IR imager A_e, f_{lens}. Carrying out the substitution $a_{ob}/L^2 = A_e/f_{lens}$ in accordance with (8.46), we transform (8.49) into

$$N_{eff} + \delta N_{eff} = \tau_{lens} \frac{D^2}{4} \frac{A_e}{f_{lens}^2} (n_{eff} + \delta n_{eff}) = K_{lens}^2 A_e (n_{eff} + \delta n_{eff}), \tag{8.50}$$

where

$$K_{lens}^2 = \tau_{lens} \left(\frac{D}{2 f_{lens}}\right)^2 . \tag{8.51}$$

We call K_{lens} the coefficient of collection of optical system of IR imager. The term is clear: if it were possible to bring the photosensitive element close to the heated body, the photosensitive element of area A_e would receive the total number of quanta $A_e(n_{eff}+\delta n_{eff})$ emitted by the object from the area $a_{ob} \equiv A_e$. The lens, however, only collects the K_{lens}^2-th part of this number (see (8.51)). The ability of a lens to collect optical radiation is characterized by its relative aperture, f-number, lens speed — that is, such quantities as $D{:}f_{lens}$, $f_{lens}{:}D$, $\pi D^2/f_{lens}^2$, or D^2/f_{lens}^2 [117]. The coefficients K_{lens}, K_{lens}^2 (8.51), introduced above, belong to the family of these parameters.

Observe that formula (8.51) can be used when $D \lesssim f_{lens}$. In this case the lens will project all the quanta that came from the object area a_{ob} on to the photosensitive element A_e, as we assumed when deriving (8.51). If, however, D is much larger than f_{lens}, the image of the object element a'_{ob} is smeared. Then (8.51) does not hold for peripheral rays: they go beyond the limits of A_e. Efficient collection is performed for the central beam, going through the central portion of the lens. For calculation of K_{lens}^2 it is expedient to use the angle 2α at which the lens is seen from the photosensitive element (Fig. 8.4). In geometrical optics, the angle α is known as the back aperture. The ratio $D/2f_{lens}$ in (8.51) is nothing else but the tangent of this angle, $D/2f_{lens} = \tan\alpha$. Rigorous calculations indicate that it is the sine rather than the tangent of the back aperture that must be used in the formula for K_{lens}^2 [5. 117]. Since $\sin\alpha \le \tan\alpha = D/2f_{lens}$, this replacement takes care of the reduced efficiency of collection because of the peripheral rays:

$$K_{lens}^2 = \tau_{lens}\sin\alpha = \tau_{lens}\frac{(D/2f_{lens})^2}{\sqrt{1+(D/2f_{lens})^2}} \,. \tag{8.52}$$

The condition $D \le 2f_{lens}$ is usually satisfied for real lenses, so $\sin\alpha \approx \tan\alpha = D/2f_{lens}$, and we return to (8.51). At $D = f_{lens}$ the value given by (8.51) differs from the exact value (8.52) by 12%, and the error decreases rapidly with the decreasing relative aperture. In the hypothetical case of infinitely large relative aperture, when $D/f_{lens} \to \infty$ and $\tau_{lens} = 1$, from (8.52) we get $K_{lens}^2 = 1$ — in other words, the irradiance of photosensitive element tends to its highest possible value (equal to that at the object), whereas (8.51) gives an unrealistic value of $K_{lens}^2 \to \infty$.

Since we are considering an idealized model, the thermal flux from the object is predominant; all the other thermal fluxes, including those from parts of the device, are negligibly small. The photodetector in this model is also perfect: it operates in the radiation-limited mode and counts all quanta in the range from λ' to λ'' (curve 2 in Fig. 1.8). Therefore, its total current is equal to the background current, and is uniquely determined by the number of quanta N_{eff} (8.50):

$$I_0 = I_{db} = qN_{eff} = qK_{lens}^2 A_e n_{eff} \,. \tag{8.53}$$

The additional number of quanta δN_{eff} because of overheating of the object causes the signal photocurrent I_s:

$$I_s = q\delta N_{\text{eff}} = qK_{\text{lens}}^2 A_e \delta n_{\text{eff}} = K_{\text{lens}}^2 A_e\left(q\frac{dn_{\text{eff}}}{dT}\Delta T\right). \tag{8.54}$$

Further on the signal is processed as described in Sect. 7.1, 7.2. In the IR imager we deal with the classical case of the detection of rectangular signal (against the background of white noise. It is clear why the signal may be considered rectangular: when a non-scanning matrix is used, the exposure of each element remains the same for the entire frame time T_f. Even when the image is scanned with a linear array or one-element photodetector, the signal between two interrogations is constant, because the element displacement l is smaller than the linear size of the image. The interval between interrogations is equal to $T_f M_e / \mathcal{M}$, where $\mathcal{M}$ is the number of pixels (image resolution elements), and M_e is the number of photosensitive elements of the photodetector (see (7.63). This interval may be taken for the signal length T_s. The ideal filter for this arrangement is the integrator with the integration time $T_i = T_s = T_f M_e / \mathcal{M}$ (Sect. 7.2). Accumulated over the time T_i are the charges of the signal $Q_s = I_s T_i$ and the background $Q_0 = I_0 T_i$. The latter is responsible for the noise, and so the sought-for signal-to-noise ratio, with due account for (7.7), (8.53), (8.54), is

$$\text{SNR} = \frac{Q_s}{Q_n} = \frac{I_s T_i}{\sqrt{qI_0 T_i}} = \frac{K_{\text{lens}}^2 A_e T_i q \dfrac{dn_{\text{eff}}}{dT}\Delta T}{\sqrt{K_{\text{lens}}^2 A_e T_i q^2 n_{\text{eff}} \mathcal{F}}} = K_{\text{lens}}\sqrt{2A_e T_i}\,\frac{\dfrac{dn_{\text{eff}}}{dT}}{\sqrt{2n_{\text{eff}}\mathcal{F}}}\Delta T. \tag{8.55}$$

Noise equivalent temperature difference for ideal IR imager. From the last formula for the signal-to-noise ratio we get the following expression for the noise equivalent temperature difference for the ideal IR imager [118]:

$$\text{NETD}_{\text{BLIP}} = \frac{\Delta T}{\text{SNR}} = \frac{1}{K_{\text{lens}}} \cdot \frac{1}{\sqrt{A_e (2T_i)}} \cdot \frac{\sqrt{2n_{\text{eff}}\mathcal{F}}}{\dfrac{dn_{\text{eff}}}{dT}}. \tag{8.56}$$

Subscript BLIP indicates that this equation is written for the background-limited infrared photodetector, when $\text{NETD} = \text{NETD}_{\text{BLIP}}$.

Now we are through with our homework in algebra, and may take pleasure in the term-by term presentation of the resulting expression.

The first term, $1/K_{\text{lens}} \sim 2f_{\text{lens}}/D$ (8.52). Since the original optical signal for the IR imager is defined at the object rather than at the photodetector, the chain of transformations of optical input into electrical output now involves the lens, which is represented in (8.56) by the relative aperture D/f_{lens}. The larger it is, the smaller is the temperature difference the IR imager is able to discern.

The second term, $1/\sqrt{A_e(2T_i)}$ characterizes the IR imager as a three-dimensional filter-integrator: in space (over the coordinates x and y) and in time (along the t-axis). Accordingly, the formula for $NETD_{BLIP}$ involves the integration intervals in space A_e and in time T_i (more precisely, the time-domain integration is represented by the effective frequency band of filter-integrator $1/2T_i$; Table 5.1).

As a matter of fact, the first term may also be regarded as an integration characteristic. Namely, combining the first term expressed via $\sin\alpha$ (8.52) with the second term, we get:

$$\frac{1}{K_{lens}} \cdot \frac{1}{\sqrt{A_e(2T_i)}} = \frac{1}{\sqrt{\sin^2\alpha \cdot A_e \cdot 2T_i}} . \tag{8.57}$$

In other words, the integration intervals A_e and T_i are supplemented by the angle of view $\pm\alpha$: the photosensitive element integrates the flux within this angle.

The second term $1/\sqrt{A_e(2T_i)}$ may be modified to taste by expressing the integration time via the frame time $T_i = T_f M_e / \mathcal{M}$ (7.57), the PSE area via the frame area $A_e = A/\mathcal{M}$ (7.59) (when the matrix photodetector occupies the entire frame area, A is also the area of the photodetector), and by introducing the frame frequency $f_f = 1/T_f$:

$$\frac{1}{\sqrt{2T_i A_e}} = \sqrt{\frac{\mathcal{M}}{2T_f M_e A_e}} = \frac{\mathcal{M}}{\sqrt{2T_f M_e A}} = \mathcal{M}\sqrt{\frac{f_f}{2M_e A}} . \tag{8.58}$$

These expressions indicate that the lowest detectable temperature difference (NETD) increases (deteriorates) with increasing resolution and frame frequency and decreasing PSE area, and decreases (improves) with increasing number of elements (when $M_e \to \mathcal{M}$). All these trends find a straightforward explanation in the dependence of the integration of thermal radiation over time and space on these parameters.

At the limit of resolution, when the image area of overheated object is equal to the area of one PSE, the energy may be split between adjacent elements and frames (Sect. 7.1, 7.2), and the value of NETD may be up to four times as large (that is, worse) as that given by (8.56), (8.58). In the pyrometer mode, when the task consists in measuring the temperature of the object with the accuracy of about NETD, while the geometry and the area of the object A_{ob} are known in advance, the signal may be accumulated over all illuminated elements. Then A_e in (8.56) will be replaced with the object image area a'_{ob}.

Observe also that (8.56) can be used for calculating the error of temperature measurement with the radiation pyrometer based on one-element photodetector. In this case A_e is the area of photosensitive element.

The designer of IR imager may find it helpful to know the linkage between NETD and angular resolution (the angle at which one element sees the object)

$2\beta = l/f_{\text{lens}} = \sqrt{A_e}/f_{\text{lens}}$, and the entire field of view of the IR imager $2\gamma = \sqrt{A}/f_{\text{lens}}$ (Fig. 8.4). This linkage can be found by combining once again the two first terms in (8.56), and setting $K_{\text{lens}} = D/2f_{\text{lens}}$:

$$\frac{2f_{\text{lens}}}{D}\frac{1}{\sqrt{A_e(2T_i)}} = \frac{1}{\beta D}\frac{1}{\sqrt{2T_i}}\ , \tag{8.59}$$

$$\frac{2f_{\text{lens}}}{D}\mathcal{M}\sqrt{\frac{f_f}{2M_e A}} = \frac{1}{\gamma D}\mathcal{M}\sqrt{\frac{f_f}{2M_e}}\ . \tag{8.60}$$

These and similar manipulations allow one to analyze the dependence of NETD on the parameters (area, field of view, number of cells, etc.) which may be collectively called the parameters of IR imager as a cellular antenna for receiving electromagnetic radiation in the optical (thermal IR) range.

***The third term in* (8.56).** Since the first two terms in (8.56) characterize the ability of IR imager to collect the thermal flux coming from the object, integrating it over the field of view of PSE 2α, the PSE area A_e, and the time T_i, the meaning of the third term is obvious. Namely, it defines the noise equivalent temperature difference for a hypothetical IR imager with the unit PSE area $A_e = 1\,\text{cm}^2$, with the unit frequency band of the integrator $1/2T_i = 1\,\text{s}^{-1}$, and with the unit value of K_{lens} (which means that the photodetector receives thermal flux from the hemisphere 2π steradian, $\alpha = \pi/2$, $K_{\text{lens}} = \sin\pi/2 = 1$ if $\tau_{\text{lens}} = 1$). In other words, it is the specific noise equivalent temperature difference in BLIP mode:

$$\text{NETD}^*_{\text{BLIP}} = \frac{\text{NETD}_{\text{BLIP}}}{\dfrac{1}{K_{\text{lens}}\sqrt{A_e(2T_i)}}} = \frac{\sqrt{2n_{\text{eff}}\mathcal{F}}}{\dfrac{dn_{\text{eff}}}{d\text{T}}}\ . \tag{8.61}$$

This term is of special importance for us because it defines the connection between the temperature sensitivity of IR imager $\text{NETD}^*_{\text{BLIP}}$ with the threshold properties of the PDA.

As indicated above, the PDA incorporated in the ideal IR imager is the optimal detector for rectangular signal. The numerator $\sqrt{2n_{\text{eff}}\mathcal{F}}$ in (8.61) is the spectral noise density of such PDA, and the denominator $dn_{\text{eff}}/d\text{T}$ is the unit signal amplitude (the noise and the signal are expressed as the effective numbers of quanta). The threshold properties of this PDA are characterized by the specific background-limited detectivity D^*_{BLIP} (Sect. 3.4). Let us try to find the linkage between $\text{NETD}^*_{\text{BLIP}}$ and D^*_{BLIP}.

Fundamental equation of IR imaging. Consider first the spectral range $\lambda \leq 14\,\mu\text{m}$, when we may assume that the background noise is Poissonic, and set $\mathcal{F} = 1$

in (8.61) for $\mathrm{NETD}^*_{\mathrm{BLIP}}$. Multiplying the numerator and the denominator of (8.61) by $\sqrt{2n_{\mathrm{eff}}}$, we get

$$\mathrm{NETD}^*_{\mathrm{BLIP}} = \frac{\sqrt{2n_{\mathrm{eff}}}}{\dfrac{dn_{\mathrm{eff}}}{d\mathrm{T}}} = 2\frac{n_{\mathrm{eff}}}{\dfrac{dn_{\mathrm{eff}}}{d\mathrm{T}}}\frac{1}{\sqrt{2n_{\mathrm{eff}}}} \ . \tag{8.62}$$

Now we may take advantage of the expression for $dn_{\mathrm{eff}}/d\mathrm{T}$ obtained in the beginning of this section. For the sake of clarity, we start with the selective photodetector (8.38), setting $n_{\mathrm{eff}} = n(\lambda)$, $\lambda = \lambda''$:

$$\mathrm{NETD}^*_{\mathrm{BLIP}} = 2\frac{n_{\mathrm{eff}}}{n_{\mathrm{eff}}\cdot\dfrac{\mathrm{h}c}{\lambda''}\cdot\dfrac{1}{\kappa\mathrm{T}^2}}\frac{1}{\sqrt{2n_{\mathrm{eff}}}} = 2\kappa\mathrm{T}^2\cdot\left(\frac{\lambda''}{\mathrm{h}c}\frac{1}{\sqrt{2n_{\mathrm{eff}}}}\right) \ . \tag{8.63}$$

The term in parentheses in (8.63) is the specific background-limited detectivity (3.116),

$$\left(\frac{\lambda''}{\mathrm{h}c}\frac{1}{\sqrt{2n_{\mathrm{eff}}}}\right) = \frac{q\lambda''}{\mathrm{h}c}\frac{1}{\sqrt{2q(qn_{\mathrm{eff}})}} = \frac{S}{\sqrt{2qj_{\mathrm{db}}}} = D^*_{\mathrm{BLIP}} \ . \tag{8.64}$$

The first term in (8.63) is a wavelength-independent constant, which we denote by

$$\theta = 2\kappa\mathrm{T}^2 \ . \tag{8.65}$$

The derivative $dn_{\mathrm{eff}}/d\mathrm{T}$ was represented in the beginning of this section as the product of three terms; now we see the reason why. The term n_{eff} simply cancels out. The term $\lambda/\mathrm{h}c$, supplemented by the charge q, is the sensitivity, and is delegated to the detectivity. And it is the third term $\kappa\mathrm{T}^2$ that defines the value of the constant. In the photodetectors for a broad spectral range this term is slightly modified: it is supplemented by the binomial $(1+1/x'')$ (recall (8.38), (8.43). The coefficient θ must be corrected accordingly:

$$\lambda < 14\mu\mathrm{m} \ , \quad x'' >> 1 \ ,$$

$$\theta \approx \frac{2\kappa\mathrm{T}^2}{1+\dfrac{1}{x''}} \approx 2\kappa\mathrm{T}^2\left(1-\frac{1}{x''}\right) = 2\kappa\mathrm{T}^2\left[1-\frac{1}{3.51}\left(\frac{\lambda}{14}\right)\right] \approx 2\kappa\mathrm{T}^2\left(1-0.3\frac{\lambda}{14}\right) \ . \tag{8.66}$$

Here we have noted that $x'' >> 1$, and therefore $1/(1+1/x'') \approx 1-1/x''$. We have also used the value of $x'' = 3.51$ at $\lambda = 14\,\mu\mathrm{m}$ as calculated in Sect. 1.2. Since $x'' \propto 1/\lambda$, for arbitrary wavelength we have $x'' = 3.51(\lambda/14)$.

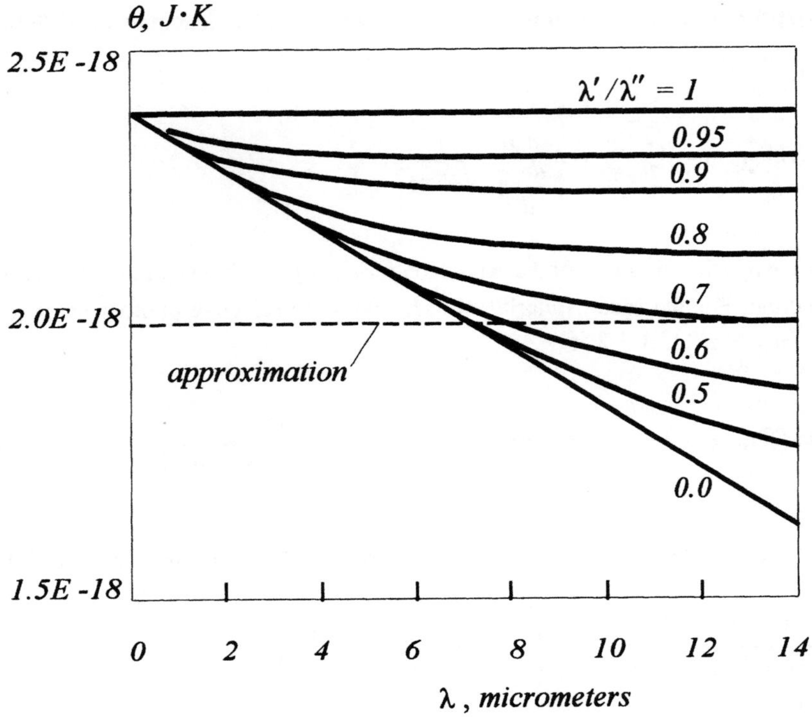

Figure 8.5: Coefficient θ weakly depends on the long-wave (λ'') and short-wave (λ') limits of spectral band of photodetector.

Expression (8.66) indicates that the coefficient θ depends on the wavelength λ, but this dependence is weak: the value of θ falls below its nominal value of $2\kappa T^2$ by 6% at $\lambda'' = 3\,\mu m$, and by 30% at $\lambda'' = 14\,\mu m$. Because of this, for most practical purposes this coefficient may be assumed to be constant and equal to its average value in the range 0...14 μm:

$$\theta \approx \overline{\theta} = 2\kappa T^2 \overline{\left(1 - 0.3\frac{\lambda}{14}\right)} \approx 2\kappa T^2 \cdot 0.85 = 2 \cdot 1.38 \cdot 10^{-23} \cdot (293)^2 \cdot 0.85$$

$$= 2 \cdot 10^{-18} \left(\frac{T}{293}\right)^2 \text{J} \cdot \text{K}. \tag{8.67}$$

If necessary, it is possible to calculate the exact value of θ for the ideal photodetector whose spectral range is limited both on the right (λ'') and on the left (λ'). The guidelines for calculating $n_{\text{eff}}(\lambda', \lambda'')$ were set forth earlier in Sect. 1.2. This technique was used for calculating $\theta(\lambda', \lambda'')$, Fig. 8.5. Naturally, when $\lambda' \to 0$ the spectral range of the photodetector is broad, and the value of θ corresponds to the straight line (8.66). When $\lambda' \to \lambda''$, the photodetector is selective, and θ tends to its constant nominal value of $2\kappa T^2$ (8.65).

So, we have coped with our task, and found the linkage between $\mathrm{NETD}^*_{\mathrm{BLIP}}$ and D^*_{BLIP}. According to (8.63), (8.64), (8.66), we get

$$\boxed{\mathrm{NETD}^*_{\mathrm{BLIP}} = \theta D^*_{\mathrm{BLIP}}} \tag{8.68}$$

We have derived yet another law of the detection theory, the IR imaging equation (8.65) – (8.68) [118]:

The specific noise equivalent temperature difference of ideal IR imager $\mathrm{NETD}^*_{\mathrm{BLIP}}$ *is equal to the specific background-limited detectivity of PDA* D^*_{BLIP} *times the coefficient* θ. *The coefficient* θ *is practically independent of the spectral range: its value is constant and equal to* $2\kappa \mathrm{T}^2$ *for a selective photodetector, and its decrease does not exceed 30% of the value for real broad spectral range photodetector.*

The restrictions on $\mathrm{NETD}^*_{\mathrm{BLIP}}$, D^*_{BLIP} are imposed neither by the IR imager nor by the PDA: they are germane to the object, being related to the fluctuations of thermal radiation emitted by the object. Trying to improve the sensitivity above $\mathrm{NETD}^*_{\mathrm{BLIP}}$ is trying to build a perpetuum mobile (this is what we said in Sect. 3.4 about trying to improve on the value of D^*_{BLIP} for the PDA). According to the IR imaging relation, the ultimate temperature sensitivity $\mathrm{NETD}^*_{\mathrm{BLIP}}$ depends only on two parameters: the selected spectral range $\lambda'...\lambda''$, and the background temperature T. Let us consider these two factors separately.

Specific noise equivalent temperature difference and spectral range. From the IR imaging relation (8.65) – (8.68) it follows that the function $\mathrm{NETD}^*_{\mathrm{BLIP}}(\lambda'')$ almost imitates the function $D^*_{\mathrm{BLIP}}(\lambda'')$. As the spectral range of the photodetector is extended from 3μm to 9...10 μm, the temperature sensitivity is improved dramatically (the value of D^*_{BLIP} is reduced by the factor of 30; see (8.68), (3.116), and Table 3.3). This is explained by the rapid increase in the number of effective quanta n_{eff} registered by the photodetector. The unit signal increases in proportion with this number, $dn_{\mathrm{eff}}/d\mathrm{T} \propto n_{\mathrm{eff}}$. The noise increases as $\sqrt{n_{\mathrm{eff}}}$, and so the signal-to-noise ratio is proportional to $\sqrt{n_{\mathrm{eff}}}$.

As the spectral range is extended further (to 10...14 μm), the function $n_{\mathrm{eff}}(\lambda'')$ and hence the function $\mathrm{NETD}^*_{\mathrm{BLIP}}(\lambda'')$ become not so steep. From the qualitative estimate $\mathrm{NETD}^*_{\mathrm{BLIP}} \propto 1/\sqrt{n_{\mathrm{eff}}}$ one might deduce that the ultimate temperature sensitivity is achieved when the photodetector registers all the photons of the blackbody radiation — that is, when $\lambda'' \to \infty$. Then, however, the IR imaging relation (8.68) no longer holds, because its derivation was based on equation (8.43) for $dn_{\text{eff}}/d\text{T}$, which is only valid when $\lambda'' \le 14\mu\text{m}$. Now we must take the values of n_{eff}, $dn_{\text{eff}}/d\text{T}$ in accordance with the Stefan – Boltzmann law (1.48), and take into account the over-Poissonic noise associated with the grouping of photons, setting $\mathfrak{F} = 1.37$ (Sect. 2.3). On the strength of these arguments, from (8.62) we get:

$$\text{NETD}^*_{\text{BLIP}} = \frac{\sqrt{2n_{\text{eff}}\mathfrak{F}}}{\dfrac{dn_{\text{eff}}}{d\text{T}}} = \frac{\sqrt{2\sigma'\text{T}^3\mathfrak{F}}}{3\sigma'\text{T}^2} = \frac{1}{3}\sqrt{\frac{2\mathfrak{F}}{\sigma'\text{T}}}$$

$$= \frac{1}{3}\sqrt{\frac{2\cdot 1.37}{1.59\cdot 10^{11}\cdot 293}} = 8.3\cdot 10^{-8}\sqrt{\frac{293}{\text{T}}}\ \text{K}\cdot\text{cm}\cdot\text{s}^{1/2}\ . \tag{8.69}$$

The two expressions for $\text{NETD}^*_{\text{BLIP}}$ (8.68), (8.69) may be regarded as two approximations which describe the function $\text{NETD}^*_{\text{BLIP}}(\lambda'')$ each in its own range of λ'': the first holds for $\lambda'' \le 14\mu\text{m}$, and the second for $\lambda'' > 14\mu\text{m}$ $(\text{T} = 293\text{K})$. Of course, we can also write the more general expression for $\text{NETD}^*_{\text{BLIP}}$, which holds for $\lambda'' = (0\ldots\infty)$ and follows from (8.62):

$$\text{NETD}^*_{\text{BLIP}} = \frac{\sqrt{2n_{\text{eff}}\mathfrak{F}}}{\dfrac{dn_{\text{eff}}}{d\text{T}}} = 2\sqrt{\mathfrak{F}}\,\frac{\sqrt{n_{\text{eff}}}}{\dfrac{dn_{\text{eff}}}{d\text{T}}} = \frac{2\sqrt{\mathfrak{F}}}{\xi\sqrt{n_{\text{eff}}}}\ . \tag{8.70}$$

Here we have used the definition of the relative temperature coefficient for the flux of quanta $\xi = (1/n_{\text{eff}})(dn_{\text{eff}}/d\text{T})$ (1.57). For all three quantities $\mathfrak{F}(\lambda'')$, $\xi(\lambda'')$, $n_{\text{eff}}(\lambda'')$ we have already given both the exact integral expressions (2.108), (1.61), (1.31) and the approximations (2.108), (1.65), (1.47); recall Sect. 1.2, 2.3.

Equation (8.70) reveals an interesting feature: strictly speaking, the function $\text{NETD}^*_{\text{BLIP}}(\lambda'')$ is nonmonotonous, and exhibits a minimum. This minimum occurs at 28 μm and is rather shallow: 9% less than the ultimate value given by (8.69) when $\lambda'' \to \infty$:

$$\lambda'' = 28\left(\frac{293}{\text{T}}\right)\mu\text{m}\ ;\quad \min\left[\text{NETD}^*_{\text{BLIP}}\right] = 7.3\cdot 10^{-8}\sqrt{\frac{293}{\text{T}}}\ \text{K}\cdot\text{cm}\cdot\text{s}^{1/2}\ . \tag{8.71}$$

This is the last relation of the detection theory in this book. It defines the absolute limit of the specific noise equivalent temperature difference of IR imager. *The absolute limit of the specific noise equivalent temperature difference is a function of only one physical quantity, the background temperature* [118].

Equation (8.71) does not mean that the designer must aspire to conquer the far infrared region at 28μm. As indicated above, at $\lambda'' > 10\mu\text{m}$ the dependence of $\text{NETD}^*_{\text{BLIP}}$ on λ''

becomes weak, and the values of $\mathrm{NETD}^*_{\mathrm{BLIP}}$ are not far from the minimum. It is quite sufficient to use photodetectors of spectral range V with the long-wave limit at 10, 12 or 14 μm. Numerical calculations show that IR imagers equipped with such photodetectors fall short of the absolute sensitivity limit by, respectively, 35, 20 and 12%.

Since the temperature sensitivity improves as the spectral range is extended (which is especially noticeable when we go from 2...5 to 8...14 μm), the engineers strive to extend the long-wave limit, to master the 8...14 μm range — the more so since this range is less hampered by atmospheric absorption. In practice, however, the extension of the long-wave limit meets with a number of problems: deeper cooling is required, the technology of photodetectors becomes more sophisticated. Signal processing also becomes more complicated, because of the drastic increase in the thermal background current I_{db}, see Sect. 7.1, 7.2. The limitations caused by I_{db} will be discussed in greater detail shortly in connection with the analysis of NETD in real IR imagers. This explains why the IR imagers employ photodetectors for the range 8...14 μm, as well as 2...5 μm.

Dependence of $\mathrm{NETD}^*_{\mathrm{BLIP}}$ on background temperature. This dependence is most readily deduced from the last expression for $\mathrm{NETD}^*_{\mathrm{BLIP}}$ (8.70). Since the range of variation of the noise factor ($\mathcal{F} = 1...1.37$) is negligibly narrow compared with the range of variation of n_{eff}, we may simplify our calculations by assuming that $d\mathcal{F}/d\mathrm{T} = 0$. Taking first the logarithm of (8.70)

$$\ln \mathrm{NETD}^*_{\mathrm{BLIP}} = -\ln \xi - \frac{1}{2}\ln n_{\mathrm{eff}} + \ln 2\sqrt{\mathcal{F}} \;, \tag{8.72}$$

and then the derivative with respect to temperature, we find the desired expression for the temperature coefficient of $\mathrm{NETD}^*_{\mathrm{BLIP}}$:

$$\xi_{\Delta \mathrm{T}} = \frac{1}{\mathrm{NETD}^*_{\mathrm{BLIP}}}\frac{d\,\mathrm{NETD}^*_{\mathrm{BLIP}}}{d\mathrm{T}} = -\frac{1}{\xi}\frac{d\xi}{d\mathrm{T}} - \frac{1}{2n_{\mathrm{eff}}}\frac{dn_{\mathrm{eff}}}{d\mathrm{T}} = -\frac{1}{\xi}\frac{d\xi}{d\mathrm{T}} - \frac{\xi}{2}. \tag{8.73}$$

We see that the relative temperature coefficient of $\mathrm{NETD}^*_{\mathrm{BLIP}}$ is uniquely defined by the relative temperature coefficient for the flux of photons ξ and its derivative. Making use of the simplified approximation for $\xi(\lambda'')$ (1.61) and assuming the correction ζ to be independent of the temperature, we find the derivative

$$\frac{d\xi}{d\mathrm{T}} = \frac{d}{d\mathrm{T}}\left(\frac{x''}{\mathrm{T}} + \frac{1+\zeta}{\mathrm{T}}\right) = \frac{d}{d\mathrm{T}}\left(\frac{\mathrm{h}c}{\kappa \mathrm{T}^2 \lambda''} + \frac{1+\zeta}{\mathrm{T}}\right)$$
$$= -\frac{2\mathrm{h}c}{\kappa \mathrm{T}^3 \lambda''} - \frac{1+\zeta}{\mathrm{T}^2} = -\frac{2x''+1+\zeta}{\mathrm{T}^2}\;. \tag{8.74}$$

Substituting this derivative and the value of ξ itself into (8.73), we get:

$$\xi_{\Delta T} = \frac{T}{x''+1+\zeta}\frac{2x''+1+\zeta}{T^2} - \frac{x''+1+\zeta}{2T}$$

$$= -\frac{1}{2T}\left(x''+\zeta-1-\frac{2x''}{x''+1+\zeta}\right) . \qquad (8.75)$$

Let us now obtain numerical evaluations for $\xi_{\Delta T}$. Up to $\lambda''=5$ μm we may assume that $x''>>1$, $\zeta \approx 0$ (1.63), and the expression for $\xi_{\Delta T}$ is simplified:

$$\lambda''=5\ \mu\text{m} , \quad \xi_{\Delta T} \approx -\frac{x''-3}{2T} = -\frac{9.83-3}{2\cdot 293} = -1.2\,\%/\text{K} . \qquad (8.76)$$

In the range 8...14 μm one must use the general expression (8.75):

$\lambda''=14$ μm ,

$$\xi_{\Delta T} = -\frac{1}{2\cdot 293}\left(3.51+0.48-1-\frac{2\cdot 3.51}{3.51+1+0.48}\right) = -0.27\,\%/\text{K} . \qquad (8.77)$$

Finally, for the absolute specific NETD limit and for its neighborhood ($\lambda''\geq 28$ μm, $x''\to 0$, $\zeta=2$) we get:

$$\xi_{\Delta T} = -\frac{1}{2T} = -\frac{1}{2\cdot 293} = -0.17\,\%/\text{K} . \qquad (8.78)$$

Correction ζ in all three formulas is calculated according to (1.63).

The last formula $\xi_{\Delta T} = -1/2T$ can be directly derived from the expression for the absolute limit of temperature sensitivity (8.71). Taking the logarithm of both sides of this equation,

$$\ln \min\left[\text{NETD}^*_{\text{BLIP}}\right] = \ln 7.3\cdot 10^{-8}\sqrt{293} - \frac{1}{2}\ln T , \qquad (8.79)$$

and then its derivative, we find

$$\xi_{\Delta T} = \frac{d}{dT}\ln \min\left[\text{NETD}^*_{\text{BLIP}}\right] = \frac{1}{\text{NETD}^*_{\text{BLIP}}}\frac{d\,\text{NETD}^*_{\text{BLIP}}}{dT} = -\frac{1}{2T} . \qquad (8.80)$$

The above analysis leads us to two conclusions.

1. The lower the long-wave limit of the photodetector, the stronger the dependence of $NETD^*_{BLIP}$ on the background temperature. This is because n_{eff} grows faster with increasing λ'' at low values of λ''.

2. Coefficient $\xi_{\Delta T}$ for all spectral ranges is negative, and the specific noise equivalent temperature difference always improves (the value of $NETD^*_{BLIP}$ decreases) as the background temperature increases. For PDA engineers this is a kind of the "IR imaging paradox": they grew accustomed to the fact that the parameters of PDA deteriorate as the temperature T increases: the background current and the noise increase, the integrator of the matrix cell is overcharged. The improvement of the sensitivity of IR imager with increasing T is explained by the fact that thermal radiation for the IR imager is signal rather than noise — more precisely, the unit signal is the increment of the number of quanta as the temperature is raised by 1K, and this increment is the larger, the higher the temperature of the body.

Noise equivalent temperature difference for IR imager with real PDAs. The temperature sensitivity of real IR imagers under real conditions depends on a large number of factors: the quality of the optical system, scanning and imaging systems, atmospheric absorption and scattering, the contrast of the object. In this book we are concerned with the deterioration of the noise equivalent temperature difference because of imperfections of the PDA.

The main difficulty in the calculation of $NETD^*_{BLIP}$ for real photodetectors arises because the quantum yield η is a function of the wavelength λ. As a consequence, the exact calculation of n_{eff} calls for numerical integration of the density of photoactive pairs $\eta(\lambda)n(\lambda)$. In Sect. 1.2 we demonstrated, however, that this numerical integration can be avoided, and the value of n_{eff} for a real photodetector (having a typical spectral curve, like curve 3 in Fig. 1.8) may be found from analytical expressions for the ideal photodetector (1.47), (1.53). Since both θ and $D^*_{BLIP}(\lambda'')$ are defined via the value of n_{eff}, it is possible to avoid this complication in the calculation of $NETD^*_{BLIP}$ as well. The calculation procedure is as follows.

1. Using the relative spectral response of the real photodetector, one determines the short-wave (λ') and the long-wave (λ'') sensitivity limits at half-maximum. Then the analytical expressions for the ideal photodetector (1.47) – (1.53) are employed for calculating $n_{eff}(\lambda',\lambda'')$.

2. The next step consists in calculating the specific detectivity. In case of PDA this parameter is usually defined at the sensitivity maximum, $D^*_{BLIP}(\lambda_m)$. The IR imaging equation, however, formally involves the quantity $D^*_{BLIP}(\lambda'')$, and $\lambda_m < \lambda''$. Therefore,

into (8.68) we must substitute $D^*_{BLIP}(\lambda'')=(\lambda''/\lambda_m)D^*_{BLIP}(\lambda_m)$, calculating $D^*_{BLIP}(\lambda_m)$ by the following formula:

$$D^*_{BLIP}(\lambda_m)=\frac{\lambda_m}{hc\sqrt{2n_{eff}(\lambda',\lambda'')}} \; . \tag{8.81}$$

3. The calculation of θ follows the same guidelines as before: one may use either its average value (8.67), or its more precise value calculated by (8.66) or taken from Fig. 8.4.

4. Of course, the deteriorated sensitivity of the PDA, the additional noise reduce the signal-to-noise ratio at the PDA output (given that the input signal is the same). The deterioration of SNR entails a proportional deterioration of both the specific detectivity of PDA D^*, and the noise equivalent temperature difference of IR imager NETD*. In case of the detectivity, however, "*deterioration*" means a *decrease* in its numerical value, and in case of the noise equivalent temperature difference "*deterioration*" means that its numerical value *increases*. The fact that D^* varies in sympathy with SNR, and NETD* in antipathy, follows directly from definitions of these parameters (3.117) and (8.56):

$$D^* \propto \frac{\mathrm{SNR}}{E_{sA}} \; , \quad \mathrm{NETD}^* \propto \frac{\Delta T}{\mathrm{SNR}} \; . \tag{8.82}$$

Thus, if the specific detectivity D^* of real PDA is worse (less) than that of the ideal one by the factor of D^*_{BLIP}/D^*, the specific noise equivalent temperature difference NETD* of IR imager based on such PDA will be worse (greater) by the same factor of D^*_{BLIP}/D^*:

$$\mathrm{NETD}^* = \mathrm{NETD}^*_{BLIP}\frac{D^*_{BLIP}(\lambda_m)}{D^*(\lambda_m)} = \theta\frac{\lambda''}{\lambda_m}D^*_{BLIP}(\lambda_m)\frac{D^*_{BLIP}(\lambda_m)}{D^*(\lambda_m)}$$

$$= \theta\frac{\lambda''}{\lambda_m}\frac{\left[D^*_{BLIP}(\lambda_m)\right]^2}{D^*(\lambda_m)} \; . \tag{8.83}$$

The value of NETD* depends on the ability of each separate channel of the PDA to detect weak signals. As indicated in the preceding sections, the specific detectivity of modern PDAs is close to the background limit: the loss is as small as 10...30%. Photoresistor is always inferior to the ideal photodetector: because of the generation-recombination noise, we have $D^*_{BLIP}/D^* = \sqrt{2}$ even if $\eta = 1$ (Sect. 2.3).

Equation (8.83) may also include coefficient $\mathcal{L}$ (the loss in the signal-to-noise ratio due to non-optimal filtering). For example, IR imagers, in which the scene is scanned by a single-element photodetector, use an amplifier instead of the optimal filter (integrator). Depending on the passband of the amplifier, the typical loss coefficient is $\mathcal{L} \approx 0.8...0.95$ (Sect. 5.1). We see that the deterioration of the specific noise equivalent temperature difference of real IR imagers because of non-perfect PDAs is within acceptable limits.

The designers of IR imagers are, however, interested not only in the performance of one single channel of the PDA, not in the value of NETD* which characterizes the quality of this channel, but rather in the overall noise equivalent temperature difference of the IR imager. This means that we must go over from the specific value of NETD* to the integral value of NETD, taking into account all three factors in (8.56), including the integration time T_i, and the number of elements (channels) M_e. With $D^*/D^*_{BLIP} = \sqrt{\eta}$, according to (8.56), (8.58), (8.83), we get:

$$\mathrm{NETD} = \frac{1}{K_{lens}\sqrt{2T_i A_e}}\mathrm{NETD}^* = \frac{1}{K_{lens}}\sqrt{\frac{\mathcal{M}}{2T_f M_e A_e}}\frac{\theta D^*_{BLIP}}{\sqrt{\eta}} \propto \frac{1}{\sqrt{T_f M_e \eta}} \, . \tag{8.84}$$

There is nothing to surprise us in this expression: once again we encounter the product $T_f M_e \eta$ which is definitive for the detective properties of matrices (see Sect. 7.2). To make NETD as small as possible, we must maximize the product $T_f M_e \eta$. This means that the matrix PDA must ensure perfect space-time integration without loss of photons ($\eta = 1$), over the entire available frame time ($T_i = T_f$), and over the entire frame pixels ($M_e = \mathcal{M}$). So, equation (8.84) has once again confirmed our main thesis: to cope with the task set before the optoelectronic system in the best possible way, we must first of all accomplish optimal detection of the low-level optical signal. In the language of football commentators, if the "individual technique" of each separate element of the PDA is good, then the element stands up to the requirements of optimal detection of low-level optical signal, and the specific values of NETD* for ideal and real IR imagers are close enough to each other. The "team play" of the PDA, however, is not as good as we would like it to be. As shown in Sect. 7.2, it has not yet been possible to implement all the desired conditions at once ($\eta = 1$, $T_i = T_f$, $M_e = \mathcal{M}$), and so the NETD value in real IR imagers is inferior to NETD calculated for perfect IR imagers based on the ideal matrix PDA. The fundamental limitations are due to the large background current. If, because of the large background current I_0, it is not possible to ensure integration over the entire frame time in the matrix PDA and in external units, and if the integration time T_i is limited by the capacitance of the integrator (by its maximum charge $Q_{max} = I_0 T_i$), then, according to (8.84), (3.116), we get

$$\mathrm{NETD} = \frac{1}{K_{lens}\sqrt{2T_i A_e}}\frac{\theta S(\lambda'')}{\sqrt{2qj_0}} \propto \frac{S(\lambda'')}{\sqrt{T_i(A_e j_0)}} \propto \frac{S(\lambda'')}{\sqrt{Q_{max}}} \propto S(\lambda'') \propto \lambda'' \, . \tag{8.85}$$

We see that in case of such non-optimal signal processing the transition to shorter wavelengths is even advantageous.

Now let us give numerical evaluation of NETD for two typical cases: a CCD matrix based on Shottky barriers (spectral range 3...5 μm), and single-element MCT-based PD (8...12 μm). The specific background-limited detectivity of these two devices is, respectively, $1.77\cdot10^{11}\ \mathrm{W^{-1}\cdot cm\cdot Hz^{1/2}}$ and $5.81\cdot10^{10}\ \mathrm{W^{-1}\cdot cm\cdot Hz^{1/2}}$ (Table 3.3). Assume that the relative aperture is $D{:}f_{\mathrm{lens}}=1{:}1$, the PSE size is $l=\sqrt{A_{\mathrm{e}}}=50\ \mu\mathrm{m}$, $A_{\mathrm{e}}=2.5\cdot10^{-5}\ \mathrm{cm}^2$, the frame format is $\mathcal{M}=300\times400=1.2\cdot10^5$ pixels, the frame time corresponds to the TV standard, $T_{\mathrm{f}}=0.04\,\mathrm{s}$. Then for the CCD matrix based on Shottky barriers, with $M_{\mathrm{e}}=\mathcal{M}$, $\eta\approx10^{-3}$, $D^*/D^*_{\mathrm{BLIP}}=\sqrt{\eta}\approx0.03$, $K_{\mathrm{lens}}=D/2f_{\mathrm{lens}}$, we get

$$\mathrm{NETD}=\frac{2f_{\mathrm{lens}}}{D}\frac{\theta D^*_{\mathrm{BLIP}}}{\sqrt{2T_{\mathrm{f}}A_{\mathrm{e}}}}\frac{D^*_{\mathrm{BLIP}}}{D^*}=2\frac{2\cdot10^{-18}\cdot1.77\cdot10^{11}}{\sqrt{2\cdot0.04\cdot2.5\cdot10^{-5}}}\frac{1}{0.03}\approx0.015\,\mathrm{K}\ ; \qquad (8.86)$$

and for the one-element MCT-based PD with $\eta\approx1$

$$\mathrm{NETD}=\frac{2f_{\mathrm{lens}}}{D}\sqrt{\frac{\mathcal{M}}{2T_{\mathrm{f}}A_{\mathrm{e}}}}\theta D^*_{\mathrm{BLIP}}$$

$$=2\sqrt{\frac{1.2\cdot10^5}{2\cdot0.04\cdot2.5\cdot10^{-5}}}2\cdot10^{-18}\cdot5.8\cdot10^{10}\approx0.06\,\mathrm{K}\,. \qquad (8.87)$$

The large number of elements in the matrix based on Shottky barriers makes up for the low quantum yield and poor detectivity, and ensures, after all, a better noise equivalent temperature difference: the detectable temperature difference is four times less than when a single-element MCT PD is used. However, it would be sufficient to use an MCT PD with $4^2=16$ to achieve the same calculated value of NETD as that for the matrix based on Shottky barriers and composed of 300×400 elements.

These NETD values (hundredths or tenths of a kelvin) are typical of the modern high-quality IR imagers. It would be interesting to note that the absolute limit of NETD value for the above values of $D{:}f_{\mathrm{lens}}=1{:}1$, $A_{\mathrm{e}}=2.5\cdot10^{-5}\,\mathrm{cm}^2$, $T_{\mathrm{f}}=0.04\,\mathrm{s}$, $\mathrm{T}=293\,\mathrm{K}$ is

$$\min[\mathrm{NETD}]=\frac{2f_{\mathrm{lens}}}{D}\frac{\min\left[\mathrm{NETD}^*_{\mathrm{BLIP}}\right]}{\sqrt{2T_{\mathrm{f}}A_{\mathrm{e}}}}=2\frac{7.3\cdot10^{-8}}{\sqrt{2\cdot0.04\cdot2.5\cdot10^{-5}}}\approx10^{-4}\,\mathrm{K}\,. \qquad (8.88)$$

The above examples make it clear why the absolute limit has not been achieved: if $\eta=1$, we cannot secure $T_{\mathrm{i}}=T_{\mathrm{f}}$; and if $T_{\mathrm{i}}=T_{\mathrm{f}}$, we get $\eta<<1$. So, $10^{-4}\,\mathrm{K}$ is the limit of noise equivalent temperature difference of IR imagers which is not yet achieved (and, as a matter of fact, not yet demanded by) the IR technology.

Summary. *We used the example of IR imager to prove our old thesis: to cope with the task set before the optoelectronic system in the best possible way, one must first of all perform optimal detection of low-level optimal signal. The task of IR imager is to register the lowest possible (noise-equivalent) temperature difference* NETD. *This task is perfectly accomplished by the matrix PDA: the optimal detection system, which ensures ideal space-time integration of the signal over the entire frame time and entire frame area (without scanning). The performance of IR imager is characterized by the value of specific noise equivalent temperature difference* $NETD^*$ *calculated for IR imager with unit passband and unit PSE area, and equipped with a hypothetical optical system which collects thermal radiation from the entire hemisphere. To calculate* $NETD^*$ *of the ideal IR imager one must simply multiply the specific detectivity of the PDA* D^* *by the coefficient* $2\kappa T^2$ *(the more precise calculations must take into account the possible reduction (by no more than* 30%) *of this coefficient as the long-wave limit varies from* 0 *to* 14μm).

The specific noise equivalent temperature difference of ideal IR imager is improved as the spectral range broadens, and achieves its absolute limit (the lowest detectable temperature difference) when the PDA registers almost all thermal photons. The absolute limit depends on the background temperature T *(decreasing as* $1/\sqrt{T}$), *and at* $T = 293\,K$ *is equal to* $7.3 \cdot 10^{-8}\,K \cdot cm \cdot s^{1/2}$. *For typical parameters of IR imager this translates into the specific noise equivalent temperature difference of* $\sim 10^{-4}\,K$.

In real PDAs with the long-wave limit at 10...14μm *the deterioration of the noise equivalent temperature difference because of the limited spectral range is not considerable, and constitutes about* 12..35%. *The fundamental restrictions arise because of the insufficient charge capacity of the matrix cell, leading to the loss of integration and deterioration of* NETD *of real IR imagers by two or three orders of magnitude as compared to the absolute limit. Even with these limitations, the temperature sensitivity of modern IR imagers is as good as* 0.01...0.1K.

Conclusion

or Praise to Photodetector and Photodetecting Assembly

This is the end of our story about PDA and the detection of ultimately weak optical signal. The diversity of modern PDAs reveals also their unity: by their functional composition they all are detection systems, and incorporate a photodetector, a filter-amplifier (integrator), sometimes a discriminator. Multielement PDAs may also include a commutator (multiplexer). The theory of detection conducted the orchestra throughout the book: it set the guidelines for synthesis of the PDA, prompted the ways to minimize its noise, defined the optimal passband of the filter and the rules for making the decision concerning the presence of the signal. All of the signal processing stages are responsible for the threshold characteristics of the device, but the first violin is always played by the photodetector. The input circuits are designed so as to match its impedance and noise level, the cooling system and the power supply maintain the desired temperature and operating regime. Under these conditions the photodetector performs its mission: it detects the optical signal almost without loss, often without introducing any additional noise. Because of this, the thermal background noise prevails in high-quality IR PDAs operating at wavelengths above 2 μm, and the amplifier noise prevails in the short-wave PDAs operating at wavelengths below 1 μm. The noisy amplifier is rescued by the photodetector (avalanche photodiode): it reinforces the signal, and, by avalanche multiplication, drags out both the signal and its own noise from the noise of the input stage. Initiative is persecuted, however: we accuse the APD of introducing the additional multiplication noise. The penitent APD then corrects the situation, and goes to the photon counting mode, registering individual quanta. This mode takes full advantage of the counting principle inherent in the intrinsic photoeffect. Thus, the performance of modern PDAs is close to the theoretical limit. The lowest detectable optical power is a fundamental quantity: it depends neither on the noise of the photodetector, nor on the noise of the PDA, but is instead delimited either by the fluctuations of thermal background or by the quantum nature of radiation.

We intend this conclusion to be a praise, not an epitaph, to photodetectors and PDAs. Let us believe in the future progress of photoelectronics and high technologies even despite the consequences of reconversion. Research and development continue in the field of new photosensitive structures (multiple compounds, superlattices), low-noise avalanche structures with the multiplication of one type of carriers (for the range 1.3...1.6 μm and higher). Complicated problems of detection of low-level signals with IR linear arrays and matrices are being solved under conditions of both weak and strong backgrounds. Improved is the design, extended are the functions of the PDAs (using digital techniques, image processing methods).

Working in this field, taking part in the detection of ultimately weak optical signals is an onerous but honorable challenge.

References

1. M.M. Miroshnikov, “Theoretical foundations of optoelectronic devices" [in Russian], Mashinostroenie, Leningrad (1983).
2. M.D. Aksenenko, M.L. Baranochnikov, O.V. Smolin, “Photodetecting assembly” [in Russian], Energoatomizdat, Moscow (1984).
3. M.D. Aksenenko, M.L. Baranochnikov, “Receivers of optical radiation. Handbook” [in Russian], Radio i Svyaz', Moscow (1987).
4. N.S. Shestov, “Separation of optical signals from random noise” [in Russian], Sovetskoye Radio, Moscow (1967).
5. W.L. Wolfe, G.J. Zissis (Eds.), “The infrared handbook”, 4th revised edition, ERIM-USA (1993).
6. R.D. Hudson, “Infrared system engineering”, Wiley, New York (1969).
7. R.D. Hudson, “Military applications of remote sensing”, *Proc. IEEE*, **67**(1), 60–128 (1976).
8. B.F. Andersen, M.S. Scholl (Eds.), “Infrared technology and applications XXII”, *Proc. SPIE*, **2744**, 762 (1996).
9. G. Gaussorgues, “La thermographie infrarouge. Principes – Technologie – Applications”, Technique et Documentation (Lavoisier), Paris (1984).
10. M.J. Howes, D.V. Morgan (Eds.), “Optical fibre communications. Devices, circuits and systems”, Wiley, New York (1980).
11. T.J. Brown, “Development of an earth resource pushbroom scanner utilizing a 90-element 8-14 μm (HgCd)Te array”, *Proc. SPIE*, **226**, 18-38 (1980).
12. R.A. Ballingall, I.D. Blenkinsop, “Two dimensional IR FPA utilizing a direct inject input scheme”, *Proc. SPIE*, **685**, 115-122 (1986).
13. C.H. Sequin, M.F. Tompset, “Charge transfer devices”, Academic Press, New York (1975).
14. M.J. Howes, D.V. Morgan (Eds.), “Charge-coupled devices and systems”, Wiley, New York (1979).
15. D.F. Barbe (Ed.), “Charge-coupled devices”, Springer, Heidelberg (1980).
16. G.S. Holson, “Charge transfer devices”, Arnold, London (1978).
17. J.E. Beynon, D.L. Lamp, “Charge-coupled devices and their applications”, McGraw-Hill, London (1980).
18. P.G. Jespers (Ed.), “Solid state imaging”, Noordhoff, Leyden (1976).
19. D.S. Glad, J.S. Lorenzo, G.M. Parry, J. Moriam, R.K. Bentley, R.H. Weber, “Pulsebias modulation of a PbS mosaic array”, *Proc. SPIE*, **197**, 9–18 (1979).
20. G.T. Elliott, “New detector for thermal imaging systems”, *Electron. Lett.* **17**(8), 312-313 (1981).
21. A. Fowler, R. Youce, I. Gatley, “Evaluation of a 256 × 256 PtSi Array for Infrared Ground-Based Astronomy”, *Proc. SPIE*, **1107**, 22-28 (1989).
22. M.B. Barron, “Low-level currents in insulated-gate FETs”, *Solid State Electron.* **15**(3), 293-302 (1972).

23. Yu.R. Vinetskiy, M.A. Trishenkov, "Potential input arrangement of charge-coupled device in the weak-signal regime", *Radio Eng. Electron. Phys.* **26**(4), 130-142 (1981).
24. M.A. Trishenkov, Yu.R. Vinetskiy, "Input circuits of charge-coupled devices", *Radio Eng. Electron. Phys.* **27**(12) 1-30 (1982).
25. A. Tebo, "Speaker report PtSi and other detector advances", *Laser Focus*, **36**(7), 51–54 (1990).
26. D.A. Schibner, M.R. Kruer, J.M. Killiany, "Infrared focal plane array technology", *Proc. IEEE*, **79**(1), 66-85 (1991).
27. M. Kimata, N. Yutani, N. Tsubouchi, "High performance 1040×1040 element PtSi Schottky barrier image sensor", *Proc. SPIE*, **1762**, 350-360 (1992).
28. L.J. Kozlowsky, R.B. Bailey, S.C. Cabelli, "640×480 Pace HgCdTe Focal Plane Array", *Proc. SPIE*, **1735**, 163–174 (1992).
29. R.P. Leftwich, D.R. Ward, "Latest developments in SPRITE detector technology", *Proc. SPIE*, **930**, 76–86 (1988).
30. S. Wicker, "New technologies for FPA dewars", *Proc. SPIE*, **1683**, 102-112 (1992).
31. F. Mongellaz, A. Fillot, J. de Lallee, "Thermoelectric cooler for infrared detector", *Proc. SPIE*, **2227**, 156-165 (1994).
32. K.I. Britsyn, V.S. Vavilov, "On the photoionization process in silicon", *Optics and Spectroscopy*, **8**(6), 452-455 (1960).
33. A. Rogalski (Ed.) "Infrared photon detectors", SPIE Optical Engineering Press, USA (1995).
34. R.J. Keyes (Ed.), "Optical and infrared detectors", Springer, Heidelberg (1980).
35. R.M. Mindock, I.B. Horak, "Temperature effects upon partially and fully depleted silicon photodetectors", *EOSD West Conference Proceedings*, May 1971.
36. J.J. Chang, "Frequency response of *pin* avalanche photodiodes", *IEEE Trans. on Electron. Dev.* **ED–14**(3), 139-145 (1967).
37. H. Melchior, A.R. Hartman, D.P. Schinke, T.E. Seidel, "Planar epitaxial silicon avalanche photodiodes", *Bell Syst. Tech. J.* **57**(6/1), 1791-1807 (1978).
38. P.P. Webb, R.J. McIntyre, "Multielement Reachthrough Avalanche Photodiodes", *IEEE Trans. on Electron Dev.* **ED–31**(9), 1206-1212 (1984).
39. R.K. Williardson, A.C. Beer (Eds.), "Semiconductors and semimetals", v. 22: Lightwave communications technology. Part D. Photodetectors (ed. by W.T. Tsang), Academic Press, New York (1985).
40. D.E. Ackley, J. Hladky, M.J. Lange, S. Mason, G. Erickson, G.H. Olsen, V.S. Ban, J.Y. Lin, S.R. Forrest, "InGaAs/InP floating guard ring avalanche photodiodes fabricated by double diffusion", *IEEE Photonics Tech. Letters*, **2**(8), 571–573 (1990).
41. L.E. Tarof, "Planar InP/InGaAs avalanche photodetectors with partial charge sheet in device periphery", *Appl. Phys. Letters*, **57**(7), 670-672 (1990).
42. S.M. Sze, "Physics of semiconductor devices", 2nd edition, Wiley, New York (1981).
43. P.I. Baranskiy, V.P. Klochkov, I.V. Potykevich, "Semiconductor electronics. Handbook" [in Russian], Naukova Dumka, Kiev (1975).

44. M.B. Reine, "Status of HgGdTe detector technology", *Proc. SPIE*, **443**, 2–10 (1985).
45. M.E. Creiner, C.G. Martin, "Indium arsenide photovoltaic detector", *Proc. SPIE*, **686**, 21–22 (1986).
46. M.A. Bramson, "Thermal radiation of heated bodies" [in Russian], Nauka, Moscow (1964).
47. P.W. Kruse, L.D. McGlauchlin, R.B. McQuistan, "Elements of infrared technology", Wiley, New York, (1962).
48. H.B. Dwight, "Tables of integrals and other mathematical data", Macmillan, New York, (1961).
49. P.H. Wendland, "Solid state combo senses light well enough to vie with tubes", *Electronics*, **44**(11), 50–54 (1971).
50. Jiri Dostall, "Operational amplifiers", Elsevier, Amsterdam (1981).
51. L.M. Faulkenberry, "An introduction to operational amplifiers with linear IC applications", 2nd edition, Wiley, New York (1982).
52. A. Barna, D.I. Portal, "Operational amplifiers", Wiley, New York (1989).
53. H.J. Weaver, "Theory of discrete and continuous Fourier analysis", Wiley, New York (1989).
54. W.M. Siebert, "Circuits, signals and systems", MIT Press, Cambridge; McGraw-Hill, New York (1986).
55. B.R. Levin, "Fondements theoriques de la radiotechnique statistique", Mir, Moscow (1973).
56. R. Barlow, "Statistics", Wiley, Chichester (1989).
57. M.J. Buckingham, "Noise in electronic devices and systems", Ellis Horwood, New York (1983).
58. R.L. Batdorf, A.G. Chynoweth, G.C. Dacey, P.W. Fay, "Uniform silicon *pn* junction. Part I. Broad area breakdown", *J. Appl. Opt.* **31**(7), 1153-1168 (1960).
59. A. Goetsberger, B. McDonald, R.H. Haitz, R.M. Scarlett, "Avalanche effects in silicon *pn* junctions. Part II. Structurally perfect junctions", *J. Appl. Phys.* **34**(6), 1591-1601 (1960).
60. C. Donolato, P.G. Meril, I. Vecchi, "Characterization of microplasma sites in silicon *np*–junctions", *J. Electrochem. Soc.* **124**(3), 473–474 (1977).
61. R. Muller, "Rauschen (Halbleiter Elektronik 15)", Springer, Heidelberg (1979).
62. A. Van der Ziel, "Noise in measurements", Wiley, New York (1977).
63. P.O. Lauritzen, "Noise due to generation and recombination of carriers in *pn*–junction transition region", *IEEE Trans. on Electron Dev.* **ED–15**, 770–776 (1968).
64. G. Cavalleri, G. Fabri, E. Gatti, V. Svelto, "On the induced charge in semiconductor", *Nuclear Instruments and Methods*, **21**(1), 177-178 (1963).
65. J. Perina, "Quantum statistics of linear and nonlinear optical phenomena", D. Reidel, Dordrecht (1984).
66. K.M. Van Vliet, L.M. Rucker, "Theory of carrier multiplication and noise in avalanche devices. Part I. One-carrier process. Part II. Two-carrier processes", *IEEE Trans. on Electron. Dev.* **ED-26**(5), 746-764 (1979).

67. A.S. Tager, "Current fluctuations in a semiconductor (dielectric) under the conditions of impact ionization and avalanche breakdown", *Soviet Phys.-Solid State*, **6**(8), 1919-1925 (1965).
68. R.Y. McIntyre, "Multiplication noise in uniform avalanche diodes", *IEEE Trans. on Electron Dev.* **ED-13**(1), 164–170 (1966).
69. K. Seeger, "Semiconductor physics. An introduction", 4th edition, Springer, Heidelberg (1989).
70. F. Schlögl, "Probability and heat: Fundamentals of thermostatistics", Wieweg, Braunschweig, Wiesbaden (1989).
71. N.W. Ashcroft, N.D. Mermin, "Solid State Physics", Holt, Rinehart and Winston, New York (1976).
72. E.S. Oxner, "FET technology and application: An introduction", Dekker, New York, Basel (1989).
73. K. Kandiah, F.B. Whiting, "Low frequency noise in junction field effect transistor", *Solid State Electronics*, **21**(8), 1079-1088 (1978).
74. V.M. Korenevskiy, "Noise electromotive force at high frequencies for a field-effect transistor with a *pn*–junction", *Sov.J. Commun. Technol. Electron.* **34**(2), 91-97 (1989).
75. A. Ambrozy (Ed.), "Noise in physical systems. Proc. 10th Int. Conf. August 21-25, 1989, Budapest, Hungary", Akademiai Kiado, Budapest (1990).
76. M.A. Trishenkov, "Avalanche photodiode operating as a detector of pulsed radiation", *Radio Eng. Electron. Phys.* **24**(8), 116-124 (1979).
77. N.N. Goryunov (Ed.), "Semiconductor devices. Transistors. Handbook" [in Russian], Energoatomizdat, Moscow (1985).
78. K. Hyyppä, K. Ericson, "Low-noise photodiode amplifier circuit", *IEEE J. Solid State Circuits*, **29**(3), 362-365 (1994).
79. B.I. Segal, K.I. Semendyaev, "Five-digit mathematical tables" [in Russian], Izdatel'stvo AN SSSR, Moscow, Leningrad (1948).
80. D.O. North, "Analysis of factors which determine signal-to-noise discrimination in pulsed carrier systems", Rep.PTR-GC, RCA, Princeton (1943).
81. B.M. Dwork, *Proc.IRE*, **31**, 771 (1950).
82. I.S. Gradshteyn, I.M. Ryzhik, "Tables of integrals, sums, series and products" [in Russian], Fizmatgiz, Moscow (1963).
83. R.C. Jones, "A method of describing the detectivity of photoconductive cells", *Rev. Sci. Instr.* **24**, 1035-1040 (1953).
84. M.H. Weik, "Fiber Optics Standard Dictionary". 2nd edition, Van Nostrand, New York (1989).
85. Yu.S. Lezin, "Optimal filters and integrators of optical signals" [in Russian], Sovetskoye Radio, Moscow (1969).
86. B.D. Anderson, J.B. Moore, "Optimal filtering", Prentice Hall, Englewood Cliffs (1979).
87. V.I. Siforov, "On the effects of noise on reception of pulsed radio signals" [in Russian], *Radiotekhnika*, **1**(1), 5-19 (1946).
88. A.P. Belousov, "On the highest real sensitivity of pulsed [signal] receiver" [in Russian], *Radiotekhnika*, **1**(5), 18-27 (1946).

89. L.N. Averin, "Reception of short light pulses with a PD equipped with quasi-optimum filtering", *Sov. J. Opt. Technol.* **41**(10), 53-55 (1974).
90. Yu.S. Mikheev, A.S. Petrov, "Quasioptimum filtering in a photodiode receiver for short signal pulses", *Sov. J. Opt. Technol.* **45**(2), 77-80 (1978).
91. S.D. Personick, "Receiver design for digital fiber optic communication systems", *Bell Syst. Tech. J.* **52**(6), 843-886 (1973).
92. Yu.M. Kazarinov (Ed.) "Search, detection and measurement of parameters of signals in radio navigation systems" [in Russian], Sovetskoye Radio, Moscow (1975).
93. R.I. McIntyre, "The distribution of gains in uniformly multiplying avalanche photodiodes: Theory", *IEEE Trans. on Electron Dev.* **ED-19**(6), 703–713 (1972).
94. R.B. Emmons, G. Lucovsky, "Frequency response of avalanche photodiodes", *IEEE Trans. on Electron. Dev.* **ED-13**(3), 297 -305 (1966).
95. S. Barber, "Photon counting with avalanche photodiodes", *Electronic Engineering*, **56**(689), 63–70 (1984).
96. A. Lacaita, S. Cova, M. Ghioni, "Four-hundred-picosecond single-photon timing with commercially available avalanche photodiodes", *Rev. Sci. Instr.* **59**(7), 1115-1121 (1988).
97. D.L. Robinson, B.D. Metscher, "Cooled avalanche photodiodes used for photon detection", *Proc. SPIE (Infrared Systems and Components)*, **750**, 32–36 (1987).
98. K. Shimizu, M. Fujise, M. Nunokawa, "Synchronous single-photon counting using a Si avalanche photodiode of room temperature", *Electron. Lett.* **23**(24), 1307- 1308 (1987).
99. P.P. Webb, R.J. McIntyre, J. Conradi, "Properties of avalanche photodiodes", *RCA Review*, **35**(6), 234–278 (1974).
100. G.P. Weckler, "Operation of *pn*–junction photodetectors in a photon flux integrating mode", *IEEE J. Solid State Circuits*, **SC-2**(3), 65-73 (1967).
101. Yu.R. Vinetskiy, "Frequency filtering of a signal by input charge-coupled device units", *Sov.J. Commun. Technol. Electron.* **32**(5), 77-85 (1987).
102. R.A. Ballingall, "2-Dimensional Random Access IR Arrays", in: Proc. 2nd Int. Conf. Advanced IR Detectors and Systems, London, 1983; New York (1983) pp. 6–11.
103. M.A. Trishenkov, "Excess noise suppression in the frame subtracting regime", *Radio Eng. Electron. Phys.* **24**(7), 82-91 (1979).
104. G. Reimbold, "Noise associated with charge injection into CCD by current integration through a MOS transistor", *IEEE Trans. on Electron Dev.* **ED-32**(5), 871–873 (1985).
105. A. Van der Ziel "1/f noise characterization of *np* and *nip* HgCdTe detectors", *J. Vacuum Sci. Tech. A*, **7**(2), 550–554 (1989).
106. P. Nicolas, "Readout of 3-5 and 8-14 μm hybrid arrays", in: Proc. 2nd Int. Conf. Advanced IR Detectors and Systems, London, 1983; New York (1983) pp. 17–22.
107. D.D. Wen, P.J. Salsbury, "Analysis and design of a single–stage floating gate amplifier", Int. Solid State Circuits Conference, Philadelphia, 1975; Digest of Tech. Papers, pp. 24–25.

108. G. Bailey, "A 256×256 HgGdTe hybrid plane for low background Earth resources and astronomy applications", *Proc. SPIE*, **972**, 122–126 (1988).
109. C.L. Carrison, N.A. Foss, "Fixed pattern noise compensation techniques for starring infrared focal planes", *Opt. Engineering*, **19**(5), 753–757 (1980).
110. J. Mosley, "CCD Imaging Arrays", *EDN*, **35**(6), 116–126 (1990).
111. E. Oda, K. Nagano, T. Tanaka, N. Muton, K. Orihara, "A 1920(H)×1035(V) pixel high-definition CCD image sensor", *IEEE J. Solid State Circuits*, **24**(3), 711-716 (1989).
112. M. Denda, M. Kimata, S. Iwade, N. Yutani, T. Kondo, N. Toubouchi, "4-band × 4096-element Schottky barrier infrared linear image sensor", *IEEE Trans. on Electron Dev.* **ED-34**(5), (1991).
113. A. Rodalski, J. Piotrowski, "Intrinsic Infrared Detectors", *Progress in Quantum Electronics*, **12**(1/2), 87–299 (1988).
114. M.M. Blanke "Current status of the 800×800 CCD", *Opt. Engineering*, **26**(9), 864-874 (1987).
115. B.E. Burke, "420×420 CCD imager and 4-chip hybrid focal plane", *Opt. Engineering*, **26**(9), 890-896 (1987).
116. J.M. Lloyd, "Thermal imaging systems", Plenum Press, New York (1975).
117. M. Born, E. Wolf, "Principles of optics. Electromagnetic theory of propagation, interference and diffraction of light", Pergamon Press, Oxford (1975).
118. I.I. Taubkin, M.A. Trishenkov, N.V. Vasil'chenko, "Minimum temperature difference detected by the thermal radiation of objects", *Infrared Phys. Tech.* **35**(5), 715-732 (1994).

9 780792 346913

GPSR Compliance

The European Union's (EU) General Product Safety Regulation (GPSR) is a set of rules that requires consumer products to be safe and our obligations to ensure this.

If you have any concerns about our products, you can contact us on ProductSafety@springernature.com

In case Publisher is established outside the EU, the EU authorized representative is:

Springer Nature Customer Service Center GmbH
Europaplatz 3
69115 Heidelberg, Germany

Batch number: 10607204

Printed by Printforce, the Netherlands